Surface Chaos and Its Applications

Shu Tang Liu · Li Zhang

Surface Chaos and Its Applications

Springer

Shu Tang Liu
School of Control Science and Engineering
Shandong University
Jinan, Shandong, China

Li Zhang
School of Control Science and Engineering
Shandong University
Jinan, Shandong, China

Business School
Shandong University of Political Science
and Law
Jinan, Shandong, China

ISBN 978-981-16-8228-5   ISBN 978-981-16-8229-2 (eBook)
https://doi.org/10.1007/978-981-16-8229-2

This Springer imprint is published by the registered company Springer Nature Singapore Pte Ltd.
The registered company address is: 152 Beach Road, #21-01/04 Gateway East, Singapore 189721, Singapore

Preface

Throughout the contents, we can see that both the plane and spatial curve chaos have been relatively mature.

Surface chaos and its applications by control theory can be regarded as a fertile land, and there are a lot of open problems to be solved there. Most existing studies of the discrete dynamical systems derived by partial differential equations of order one or two did not consider the high-dimensional problems directly. These studies bypassed the high-dimensional difficulties and transformed the complex system into a form of nonlinear ordinary differential equations which is usually easy to solve. In fact, the study of high-dimensional discrete dynamical systems has recently turned into a hot topic, and there are currently a lot of open problems to be solved in this field.

Surface chaos, as a space body involving the nonlinear behaviors of the motion of stars in astrophysics, is still a brand new research topic.

With the improvement of the plane and spatial curve chaos and, more importantly, with the development of the surface chaos theory, this book can be regarded as a small dividing line or a watershed in the field of chaos in nonlinear science.

At the same time, it is further clear for the researchers in this field that

I. the research of plane curve and space-related chaos is relatively mature;

II. the research of surface chaos in space has just begun;

III. in this monograph, we show that the basic theories of continuity, differentiability, and differentiability of space surface chaos involved with multivariate function are similar to those with univariate function, but there are still many differences between the two. It can be seen that the motion behavior of a nonlinear high-dimensional system is more complex due to issues resulting from multiple dimensions in mathematical analyses. Nonlinear spatiotemporal behavior has more complex categories, and it will have more extensive applications in all aspects of human life. Therefore, we would cordially invite more and more researchers who would pay attention to and devote themselves to the research of this multi-dimensional nonlinear field. We hope that the research in this field will be as vigorous as bamboo shoots after a spring rain!

Jinan, China

Shu Tang Liu
Li Zhang

Acknowledgements

We would like to express our appreciation for the great efforts of Drs. Jian Liu, Chunhua Yuan, Yongpin Zhang, Fuyan Sun, Fangfang Zhang, Quan Hai, Pei Wang, Dadong Tian, Qiuyue Zhao, Zhiping Liu, and graduate student Yang Zheng. Additionally, we are grateful for the sequential support of the Key Program of the National Natural Science Foundation of China (No. 61533011), the National Natural Science Foundation of China-Shandong Joint Fund (No. U1806203), the Foundation for the Author of National Excellent Doctoral Dissertation of China (No. 200444), the National Natural Science Foundation of China (No. 60472112, No. 60372028, No. 600874009, No. 10971120, No. 61273088), and the Shandong Province Natural Science Foundation (No. ZR2010FM010, No. Y98A02005), which ensure the smooth developments of this research project.

Shu Tang Liu
Li Zhang

Contents

Abbreviations

$\mathbb{R}$	Set of real numbers
$\mathbb{R}^n$	Set of n-dimensional real vectors
$det(A)$	Determinant of matrix A
$\vec{r}$	Vector of r
$\lambda(A)$	Eigenvalue of matrix A
$\sum_{i=1}^{n} A(i)$	Sum of $A(i)$ $i = 1, 2, ...n.$
$f^k(x)$	The $k-th$ mapping of $f(x)$
$supf(x)$	Super $f(x)$
$inff(x)$	Inf $f(x)$
$\frac{\partial v(x,y)}{\partial x}$	Partial differential of $v(x, y)$ about x
$\frac{\partial^2 v(x,y)}{\partial x^2}$	The second partial differential of $v(x, y)$ about x
$\frac{\partial^n v(x,y)}{\partial x^n}$	The $n-th$ partial differential of $v(x, y)$ about x
$\frac{\partial^2 v(x,y)}{\partial x \partial y}$	The second partial differential of $v(x, y)$ about x and y
F'_μ	The partial differential of $F(\mu, v)$ about μ
∇D_g	Gradient of D_g
$\Delta_1 y_{m,n}$	$y_{m+1,n} - y_{m,n}$
$\Delta_2 y_{m,n}$	$y_{m,n+1} - y_{m,n}$
$\nabla_1 y_{m,n}$	$y_{m-1,n} - y_{m,n}$
$\nabla_2 y_{m,n}$	$y_{m,n-1} - y_{m,n}$
Inf	Infimum
lim	Limit
max	Maximum
min	Minimum
sup	Supremum
CML	Coupled map lattice
$\vert \cdot \vert$	Absolute value (or modulus)
$\vert\vert \cdot \vert\vert$	Euclidean norm of a vector or spectral norm of a matrix
$\vert\vert \cdot \vert\vert_n$	Included l_n−norm
$\mathcal{C}([a, b], \mathbb{R}^n)$	Family of continuous functions ϕ from $[a, b]$ to $\mathbb{R}^n$
$\vert\vert\phi\vert\vert_c$	Continuous norm $\sup_{a \le t \le b} \vert\vert\phi t\vert\vert$ for $\phi \in \mathcal{C}([a, b], \mathbb{R}^n)$

$\mathcal{E}$	Mathematical expectation
$\left[\begin{array}{c\|c} A & B \\ \hline C & D \end{array}\right]$	Shorthand for state space realization $C(sI - A)^{-1}B + D$ for a continuous-time system or $C(zI - A)^{-1}B + D$ for a discrete-time system
$\forall$	For all
$\in$	Belongs to
$\exists$	There exists
$\subseteq$	Is a subset of
$\cup$	Union
$\cap$	intersection
$\rightarrow$	Tends toward or is mapped into (case sensitive)
$\Rightarrow$	Implies
$\prod_{i=1}^{n} \lambda_i$	Product of $\lambda_1, \lambda_2, \ldots, \lambda_n$
$\Delta P(x, y)$	Laplace operator of $P(x, y)$
$\square$	End of proof

Chapter 1
Introduction

Chaos is one of the most important research areas in nonlinear science. There are currently extensive achievements in chaos theory and its applications. For example, chaos has been widely used in control systems, engineering optimization, biomedicine, etc. [1–242]. For instance, $t \in (a, b) \subseteq \mathbb{R}$(real axis), having the following condition:

$$t \to (x_1(t), x_2(t), \ldots, x_n(t)),$$

in which the number of independent variables contained in the unknown function is 1, the state space is n-dimensional where the curve chaos and the surface chaos can be defined according to the domain of the unknown function and the number of unknown functions of independent variables is one or two. For example, the surface chaos is generated from a mathematical system equation formed by $\mu(x, y)$, two independent variables are x and y. In fact, its definition domain is a region on the plane $\mathbb{R}^2$, so its state behavior is the motion behavior of a surface in space $\mathbb{R}^3$. And if the surface is a nonlinear system with $U(x, y)$, chaos and bifurcation will occur under certain parameters. Naturally, we regard the nonlinear chaos as surface chaos and surface bifurcation [243–250, 250–267, 267, 268, 270–280, 282–293]. If you want to consider time to change in it, the variable x or y can be used as a time variable.

Hence the concepts of curve chaos and surface chaos can be treated as a new area of nonlinear behaviors, which motivates us to pay more attention to the relationship and difference between them.

The curve chaos and the surface chaos are different in structure because they involve the unknown univariate or multivariate functions in the corresponding nonlinear systems. Furthermore, some of the analytical properties of functions of one variable and those of functions of multiple variables can be derived from each other, but some others can not, especially for multivariate functions, which involve the increase of independent variables and the change of definition domain, partial derivatives, differentiability, continuity and so on. So its analytical properties are not fully

S. T. Liu and L. Zhang, *Surface Chaos and Its Applications*,
https://doi.org/10.1007/978-981-16-8229-2_1

generalized, which makes the surface chaos more complicated. Therefore, it is of great significance to study this new area.

Feigenbaum problem of the Lyapunov exponent, static bifurcation, other related wave variation of the surface in 2-D discrete system, the different control methods for the surface by static bifurcation, generalized synchronization, nonlinear feedback-controlled synchronization, prediction-based feedback control and envelope curve method are studied. These results can also be used in spatial physics kinematic system, molecular orbit and growth of harmful algal bloom.

Our work about the understanding of curve chaos and introduction of the concept of surface chaos have established some solid outcomes. However, these preliminary results are just the tip of the iceberg of surface chaos research. The rationality of such a new definition needs to be further investigated, which can also be regarded as an open problem. The researchers are warmly welcome to join this new area and any discussion or collaboration with us is much appreciated. Our goal is to better grasp the nonlinear characteristics of surface chaos and apply them to various fields of nonlinear system science.

Part I
Surface Chaos and Surface Bifurcation

Chapter 2
Basic Knowledge

2.1 Curve

It is well known that curve definitions can be understood through the image of a homeomorphism function defined on an open straight-line segment in the three-dimensional Euclidean space $\mathbb{R}^3$. In this book, we set the curve as the graph $G = \{(x, y) \mid y = f(x), x \in D\}$ defined by the following mapping:

$$f : D \to R, \tag{2.1}$$

where $D \subseteq \mathbb{R}^j$, $j = 1, 2$.

2.1.1 Plane Curve

2.1.1.1 Definition of Plane Curve

There are different definitions about plane curve such as connected set, complete set, nowhere dense set, and set of points as graph $G = \{(x, y) \mid x = \phi(t), y = \psi(t), t \in [0, 1]\}$ where ϕ and ψ are continuous functions. In the following content, the plane curve is the generalization of the Jordan plane curve as

$$G = \{(x, y) \mid x = \phi(a), y = \psi(a), a \in D \subseteq \mathbb{R}\}, \tag{2.2}$$

where ϕ and ψ are continuous or discrete real functions and a is time or other variable in different situations.

S. T. Liu and L. Zhang, *Surface Chaos and Its Applications*,
https://doi.org/10.1007/978-981-16-8229-2_2

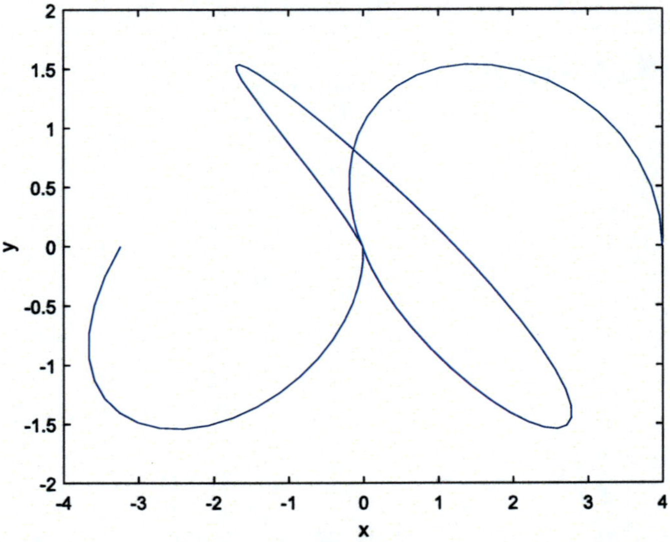

Fig. 2.1 Continuous plane curve by (2.3) where $a = [0, 4\pi]$

2.1.1.2 Continuous Plane Curve

In general, the continuous plane curve is defined when ϕ and ψ are continuous functions on $D \subseteq \mathbb{R}$.

Example 2.1 As shown in Fig. 2.1:

$$y = r sin(0.5a), x = r cos(0.8a), \tag{2.3}$$

where $r = 2(1 + cos(a))$, $a \in (0, 4n\pi)$, $n = 1, 2, \ldots$.

2.1.1.3 Discrete Plane Curve

Similarly, the discrete plane curve is defined when ϕ and ψ are discrete functions on $D \subseteq \mathbb{R}$. We consider the discrete plane curve as a graph $G = \{(n, x_n) \mid x_{n+1} - f(x_n, c) = 0\}$ where f is a nonlinear function and c is a real parameter.

Example 2.2 For the Logistic system [294]:

$$x_{n+1} = f(x_n, \mu) = \mu x_n(1 - x_n), \ \ \mu > 0, x \in [0, 1],$$

it can be rewritten as:

$$x_{n+1} = f(x_n, \mu) = 1 - \mu x_n^2, \tag{2.4}$$

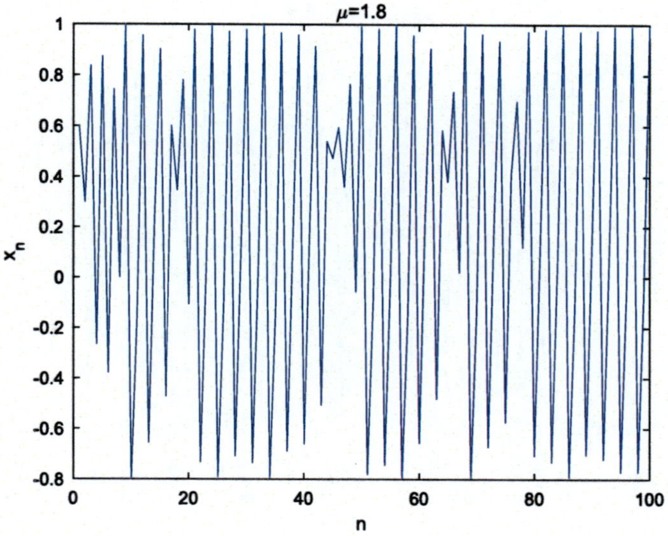

Fig. 2.2 Discrete plane curve of (2.4)

illustrated by Fig. 2.2 where μ is real parameter.

Example 2.3 And for the Hénon system [358]:

$$x_{n+1} = f(x_n, \alpha, \beta) = 1 - \alpha x_n^2 + \beta x_{n-1}, \tag{2.5}$$

illustrated by Fig. 2.3 where α and β are all real parameters, it can be replaced as the following form if we take $y_{n+1} = x_n$:

$$\begin{cases} x_{n+1} = 1 - \alpha x_n^2 + \beta y_n, \\ y_{n+1} = x_n. \end{cases}$$

Actually, the complexity behaviors of the discrete systems (2.4) and (2.5) are produced by the iteration process of the map f about independent variable n and have a lot of related theoretical and application results [295–310].

Moreover, the chaotic features of the discrete plane curve have generated a large number of results in control theory, communication field and other related technology fields [311–350, 352–354].

Noticeably, the famous Logistic and Henon mapping are examples of the systems (2.4) and (2.5), respectively.

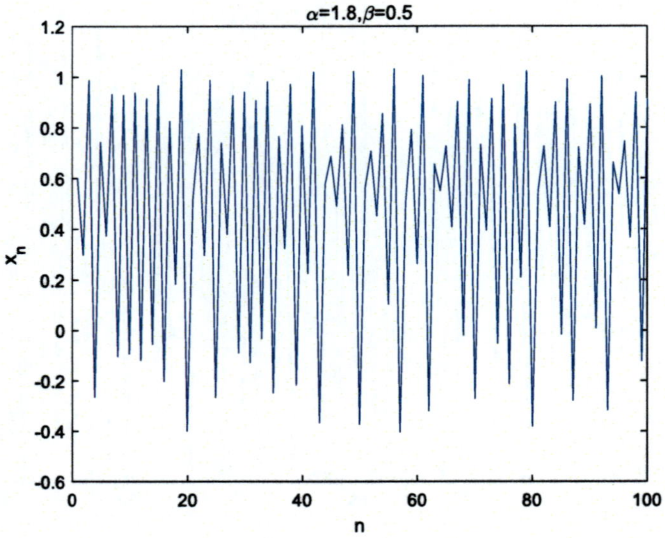

Fig. 2.3 Discrete plane curve of (2.5)

2.1.2 *Spatial Curve*

2.1.2.1 Definition of Spatial Curve

Generally, spatial curve can be defined as the graph

$$G = \{(x, y, z) \mid x = x(a), y = y(a), z = z(a), a \in D \subseteq \mathbb{R}\}.$$

2.1.2.2 Spatial Continuous Curve

The spatial continuous curve can thus be generated by following:

$$G = \{(x, y, z) \mid F(\frac{dx}{dt}, \frac{dy}{dt}, \frac{dz}{dt}, x(t), y(t), z(t)) = 0, t \in D \subseteq \mathbb{R}\}.$$

Example 2.4 Lorenz system [64]:

$$\begin{cases} \frac{dx}{dt} = a(y - x), \\ \frac{dy}{dt} = cx - xz - y, \\ \frac{dz}{dt} = xy - bz, \end{cases} \tag{2.6}$$

and its spatial continuous curve illustrated by Fig. 2.4.

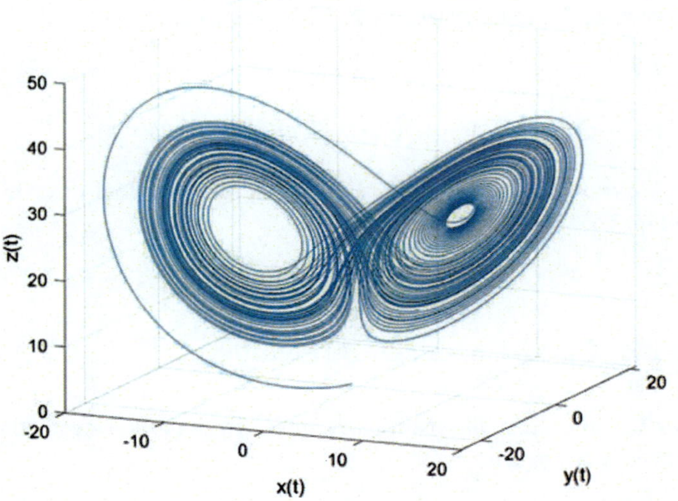

Fig. 2.4 Spatial continuous curve of Lorenz system, where $(a, b, c) = (18, \frac{7}{3}, 26)$

2.1.2.3 Spatial Discrete Curve

The spatial discrete curve is defined as

$$G = \{(m, n, x(m, n)) \mid F(x(m, n), m, n, c) = 0, m, n \in N_r\}.$$

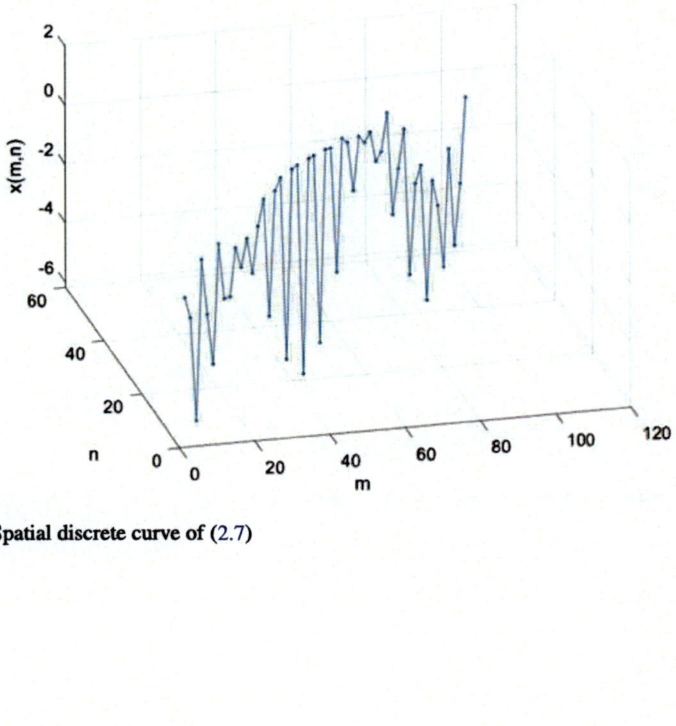

Fig. 2.5 Spatial discrete curve of (2.7)

where the form of F can be generated by other functions.

Example 2.5

$$x(m,n) = 1 - 1.8(1 - 1.89m \cdot sin^2 m + cos^2 n)^2, \tag{2.7}$$

where $m = a$, $n = 2a$, $a = 1, 2, \ldots, N$. We can see its spatial discrete curve as shown in Fig. 2.5.

2.2 Spatial Curve Chaos and Bifurcation

Definition 2.1 In general, the spatial curve given by the following continuous differential dynamical system:

$$\begin{cases} x' = f_1(x, y, z), \\ y' = f_2(x, y, z), \\ z' = f_3(x, y, z), \end{cases} \tag{2.8}$$

produces chaotic behavior under some changes of the parameters and other disturbance, which is defined as **spatial curve chaos**.

Example 2.6 Rossler system [204]:

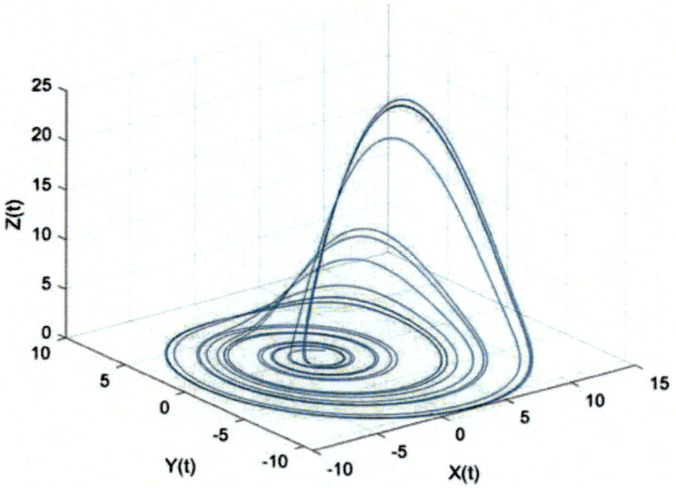

Fig. 2.6 Spatial attractive behavior of Rossler system (2.9), where $(a, b, c) = (0.25, 0.46, 5.78)$

$$\begin{cases} \frac{dx}{dt} = -(y + x), \\ \frac{dy}{dt} = x + ay, \\ \frac{dz}{dt} = b + xz - cz, \end{cases} \tag{2.9}$$

where we can see its spatial continuous curve as attractive behavior as shown in Fig. 2.6.

2.3 Surface

2.3.1 *Definition of Surface*

Firstly, the closed curve which does not intersect in plane is named by Jordan curve. And the Jordan curve departs the plane $\mathbb{R}^2$: the inner part that is finite can be called as the elementary region, while the outer part is infinite.

Definition 2.2 If the relationship from the elementary region to the Euclid space is one to one full mapping, the surface generated by the map is defined as the simple surface.

The surface we study in this book is simple surface.

Based on the Definition (2.2), we set up a surface equation:

$$z = f(x, y), \tag{2.10}$$

where $(x, y) \in G \subset \mathbb{R}^2$ or an implicit function:

$$F(x, y, z) = 0, \tag{2.11}$$

The figure of the spatial surface S is shown as listed below (see examples in Fig. 2.7).

Similarly, the parametric equation and the vector equation can be shown as following:

$$\begin{cases} x = x(u, v), \\ y = y(u, v), \\ z = z(u, v), \end{cases} \tag{2.12}$$

and

$$\vec{r} = \vec{r}(u, v) = \vec{r}(x(u, v), y(u, v), z(u, v)),$$

respectively (Fig. 2.8).

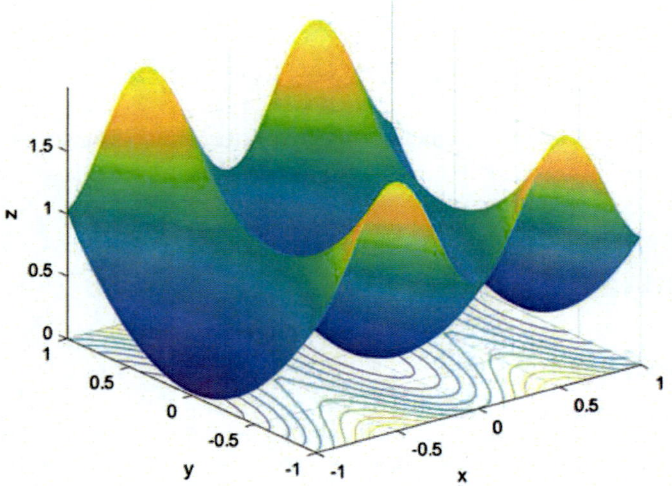

Fig. 2.7 Surface S in $\mathbb{R}^3$ where $z = sin(3x)^2 + y^2$

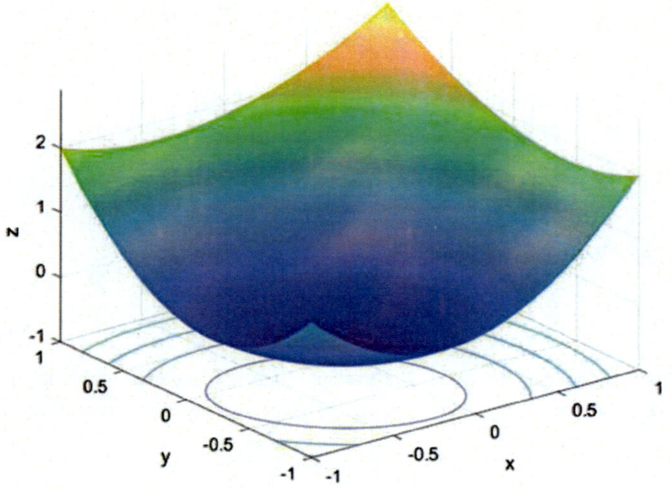

Fig. 2.8 Surface S in $\mathbb{R}^3$ of (2.13)

2.3.2 Continuous Surface

Definition 2.3 Generally, the surface generated by the system (2.10) or (2.11) is named by continuous surface.

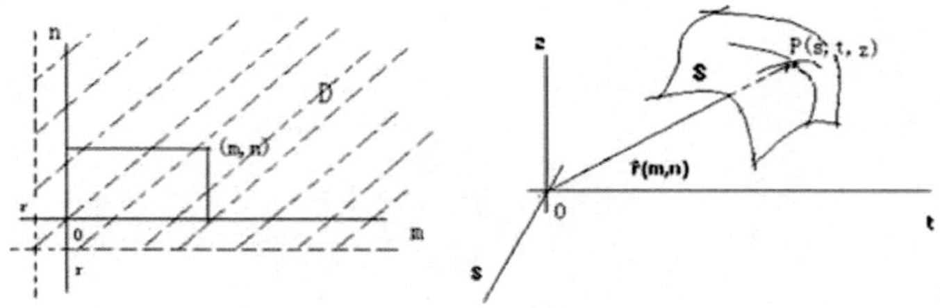

Fig. 2.9 Elementary G and Surface S in $\mathbb{R}^3$

Example 2.7

$$z = f(x, y) = x^2 + y^2 + sin(x + y). \tag{2.13}$$

2.3.3 *Discrete Surface*

Let G to be a discrete set, for example,

$$G = D \times N_r,$$

where $D = N_r \times N_r$,

$$N_r = \{r, r+1, r+2, \ldots, |r \text{ } is \text{ } an \text{ } integer\}.$$

Then we get the corresponding discrete surface according to system (2.11):

$$F(x_{m+1,n}, x_{m,n+1}, x_{m,n}) = 0. \tag{2.14}$$

For arbitrary $(m, n) \in D$, one may consider a spatial surface S, defined by system (2.14) via the implicit function x_{mn}, that is,

$$S : \overrightarrow{r}(m, n) = (s(m, n)), t(m, n), z(m, n)),$$

which is the vector form of the surface S, as shown by Fig. 2.9.

Example 2.8

$$x_{m+1,n} + wx_{m,n+1} = \alpha + \beta x_{m,n}^2 + sin(m^2 + 1) \tag{2.15}$$

is illustrated by Fig. 2.10.

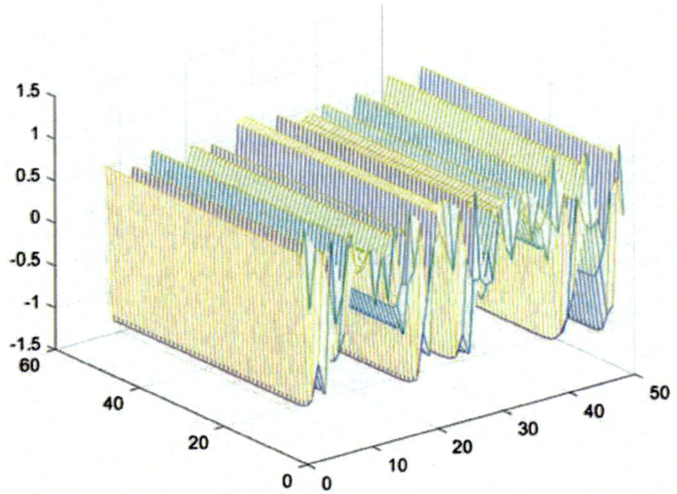

Fig. 2.10 Discrete surface S in $\mathbb{R}^3$ of system (2.15)

2.4 Chaotic Behavior and Bifurcation of the Surface

Definition 2.4 Especially, the chaotic and bifurcation behavior generated by the discrete system (2.14) under some parameter conditions are defined as the surface chaos and surface bifurcation.

Generally, the surface equation of 2-dimensional (2-D) discrete system is given by

$$x_{m+1,n} + a_1 x_{m,n+1} + a_2 x_{m-1,n} + a_3 x_{m,n-1} = f(b, x_{m,n}), \tag{2.16}$$

and the Eq. (2.16) is actually the nonlinear Laplace equation of continuous system:

$$\frac{\partial^2 \phi}{\partial x^2} + a\frac{\partial^2 \phi}{\partial y^2} = f(c, \phi(x, y)), \tag{2.17}$$

where f is a nonlinear function, a_i, $i = 1, 2, 3$ and c are real parameters. In particular, when $f = 1 - \mu x_{m,n}^2$, the system (2.16) is transformed into:

$$x_{m+1,n} + a_1 x_{m,n+1} + a_2 x_{m-1,n} + a_3 x_{m,n+1} = 1 - \mu x_{m,n}^2 \tag{2.18}$$

named as a generalized 2-D logistic system. When $a_i = 0,\ i = 1, 2, 3$, system (2.18) can be replaced by:

$$x_{m+1} = 1 - \mu x_m^2, \tag{2.19}$$

which is a 1-D logistic system that we are familiar with. For (2.16), if different values are given, their surface chaotic behaviors are quite different as shown in the following

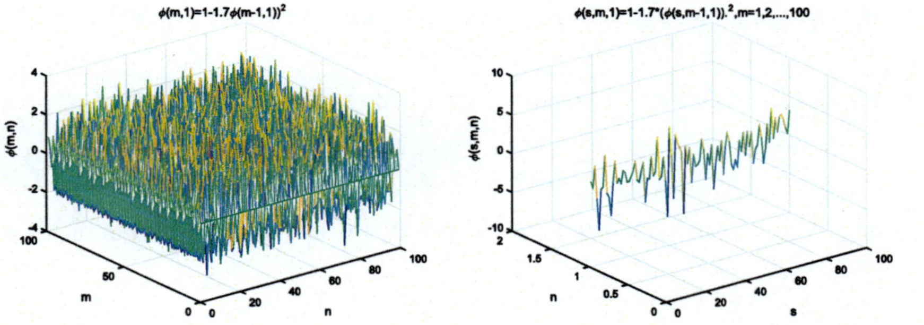

(a) (b)

Fig. 2.11 Surface and slicing chaotic behavior of (2.18)

Fig. 2.11a. In the same way, we can also look at the slicing behavior of their surface chaotic behavior in Fig. 2.11b.

To sum up, it is shown that the chaos and bifurcation behavior of 2-D surface have different characteristics due to different quantity changes. In particular, the chaotic and bifurcation problems of 2-D surface will be discussed in this book.

In addition, we will generalize the theoretical results and applications of surface chaos and bifurcation into wider fields because the surface chaos and bifurcation involve nonlinear partial differential equations, its corresponding discrete systems, 3-dimensional space volume and hyperspace with more independent variables.

The differences and commonness between higher and 1-D or 2-D dimensional discrete chaos and bifurcations are also worth exploring such as the number of the independent variables are larger than 2. It is expected to attract the wide attention of scientists and engineers to obtain good results.

2.5 Related Definitions and Theorems

2.5.1 *About Lyapunov Exponent and Marotto Theorem*

For the discrete system (2.16), the 2-D Lyapunov exponent is defined as follows:

$$\lambda = \frac{1}{N} \lim_{N \to \infty} \sum_{j=1}^{N} \log \left| \frac{||dx_j||}{||dx_{j-1}||} \right|, \tag{2.20}$$

which means that two trajectories would depart exponentially when the two are close at initial time as time increases, where $||dx_j|| = \sqrt{\sum_{j=1}^{N}[dx_k(i)]^2}$, $dx_k = [dx_k(1), dx_k(2), \ldots, dx_k(N)]^T$,

In this section, we first establish some useful preliminary results, which will be needed for the proofs of the main results in the next section.

Definition 2.5 (*Marotto Definition* [339]) A fixed point u^* of the system

$$u_{i+1} = f(\mu, u_i) \tag{2.21}$$

is said to be a snap-back repeller if we can find a solution u_k with not all $u_k = u^*$satisfying:
(i) $u_k = u^*$ for all $k \geq M$ for some positive integer M,
(ii) $u_k \to u^*$ as $k \to -\infty$,
(iii) $f'(\mu, u_k) \neq 0$ for all k,
where, the condition $f'(\mu, u_k) \neq 0$ is the Jacobian condition required of the u'_k in the definition of a snap-back repeller.

Theorem 2.1 (Marotto Theorem [342]) *If f possesses a snap-back repeller then (2.21) is chaotic.*

2.5.2 About Feigenbaum Constant

Feigenbaum constant generated by the bifurcation behavior from the chaotic process for the system (2.21) shows the constant rule of the nonlinear feature from the chaotic behavior which is calculated as:

$$\delta = \lim_{n\to\infty} \frac{\mu_n - \mu_{n-1}}{\mu_{n+1} - \mu_n} = 4.669201609\cdots. \tag{2.22}$$

2.5.3 About Sarkovskii Sequence

Definition 2.6 ([350]) The sequence listed below is defined as the Sarkovskii sequence:

$$\begin{aligned}
&3, 5, 7, 9, 11, \ldots,\\
&3\times 2, 5\times 2, 7\times 2, 9\times 2, 11\times 2, \ldots,\\
&3\times 2^2, 5\times 2^2, 7\times 2^2, 9\times 2^2, 11\times 2^2, \ldots,\\
&\cdots\cdots\\
&\cdots, 2^5, 2^4, 2^3, 2^1, 2^0.
\end{aligned}$$

Theorem 2.2 ([350] Sarkovskii Theorem) *Suppose that $f : R \to R$ is continuous and has a periodic point of prime period k. If $k \rhd l$ in the above ordering, then f also has a periodic point of period l.*

The following period-doubling bifurcation and saddle-node bifurcation theorems are useful.

Lemma 2.1 ([350]) *Period-doubling bifurcation [371].*

Suppose that

1. $f(\mu, 0) = 0$ *for all* μ *in an interval containing* μ_0,
2. $f'_x(\mu_0, 0) = -1$,
3. $\left.\frac{\partial(f^2(\mu,\,(1+\omega)x))}{\partial\mu}\right|_{\mu=\mu_0}(0) \neq 0$.

Then, there is an interval I, *containing 0, and a function* $p((1+\omega)x) : I \to R$, *such that* $f(\mu, (1+\omega)x) \neq (1+\omega)x$ *but* $f^2(\mu, (1+\omega)x) = (1+\omega)x$, *where* $f^2(x) = f(f(x))$ *and* $f'_x(\mu, x) = \frac{\partial f(\mu,x)}{\partial x}$.

Denote

$$f'_\mu(\mu, x) = \frac{\partial f(\mu, x)}{\partial \mu}, \quad f''_x(\mu, x) = \frac{\partial^2 f(\mu, x)}{\partial x^2}.$$

Lemma 2.2 ([350] Saddle-node bifurcation)

Suppose that

1. $f(\mu_0, 0) = 0$;
2. $f'_x(\mu_0, 0) = 1$;
3. $f''_x(\mu_0, 0) \neq 0$;
4. $\left.\frac{\partial(f(\mu,\,(1+\omega)x))}{\partial\mu}\right|_{\mu=\mu_0}(0) \neq 0$. *Then, there exists an interval* I, *containing 0, and a smooth function* $p((1+\omega)x) : I \to R$ *satisfying* $p(0) = \mu_0$, *such that*

$$f_{p((1+\omega)x)}(\mu, (1+\omega)x) = (1+\omega)x.$$

Moreover, $p'(0) = 0$ *and* $p''(0) \neq 0$.

Chapter 3
Spatial Periodic Orbits and Surface Chaos

3.1 Spatial Periodic Orbits

It has been observed that there exist complex spatial, temporal, and spatiotemporal dynamical behaviors in turbulent flows, coupled map lattices (CMLs), circuit arrays, optical waves, etc.:

$$x_{m+1,n} = (1-\varepsilon) f_0(x_{mn}) + \frac{\varepsilon}{2}\left[f_0(x_{m,n-1}) - f_0(x_{m,n+1})\right], \tag{3.1}$$

where m is the discrete time index, n is the lattice site index ($n = 1, 2, \ldots, L$, L is the system size), and ε the coupling constant. In general, the periodic boundary condition $x_{m,n+L} = x_{mn}$ is assumed, and the mapping function $f_0(x)$ is chosen to be the logistic map $f_0(x) = a\,x(1-x)$ in this study, with a being a constant parameter.

Dynamical behaviors such as chaos and synchronization of system (3.1) have been a hot topic of intensive investigation, however the research was only performed experimentally or semi-analytically. Moreover, focus has been given on spatiotemporal dynamics rather than spatial ones, and it is known that they are not the same and one cannot include the other as a special case for study.

It is noted that system (3.1) has some special features that significantly restrict the current investigations:

1. It is a very special lattice model built on the logistic map $f_0(x)$, and its spatial behavior has not been carefully studied.
2. A periodic boundary condition, $x_{m,n+L} = x_{mn}$, is given on n, where the lattice site index n has a finite boundary, that is, $L < \infty$.

On the other hand, it is noted that the following 2-D delay model has also been frequently discussed:

$$x_{m+1,n} + c x_{m,n+1} - x_{m,n} + \sum_{i=1}^{u} p_i(m,n) f(x_{m-\sigma_i, n-\tau_i}) = 0, \tag{3.2}$$

S. T. Liu and L. Zhang, *Surface Chaos and Its Applications*,
https://doi.org/10.1007/978-981-16-8229-2_3

where $p_i(m, n)$ are functions of m and n, and c, σ_i, and $\tau_i, i = 1, \ldots, u$, are constants, with $m, n \in N_r$:

$$N_r = \{r, r+1, r+2, \ldots \mid r \textit{ is an integer and } r \leq 0\},$$

and the following conditions are assumed:

(c_1) $u \in N_1, \sigma_i, \tau_i \in N_0, p_i(m, n) \in [N_r^2, (0, \infty)], i = 1, 2, \ldots, u$.

(c_2) $f : \mathbb{R}^2 \to \mathbb{R}$ and $f(\xi, *) > 0 \; for \; \xi \neq 0$.

Here:

1. The lattice site index n can have either a finite or an infinite boundary.
2. The periodic boundary condition can be either on n, $x_{m,n+L} = x_{mn}$, or on m, $x_{m+L,n} = x_{mn}$, or on (m, n), $x_{m+L,n+L} = x_{mn}$, and it can also be a non-periodic boundary.
3. The mapping function $f_0(x)$ is not necessarily restricted to be the logistic map.

Obviously, system (3.2) is more general than system (3.1), and both of them are interesting with different characteristics.

In this chapter, an even more general model is considered. Precisely, the aim of this study is to introduce a constructive technique for generating spatial periodic orbits and then to give a criterion of surface chaos for the following 2-D nonlinear system:

$$x_{m+1,n} + a x_{m,n+1} = f\left[(1+a)x_{mn}\right], \tag{3.3}$$

where, a is a real constant, and $f(x)$ is a nonlinear continuous function, $m, n \in N_r$ with N_r defined in (3.2), and both $m, n = 0, 1, 2, \ldots$. Let

$$\Omega = N_r \times N_r$$

First, observe that for a given function $\varphi(i, j)$ defined on Ω, it is easy to construct, by induction, a double sequence $\{x_{ij}\}$ that equals $\varphi(i, j)$ on Ω and satisfies system (3.3) for $i, j = 0, 1, 2, \ldots$. Indeed, one can rewrite system (3.3) as

$$x_{m,n+1} = f\left[(1+a)x_{mn}\right] - a x_{m+1,n}$$

and then use it, with initial condition

$$x_{ij} = \varphi(i, j), (i, j) \in \Omega$$

to successively calculate

$$x_{01},\ x_{11},\ x_{02},\ x_{21},\ x_{12},\ x_{03},\ \ldots.$$

Such a double sequence is unique and is a solution of system (3.3).

In general, the following 2-D nonlinear system can be studied:

$$x_{m+1,n} + ax_{m,n+1} = f(bx_{mn}, cx_{m,n-1}, dx_{m,n+1}), \tag{3.4}$$

where, b, c, d are real constants, and one can see that in a special case of

$$\begin{cases} a = 0 \; and \; b = c = d = 1, \\ f(x_{m,n}, x_{m,n-1}, x_{m,n+1}) \\ = (1-\varepsilon) f_0(x_{m,n}) + \frac{\varepsilon}{2}(f_0(x_{m,n-1}) - f_0(x_{m,n+1})), \end{cases}$$

system (3.4) reduces to (3.1). Also when

$$b = 1 + a \; and \; c = d = 0,$$

system (3.4) reduces to (3.3). However, system (3.4) will be studied elsewhere, and the focus of this chapter is on system (3.3).

Another example is that, system (3.3) can be regarded as a discrete analog of the functional partial differential system:

$$\frac{\partial v}{\partial x} + a\frac{\partial v}{\partial y} = f\left[(1+a)v(x, y)\right]. \tag{3.5}$$

In fact, system (3.5) is a Convection Equation with a forced term, quite classical in physics. Therefore, qualitative properties of system (3.3) should provide some useful information for analyzing this companion partial differential system.

3.1.1 Definitions of the Spatial Periodic Orbits

First, some basic concepts are introduced.

Definition 3.1 Let $V \subseteq \mathbb{R}^3$, V_0 be a nonempty subset of V, and take $I \subseteq V_0$, and $I \subset \mathbb{R}$. Assume that $f : I \to I$ is a continuous map, with $f^k(x) = f\left(f^{k-1}(x)\right)$ and $f^0(x) = x$. Then, f is said to be a continuous self-map in I if $f \in C^0(I, I)$ and $f(I) \subset I$. Also, x_0 is said to be a spatial periodic point of period k if $x_0 \in I$ such that

$$\begin{cases} f^k(x_0) = x_0, \\ f^s(x_0) \neq x_0, \; for \; 1 \leq s < k, \end{cases} \tag{3.6}$$

where k is the spatial period of x_0. If it satisfies (3.6) and is the smallest such positive integer, the sequence

$$x_0, x_1, x_2, \ldots, x_k, x_0, x_1, \ldots, \tag{3.7}$$

is called a spatial period orbit of $f(x)$, with period k.

For the sake of symmetry and unification, we shall use the convention

$$(1+a)x_{mn} = x_{mn} + ax_{mn}.$$

3.1.2 Theorems of the Spatial Periodic Orbits

Theorem 3.1 *For any given sequence of distinct nonzero real functions in* $m, n \in N_0$ *in the form of*

$$x_{mn} + ax_{mn},\ x_{m+1,n} + ax_{m,n+1},\ x_{m+2,n} + ax_{m,n+2},\ \ldots,\ x_{m+(k-1),n} + ax_{m,n+(k-1)}, \tag{3.8}$$

the map

$$f(x+y) = a_1 (x+y)^k + a_2 (x+y)^{k-1} + \cdots + a_k (x+y) \tag{3.9}$$

has a periodic point $x_{mn} + x_{mn}$ *with period* k*, where* $a_i = \frac{\Delta^{(i)}}{\Delta}$, $i = 1, 2, \ldots, k$, $\Delta =$

$$\begin{vmatrix} (x_{mn} + ax_{mn})^k & (x_{mn} + ax_{mn})^{k-1} & \cdots & x_{mn} + ax_{mn} \\ (x_{m+1,n} + ax_{m,n+1})^k & (x_{m+1,n} + ax_{m,n+1})^{k-1} & \cdots & x_{m+1,n} + ax_{m,n+1} \\ \cdots & \cdots & \cdots & \cdots \\ (x_{m+(k-1),n} + ax_{m,n+(k-1)})^k & (x_{m+(k-1),n} + ax_{m,n+(k-1)})^{k-1} & \cdots & x_{m+(k-1),n} + ax_{m,n+(k-1)} \end{vmatrix}.$$

Here, determinant $\Delta^{(i)}$ *is obtained from* Δ *by replacing the* i*th column of* Δ *with the following vector:*

$$\left(x_{m+1,n} + ax_{m,n+1},\ x_{m+2,n} + ax_{m,n+2},\ \ldots,\ x_{m+(k-1),n} + ax_{m,n+(k-1)},\ x_{mn} + ax_{mn}\right)^T,$$

$i = 1, 2, \ldots, k.$

Proof Given a nonzero real sequence (3.8), satisfing

$$x_{m+i,n} + ax_{m,n+i} \neq x_{m+j,n} + ax_{m,n+j},\ i \neq j, \quad i, j = 1, 2 \ldots, k, \tag{3.10}$$

for any $m, n \in N_r$, suppose that the map

$$f(x+y) = a_1 (x+y)^k + a_2 (x+y)^{k-1} + \cdots + a_k (x+y)$$

satisfies

$$\left.\begin{aligned} f(x_{mn}+ax_{mn}) &= x_{m+1,n}+ax_{m,n+1}, \\ f(x_{m+1,n}+ax_{m,n+1}) &= x_{m+2,n}+ax_{m,n+2}, \\ \cdots \qquad & \cdots \qquad \cdots \\ f(x_{m+(k-1),n}+ax_{m,n+(k-1)}) &= x_{mn}+ax_{mn}. \end{aligned}\right\} \tag{3.11}$$

For notational simplicity, let $x_{m+i,n}+ax_{m,n+i}=r_i(m,n)=r_i$, $i=0,1,2,\ldots,k-1$. Then, (3.11) is equivalent to the following system:

$$\left.\begin{aligned} a_1r_0^k+a_2r_0^{k-1}+\cdots+a_kr_0 &= r_1, \\ a_1r_1^k+a_2r_1^{k-1}+\cdots+a_kr_1 &= r_2, \\ \cdots \qquad \cdots \qquad & \cdots \\ a_1r_{k-2}^k+a_2r_{k-2}^{k-1}+\cdots+a_kr_{k-2} &= r_{k-1}, \\ a_1r_{k-1}^k+a_2r_{k-1}^{k-1}+\cdots+a_kr_{k-1} &= r_0, \end{aligned}\right\} \tag{3.12}$$

which determines the unknown a_i, $i=1,2,\ldots,k$.

It is easy to check that the determinant of the coefficients of system (3.12) is the kth Vandermonde determinant, and

$$\Delta=\begin{vmatrix} r_0^k & r_0^{k-1} & \cdots & r_0 \\ r_1^k & r_1^{k-1} & \cdots & r_1 \\ \cdots & \cdots & \cdots & \cdots \\ r_{k-1}^k & r_{k-1}^{k-1} & \cdots & r_{k-1} \end{vmatrix} = (-1)^{\frac{k(k-1)}{2}}\left(\prod_{i=0}^{k} r_i\right)\left[\prod_{k-1\geq i>j\geq 1}(r_i-r_j)\right].$$

Since (3.10) holds, one has $\Delta\neq 0$, so that there exists a unique solution of system (3.12), given by

$$a_i^*=\frac{\Delta^{(i)}}{\Delta}, \quad i=1,\,2,\,\ldots,\,k,$$

where the determinant $\Delta^{(i)}$ is obtained from Δ by replacing the ith column of Δ with the following vector:

$$\big(x_{m+1,n}+ax_{m,n+1},\ x_{m+2,n}+ax_{m,n+2},\ \cdots,\\ x_{m+(k-1),n}+ax_{m,n+(k-1)},\ x_{mn}+ax_{mn}\big)^T,$$

which is the vector on the right-hand side of (3.11).

Now, substituting $a_i=a_i^*$ into (3.9), $i=1,2,\ldots,k$, one has

$$f(x+y)=a_1^*(x+y)^k+a_2^*(x+y)^{k-1}+\cdots+a_k^*(x+y). \tag{3.13}$$

It is easy to see that the function in (3.13) is continuous and

$$f^k(x_{mn} + ax_{mn}) = f^k(r_0) = r_0 = x_{mn} + ax_{mn},$$

but

$$x_{mn} + ax_{mn} \neq f^s(x_{mn} + ax_{mn}), \quad \text{for } 1 \leq s < k.$$

Hence, the function in (3.13) is a continuous map with a spatial periodic point $x_{mn} + ax_{mn}$ of period k.

Corollary 3.1 *For any positive integer k, there exists a kth order polynomial with period k.*

3.2 Surface Chaos

First, recall the concept of discrete chaos in the sense of Li and Yorke.

3.2.1 Definition of the Surface Chaos

Lemma 3.1 ([351]) *Let $I \subset \mathbb{R}$ be an interval and $f: I \to I$ be a continuous map. Assume that there is a point $a \in I$, satisfying*

$$f^3(a) \leq a < f(a) < f^2(a) \quad \text{or} \quad f^3(a) \geq a > f(a) > f^2(a).$$

Then:

(1) *For every $i = 1, 2, \ldots$, there is a periodic point of f^i with period k in I.*
(2) *There are an uncountable set $S \subset I$ (containing no periodic points) and an uncountable subset $S_0 \subset S$, such that*

(2.1) *for every $p, q \in S_0$ with $p \neq q$,*

$$\lim_{k\to\infty} \sup \left| f^k(p) - f^k(q) \right| > 0$$

and

$$\lim_{k\to\infty} \inf \left| f^k(p) - f^k(q) \right| = 0,$$

(B) *for every $p \in S$ and periodic point $q \in I$ with $p \neq q$*

$$\lim_{k\to\infty} \sup \left| f^k(p) - f^k(q) \right| > 0.$$

The 1-D dynamical system $x_{i+1} = f(x_i)$ that satisfies the above conditions is said to be chaotic in the sense of Li and Yorke.

Definition 3.2 Let $V \subseteq \mathbb{R}^3$, V_0 be a nonempty subset of V, and take $I \subseteq V_0$, and $I \subset \mathbb{R}$. Then, f is said to be chaotic on V_0 if it is chaotic on I, and f is said to be surface chaotic on V if it is chaotic on V_0, both in the sense of Li and Yorke.

3.2.2 Theorems of the Surface Chaos

Theorem 3.2 *Let $V \subseteq \mathbb{R}^3$, V_0 be a non-empty subset of V, and I be an interval in V_0, and I is one dimension. Denote*

$$r_i(m,n) = x_{m+i,n} + ax_{m,n+i}, \quad i = 1,\ 2,\ 3, \tag{3.14}$$

and assume the following conditions:

1^0 $r_i(m,n) \neq 0$, $i = 1, 2, 3$, *and* $r_i(m,n) \neq r_j(m,n)$ *if* $i \neq j$ *for all* $m, n \in N_0$, $i,\ j = 1,\ 2,\ 3$.
2^0 $r_1(m,n) < r_2(m,n) < r_3(m,n)$ *or* $r_1(m,n) > r_2(m,n) > r_3(m,n)$ *for all* m, $n \in N_0$.
3^0 *Let*

$$f^*(x+y) = a_1^*(x+y)^3 + a_2^*(x+y)^2 + a_3^*(x+y), \tag{3.15}$$

where $a_i^* = \frac{D^{(i)}}{D}$, $i = 1, 2, 3$,

$$D = \begin{vmatrix} r_1^3 & r_1^2 & r_1 \\ r_2^3 & r_2^2 & r_2 \\ r_3^3 & r_3^2 & r_3 \end{vmatrix},$$

and the determinant $D^{(i)}$ is obtained from D by replacing the ith column of D with the following vector:

$$\left(x_{m+1,n} + ax_{m,n+1},\ x_{m+2,n} + ax_{m,n+2},\ x_{mn} + ax_{mn},\right)^T, \quad i = 1, 2, \ldots, k.$$

4^0 $f^*(I) \subset I$.

Then, system

$$x_{m+1,n} + ax_{m,n+1} = f^*((1+a)x_{mn}) \tag{3.16}$$

is surface chaotic on V in the sense of Li and Yorke.

Proof Without loss of generality, consider system

$$x_{s+1,n} + ax_{m,t+1} = f^*(x_{sn} + ax_{mt}),$$

where $s, t, m, n \in N_r$, and take $s, t = 0, 1, 2, 3$. Then, one can compute and obtain the above $r_i(m, n), i = 0, 1, 2, 3$, where $r_0(m, n) = x_{0n} + x_{m0}$. Next, applying Theorem 3.1 yields

$$x_{s+1,n} + ax_{m,t+1} = a_1^*(x_{sn} + ax_{mt})^3 + a_2^*(x_{sn} + ax_{mt})^2 + a_3^*(x_{sn} + ax_{mt}). \quad (3.17)$$

Since $s, t, m, n \in N_r$, leting $s = m$ and $t = n$ gives

$$\begin{aligned} x_{m+1,n} + ax_{m,n+1} &= a_1^*(x_{mn} + ax_{mn})^3 + a_2^*(x_{mn} + ax_{mn})^2 + a_3^*(x_{mn} + ax_{mn}) \\ &= f^*\left[(1+a)x_{mn}\right]. \end{aligned} \quad (3.18)$$

Note that

$$\begin{aligned} &f^{*3}(x_{mn} + ax_{mn}) = f^{*3}(r_1(m, n)) = r_1(m, n) = x_{mn} + ax_{mn}, \\ &f^{*i}(r_1(m, n)) \neq r_1(m, n), \quad 1 \le i < 3. \end{aligned}$$

Therefore, the map (3.16) has a periodic point $r_1(m, n)$ of period 3.

Since $r_i(m, n)$ are distinct points and $r_i(m, n) \neq 0$ for $m, n \in N_0$, $i = 1, 2, 3$, it follows immediately from Theorem 3.1 that $D \neq 0$, that is, a_i^* are uniquely determined. Therefore, $f^*(x_{mn} + ax_{mn}) = f^*[(1+a)x_{mn}]$ is uniquely determined. In addition, since $r_i(m, n) \in I$ and $f^*(I) \subset I$, one concludes that $f^*((1+a)x_{mn})$ is a continuous self-map in $I \subset V_0$.

On the other hand, it follows from 2^0 that

$$\begin{aligned} f^{*3}(r_1(m, n)) = r_1(m, n) < r_2(m, n) &= f^*(r_1(m, n)) \\ &< f^{*2}(r_1(m, n)) = r_3(m, n), \end{aligned}$$

for $m, n \in N_0$.

Thus, applying Lemma 3.1 leads to the conclusion that system (3.16) is chaotic on I, so that Definition 3.2 implies that system (3.16) is surface chaotic on V, in the sense of Li and Yorke.

Example 3.1 Let $a = 1$, $V_0 = V = [0, 50]^3$, and take $I = [\frac{3}{2}, \frac{17}{4}]$, then $I \subset V_0$ and $I \subset R$. For the boundary condition of V_0, $r_0(m, n) = x_{0n} + x_{m0}$, without loss of generality one may take $r_0(m, n) = x_{0n} + x_{m0} = 2$, $r_1(m, n) = x_{1n} + x_{m1} = 3$, $r_2(m, n) = x_{2n} + x_{m2} = 4$, $r_3(m, n) = x_{3n} + x_{m3} = 2$.

By applying Theorem 3.2, one has

$$\left.\begin{aligned} f(x_{0n} + x_{m0}) &= r_1(m, n) = x_{m+1,n} + x_{m,n+1} = 3, \\ f(x_{m+1,n} + x_{m,n+1}) &= r_2(m, n) = x_{m+2,n} + x_{m,n+2} = 4, \\ f(x_{m+2,n} + x_{m,n+2}) &= r_3(m, n) = x_{3n} + x_{m3} = r_0(m, n) = 2, \end{aligned}\right\}$$

which yields

$$\left.\begin{aligned} 2a_3 + 4a_2 + 8a_1 &= 3, \\ 3a_3 + 9a_2 + 27a_1 &= 4, \\ 4a_3 + 16a_2 + 64a_1 &= 2. \end{aligned}\right\} \tag{3.19}$$

Solving system (3.19) gives $a_3 = -\frac{1}{6}$, $a_2 = \frac{3}{2}$, $a_1 = -\frac{1}{3}$, and, substituting a_i, $i = 1, 2, 3$, into (3.53) for this example yields

$$x_{m+1,n} + x_{m,n+1} = -\frac{1}{3}(x_{mn} + x_{mn})^3 + \frac{3}{2}(x_{mn} + x_{mn})^2 - \frac{1}{6}(x_{mn} + x_{mn}), \tag{3.20}$$

for all $m, n \in N_0$. Let

$$f(2x_{mn}) = -\frac{1}{3}(2x_{mn})^3 + \frac{3}{2}(2x_{mn})^2 - \frac{1}{6}(2x_{mn}), \tag{3.21}$$

where $m, n \in N_0$. Then, it follows from (3.20) that

$$\begin{cases} 3 = f(2), \\ 4 = f(3) = f(f(2)) = f^2(2), \\ 2 = f(4) = f(f(f(2))) = f^3(2). \end{cases} \tag{3.22}$$

In view of (3.22), it is easy to see that $r_0(m, n) = x_{0n} + x_{m0} = 2$ is a periodic point of peniodc 3. On the other hand, notice that

$$f^3(2) = 2 < 3 = f(2) < 4 = f^2(2),$$

for $m, n \in N_0$. Now, by letting $\xi = 2x$ in (3.21), one has $f(\xi) = -\frac{1}{3}\xi^3 + \frac{3}{2}\xi^2 - \frac{1}{6}\xi$.

Solving equation $f'(\xi) = 0$ gives $\xi_1 \approx -0.85$, $\xi_2 \approx -2.15$, since another pair $\xi_1 \approx -0.85$ and $\xi_2 \approx -2.15$ are not in I. In addition, note that $f(\frac{3}{2}) = 2$, $f(\frac{17}{4}) = 3.354$. Hence,

$$\min\left\{f\left(\frac{3}{2}\right), f\left(\frac{17}{4}\right)\right\} < f(\xi) < \max\left\{f\left(\frac{3}{2}\right), f\left(\frac{17}{4}\right)\right\},$$

implying that $f(I) \subset I$.

It follows from Theorem 3.2 that this system is chaotic on I, so also on V_0 by Definition 3.2, in the sense of Li and Yorke.

The surface chaotic behavior of this system is demonstrated by Fig. 3.1.

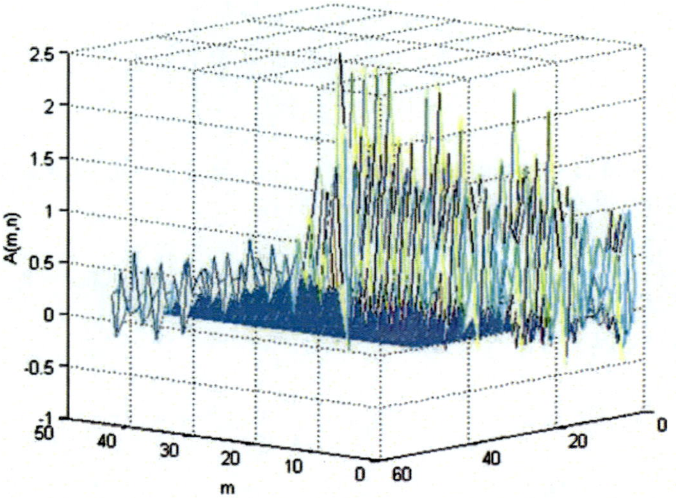

Fig. 3.1 Surface chaotic behavior of the Example 3.1

3.3 Construction of Surface Chaos of Generalized 2-D Discrete Logistic Systems

For the following 2-D discrete logistic system:

$$\begin{aligned}x_{m+1,n} + x_{m,n+1} = \mu(2x_{mn})^r(\lambda_1(2x_{mn})^{k+\alpha} + \lambda_2(2x_{mn})^{k+\alpha-1}\\ + \cdots + \lambda_v(2x_{mn})^{\alpha+1}),\end{aligned} \tag{3.23}$$

there also exists theorem about the surface chaos, where μ are a non-zero real, $m, n \in N_u = \{\ u, u+1, u+2, ..., \mid u$ is a integer and $u \leq 0\ \}$, $\lambda_i, i = 1, 2, \ldots \nu$, α is real parameter. Especially, when $\mu = 1$, $k = 2$, $r = \alpha = 0$, $\lambda_1 = -\frac{1}{4}\mu$ and $\lambda_2 = \frac{1}{2}\mu$, then system (3.23) is quite general.

Moreover,

$$\begin{aligned}x_{m,n+1} = \mu(2x_{mn})^r(\lambda_1(2x_{mn})^{k+\alpha} + \lambda_2(2x_{mn})^{k+\alpha-1}\\ + \cdots + \lambda_v(2x_{mn})^{\alpha+1}) - x_{m+1,n}\end{aligned}$$

can be regarded as a discrete analog of the following functional partial differential system:

$$\begin{aligned}\frac{\partial u}{\partial x} + \frac{\partial u}{\partial y} = \mu\,(2u(x.y))^r\,[\lambda_1(2u(x.y))^{k+\alpha} + \lambda_2\,(2u(x.y))^{k+\alpha-1}\\ + \cdots + \lambda_v\,(2u(x.y))^{\alpha+1}].\end{aligned} \tag{3.24}$$

Theorem 3.3 *For any given distinct nonzero real function sequences for any* $m, n \in N_0$ *of the form:*

$$x_{mn} + x_{mn},\ x_{m+1,n} + x_{m,n+1},\ x_{m+2,n} + x_{m,n+2},\ \ldots,\ x_{m+(v-1),n} + x_{m,n+(v-1)}, \tag{3.25}$$

the map

$$f\,(x+y) = \mu(x+y)^r \left(\lambda_1 (x+y)^{k+\alpha} + \lambda_2 (x+y)^{k+\alpha-1} + \cdots + \lambda_v (x+y)^{\alpha+1}\right), \tag{3.26}$$

have a periodic point $x_{mn} + x_{mn}$ *with period* v, *where* $\lambda_i = \frac{\Delta^{(i)}}{\Delta}$, $i = 1, 2, \ldots, v$, α *is real parameter, and* $\Delta =$

$$\begin{vmatrix} \mu(x_{m,n} + x_{m,n})^{k+\alpha+r} & \mu(x_{m,n} + x_{m,n})^{k+\alpha+r-1} & \cdots & \mu(x_{m,n} + x_{m,n})^{\alpha+r-1} \\ \mu(x_{m+1,n} + x_{m,n+1})^{k+\alpha+r} & \mu(x_{m+1,n} + x_{m,n+1})^{k+\alpha+r-1} & \cdots & \mu(x_{m+1,n} + x_{m,n+1})^{\alpha+r-1} \\ \cdots & \cdots & \cdots & \cdots \\ \mu(x_{m+(v-1),n} + x_{m,n+(v-1)})^{k+\alpha+r} & \mu(x_{m+(v-1),n} + x_{m,n+(v-1)})^{k+\alpha+r-1} & \cdots & \mu(x_{m+(v-1),n} + x_{m,n+(v-1)})^{\alpha+r-1} \end{vmatrix}.$$

Moreover, the determinant $\Delta^{(i)}$ *is obtained from* Δ *by replacing* i*th column of* Δ *with the vector*

$$\left(x_{m+1,n} + x_{m,n+1},\ x_{m+2,n} + x_{m,n+2},\ \ldots,\ x_{m+(v-1),n} + x_{m,n+(v-1)},\ x_{mn} + x_{mn}\right)^T,$$

$i = 1, 2, \ldots, v$.

Proof Given a nonzero real function sequences (3.25), which satisfies

$$x_{m+i,n} + x_{m,n+i} \neq x_{m+j,n} + x_{m,n+j},\ i \neq j,\ i,\ j = 1, 2 \ldots, k,\ \text{for any } m, n \in N_u. \tag{3.27}$$

Suppose that the map

$$f\,(x+y) = \mu(x+y)^r \left(\lambda_1 (x+y)^{k+\alpha} + \lambda_2 (x+y)^{k+\alpha-1} + \cdots + \lambda_v (x+y)^{\alpha+1}\right)$$

satisfies

$$\left.\begin{aligned} f\,(x_{mn} + x_{mn}) &= x_{m+1,n} + x_{m,n+1}, \\ f\,(x_{m+1,n} + x_{m,n+1}) &= x_{m+2,n} + x_{m,n+2}, \\ \cdots \qquad & \cdots \qquad \cdots \\ f\,(x_{m+(v-1),n} + x_{m,n+(v-1)}) &= x_{mn} + x_{mn}. \end{aligned}\right\} \tag{3.28}$$

For notational simplicity, let $x_{m+i,n} + x_{m,n+i} \overset{def.}{=} r_i = (m, n) \overset{\text{def.}}{=} r_i$, $i = 0, 1, 2, \ldots,$ $v - 1$.

Then (3.28) reduces to the following system

$$\left.\begin{aligned} \mu r_0^r(\lambda_1 r_0^{k+\alpha} + \lambda_2 r_0^{k+\alpha-1} + \cdots + \lambda_v r_0^{\alpha+1}) &= r_1, \\ \mu r_1^r(\lambda_1 r_1^{k+\alpha} + \lambda_2 r_1^{k+\alpha-1} + \cdots + \lambda_v r_1^{\alpha+1}) &= r_2, \\ \cdots \qquad \cdots \qquad \cdots \\ \mu r_{v-2}^r(\lambda_1 r_{v-2}^{v+\alpha} + \lambda_2 r_{v-2}^{v+\alpha-1} + \cdots + \lambda_v r_{v-2}^{\alpha+1}) &= r_{v-1}, \\ \mu r_{v-1}^r(\lambda_1 r_{v-1}^{k+\alpha} + \lambda_2 r_{v-1}^{k+\alpha-1} + \cdots + \lambda_v r_{v-1}^{\alpha+1}) &= r_0, \end{aligned}\right\} \tag{3.29}$$

which determine the unknown λ_i, $i = 1, 2, \ldots, v$.

It is easy to check that the determinant of the coefficients of system (3.29) is the kth Vandermonde determinant, and

$$\Delta = \begin{vmatrix} \mu r_0^{k+r+\alpha} & \mu r_0^{k+r+\alpha-1} & \ldots & \mu r_0^{r+\alpha+1} \\ \mu r_1^{k+r+\alpha} & \mu r_1^{k+r+\alpha-1} & \ldots & \mu r_1^{r+\alpha+1} \\ \ldots & \ldots & \ldots & \ldots \\ \mu r_{v-1}^{k+r+\alpha} & \mu r_{v-1}^{k+r+\alpha-1} & \ldots & \mu r_{v-1}^{r+\alpha+1} \end{vmatrix}$$

$$= \mu^v (-1)^{\frac{k(k-1)}{2}} \left(\prod_{i=0}^{v} r_i\right)^{r+\alpha} \left[\prod_{v-1 \geq i > j \geq 1} (r_i - r_j)\right].$$

Since (3.27) holds, one has $\Delta \neq 0$. So there exists a unique solution of system (3.29), given by

$$\lambda_i^* = \frac{\Delta^{(i)}}{\Delta}, \quad i = 1, 2, \ldots, v,$$

where the determinant $\Delta^{(i)}$ is obtained from Δ by replacing ith column of Δ with the following vector

$$(x_{m+1,n} + x_{m,n+1}, x_{m+2,n} + x_{m,n+2}, \ldots, \\ x_{m+(v-1),n} + x_{m,n+(v-1)}, x_{mn} + x_{mn})^T,$$

which is the vector of the values on the right-hand side of (3.29).

Now, substituting $\lambda_i = \lambda_i^*$ into (3.26), $i = 1, 2, \ldots, v$, one has

$$f(x+y) = \mu(x+y)^r(\lambda_1^*(x+y)^{k+\alpha} + \lambda_2^*(x+y)^{k+\alpha-1} \\ + \cdots + \lambda_v^*(x+y)^{\alpha+1}). \tag{3.30}$$

It is easy to see that the function in (3.30) is continuous in the space $\mathbb{R}^3$ and

$$f^v(x_{mn} + x_{mn}) = f^v(r_0) = r_0 = x_{mn} + x_{mn},$$

but

$$x_{mn} + x_{mn} = f^s(x_{mn} + x_{mn}) \ for\ 1 \leq s < v.$$

Hence, the function in (3.30) is a continuous map with a periodic point $x_{mn} + x_{mn}$ with *period* v.

Corollary 3.2 *There must exist a vth polynomial with period v for any positive integer v.*

Theorem 3.4 *Let $V \subseteq \mathbb{R}^3 = (-\infty, \infty)^3$, V_0 be a non-empty subset of V, let I be an interval such that $I \subset V_0$, assume that*

$$r_i(m,n) = x_{m+i,n} + x_{m,n+i},\ i = 1,\ 2,\ 3, \tag{3.31}$$

and the following conditions holds:

1^0 $r_i(m,n) \neq 0$, $i = 1, 2, 3$, and $r_i(m,n) \neq r_j(m,n)$ for $m, n \in N_0$, $i, j = 1, 2, 3$.

2^0 $r_1(m,n) < r_2(m,n) < r_3(m,n)$ or $r_1(m,n) > r_2(m,n) > r_3(m,n)$ for $m, n \in N_0$.

3^0 Let

$$f^*(x+y) = \mu(x+y)^r \left(\lambda_1^*(x+y)^{k+\alpha} + \lambda_2^*(x+y)^{k+\alpha-1} + \lambda_3^*(x+y)^{\alpha+1}\right) \tag{3.32}$$

where $\lambda_i^* = \frac{D^{(i)}}{D}$, $i = 1, 2, 3$,

$$D = \begin{vmatrix} \mu r_1^{3+r+\alpha} & \mu r_1^{2+r\alpha} & \mu r_1^{1+r+\alpha} \\ \mu r_2^{3+r+\alpha} & \mu r_2^{2+r\alpha} & \mu r_2^{1+r+\alpha} \\ \mu r_3^{3+r+\alpha} & \mu r_3^{2+r\alpha} & \mu r_3^{1+r+\alpha} \end{vmatrix},$$

and the determinant $D^{(i)}$ is obtained from D by replacing ith column of D with the vector

$$\left(x_{m+1,n} + x_{m,n+1},\ x_{m+2,n} + x_{m,n+2},\ x_{m,n} + x_{m,n}\right)^T,\ i = 1, 2, \ldots, k.$$

4^0 $f(x) \in C^0(I, I)$ and $f(I) \subset I$.
Then, system

$$x_{m+1,n} + x_{m,n+1} = f^*(2x_{mn}) \tag{3.33}$$

is chaotic on V_0 in the sense of Li -Yorke.

Proof Without loss of generality, consider system

$$x_{s+1,n} + x_{m,t+1} = f^*(x_{s,n} + x_{m,t}),$$

where $s, t, m, n \in N_u$, and take $s, t = 0, 1, 2, 3$, then one can compute and obtain the above $r_i(m,n)$, $i = 0, 1, 2, 3$, where, $r_0(m,n) = x_{0,n} + x_{m,0}$. Next, applying Theorem 3.1, we get that

$$x_{s+1,n} + x_{m,t+1} = \lambda_1^*(x_{sn} + x_{mt})^{k+\alpha} + \lambda_2^*(x_{sn} + x_{mt})^{k+\alpha-1} + \lambda_3^*(x_{sn} + x_{mt})^{\alpha+1}. \tag{3.34}$$

Since $s, t, m, n \in N_u$, in particular, let $s = m$, $t = n$, to obtain

$$x_{m+1,n} + x_{m,n+1} = \lambda_1^*(2x_{mn})^{k+\alpha} + \lambda_2^*(2x_{mn})^{k+\alpha-1} + \lambda_3^*(2x_{mn})^{\alpha+1} = f^*(2x_{mn}). \quad (3.35)$$

Note that

$$f^{*3}(x_{mn} + x_{mn}) = f^{*3}(r_1(m,n)) = r_1(m,n) = x_{mn} + x_{mn},$$
$$f^{*i}(r_1(m,n)) \neq r_1(m,n),\ 1 \leq i < 3.$$

Therefore, the map (3.34) have a periodic point $r_1(m,n)$ of period 3.

Since $r_i(m,n)$ are distinct points and $r_i(m,n) \neq 0$ in $m, n \in N_0$, $i = 1, 2, 3$, it follows immediately from Theorem 1 that $D \neq 0$, that is, λ_i^* is uniquely determined. Therefore $f^*(x_{mn} + x_{mn}) = f^*(2x_{mn})$ is uniquely determined. In addition, since $r_i(m,n) \in I$ and $f^*(I) \subset I$ from 4^0, one concludes that $f^*(2x_{mn})$ is a continuous self-map in $I \subset V_0$. On the other hand, it follows from 2^0 that

$$f^{*3}(r_1(m,n)) = r_1(m,n) < r_2(m,n) = f^*(r_1(m,n)) < f^{*2}(r_1(m,n)) = r_3(m,n),$$

for $m, n \in N_0$. Thus, applying Lemma 3.1, one concludes that system (3.33) is chaotic on I, thus, using Definition 3.2, one can further conclude that system (3.33) is chaotic on V_0, in the sense of Li and Yorke.

Example 3.2 Let $V_0 = V = [0, 50]^3 = [0, 50] \times [0, 50] \times [0, 50] \subseteq R^3, \alpha = -\frac{2}{3}$, $\mu = 1, r = 0$ and $I = [1, 3.1] \subset V_0$. Since $r_0(m,n) = x_{0n} + x_{m0}$ is boundary values of V_0, therefore, without loss of generality, one may take $r_0(m,n) = x_{0n} + x_{m0} = 1$, and $r_1(m,n) = x_{1n} + x_{m1} = 2.1$, $r_2(m,n) = x_{2n} + x_{m2} = 3.1$, $r_3(m,n) = x_{3n} + x_{m3} = 1$. By using Theorem 3.4, then one has

$$\left.\begin{aligned} f(x_{0n} + x_{m0}) &= r_1(m,n) = x_{m+1,n} + x_{m,n+1} = 2.1, \\ f(x_{m+1,n} + x_{m,n+1}) &= r_2(m,n) = x_{m+2,n} + x_{m,n+2} = 3.1, \\ f(x_{m+2,n} + x_{m,n+2}) &= r_3(m,n) = x_{3n} + x_{m3} = r_0(m,n) = 1, \end{aligned}\right\}$$

which yields

$$\left.\begin{aligned} \lambda_1 + \lambda_2 + \lambda_3 &= 2.1, \\ 2.1^{\frac{7}{3}}\lambda_1 + 2.1^{\frac{4}{3}}\lambda_2 + 2.1^{\frac{1}{3}}\lambda_3 &= 3.1, \\ 3.1^{\frac{7}{3}}\lambda_1 + 3.1^{\frac{4}{3}}\lambda_2 + 3.1^{\frac{1}{3}}\lambda_3 &= 1. \end{aligned}\right\} \quad (3.36)$$

Solving system (3.36) gives $\lambda_1 = -0.9650$, $\lambda_2 = 3.2832$, $\lambda_3 = -0.2181$, and, substituting a_i, $i = 1, 2, 3$ into (3.35) yields

$$x_{m+1,n} + x_{m,n+1} = -0.9650(2x_{mn})^{\frac{7}{3}} + 3.2832(2x_{mn})^{\frac{4}{3}} - 0.2181(2x_{mn})^{\frac{7}{3}}, \quad (3.37)$$

for all $m, n \in N_t$. Let

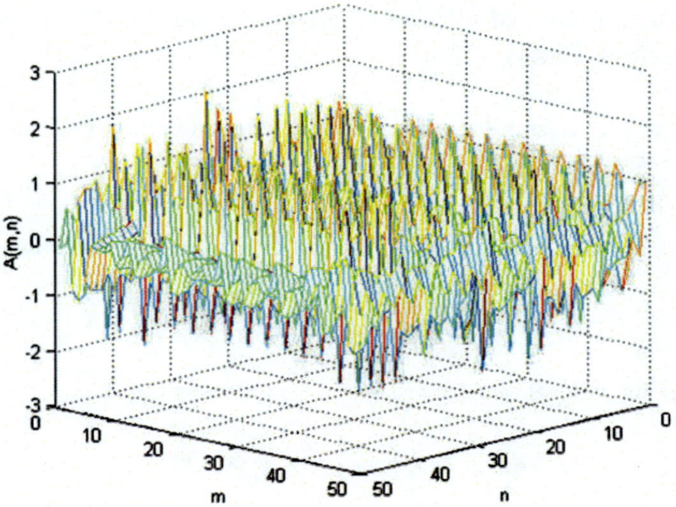

Fig. 3.2 The surface chaos of system (3.37)

$$f(2x_{mn}) = -0.9650(2x_{mn})^{\frac{7}{3}} + 3.2832(2x_{mn})^{\frac{4}{3}} - 0.2181(2x_{mn})^{\frac{7}{3}}, \tag{3.38}$$

for all $m, n \in N_t$. Then, it follows from (3.37) that

$$\begin{cases} 2.1 = f(1), \\ 3.1 = f(2.1) = f(f(1)) = f^2(1), \\ 1 = f(3.1) = f(f(f(1))) = f^3(1). \end{cases} \tag{3.39}$$

In view of (3.39), it is easy to see that $r_0(m, n) = x_{0n} + x_{m0} = 1$ is a 3-iterative periodic point. On the other hand, notice that

$$f^3(1) = 1 < 2.1 = f(1) < 3.1 = f^2(1), \text{ for } m, n \in N_0.$$

So, by letting $\xi = 2x$ in (3.38), one has $f(\xi) = -0.9650\xi^{\frac{7}{3}} + 3.2832\xi^{\frac{4}{3}} - 0.2181\xi^{\frac{1}{3}}$. Now, solving equation $f'(\xi) = 0$ gives $\xi_1 \approx 1.93^{\frac{1}{3}}$, $\xi_2 \approx 0.017^{\frac{1}{3}}$, in addition, note that $f(1) = 2.1$, $f(3.1) = 1$, hence,

$$\min\{f(1), f(\xi_1), f(\xi_2), f(3.1)\} < f(\xi) < \max\{f(1), f(\xi_1), f(\xi_2), f(3.1)\},$$

implying that $f(I) \subset I$. It follows from Theorem 3.4 that system (3.37) is chaotic on I, so also on V_0, in the sense of Li and Yorke. Finally, the chaotic behavior of system (3.37) is demonstrated by Fig. 3.2.

3.4 Construction of a Class of Space Periodic Orbits and Surface Chaos

A model of atom arranged in order in fixed physics and a model of light turbulent flow in laser physics are expressed the following of the form

$$x_{n+1} + x_{n-1} = \frac{\mu}{2\pi}\sin(2\pi x_n),$$

and

$$\varphi_{n+1} = A^2[1 + 2B\cos(\varphi_n - \varphi_0)].$$

Their two-dimensional case can be showed the following form:

$$x_{m+1,n} - x_{m,n} + x_{m,n-1} = \frac{\mu}{2\pi}\sin(2\pi x_{mn}),$$

and

$$\varphi_{m+1,n} + \varphi_{m,n+1} - \varphi_{mn} = A^2[1 + 2B\cos(\varphi_{mn} - \varphi_{00})].$$

Therefore, regarding iterative of triangle polynomial has rich dynamics. In the part, we mainly study a spatio-temporal behavior for 2-D discrete system as the following form:

$$x_{m+1,n} + x_{m,n-1} = \mu_1\sin(x_{mn}) + \mu_2\sin(2x_{mn}) + \cdots + \mu_k\sin(kx_{mn}). \quad (3.40)$$

In addition, we have a lot of problems and will deal with the above system (3.40). Our aims are to consider the construction of the space periodic orbits and to give criterion of surface chaos of the system (3.40). And let

$$\Omega = [r,\infty) \times [r,\infty)\backslash[1,\infty) \times [1,\infty).$$

On the other hand, given a function $\varphi(i, j)$ defined on $\Omega = [r,\infty] \times [r,\infty]\backslash[1,\infty] \times [1,\infty]$, it is easy to construct by induction a double sequence $\{x_{ij}\}$ that equals $\varphi(i, j)$ on Ω and satisfies Eq. (3.40) for $i, j = 0, 1, 2, \ldots$. Indeed, we can rewrite Eq. (3.40) as

$$x_{m,n+1} = \sum_{i=1}^{k}\mu_i\sin(ix_{mn}) - x_{m+1,n},$$

and then use it to calculate, successively,

$$x_{01}, x_{11}, x_{02}, x_{21}, x_{12}, x_{03}, \ldots.$$

Such a double sequence is unique and is a solution of Eq. (3.40) subject to the initial condition:

$$x_{ij} = \varphi(i, j),\ (i, j) \in \Omega.$$

We can see that system (3.40) can be regarded as a discrete analog of the following functional partial differential system:

$$\frac{\partial u}{\partial x} + \frac{\partial u}{\partial y} = \sum_{i=1}^{k} \mu_i \sin\left(iu(x, y)\right). \tag{3.41}$$

In fact, Eq. (3.41) is a Convection Equation with a forecd term in physics. Therefore, qualitative properties of system (3.40) may lead to some useful information for analyzing this companion partial differential system.

Theorem 3.5 *For any given distinct non-zero real function sequences for any* $s, m, n \in N_0$ *of the form:*

$$x_{m,n} + ax_{m,n}, x_{m+1,n} + ax_{m,n+1}, x_{m+2,n} + ax_{m,n+2}, \\ \ldots, x_{m+(k-1),n} + ax_{m,n+(k-1)},$$

and which satisfy

$$r_i \stackrel{det.}{=} r_i(m, n) \stackrel{det.}{=} x_{m+i,n} + ax_{m,n+i} \neq p\pi,\ \textit{for any}\ m, n \in N_0,\ k \geq i \geq 1 \tag{3.42}$$

and

$$r_i + r_j \neq 2t\pi, r_i - r_j \neq 2l\pi,\ \textit{for any}\ m, n \in N_0,\ k \geq i > j \geq 1, \tag{3.43}$$

where $p, t, l \in Z$, *there exists a continuous self-map of the form*

$$S(x) = \mu_1 \sin x + \mu_2 \sin(2x) + \cdots + \mu_n \sin(kx) \tag{3.44}$$

having a periodic point $r(m, n) = x_{mn} + x_{mn}$ *with period* k, *where* $a_i = \frac{\Delta^{(i)}}{\Delta}$, $i = 1, 2, \ldots, k$,

$$\Delta = \begin{vmatrix} \sin r_1 & \sin 2r_1 & \cdots & \sin nr_1 \\ \sin r_2 & \sin 2r_2 & \cdots & \sin nr_2 \\ \cdots & \cdots & \cdots & \cdots \\ \sin r_n & \sin 2r_n & \cdots & \sin nr_n \end{vmatrix},$$

and the determinant $\Delta^{(i)}$ *is obtained from* Δ *by replacing column* i*th column of* Δ *by the vector*

$$(x_{m+1,n} + ax_{m,n+1},\ x_{m+2,n} + ax_{m,n+2}, \ldots, \\ x_{m+(k-1),n} + ax_{m,n+(k-1)}, x_{mn} + ax_{mn})^T,$$

$$i = 1, 2,\ \ldots,\ k.$$

Proof Given non-zero real function sequences (3.43) which satisfy

$$x_{m+i,n} + ax_{m,n+i} \neq x_{m+j,n} + ax_{m,n+j},\ i \neq j, i, j = 1, 2, \ldots, k, \text{ for any } m, n \in N_r. \tag{3.45}$$

Suppose that the map $S(x) = \mu_1 \sin x + \mu_2 \sin(2x) + \cdots + \mu_n \sin(kx)$ satisfies

$$\left.\begin{aligned} S(x_{mn} + ax_{mn}) &= x_{m+1,n} + ax_{m,n+1}, \\ S(x_{m+1,n} + ax_{m,n+1}) &= x_{m+2,n} + ax_{m,n+2}, \\ \cdots \qquad & \cdots \qquad \cdots \\ S(x_{m+(k-1),n} + ax_{m,n+(k-1)}) &= x_{mn} + ax_{mn}. \end{aligned}\right\} \tag{3.46}$$

For simplicity, we shall use the following notation. Let $x_{m+i,n} + ax_{m,n+i} = r_i(m,n) = r_i$, $i = 0, 1, 2, \ldots, k-1$.

Then (3.46) reduces to the following system:

$$\left.\begin{aligned} \mu_1 \sin r_0 + \mu_2 \sin 2r_0 + \cdots + \mu_k \sin kr_0 &= r_1, \\ \mu_1 \sin r_1 + \mu_2 \sin 2r_1 + \cdots + \mu_k \sin kr_1 &= r_2, \\ \cdots \qquad \cdots \qquad \cdots \\ \mu_1 \sin r_{k-2} + \mu_2 \sin 2r_{k-2} + \cdots + \mu_k \sin kr_{k-2} &= r_{k-1}, \\ \mu_1 \sin r_{k-1} + \mu_2 \sin 2r_{k-1} + \cdots + \mu_k \sin kr_{k-1} &= r_1. \end{aligned}\right\} \tag{3.47}$$

Hence we obtain Eq. (3.47) and which determine the unknown μ_i, $i = 1, 2, \ldots, k$.

It is easy to check that the determinant of the coefficients of Eq. (3.47) is the kth triangle determinant and

$$\begin{aligned} \Delta &= \begin{vmatrix} \sin r_0 & \sin 2r_0 \cdots & \sin kr_0 \\ \sin r_1 & \sin 2r_1 \cdots & \sin kr_1 \\ \cdots & \cdots\cdots & \cdots \\ \sin r_{k-1} & \sin 2r_{k-1} \cdots & \sin kr_{k-1} \end{vmatrix} \\ &= (-1)^{\frac{k(k-1)}{2}} 2^{k(k-1)} \left(\prod_{i=1}^{k} \sin ir_i \right) \left(\prod_{k \geq i > j \geq 1} \sin \frac{r_i + r_j}{2} \cdot \sin \frac{r_i - r_j}{2} \right) \neq 0. \end{aligned}$$

Since (3.42) and (3.43) hold, hence, we have, $\Delta \neq 0$. Then, there exists the unique solution of Eq. (3.47):

$$\mu_i^* = \frac{\Delta^{(i)}}{\Delta}, \quad i = 1, 2, \ldots, k,$$

where the determinant $\Delta^{(i)}$ is obtained from Δ by replacing column ith column of Δ by the vector

$$(x_{m+1,n} + ax_{m,n+1},\ x_{m+2,n} + ax_{m,n+2},\ \ldots, x_{m+(k-1),n} + ax_{m,n+(k-1)}, x_{mn} + ax_{mn})^T,$$
$$i = 1, 2, \ldots, k,$$

that is, the vector of the values on the right-hand side of (3.47).

Substituting $\mu_i = \mu_i^*$ into (3.44), $i = 1, 2, \ldots, k$, we obtain

$$S(x) = \mu_1^* \sin x + \mu_2^* \sin(2x) + \cdots + \mu_n^* \sin(kx). \tag{3.48}$$

It is easy to see that the function form (3.48) is continuous in the space $\mathbb{R}^3$ and $S^k(x_{mn} + ax_{mn}) = S^k(r_0) = r_0 = x_{mn} + ax_{mn}$, but $x_{mn} + ax_{mn} = S^s(x_{mn} + ax_{mn})$ for $1 \le s < k$. Hence the function of the form (3.48) is a continuous map which have a periodic point $x_{mn} + ax_{mn}$ with period k.

Theorem 3.6 *Let $V \subseteq \mathbb{R}^3 = (-\infty, \infty)^3$, V_0 be a non-empty subset of V, let I be an interval such that $I \subset V_0$, assume that*

$$r_i(m, n) = x_{m+i,n} + x_{m,n+i},\ i = 1, 2, 3, \tag{3.49}$$

and the following conditions holds:
1^0 *$r_i(m, n) \neq 0$, $i = 1, 2, 3$, and $r_i(m, n) \neq r_j(m, n)$ for $m, n \in N_0$, $i, j = 1, 2, 3$, which satisfy (3.42) and (3.43).*
2^0 *$r_1(m, n) < r_2(m, n) < r_3(m, n)$ or $r_1(m, n) > r_2(m, n) > r_3(m, n)$ for $m, n \in N_0$.*
3^0 *let*

$$S^*(x) = \mu_1^* \sin x + \mu_2^* \sin(2x) + \mu_3^* \sin(3x), \tag{3.50}$$

where $\mu_i^* = \frac{D^{(i)}}{D}$, $i = 1, 2, 3$, $D = \begin{vmatrix} \sin x_1 & \sin 2x_1 & \sin 3x_1 \\ \sin x_2 & \sin 2x_2 & \sin 3x_2 \\ \sin x_3 & \sin 2x_3 & \sin 3x_3 \end{vmatrix}$, and the determinant $D^{(i)}$ is obtained from D by replacing column ith column of D by the vector $(x_{m+1,n} + x_{m,n+1}, x_{m+2,n} + x_{m,n+2}, x_{mn} + x_{mn},)^T$, $i = 1, 2, \ldots, k$.
4^0 *$S(x) \in C^0(I, I)$ and $S(I) \subset I$.*

Then system

$$x_{m+1,n} + x_{m,n+1} = \sum_{i=1}^{k} \mu_i^* \sin(i x_{mn}) \tag{3.51}$$

is surface chaotic on V_0 in the sense of Li -Yorke.

Proof Without loss of generality, we consider system

$$x_{s+1,n} + x_{m,t+1} = \sum_{i=1}^{3} \mu_i^* \sin[i(x_{sn} + x_{nt})],$$

where $s, t, m, n \in N_r$. And we take $s,\ t = 0, 1, 2, 3$, then we can obtain the above $r_i(m, n)$, $i = 0, 1, 2, 3$, where, $r_0(m, n) = x_{0n} + x_{m0}$. Next, applying Theorem 3.1, we get that

$$x_{s+1,n} + x_{m,t+1} = \mu_1^* \sin(x_{sn} + x_{mt}) + \mu_2^* \sin\left(2(x_{sn} + x_{mt})\right) + \mu_3^* \sin\left(3(x_{sn} + x_{mt})\right). \tag{3.52}$$

Due to that $s, t, m, n \in N_r$, in particular, let $s = m$, $t = n$, we have

$$\begin{aligned} x_{m+1,n} + x_{m,n+1} &= \mu_1^* \sin(x_{mn} + x_{mn}) + \mu_2^* \sin\left(2(x_{mn} + x_{mn})\right) \\ &+\mu_3^* \sin\left(3(x_{mn} + x_{mn})\right) = S^*(2x_{mn}). \end{aligned} \tag{3.53}$$

Note that

$$\begin{aligned} &S^{*3}\left(x_{mn} + x_{mn}\right) = S^{*3}\left(r_1(m, n)\right) = r_1(m, n) = x_{mn} + x_{mn}, \\ &S^{*i}\left(r_1(m, n)\right) \neq r_1(m, n),\ 1 \leq i < 3, \end{aligned}$$

therefore, the map (3.52) have a periodic point $r_1(m, n)$ with period 3. Since $r_i(m, n)$ is distinct from each other and $r_i(m, n) \neq 0$ in $m, n \in N_0$, $i = 1, 2, 3$, hence, we have immediately from Theorem 3.5, $D \neq 0$, that is, μ_i^* is uniquely determined. Therefore $S^*\left(x_{mn} + x_{mn}\right) = S^*(2x_{mn})$ is uniquely determined. In addition, due to that $r_i(m, n) \in I$ and $S^*(I) \subset I$ from 4^0, $S^*(2x_{mn})$ is a continuous self-map in $I \subset V_0$. On the other hand, it follows from 2^0 that

$$\begin{aligned} S^{*3}\left(r_1(m, n)\right) &= r_1(m, n) < r_2(m, n) = S^*\left(r_1(m, n)\right) \\ &< S^{*2}\left(r_1(m, n)\right) = r_3(m, n), \end{aligned}$$

for $m,\ n \in N_0$.

Applying Lemma 3.1, we obtain system (3.51) is chaotic on I in the sense of Li and Yorke. Using Definition 3.2 again, we have system (3.51) is also surface chaotic on V_0 in the sense of Li and Yorke.

Example 3.3 Let $V_0 = V = [0, 100]^3 = [0, 100] \times [0, 100] \times [0, 100] \subseteq \mathbb{R}^3$, and $I = [\frac{\pi}{4}, \frac{\pi}{2}] \subset V_0$. Since $r_0(m, n) = x_{0n} + x_{m0}$ is boundary value of V_0, therefore, without loss of generality, we will take $r_0(m, n) = x_{0n} + x_{m0} = \frac{\pi}{2}$, and $r_1(m, n) = x_{1n} + x_{m1} = \frac{\pi}{4}$, $r_2(m, n) = x_{2n} + x_{m2} = \frac{\pi}{6}$, $r_3(m, n) = x_{3n} + x_{m3} = \frac{\pi}{2}$. Using Theorem 3.6, we have

$$\left.\begin{aligned} f(x_{0n} + x_{m0}) &= r_1(m, n) = x_{m+1,n} + x_{m,n+1} = \tfrac{\pi}{4}, \\ f(x_{m+1,n} + x_{m,n+1}) &= r_2(m, n) = x_{m+2,n} + x_{m,n+2} = \tfrac{\pi}{6}, \\ f(x_{m+2,n} + x_{m,n+2}) &= r_3(m, n) = x_{3n} + x_{m3} = r_0(m, n) = \tfrac{\pi}{2}. \end{aligned}\right\}$$

When we take

$$\left.\begin{aligned} \mu_1 - \mu_3 &= \tfrac{\pi}{4}, \\ \tfrac{\sqrt{2}}{2}\mu_1 + \mu_2 + \tfrac{\sqrt{2}}{2}\mu_3 &= \tfrac{\pi}{6}, \\ \tfrac{1}{2}\mu_1 + \tfrac{\sqrt{3}}{2}\mu_2 + \mu_3 &= \tfrac{\pi}{2}, \end{aligned}\right\} \tag{3.54}$$

due to that $\Delta = \frac{3-\sqrt{6}}{2}$, $\Delta_1 = \frac{\pi}{48}\left(36 - 4\sqrt{3} - 3\sqrt{6}\right)$, $\Delta_2 = \frac{\pi}{4}\left(1 - \frac{94}{\sqrt{2}}\right)$, $\Delta_3 = \frac{\pi}{48}\left(18 + 3\sqrt{6} - 4\sqrt{3}\right)$, we have $\mu_1 \approx 5.1653$, $\mu_2 \approx -6.2259$, $\mu_3 \approx 4.3799$. And similarly, we have

$$\begin{aligned} x_{m+1,n} + x_{m,n+1} = & \; 5.1653 \sin\left((x_{mn} + x_{mn})\right) \\ & - 6.2259 \sin\left(2(x_{mn} + x_{mn})\right) \\ & + 4.3799 \left(3(x_{mn} + x_{mn})\right), \end{aligned} \tag{3.55}$$

where $\forall m,\ n \in N_t$. Let

$$\begin{aligned} S\left(2x_{mn}\right) = & \; 5.1653 \sin\left((2x_{mn})\right) \\ & - 6.2259 \sin\left(2(2x_{mn})\right) + 4.3799 \left(3(2x_{mn})\right), \end{aligned} \tag{3.56}$$

it follows from (3.55) that we obtain

$$\begin{cases} \frac{\pi}{4} = S\left(\frac{\pi}{2}\right), \\ \frac{\pi}{6} = S\left(\frac{\pi}{4}\right) = S\left(S\left(\frac{\pi}{2}\right)\right) = S^2\left(\frac{\pi}{2}\right), \\ \frac{\pi}{2} = S\left(\frac{\pi}{6}\right) = S\left(S\left(\frac{\pi}{4}\right)\right) = S\left(S\left(S\left(\frac{\pi}{2}\right)\right)\right) = S^3\left(\frac{\pi}{2}\right). \end{cases} \tag{3.57}$$

In view of (3.57), it is easy to see the $r_0(m,n) = x_{0n} + x_{m0} = \frac{\pi}{2}$ is a three iterative periodic point, on the other hand, we note that

$$S^3\left(\frac{\pi}{2}\right) = \frac{\pi}{2} > \frac{\pi}{4} = S\left(\frac{\pi}{2}\right) > \frac{\pi}{6} = S^2\left(\frac{\pi}{2}\right), \quad \text{for } m, n \in N_0.$$

Let $\xi = 2x$, we get $S(\xi) = 5.1653 \sin\xi - 6.2259 \sin(2\xi) + 4.3799 \sin(3\xi)$. Next writing $M = \max\{|\mu_i| \,|1 \le i \le 3\} = |\mu_2|$, then we obtain

$$|S(x)| = \left|\sum_{k=1}^{3} \mu_k \sin(kx)\right| < 3M.$$

Hence $S(2x)$ is a continuous self-map in $I = [-3M, 3M]$ and satisfy

$$S^3\left(\frac{\pi}{2}\right) > S\left(\frac{\pi}{2}\right) S^2\left(\frac{\pi}{2}\right), \quad \left(\frac{\pi}{2} \in I\right).$$

Using Theorem 3.2, system (3.20) is surface chaotic on I in the sense of Li and Yorke. So does V_0 by Definition 3.2. In fact, we can also calculate its Lyapunov exponent of the following form:

$$\lambda = \lim_{q \to \infty} \frac{1}{q} \sum_{i=1}^{q-1} \ln\left|S'(x)\right|\} = 0.4659 > 0.$$

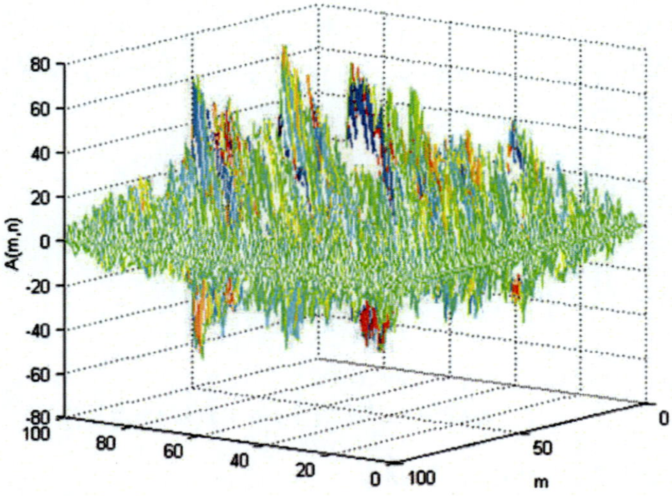

Fig. 3.3 The surface chaos of system (3.20)

On the other hand, we can also see a surface chaotic behavior of (3.20) from the following Fig. 3.3:

Remark 3.1 In general, in the above example, $r_i(m, n) \neq$ constant.

Remark 3.2 We can also construct the following surface chaotic maps of the form

$$x_{m+1,n} + x_{m,n+1} = a_0 + a_1(2x_{mn}) + a_2(2x_{mn})^2, \ \forall m, n \in N_t.$$

Remark 3.3 We can also construct the following surface chaotic maps of the form

$$\begin{aligned} x_{m+1,n} + x_{m,n+1} &= C(x) \\ &= \lambda_0 + \lambda_1 \cos 2x_{mn} + \lambda_2 \cos 2(2x_{mn}) \\ &+ \cdots + \lambda_k \cos k(2x_{mn}), \end{aligned}$$

$\forall m, \ n \in N_t$.

Finally, it should be mentioned that the concept of "snap-back repeller" with control and synchronization of surface chaos should be studied deeply.

Chapter 4
Surface Chaos and Its Spatial Lyapunov Exponent

In this chapter, we consider mainly for nonlinear 2-D discrete system:

$$x_{m+1,n} + ax_{m,n+1} = f[\mu, (1+a)x_{mn}], \tag{4.1}$$

where $f(x)$ is a nonlinear function, $m, n \in N_t$, $\mu \geq 0$ is a real parameter and let

$$\Omega = [r, \infty) \times [r, \infty)\backslash[1, \infty) \times [1, \infty).$$

First, observe that for a given function $\varphi(i, j)$ defined on Ω, it is easy to construct by inducing a double sequence $\{x_{ij}\}$ that equals $\varphi(i, j)$ on Ω and satisfies system (4.1) for $i, j = 0, 1, 2, \ldots$. Indeed, one can rewrite system (4.1) as

$$x_{m,n+1} = f[\mu, (1+a)x_{mn}] - ax_{m+1,n},$$

and then use it to calculate, successively,

$$x_{01}, x_{11}, x_{02}, x_{21}, x_{12}, x_{03}, \ldots.$$

Such a double sequence is unique and is a solution of Eq. (4.1) subject to the initial condition:

$$x_{ij} = \varphi(i, j),\ (i, j) \in \Omega.$$

One can see that system (4.1) can be regarded as a discrete analog of the following functional partial differential system:

$$\frac{\partial v}{\partial x} + a\frac{\partial v}{\partial y} + av = f[\mu, (1+a)v]. \tag{4.2}$$

S. T. Liu and L. Zhang, *Surface Chaos and Its Applications*,
https://doi.org/10.1007/978-981-16-8229-2_4

On the other hand, when $n = n_0$ (constant) and $a = 0$ in (4.1), system (4.1) becomes

$$x_{m+1,n_0} = f(\mu, x_{m,n_0}), \tag{4.3}$$

which is the familiar simple case for which asymptotic behaviors of all solutions have been discussed thoroughly. Furthermore, when $f(\mu, x_{m\,n_0}) = 1 - \mu x_{m,n_0}^2$ again, system (4.1) becomes

$$x_{m+1,n_0} = 1 - \mu x_{m,n_0}^2. \tag{4.4}$$

In particular, when $a = 1$, it follows from (4.1) that

$$x_{m+1,n} + x_{m,n+1} = f(\mu, 2x_{mn}). \tag{4.5}$$

Therefore, system (4.1) is quite general and has a more richful spatiotemporal behavior. In this chapter, the aim is to introduce a spatial fixed point, periodic orbits, spatial Lyapunov exponents and stability of spatial periodic orbits and other qualitative properties on the surface chaos by using Li-Yorke and Marotto Theorem.

4.1 A Fixed Plane of Surface

In this section, some useful preliminary results are firstly established, which will be needed for the proofs of the main results in the next section.

For the one-dimensional map (4.6), the definition is given in the following.

As is known, a zero point x^* of $x - f(\mu, x) = 0$ is said to be a fixed point of (4.6):

$$x_{m+1} = f(\mu, x_m). \tag{4.6}$$

In particular, we have also been known that the fixed point of 1-D logistic discrete system (4.6) is that

$$x^* = \frac{-1 \pm \sqrt{1 + 4\mu}}{2\mu}. \tag{4.7}$$

Definition 4.1 A fixed plane of 2-D discrete system (4.1) is defined as listed below:

$$L(2D) = \{y^* | (1 + a)y^* - f(\mu, (1 + a)y^*) = 0\},$$

where $m, n \in N_r$, $x_{m,n} = y^*$. In particular, when $f(\mu, (1 + a)x_{m,n}) = 1 - \mu\,[(1 + a)x_{mn}]^2$, 2-D discrete system (4.1) becomes

$$x_{m+1,n} + ax_{m,n+1} = 1 - \mu\,[(1 + a)x_{mn}]^2. \tag{4.8}$$

And usually, it is said to be a **general logistic spatial discrete system**. In particular, when $a = 1$, 2-D discrete system (4.1) becomes

$$x_{m+1,n} + x_{m,n+1} = 1 - 4\mu x_{mn}^2, \tag{4.9}$$

and it is said to be a **standing logistic spatial discrete system.**

Moreover, when $a = 0$, and $n = n_0$ (constant), then (4.8) reduce to (4.4), by using Definition 4.1 and equation $(1 + a)x = 1 - \mu\,[(1 + a)x]^2$. It is easy to obtain the following results for fixed plane of (4.8):

Theorem 4.1 *A fixed plane of the spatially generalized Logistic discrete system (4.8) is given by*

$$x_{m,n} = x^* = \frac{-1 \pm \sqrt{1 + 4\mu}}{2(1 + a)\mu}, \quad for\ all\ m, n \in N_r. \tag{4.10}$$

Example 4.1 For system (4.8), let $\mu = 2$, $a = \frac{1}{2}$, we have the two fixed planes: $x_{m,n} = x^* = \frac{1}{3}$ and $x_{m,n} = x^* = -\frac{2}{3}$ of the following system

$$x_{m+1,n} + \frac{1}{2}x_{m,n+1} = 1 - \frac{9}{2}x_{mn}^2,$$

shown as in Fig. 4.1.

Remark 4.1 (i) When $a = 1$, it is easy to see that a fixed plane of general 2-D logistic spatial discrete system (4.8) is:

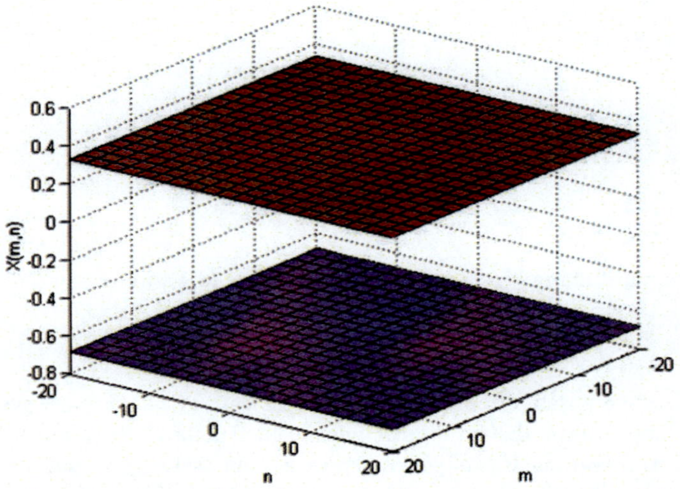

Fig. 4.1 Two spatial fixed planes

$$x_{m,n} = x^* = \frac{-1 \pm \sqrt{1+4\mu}}{4\mu}. \tag{4.11}$$

(ii) When $a = 0$, the fixed plane reduces to the fixed point:

$$x^* = \frac{-1 \pm \sqrt{1+4\mu}}{2\mu}, \tag{4.12}$$

which is a fixed point of standard 1-D logistic space discrete system (4.9).

4.2 Spatial Fixed Plane, Periodic Orbits and Their Stability

Firstly, we introduce some definitions.

Based on the definitions of spatial point and spatial periodic orbit, we have the following theorem:

Theorem 4.2 *Assume that $a \in (-\infty, -1) \cup (-1, \infty)$, the stability condition for the fixed plane of the 2-D discrete system (4.1) is*

$$|f'(\mu, (1+a)x^*)| \le \begin{cases} \frac{a-1}{1+a}, & when\ a \in (-\infty, -1), \\ \frac{1-a}{1+a}, & when\ a \in (-1, 0), \\ 1, & when\ a \in [0, \infty). \end{cases} \tag{4.13}$$

Proof Let the fixed plane $x^* \in L(2D)$. And to do so we iterate the map in the vicinity of the fixed point by writing

$$x_{mn} = x^* + \varepsilon_{mn}, \tag{4.14}$$

and one also has

$$\left.\begin{array}{l} x_{m+1,n} = x^* + \varepsilon_{m+1,n}, \\ x_{m,n+1} = x^* + \varepsilon_{m,n+1}, \end{array}\right\} \tag{4.15}$$

where ε_{mn} is a small deviation from the fixed point. Hence

$$x_{m+1,n} + ax_{m,n+1} = (1+a)x^* + (\varepsilon_{m+1,n} + a\varepsilon_{m,n+1}). \tag{4.16}$$

Moreover, substituting (4.14) and (4.16) into system (4.1), we obtain

$$(1+a)x^* + (\varepsilon_{m+1,n} + a\varepsilon_{m,n+1}) = f\left[\mu, (1+a)(x^* + \varepsilon_{mn})\right].$$

Since the x^* be a stable fixed point, then, in the case of 2-D discrete system (4.1) stability simply means that on approaching the fixed point x^* the difference between the mth iterate (without loss of generality, we assume that $m \ge n$, and when $m \le n$, we can use nth iterate to do) x_{mn} and x^* must keep decreasing. In other words, then, it follows from (4.14) and (4.15) that

$$\left|\frac{\varepsilon_{m+1,n}}{\varepsilon_{mn}}\right| < 1 \text{ and } \left|\frac{\varepsilon_{m,n+1}}{\varepsilon_{mn}}\right| < 1. \tag{4.17}$$

Therefore, when m and n sufficient large, then one must have

$$\left|\frac{\varepsilon_{m+1,n}}{(1+a)\varepsilon_{mn}}\right| = \frac{1}{|1+a|}\left|\frac{\varepsilon_{m+1,n}}{\varepsilon_{mn}}\right| < \frac{1}{|1+a|}, \tag{4.18}$$

$$\left|\frac{\varepsilon_{m,n+1}}{(1+a)\varepsilon_{mn}}\right| = \frac{1}{|1+a|}\left|\frac{\varepsilon_{m,n+1}}{\varepsilon_{mn}}\right| < \frac{1}{|1+a|}. \tag{4.19}$$

Hence

$$\begin{aligned}\left|\frac{\varepsilon_{m+1,n}+a\varepsilon_{m,n+1}}{(1+a)\varepsilon_{mn}}\right| &\leq \left|\frac{\varepsilon_{m+1,n}}{(1+a)\varepsilon_{mn}}\right| + |a|\left|\frac{\varepsilon_{m,n+1}}{(1+a)\varepsilon_{mn}}\right| \\ &= \frac{1}{|1+a|}\left|\frac{\varepsilon_{m+1,n}}{\varepsilon_{mn}}\right| + \frac{|a|}{|1+a|}\left|\frac{\varepsilon_{m,n+1}}{\varepsilon_{mn}}\right| \\ &< \frac{1}{|1+a|} + \frac{|a|}{|1+a|} \\ &= \frac{1+|a|}{|1+a|} \\ &= \begin{cases} \frac{a-1}{1+a}, & when\ a \in (-\infty, -1), \\ \frac{1-a}{1+a}, & when\ a \in (-1, 0), \\ 1, & when\ a \in [0, \infty). \end{cases}\end{aligned} \tag{4.20}$$

Next expanding f to its Taylor series for the first order in ε_{mn}, we get that

$$\begin{aligned} f(\mu, (1+a)x_{mn}) &= x_{m+1,n} + ax_{m,n+1} \\ &= (1+a)x^* + (\varepsilon_{m+1,n} + a\varepsilon_{m,n+1}) \\ &= f\left[\mu, (1+a)x^*\right] + f'\left[\mu, (1+a)x^*\right](1+a)\varepsilon_{mn} + \cdots. \end{aligned} \tag{4.21}$$

Note that x^* is a fixed point of system (4.1), one has

$$x^* + ax^* = (1+a)x^* = f\left[\mu, (1+a)x^*\right]. \tag{4.22}$$

Substituting (4.22) into (4.21) and suppressing from both sides the same items, we obtain

$$\varepsilon_{m+1,n} + a\varepsilon_{m,n+1} = f'\left[\mu, (1+a)x^*\right](1+a)\varepsilon_{mn} + \cdots,$$

which lead to the stability condition for the fixed point x^*

$$|\frac{\varepsilon_{m+1,n}+a\varepsilon_{m,n+1}}{(1+a)\varepsilon_{mn}}|=|f'(\mu,(1+a)x^*)|\le\begin{cases}\frac{a-1}{1+a}, & when\ a\in(-\infty,-1),\\ \frac{1-a}{1+a}, & when\ a\in(-1,0),\\ 1, & when\ a\in[0,\infty).\end{cases}\tag{4.23}$$

Remark 4.2 In particular, when $a=0, n=n_0$ we obtain the stability condition of (4.3)

$$|f'(\mu,x^*)|<1,$$

which just is the familiar simple case of 1-D discrete system, and also show that system (4.1) is nature extended of 1-D discrete system (4.3) from a angle of view of stability condition of fixed point.

In addition, for a general ($a\neq 0$) 2-D logistic spatial discrete system (4.8), we have the following of result.

Theorem 4.3 *The stability condition for a fixed plane $x^*$of 2-D logistic spatial discrete system (4.8) is*

$$\mu<\frac{1}{4}\left[\left(\frac{1+|a|+(a+1)^2}{(a+1)^2}\right)^2-1\right].\tag{4.24}$$

Proof For 2-D logistic spatial discrete system (4.8), note that $f(\mu,(1+a)x)=1-\mu[(1+a)x]^2$, one has $f'(\mu,(1+a)x)=-2\mu(1+a)^2x$. Using (4.24), we get

$$\left|-2\mu(1+a)^2x\right|<\frac{1+|a|}{|1+a|}.$$

Applying (4.8) again, one has

$$x_{m,n}=x^*=\frac{-1+\sqrt{1+4\mu}}{2(1+a)\mu},$$

and

$$\left|-(1+a)(-1+\sqrt{1+4\mu})\right|<\frac{1+|a|}{|1+a|},\tag{4.25}$$

in the further reduces (4.25) to obtain (4.24).

Remark 4.3 (*i*) When $a\in[0,+\infty)$, we obtain

$$\begin{aligned}|-1+\sqrt{1+4\mu}|&\le|1+a||-1+\sqrt{1+4\mu}|\\ &=\left|-(1+a)(-1+\sqrt{1+4\mu}\right|<\tfrac{1+|a|}{|1+a|},\end{aligned}$$

from (4.25), which means that

$$\left|-1+\sqrt{1+4\mu}\right| < \frac{1+|a|}{|1+a|}.$$

So we get

$$\mu < \frac{1}{4}\left[\left(\frac{1+|a|+(a+1)^2}{(a+1)^2}\right)^2 - 1\right].$$

Corollary 4.1 *(ii) In particular, when $a = 0$ and $n = n_0$ (constant), then we obtain the stability condition of 1-D logistic discrete system (4.4)*

$$\mu < \frac{3}{4}. \tag{4.26}$$

Remark 4.4 Equation (4.26) is a familiar case of 1-D, which shows again that a result of 2-D system is naturally extended of 1-D discrete system from (4.6).

Example 4.2 (i) Let $\mu = \frac{1}{4}, a = 1$. Figure 4.2 shows the fixed plane and the stability of the solution for system (4.8).

(ii) Otherwise, when $\mu = \frac{1}{2}$, $a = 1$, the fixed plane and the instability of the solution is illustrated by Fig. 4.3.

On the other hand, if $(1+a)x_i^*$, $i = 1, 2$ are two fixed points of system (4.1), then we can define a composite function of the form

$$F[2, \mu, (1+a)x] = f(\mu, (1+a)f(\mu, (1+a)x)),$$

and one has

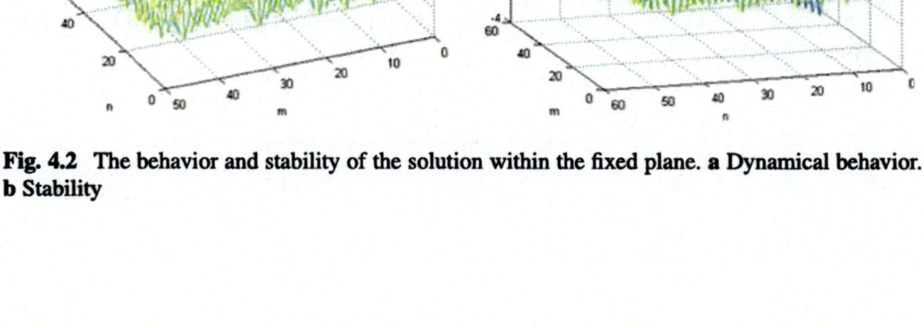

Fig. 4.2 The behavior and stability of the solution within the fixed plane. **a** Dynamical behavior. **b** Stability

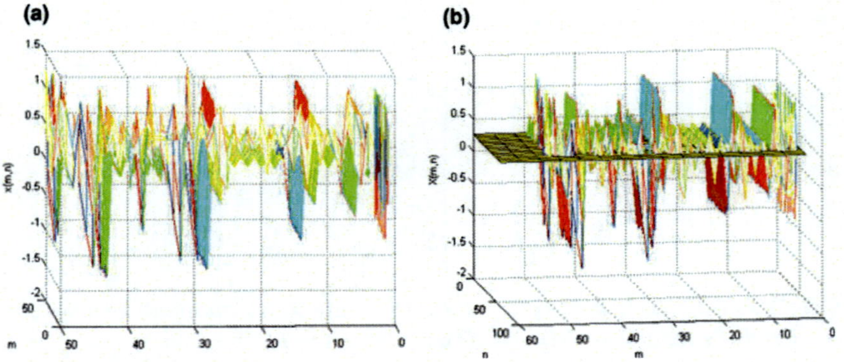

Fig. 4.3 The behavior and instability of the solution within the fixed plane. **a** Dynamical behavior. **b** Instability

$$F\left[2, \mu, (1+a)x_i^*\right] = (1+a)x_i^*, i = 1, 2. \tag{4.27}$$

For (4.27), repeating the above argument as Theorem 4.3, we obtain the following corollary.

Corollary 4.2 *When the stability condition of fixed point* $(1+a)x_i^*, i = 1, 2$ *is*

$$\left|f'(\mu, f(\mu, (1+a)x_i^*)\right| < \frac{1+|a|}{|1+a|}.$$

In general, a period k sequence emerges when the iteration loops among k points: $x_{m+k,n} + ax_{m,n+k} = x_{mn} + ax_{mn}$ for all m and n. Written done explicitly, we have

$$\left.\begin{aligned} x_{m+1,n} + ax_{m,n+1} &= f(\mu, x_{mn} + ax_{mn}), \\ x_{m+2,n} + ax_{m,n+2} &= f(\mu, x_{m+1,n} + ax_{m,n+1}), \\ &\cdots\cdots\cdots \\ x_{mn} + ax_{mn} &= f(\mu, x_{m+(k-1),n} + ax_{m,n+(k-1)}). \end{aligned}\right\} \tag{4.28}$$

These k points form a k periodic orbit:

$$\left\{x_{m,n} + ax_{m,n}, x_{m+1,n} + ax_{m,n+1}, \cdots, x_{m+k-1,n} + ax_{m,n+k-1}, x_{m,n} + ax_{m,n}, \cdots\right\},$$

and let $r_i(a, m, n) = x_{m+i,n} + ax_{m,n+i}, i = 0, 1, \cdots, k-1$. Obviously, the k periodic orbit $r_0(a, m, n), r_1(a, m, n), \cdots, r_{k-1}(a, m, n)$ is the fixed point of the following composite function

$$f^k(\mu, a, y) = \underbrace{f(\mu, a, f(\mu, a, \cdots, f(\mu, a, y)\cdots))}_{k\ times}.$$

Let

$$\begin{aligned} F[k,\mu,a,y] &= \underbrace{f(\mu,a,f(\mu,a\cdots,f(\mu,a,y)\cdots))}_{k\ times} \\ &= \underbrace{f\circ f\circ\cdots f(\mu,a,y)}_{k\ times} \\ &= f^k(\mu,a,y). \end{aligned}$$

We can present each point of the $k-cycles$ as a fixed point of F:

$$r_i(a,m,n) = F[k,\mu,a,r_i(a,m,n)], i = 0,1,\ldots,k-1.$$

The last statement allows us to carry over the stability condition (4.13) for a fixed point to a period k orbit. In order to do so, we recal the chain rule of differentiation $\frac{d}{dx}f^k(\mu,a,y) = \prod_{i=0}^{k-1} f^{(i)}(\mu,a,r_i(a,m,n))$. Using the iterations (4.28), the condition

$$\left|F'[k,\mu,a,r_i(a,m,n)]\right| < \frac{1+|a|}{|1+a|}, i = 0,1,\ldots,k-1,$$

becomes

$$\left|\prod_{i=0}^{k-1} f'(\mu,a,r_i(a,m,n))\right| < \frac{1+|a|}{|1+a|}. \tag{4.29}$$

Remark 4.5 In particular, when $a = 0$ in (4.29), we obtain the stability condition period k orbit of (4.3).

$$\left|\prod_{i=0}^{k-1} f'(\mu,x_i)\right| < 1,$$

which is just the familiar simple case of 1-D discrete system.

4.3 Spatial Lyapunov Exponents

In order to obtain spatial Lyapunov exponents for systems (4.1), in general, we consider more extensive simulation in the following form

$$x_{p+1,n} + ax_{m,q+1} = f(\mu, x_{pn} + ax_{mq}), \tag{4.30}$$

where $p, q, m, n \in N_0$. And without loss of generality we assume that

$$p \geq q. \tag{4.31}$$

Theorem 4.4 *For system (4.1), we assume that $m \geq n$ and (4.31) holds. Then we have*

$$\begin{aligned} x_{m+1,n} + a x_{m,n+1} &= \underbrace{f(\mu, f(\mu, \cdots, f(\mu, x_{0n} + a x_{m0}) \cdots)}_{m+1 \text{ times}} \\ &= f^{m+1}(\mu, x_{0n} + a x_{m0}), \end{aligned} \tag{4.32}$$

where, because $x_{0n} + a x_{m0}$ is boundary value of N_0, therefore $x_{0n} + a x_{m0}$ is known for any $m, n \in N_0$.

Proof Since $p \geq q$ by (4.31), therefore we assume that

$$x_{mq} = x_{m,t+1},\ q \geq t + 1, \tag{4.33}$$

then we have

$$\left.\begin{array}{rl} p = 0, q = 0: & x_{1n} + a x_{m1} = f(\mu, x_{0n} + a x_{m0}), \\ p = 1, q = 1: & x_{2n} + a x_{m2} = f(\mu, x_{1n} + a x_{m1}), \\ \cdots & \cdots \quad \cdots \quad \cdots \\ p = t-1, q = t-1: & x_{tn} + a x_{mt} = f(\mu, x_{t-1,n} + a x_{m,t-1}), \\ p = t, q = t: & x_{t+1,n} + a x_{m,t+1} = f(\mu, x_{tn} + a x_{mt}), \end{array}\right\} (t+1) \text{ steps}$$

$$\left.\begin{array}{rl} p = t+1, q = t+1: & x_{t+2,n} + a x_{m,t+1} = f(\mu, x_{t+1,n} + a x_{m,t+1}), \\ p = t+2, q = t+2: & x_{t+3,n} + a x_{m,t+1} = f(\mu, x_{t+2,n} + a x_{m,t+1}), \\ \cdots & \cdots \quad \cdots \quad \cdots \\ p = s-1, q = s-1: & x_{sn} + a x_{m,t+1} = f(\mu, x_{s-1,n} + a x_{m,t+1}), \\ p = s, q = s: & x_{s+1,n} + a x_{m,t+1} = f(\mu, x_{sn} + a x_{m,t+1}). \end{array}\right\} (s-t) \text{ steps}$$

In view of above equalities, we obtain

$$x_{s+1,t} + a x_{s,t+1} = f^{s+1}(\mu, x_{0n} + a x_{m0}). \tag{4.34}$$

In particular, we take that $s = m$, $t = n$, then (4.34) reduces to (4.32). Then Theorem 4.4 proof is complete.

Similialy, we obtain the following result:

Corollary 4.3 *It is easy to see that if $m \leq n$, then (4.34) reduces to*

$$x_{m+1,n} + a x_{m,n+1} = f^{n+1}(\mu, x_{0n} + a x_{m0}). \tag{4.35}$$

Applying Theorem 4.4 and Corollary (4.3) to obtain:

Theorem 4.5 *For systems (4.1): (i) If $m \geq n$, then spatial Lyapunov exponents of systems (4.1) is*

$$\lambda_{space} = \lim_{m\to\infty} \frac{1}{m}\left[\sum_{j=0}^{n} \ln\left|f'(\mu,\, x_{jn} + ax_{mj})\right| + \sum_{j=n+1}^{m-1} \ln\left|f'(\mu,\, x_{jn} + ax_{m,n+1})\right|\right]. \tag{4.36}$$

(ii) If $m \leq n$, then spatial Lyapunov exponents of systems (4.1) is

$$\lambda_{space} = \lim_{n\to\infty} \frac{1}{n}\left[\sum_{r=0}^{m} \ln\left|f'(\mu,\, x_{mr} + ax_{r\,n})\right| + \sum_{r=m+1}^{n-1} \ln\left|f'(\mu,\, x_{m+1,n} + ax_{r\,n})\right|\right]. \tag{4.37}$$

Proof First, we prove the (i), without loss of generality, using (4.30), and we still assume that $s \geq t$. Then (4.32) holds from Theorem 4.2. We will take two nearby initial positions, $P_0 = x_{0n} + ax_{m,0}$ and $P_0' = x_{0n}' + ax_{m,0}'$. Using (4.32), one have

$$x_{mn} + ax_{m-1,n+1} = f^m(\mu, x_{0n} + ax_{m0}), \tag{4.38}$$

and, similarly,

$$x_{mn}' + ax_{m-1,n+1}' = f^m(\mu, x_{0n}' + ax_{m0}'). \tag{4.39}$$

If these positions, $x_{mn} + ax_{m-1,n+1}$ and $x_{mn}' + ax_{m-1,n+1}'$, are separated exponentially fast in iterations. Then

$$\left|(x_{mn} + ax_{m-1,n+1}) - (x_{mn}' + ax_{m-1,n+1}')\right| = \left|(x_{0n} + ax_{m0}) - (x_{0n}' + ax_{m0}')\right| e^{m\lambda},$$

$\lambda > 0$ and

$$\begin{aligned}
&\frac{1}{m}\ln\frac{\left|(x_{mn} + ax_{m-1,n+1}) - (x_{mn}' + ax_{m-1,n+1}')\right|}{\left|(x_{0n} + ax_{m0}) - (x_{0n}' + ax_{m0}')\right|}\\
&= \frac{1}{m}\ln\frac{\left|(x_{mn} + ax_{m-1,n+1}) - (x_{mn}' + ax_{m-1,n+1}')\right|}{\left|P_0 - P_0'\right|}\\
&\to \lambda,
\end{aligned}$$

when $m \to \infty$.

In the case where the trajectory motion is within a bounded region, this phenomenon of exponential separation cannot occur unless the two initial positions are extremely close to each other. Thus, by letting $\left|P_0 - P_0'\right| \to 0$ before taking the limit, $k \to \infty$, a constant is induced:

$$
\begin{aligned}
\lambda_{space} &= \lim_{m\to\infty}\frac{1}{m}\lim_{|P_0-P_0'|\to 0}\ln\left|\frac{(x_{mn}+ax_{m-1,n+1})-(x'_{mn}+ax'_{m-1,n+1})}{|P_0-P_0'|}\right| \\
&= \lim_{m\to\infty}\frac{1}{m}\lim_{|P_0-P_0'|\to 0}\ln\left|\frac{f^m(P_0)-f^m(P_0')}{|P_0-P_0'|}\right| \\
&= \lim_{m\to\infty}\frac{1}{m}\ln\left|\frac{df^m(P_0)}{dP_0}\right| \\
&= \lim_{m\to\infty}\frac{1}{m}\left(\sum_{j=0}^{n}\ln\left|\frac{df(x_j)}{dx_j}\right|+\sum_{j=n+1}^{m-1}\ln\left|\frac{df(x_j)}{dx_j}\right|\right) \\
&= \lim_{m\to\infty}\frac{1}{m}\left[\sum_{j=0}^{n}\ln\left|f'(\mu,x_{jn}+ax_{mj})\right|+\sum_{j=n+1}^{m-1}\ln\left|f'(\mu,x_{jn}+ax_{m,n+1})\right|\right].
\end{aligned}
$$

(ii) As in the (i) case, one can be shown that (4.37) hold.

Remark 4.6 It is easy to see that when $a = 0$, we have
(i) when $m \geq n$, and $n = \text{constant} = n_0$, it follows from (4.38), we have

$$
\lambda = \lim_{m\to\infty}\frac{1}{m}\sum_{j=0}^{m-1}\ln\left|f'(\mu,\ x_{jn_0})\right|.
$$

(ii) when $m \leq n$, and $m = \text{constant} = m_0$, it follows from (4.38), we have

$$
\lambda = \lim_{n\to\infty}\frac{1}{n}\sum_{r=0}^{n-1}\ln\left|f'(\mu,\ x_{m_0,r})\right|,
$$

which is just the familiar simple case of Lyapunov exponents of 1-D discrete system.

Example 4.3 The leading spatial Lyapunov exponent of system (4.1) is defined:

$$
\begin{aligned}
&\lambda(x_{0n}+ax_{m0}) \\
&= \lim_{m\to\infty}\frac{1}{m}\left(\sum_{j=0}^{n}\ln\left|f_x'(\mu,x_{jn}+ax_{mj})\right|+\sum_{j=n+1}^{m-1}\ln\left|f_x'(\mu,x_{jn}+ax_{m,n+1})\right|\right).
\end{aligned}
$$

Now, if one takes $\mu = 1.76$, $a = -0.002$, then system (4.9) reduces to

$$
x_{m+1,n} - 0.002x_{m,n+1} = 1 - 1.76[0.998x_{mn}]^2. \tag{4.40}
$$

Because $\mu = 1.76 > 1.55$ in (4.40), this system is surface chaos. This can also be confirmed by examining its spatial Lyapunov exponent via simulation. In fact, its spatial Lyapunov exponent is positive when $\mu = 1.76$, as shown in Fig. 4.4a, b.

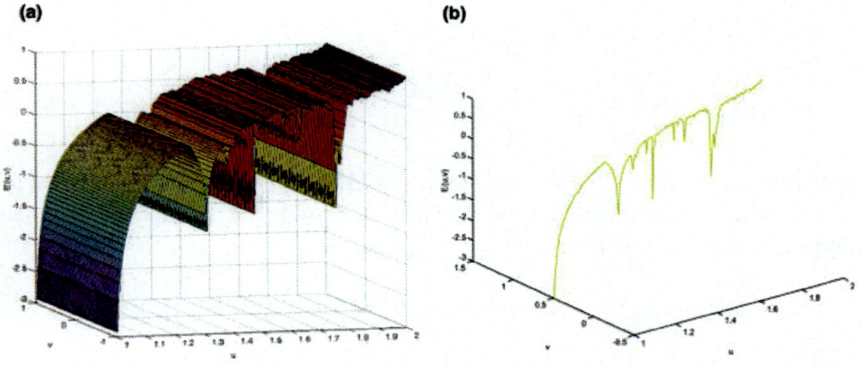

Fig. 4.4 **a** The spatial Lyapunov exponent of system (4.40). **b** A section curve of the spatial Lyapunov exponent of system (4.40)

4.4 Surface Chaos Behavior of General 2-D Logistic Discrete System

In this section, we will investigate the chaotic behavior of general 2-D logistic discrete system (4.8). That is, we will show the surface chaos in the sense of *Li* and *York* when $\mu \geq 1.55$.

Theorem 4.6 *Assume that $\mu \geq 1.55$, then general 2-D logistic discrete system (4.8) is surface chaotic in the sense of Li and York.*

Proof Let $V = R^3$, and take $V_0 = R$, then $V_0 \subset V$. First observe that $(1+a)x^* = \frac{-1+\sqrt{1+4\mu}}{2\mu}$ is an unstable fixed point of general 2-D logistic discrete system (4.8) i.e., let

$$x_{m+1,n} + ax_{m,n+1} = 1 - \mu\,[(1+a)x_{mn}]^2 = g(\mu, a, x_{mn}), \tag{4.41}$$

then we have $g'(\mu, a, x) = -2\mu(1+a)x$, hence $g'(\mu, a, (1+a)x^*) = 1 - \sqrt{1+4\mu} < -1$ for all $\mu > \frac{3}{4}$. Now if we can find a solution $\{u_{kl}\}_{k,l=-\infty,-\infty}^{+\infty,+\infty}$ with not all $u_{kl} = (1+a)x^*$ satisfying (i), (ii), (iii) of definition of Marotto sanp-back repelly in two sections. Such a sequence can be generated in the following manner. If we let $x_{0n} + ax_{m0} = (1+a)x^*$ then, since $(1+a)x^*$ is a fixed point of (4.8), $x_{k+1,n} + ax_{m,l+1} = (1+a)x^*$ for all $k \geq 0$, $l \geq 0$. In addition, $x_{k+1,n} + ax_{m,l+1}$ for all $k < 0$, $l < 0$ can be constructed by iterating the multi-valued inverse of (4.8), that is, from $x_{k+1,n} + ax_{m,l+1} = 1 - \mu(x_{kn} + ax_{ml})^2$, then

$$\mu(x_{kn} + ax_{ml})^2 = 1 - (x_{k+1,n} + ax_{m,l+1}),$$

hence we get

$$x_{k,n} + ax_{m,l} = \pm\left(\frac{1-(x_{k+1,n}+ax_{m,l+1})}{\mu}\right)^{\frac{1}{2}} = g_{\pm}^{-1}(\mu, a, x_{kn} + ax_{ml}), \quad (4.42)$$

provided $x_{kn} + ax_{ml} \le 1$. With $x_{0n} + ax_{m0} = (1+a)x^*$ we have two choices for $x_{-1,n} + ax_{m,-1}$ according to (4.42). However, it will not yield an appropriate sequence, since we would have

$$x_{-1,n} + ax_{m,-1} = g_{+}^{-1}(\mu, a, x_{0n} + ax_{m0}) = g_{+}^{-1}(\mu, a, (1+a)x^*) = (1+a)x^*.$$

Therefore define

$$x_{-1,n} + ax_{m,-1} = g_{-}^{-1}(\mu, a, x_{0n} + ax_{m0}),$$

and note that

$$x_{-1,n} + ax_{m,-1} = g_{-}^{-1}(\mu, a, x_{0n} + ax_{m0}) = g_{-}^{-1}(\mu, a, (1+a)x^*) = -(1+a)x^*.$$

Therefore let $x_{k-1,n} + ax_{m,l-1} = g_{+}^{-1}(\mu, a, x_{kn} + ax_{ml})$ in (4.42) for all $k \le -1, l \le -1$.

We shall show that this sequence $\{x_{kn} + ax_{ml}\}_{k,l=-\infty,-\infty}^{0,0}$ satisfies $x_{kn} + ax_{ml} \to (1+a)x^*$ as $k,\ l \to \infty$ (for appropriate values of μ). Note that since $x_{-1,n} + ax_{m,-1} = -(1+a)x^* < (1+a)x^*$:

$$\begin{aligned} x_{-2,n} + ax_{m,-2} &= g_{+}^{-1}(\mu, a, x_{-1,n} + ax_{m,-1}) \\ &\in g_{+}^{-1}\left[-\infty,\ (1+a)x^*\right] \\ &\subset \left(-\infty,\ (1+a)x^*\right). \end{aligned}$$

Let us find those values of a for which $x_{-2,n} + ax_{m,-2} < -1$. Since $x_{-1,n} + ax_{m,-1} = -(1+a)x^* = \frac{1-(1+4\mu)^{\frac{1}{2}}}{2\mu}$, then by (4.42)

$$x_{-2,n} + ax_{m,-2} = g_{+}^{-1}(\mu, a, x_{-1,n} + ax_{m,-1}) = \left[\frac{1+(1+a)x^*}{\mu}\right]^{\frac{1}{2}}.$$

So, $x_{-2,n} + ax_{m,-2} < 1$ implies $\left[\frac{1+(1+a)x^*}{\mu}\right]^{\frac{1}{2}} < 1$, that is $(1+a)x^* < \mu - 1$, or $\frac{-1+(1+4\mu)^{\frac{1}{2}}}{2\mu} < \mu - 1$. This is equivalent to

$$\mu^3 - 2\mu^2 + 2\mu - 2 > 0.$$

It can be shown that all values of $\mu > 1.55$ satisfy this equation, and thus $x_{-2,n} + ax_{m,-2} \in ((1+a)x^*, 1)$ for these values of μ. We restrict the remainder of the discussion to the problem when $\mu > 1.55$.

Now because $x_{-2,n} + ax_{m,-2} \in ((1+a)x^*, 1)$ for these values:

$$\begin{aligned} x_{-3,n} + ax_{m,-3} &= g_+^{-1}(\mu, a, x_{-2,n} + ax_{m,-2}) \\ &\in g_+^{-1}\left((1+a)x^*,\ 1\right) \\ &\subset \left(0,\ (1+a)x^*\right), \end{aligned}$$

and consequently $x_{-3,n} + ax_{m,-3} \in (0,\ (1+a)x^*)$. Also

$$\begin{aligned} x_{-4,n} + ax_{m,-4} &= g_+^{-1}(\mu, a, x_{-3,n} + ax_{m,-3}) \in g_+^{-1}\left(0, (1+a)x^*\right) \\ &\subset g_+^{-1}\left[x_{-1,n} + ax_{m,-1},\ (1+a)x^*\right] \\ &\subset \left((1+a)x^*,\ x_{-2,n} + ax_{m,-2}\right), \end{aligned}$$

and so $x_{-4,n} + ax_{m,-4} \in \left((1+a)x^*,\ x_{-2,n} + ax_{m,-2}\right)$. This implies:

$$\begin{aligned} x_{-5,n} + ax_{m,-5} &= g_+^{-1}(\mu, a, x_{-4,n} + ax_{m,-4}) \\ &\in g_+^{-1}\left((1+a)x^*,\ x_{-2,n} + ax_{m,-2}\right) \\ &\subset g_+^{-1}\left[x_{-1,n} + ax_{m,-1},\ (1+a)x^*\right] \\ &\subset \left(x_{-3,n} + ax_{m,-3},\ (1+a)x^*\right), \end{aligned}$$

and hence $x_{-5,n} + ax_{m,-5} \in \left(x_{-3,n} + ax_{m,-3},\ (1+a)x^*\right)$. It can also see that such a sequence generated in manner shown in Fig. 4.5: (note that we will introduce the notations $r_{-k,-l} = x_{-k,n} + ax_{m,-l}$, $r_0 = (1+a)x^*$, and only to drow the case of when $a = 1$, that is, inverse iterates of $x_{m+1,n} + x_{m,n+1} = 1 - \mu x_{mn}^2$ for $\mu > 1.55$.)

Continuing in this manner, it is apparent that the sequence $\{x_{kn} + ax_{ml}\}_{k,l=-\infty,-\infty}^{0,0}$ thus constructed satisfies the following: $x_{-2k,n} + ax_{m,-2l}$ is a decreasing sequence bounded below by $(1+a)x^*$, and $x_{-2k-1,n} + ax_{m,-2l-1}$ is an increasing sequence bounded above by $(1+a)x^*$ (Fig. 4.5). There must therefore exist a point $\alpha \in (0,\ (1+a)x^*)$ which is the limit of $x_{-2k,n} + ax_{m,-2l}$, and a point $\beta \in [(1+a)x^*, 1]$ which is the limit of $x_{-2k,n} + ax_{m,-2l}$ as $k, l \to \infty$.

We shall show $\alpha = \beta = (1+a)x^*$. Since

$$g\left(\mu, a,\ x_{-2k-1,n} + ax_{m,-2l-1}\right) = x_{-2k,n} + ax_{m,-2l},$$

and

$$g\left(\mu, a,\ x_{-2k,n} + ax_{m,-2l}\right) = x_{-2k+1,n} + ax_{m,-2l+1},$$

it must be that $g(\mu, a, \alpha) = \beta$ and $g(\mu, a, \beta) = \alpha$. Consequently, $g(g(\alpha)) = \alpha$, and α is thus a fixed point of the function $g \circ g$. But for $\mu > 1.55$ there are precisely four

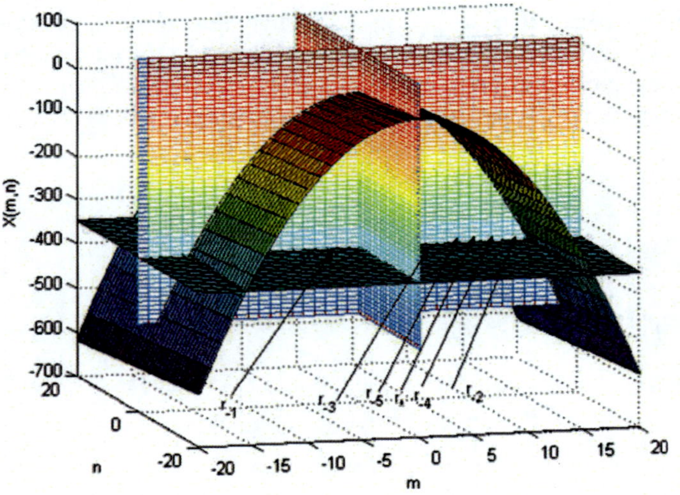

Fig. 4.5 Inverse iterates of $x_{m+1,n} + x_{m,n+1} = 1 - \mu x^2_{m,n}$ for $a = 1$, with $\mu > 1.55$

fixed points of $g \circ g$ (since this function is a quartic polynomial) each of which can be computed exactly:

$$\frac{-1 \pm (1+4\mu)^{\frac{1}{2}}}{2\mu} \text{ and } \frac{-1 \pm (4\mu - 3)^{\frac{1}{2}}}{2\mu}.$$

It is clear that, for $\mu > 1.55$, $\frac{-1\pm(1+4\mu)^{\frac{1}{2}}}{2\mu}$ and $\frac{-1\pm(4\mu-3)^{\frac{1}{2}}}{2\mu}$ are both negative, and thus neither of these can equal $\alpha \in (0, (1+a)x^*)$. Also, suppose

$$\frac{-1 \pm (4\mu - 3)^{\frac{1}{2}}}{2\mu} \leq (1+a)x^* = \frac{-1 \pm (1+4\mu)^{\frac{1}{2}}}{2\mu}.$$

Then, $(1+4\mu)^{\frac{1}{2}} \geq 2 + (4\mu - 3)^{\frac{1}{2}}$. Squaring each side of this inequality implies $4(4\mu - 3)^{\frac{1}{2}} \leq 0$ which is a contradiction for $\mu > 1.55$. This implies $\frac{-1\pm(4\mu-3)^{\frac{1}{2}}}{2\mu} \geq (1+a)x^*$. Therefore it must be that $\alpha = \frac{-1\pm(1+4\mu)^{\frac{1}{2}}}{2\mu} = (1+a)x^*$, and $\beta = g(\alpha) = (1+a)x^*$. Hence $x_{kn} + ax_{ml} \to (1+a)x^*$ as $k, l \to \infty$.

So far we have shown that the sequence $\{x_{kn} + ax_{ml}\}_{k,l=-\infty,-\infty}^{0,0}$ satisfies (i) and (ii) from above the defination of Marotto. It can easily show that (iii) is also satisfied. Since $g'(\mu, a, x) = -2\mu(1+a)x$, the only possible way for $g'(\mu, a, x_{kn} + ax_{ml}) = 0$ is if $x_{kn} + ax_{ml} = 0$ for some k and l. But based on the manner in which the sequence was constructed: $x_{kn} + ax_{ml} = (1+a)x^*$ for $k \geq 0, l \geq 0$, $x_{-1,n} + ax_{m,-1} = -(1+a)x^* < 0$; $x_{-2,n} + ax_{m,-2} > (1+a)x^* > 0$; $x_{-3,n} + ax_{m,-3} > 0$; and

$$x_{kn} + ax_{ml} \in (x_{-3,n} + ax_{m,-3}, x_{-2,n} + ax_{m,-2}) \subset (0, 1),$$

for all $k - 3$ and $l < -3$. Thus the sequence also satisfies (iii), and $(1 + a)x^*$ is therefore a sanp-back repelly of 2-D discrete system (4.8). Now applying Marotto Theorem of above in the two section, implying the 2-D discrete system (4.8) is surface chaotic in the sence of Li and York.

Remark 4.7 We only consider the Convection equation of discrete system the above content, we can also study the Laplace equation; Wave equation and Purabollic equation of discrete systems by using other methods.

4.5 Illustrative Examples

Finally, we give an illustrative example:

Example 4.4 Consider the system (4.8). Due to

$$\xi = \frac{1}{4}\left[\left(\frac{1 + |a| + (a + 1)^2}{(a + 1)^2}\right)^2 - 1\right],$$

with parameters as shown in Table 4.1.

It can be seen from Table 4.1 that a changes from -0.98 to -0.05, μ from 0.125 to 1.77285, and when a gradually changes from small to big and μ gradually changes from small to big, system (4.8) changes from stable to chaotic. For those cases shown in Table 4.1, the system dynamics are demonstrated by Figs. 4.6, 4.7, 4.8 and 4.9.

Example 4.5 Consider the system (4.8) and take $a = -\frac{18}{25}$, $\mu = \frac{375}{196}$. Then, the system (4.8) becomes

$$x_{m+1,n} - \frac{18}{25}x_{m,n+1} = 1 - \frac{3}{20}x_{mn}^2, \tag{4.43}$$

Table 4.1 Variation of Parameters and its Stabilities

$\mu(1+a)^2$	0.0005	0.0025	0.05	1.6
a	-0.98	-0.98	-0.72	-0.05
μ	0.125	0.625	0.641	1.77285
ξ	6128100	6128100	131.3	0.92
$\mu \bigvee \xi$	$<$	$<$	$<$	$>$
$\mu \bigvee 1.55$	$<$	$<$	$<$	$>$
Sanp-back repelly	No definitude	No definitude	No definitude	Existence
Figured state	Stability	From stability tend to chaos	From stability tend to chaos	Chaos
Figure	Fig. 4.6	Fig. 4.7	Fig. 4.8a, b	Fig. 4.9

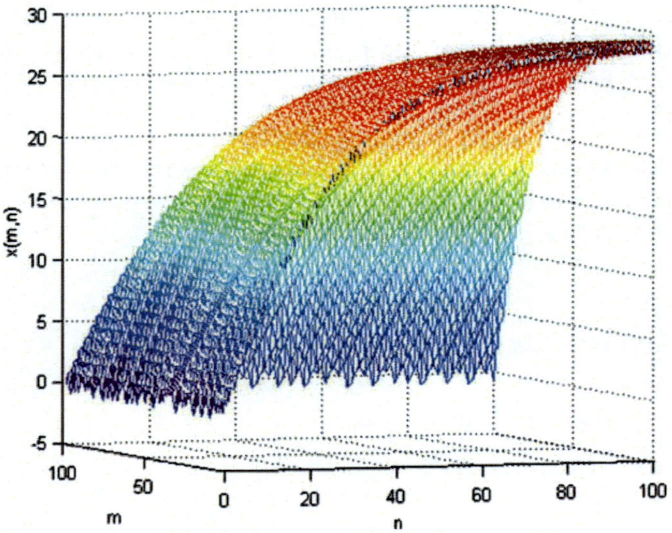

Fig. 4.6 The stable orbits of system (4.8) with $a = -0.98$, $\mu = 0.125$

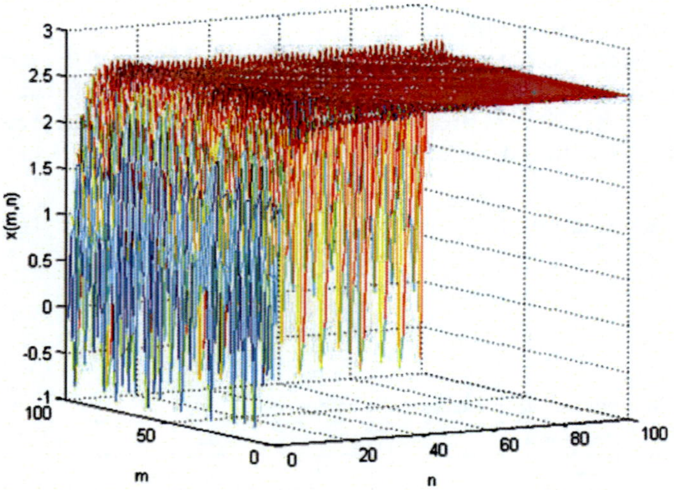

Fig. 4.7 From stability to surface chaos of system (4.8) with $a = -0.98$, $\mu = 0.125$

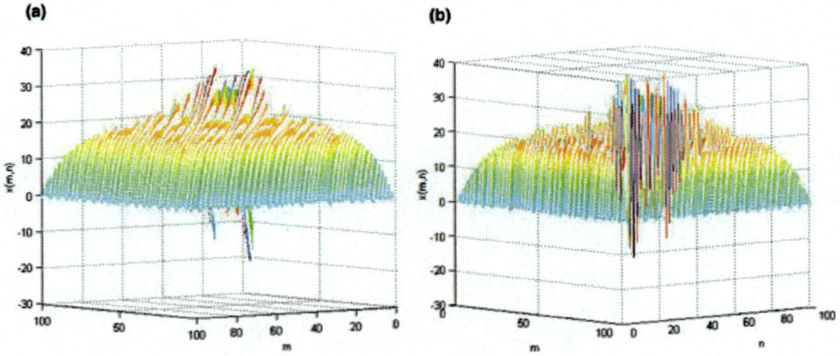

Fig. 4.8 **a** From stability to surface chaos of system (4.8) with $a = -0.72$, $\mu = 0.641$. **b** From stability to chaos of system (4.8) with $a = -0.72$, $\mu = 0.641$

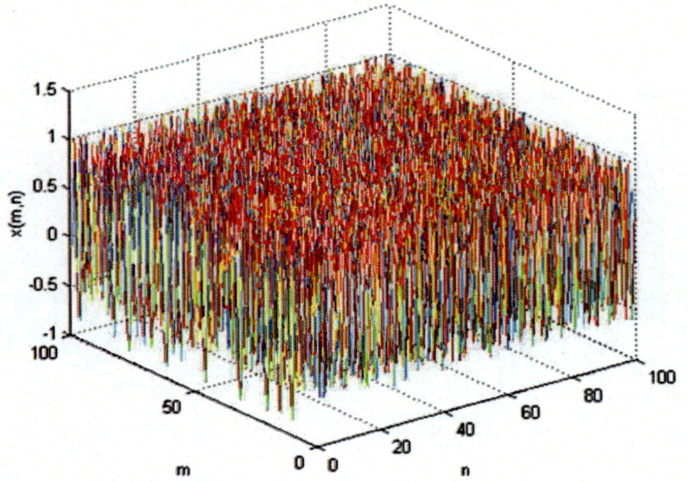

Fig. 4.9 Surface chaotic behavior of system (4.8) with $a = -0.05$, $\mu = 1.77285$

with $\mu = \frac{375}{196} > 1.55$. It is easy to see that all assumptions of Theorem 4.6 are satisfied. Therefore, Theorem 4.6 implies that system (4.43) is chaotic in the sence of *Li* and *York*. The chaotic behavior of system (4.43) is demonstrated by Fig. 4.10.

Example 4.6 In particular, when m =constant, or n = cotnstant, for example, when $m_0 = 30$ and $n_0 = 40$, Fig. 4.11 shows these two section cases.

Example 4.7 On the other hand, when $a = 0$ and $n_0 = 0$, system (4.43) reduces to

$$x_{m+1,n_0} = 1 - \frac{375}{196} x_{mn_0}^2. \tag{4.44}$$

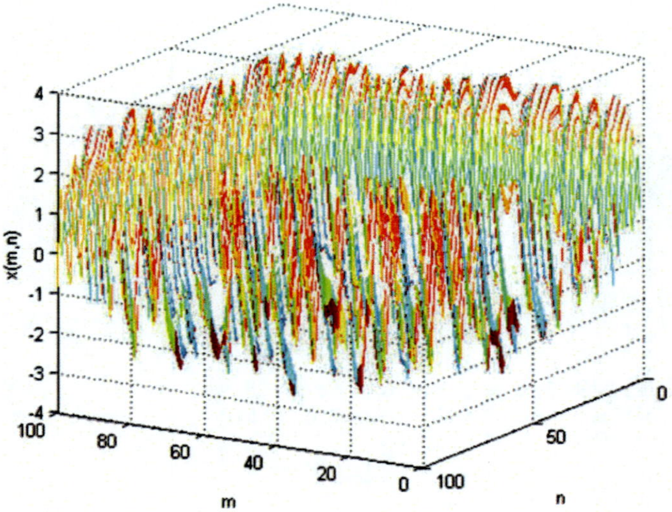

Fig. 4.10 Surface chaotic behavior of system (4.43)

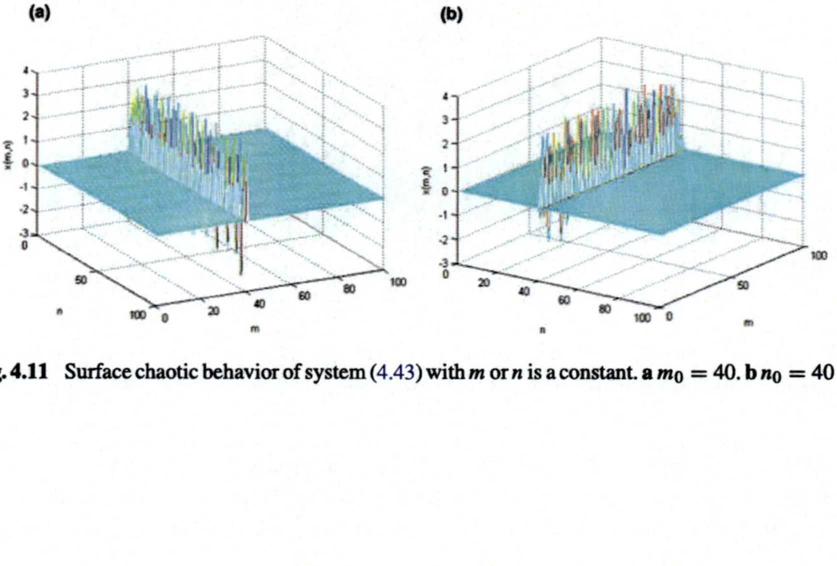

Fig. 4.11 Surface chaotic behavior of system (4.43) with *m* or *n* is a constant. **a** $m_0 = 40$. **b** $n_0 = 40$

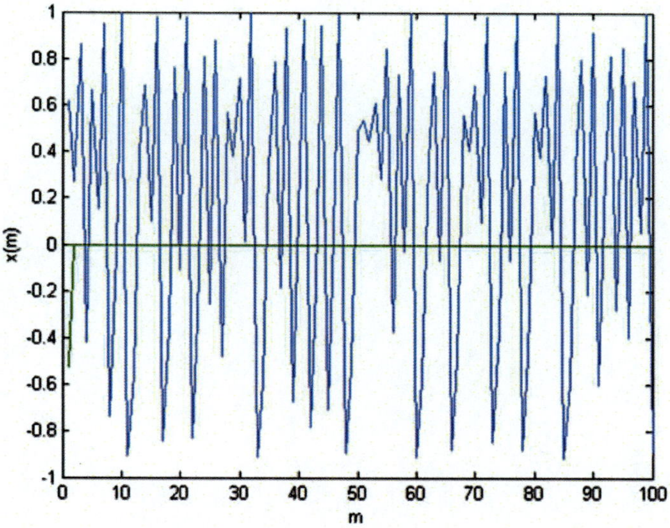

Fig. 4.12 Surface chaotic behavior of system (4.44)

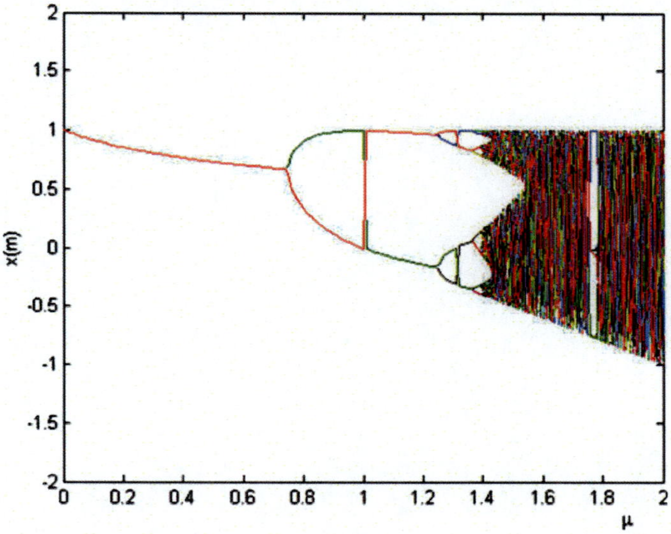

Fig. 4.13 Period-doubling bifurcations of system (4.44)

Clearly, system (4.44) is just a special simple case of 1-D Logistic system. Figure 4.12 shows its chaotic behavior. For $\mu \in [0, 2]$, Fig. 4.13 shows its peniod-doubling bifurcations.

Chapter 5
Surface Chaos and Its Associated Bifurcation and Feigenbaum Problem

5.1 Preliminaries

A surface is defined by the 2-D discrete dynamical system:

$$x_{m+1,n} + a x_{m,n+1} = G(\mu, (1+a)x_{m,n}), \tag{5.1}$$

where $G(\mu, (1+a)x_{m,n})$ is a nonlinear, implicit function, $\mu > 0$, $a \in \mathbb{R}$, $a \neq 1$. In this chapter, we aim to show that system (5.1) can lead to some significant nonlinear dynamics such as surface chaos, surface bifurcation, and the Feigenbaum problem of bifurcation about the surface chaos.

5.2 Changing the System Spatial Structures

Consider system (5.1). To carry out an in-depth investigation of the dynamic behaviors of this system on a spatial surface, one may first rewrite it in the following form:

$$x_{m+1,n} + \omega x_{m,n+1} = f(\mu, x_{mn} + \omega x_{mn}), \tag{5.2}$$

and then introduce a corresponding spatially multi-lateral variable system in the four-dimensional Euclidean space, that is,

$$x_{s+1,n} + \omega x_{m,t+1} = f(\mu, x_{sn} + \omega x_{mt}), \tag{5.3}$$

where each function on both sides are regarded as a function of the four variables s, t, m, n. Clearly, it is a system in $\mathbb{R}^4$, and one may further make a transformation by

$$C_{s+1,t+1}(m, n) = x_{s+1,n} + \omega x_{m,t+1}, \tag{5.4}$$

S. T. Liu and L. Zhang, *Surface Chaos and Its Applications*,
https://doi.org/10.1007/978-981-16-8229-2_5

so that

$$C_{st}(m,n) = x_{sn} + \omega x_{mt}. \tag{5.5}$$

By (5.4) and (5.5), Eq. (5.3) becomes

$$C_{s+1,t+1}(m,n) = f(\mu, C_{st}(m,n)). \tag{5.6}$$

If $f(\mu, C_{st}(m,n)) = 1 - \mu C_{st}^2(m,n)$, then system (5.6) reduces to

$$C_{s+1,t+1}(m,n) = 1 - \mu C_{st}^2(m,n), \tag{5.7}$$

implying that the 1-D Logistic system is

$$C_{s+1} = 1 - \mu C_s^2.$$

In addition, from (5.5) and (5.6), if one lets $s = m,\ t = n$ in (5.4), then

$$C_{m+1,n+1}(m,n) = x_{m+1,n} + \omega x_{m,n+1},\ \ C_{mn}(m,n) = (1+\omega)x_{mn}, \tag{5.8}$$

which changes (5.7) to

$$C_{m+1,n+1}(m,n) = 1 - \mu C_{mn}^2(m,n), \tag{5.9}$$

leading to the following system:

$$x_{m+1,n} + \omega x_{m,n+1} = 1 - \mu\left((1+\omega)x_{mn}\right)^2. \tag{5.10}$$

System (5.10) is called a spatially generalized Logistic system.

In view of (5.8), system (5.10) is only a special case of system (5.7). However, it is noted that system (5.10) is a 2-D system, generating a surface in $\mathbb{R}^3$, while system (5.7) is a 4-D system, generating a surface in $\mathbb{R}^5$. In fact, if one applies the transformation (5.4) such that the 2-D system (5.10) becomes the 4-D system (5.7), and considers s and t to be index-parameters while m and n to be variables, then system (5.7) can be regarded as having a surface in $\mathbb{R}^3$ spanned by s and t. Thus, one may investigate the 4-D system (5.7), subject to the changeable function $C_{st}(m,n)$, with four independent variables s, t, m, n. In particular, for system (5.7) with variables m, n, if for any (s,t), $(m,n) \in N_r \times N_r$, system (5.7) is chaotic, then system (5.10) is also chaotic when $s = m, t = n$; that is, system (5.10) is chaotic by nature. Hence, one may investigate the chaotic behaviors of system (5.7), thereby understanding the chaotic behaviors of system (5.10).

5.3 Fixed Surfaces and Fixed Curves

For the spatial Logistic system (5.10), one can easily obtain its fixed plane as follows:

$$(1+\omega)x^* = \frac{-1 \pm \sqrt{1+4\mu}}{2\mu}, \tag{5.11}$$

which is a plane consisting of all system fixed points, and is usually written in the following form:

$$x^* = \frac{-1 \pm \sqrt{1+4\mu}}{2\mu(1+\omega)}.$$

In particular, because μ, ω are two real parameters, and $\mu > 0,\ \omega \in (-1, \infty)$, formula (5.11) is a dual function of μ, ω, and represents a surface in the space.

Let

$$f(\mu, \omega) = \frac{-1 \pm \sqrt{1+4\mu}}{2\mu(1+\omega)}. \tag{5.12}$$

Then surface (5.11) has two leaves, that is,

$$f_+(\mu, \omega) = \frac{-1 + \sqrt{1+4\mu}}{2\mu(1+\omega)},\quad f_-(\mu, \omega) = \frac{-1 - \sqrt{1+4\mu}}{2\mu(1+\omega)}. \tag{5.13}$$

Since it is composed of all fixed points of system (5.10), the two leaves $f_+(\mu, \omega)$ and $f_-(\mu, \omega)$ are called fixed surfaces of system (5.10). When and only when $\mu = \mu_0$ and $\omega = \omega_0$, they change into planes in the space.

Note that, again, when $\mu > 0$ and $\omega \in (-1, \infty)$, one has

$$\begin{cases} f'_{\mu,+}(\mu, \omega) = \frac{-1+\sqrt{1+4\mu}}{2\mu(1+\omega)^2} < 0, \\ f'_{\omega,+}(\mu, \omega) = \frac{\sqrt{1+4\mu}-(1+2\mu)}{2\mu^2(1+\omega)\sqrt{1+4\mu}} < 0. \end{cases} \quad \begin{cases} f'_{\mu,-}(\mu, \omega) = \frac{(1+2\mu)+\sqrt{1+4\mu}}{2\mu^2(1+\omega)\sqrt{1+4\mu}} > 0, \\ f'_{\omega,-}(\mu, \omega) = \frac{1+\sqrt{1+4\mu}}{2\mu(1+\omega)^2} > 0. \end{cases}$$

Here, for the two leaves (fixed surfaces), one is monotonically decreasing in μ, ω, and another is monotonically increasing in μ, ω. This case is demonstrated by Fig. 5.1a.

In addition, one can obtain the spatial sections of the two fixed surfaces, and they are two spatial curves, called *fixed curves* of system (5.10), as shown in Fig. 5.1b.

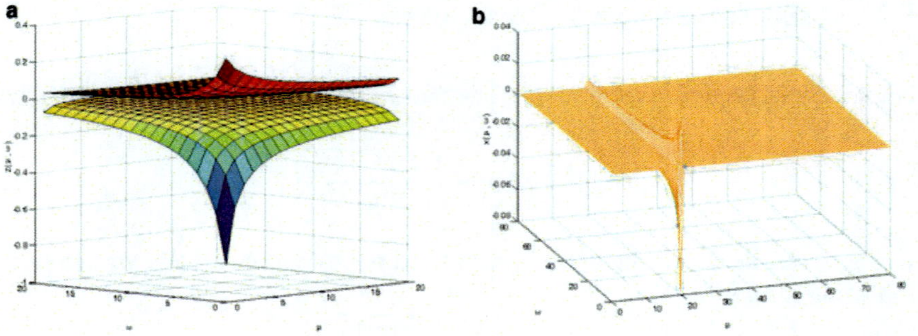

Fig. 5.1 **a** Two leaves (fixed surfaces) of system (5.10). **b** The spatial sections of the two fixed surfaces of system (5.10)

5.4 Surface Bifurcation

First, consider the following example:

$$\mu - 1 + 3\cos(Z) - (m^2 - Z^2) = 0, \tag{5.14}$$

where $\mu > 0$ is a real parameter, and $Z, m \in R$, as visualized in Fig. 5.2a, b.

One can see that the spatial surface of system (5.14) changes the two (up and down) leaves according to the plane of $z = 0$ with respect to the parameter μ.

Similarly, consider the following function:

$$z(x, y) = 2 - \mu\left[2(x-1)^2 - y^2\right]\sin y^2, \tag{5.15}$$

where $\mu > 0$ is a real parameter. System (5.15) can be considered as an iterative relation on variable x, and express the iterative relation by $z(x, y)|_x$. If, for example, one

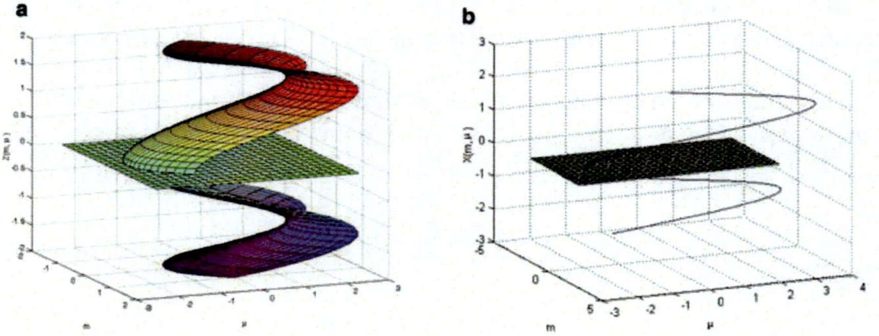

Fig. 5.2 **a** The spatial surface of system (5.14). **b** The spatial section of system (5.14)

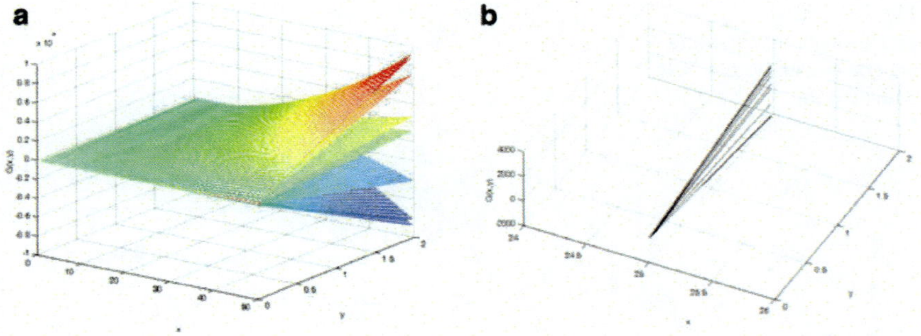

Fig. 5.3 **a** The computer plot of function $G(x, y)$. **b** The spatial section of function $G(x, y)$

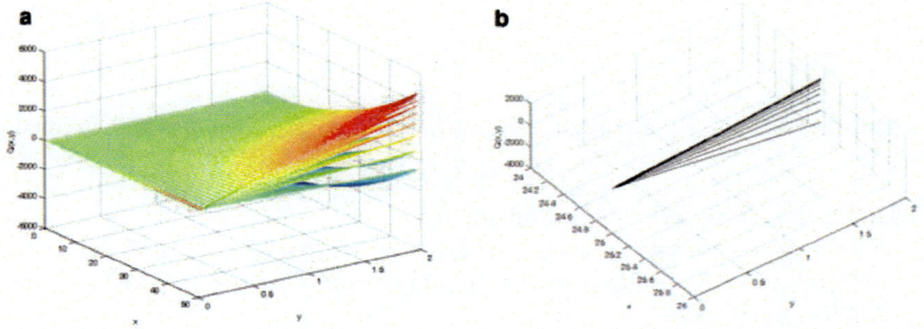

Fig. 5.4 **a** The computer plot of function $Q(x, y)$. **b** The spatial section of function $Q(x, y)$

iterates it for 10 times, then the resultant surface is given by $G(x, y) = z^{10}(x, y)\big|_x$, where $z^2(x, y)\big|_x = z(z(x, y))|_x$. Computer plots of this function are shown in Fig. 5.3a, b.

Likewise, one can examine the spatial surface of system (5.15) with, for example,

$$Q(x, y) = u^8(x, y)\big|_y,$$

where $u^2(x, y)\big|_y = u(u(x, y))|_y$ and

$$u(x, y) = 1 - \mu\left[2x^3 - (y-1)^2\right]\sin y^2, \ \mu > 0.$$

Computer plots of function Q are shown in Fig. 5.4a, b.

The above examples show that if the parameter μ and the number of iterations increase, then the spatial surface will change from leave to leave. This is the bifurcation on spatial surface, as further discussed below.

Now, consider the parametric spatial function

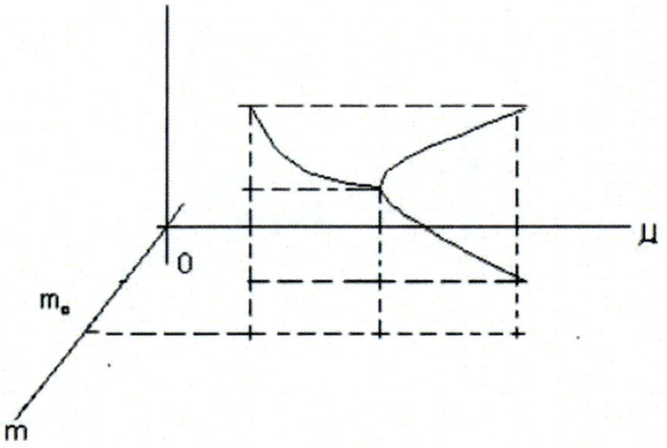

Fig. 5.5 The period-doubling bifurcation on a section of system (5.16) with $m = m_0$

$$G(\mu, \omega, x(m, n)),$$

where $\mu > 0$, $\omega \in R$, and $m, n \in Z = \{..., -1, 0, 1, ...\}$. One has the following results.

Theorem 5.1 (s-period-doubling bifurcation)
For the above function $G(\mu, \omega, x(m, n))$, suppose that for all $\omega \in R$,
1. *$G(\mu, \omega, 0) = 0$, for all μ in an interval containing μ_0,*
2. *$G'(\mu_0, \omega, 0) = -1$,*
3. *$\left.\frac{\partial(G^2(\mu, \omega, x(m,n)))'}{\partial \mu}\right|_{\mu=\mu_0} (0) \neq 0$.*

Then, system $x_{m+1,n} + \omega x_{m,n+1} = G(\mu, \omega, x(m, n))$ has a period-doubling bifurcation on the surface with respect to the parameter $\mu > 0$.

Proof For surface S: $G(\mu, \omega, x(m, n))$, one takes an arbitrary section on it. For example, let $m = m_0$, one can obtain a section of surface S :

$$S_{m_0} : G(\mu, \omega, x(m_0, n)). \tag{5.16}$$

System (5.16) has a period-doubling bifurcation in $\mathbb{R}^3$, as shown by Fig. 5.5.

Since system (5.16) has a plane curve in the plane $m = m_0$ in space, by using the conditions 1–3 of the theorem, all conditions for period-doubling bifurcation are satisfied. Thus, from Lemma 2.1, the section $G(\mu, \omega, x(m_0, n))$ has a period-doubling bifurcation. Moreover, since $m = m_0$ is arbitrarily, the surface

$$S : G(\mu, \omega, x(m, n)),$$

has an s-bifurcation in $\mathbb{R}^3$.

Theorem 5.2 (saddle-node s-bifurcation)
For system

$$x_{m+1,n} + \omega\, x_{m,n+1} = G(\mu, \omega, x(m, n)),$$

if a surface $S : G(\mu, \omega, x(m, n))$ satisfies

1. $G(\mu_0, \omega, 0) = 0,$

2. $G'(\mu_0, \omega, 0) = 1,$

3. $G''(\mu_0, \omega, 0) \neq 1,$

4. $\left.\frac{\partial G^2(\mu,\omega,x(m,n))}{\partial \mu}\right|_{\mu=0} (0) \neq 0.$

Then, the system has a saddle-node surface bifurcation on the surface S with respect to the parameter $\mu > 0$.

Theorem 5.2 can be similarly proved.

5.5 Surface Bifurcation and Surface Chaos in Spatial Logistic System

5.5.1 2^r-Period-Doubling Bifurcation on Surface

Consider again the 2-D Logistic system (5.10), namely, $x_{m+1,n} + \omega\, x_{m,n+1} = G(\mu, \omega, x(m, n))$, with

$$G(\mu, \omega, x) = 1 - \mu[(1 + \omega)x]^2. \tag{5.17}$$

Its fixed-point equation is

$$(1 + \omega)x = 1 - \mu[(1 + \omega)x]^2. \tag{5.18}$$

This is the 0-iterative fixed-point equation, as further discussed below.

Consider the cases of the surface with 2^r-period-doubling bifurcation in space in the 2-D Logistic system (5.10), $r = 0, 1, 2 \ldots$. For $\mu > 0$ with 2^r times of iterations on the system, $r = 0, 1, 2, \ldots$, one has results shown in the following Table 5.1:

In the table, to determine the 2^rth critical and bifurcating fixed curve on surface $\mu_r = \mu_r(\omega)$, one performs on the r-iterative fixed-point equation and obtains the stable boundary condition for the r-times of iterations, as follows:

$$\begin{cases} G^r(\mu, \omega, x) = (1 + \omega)x, \\ \left| \prod_{i=1}^{r} G'_x(\mu, \omega, x_s) \right| \leq \frac{1+|\omega|}{|1+\omega|}. \end{cases} \tag{5.19}$$

For example, for $r = 0$ one has $\mu_1 = \mu_1(\omega)$ in the 0-iteration, obtaining

$$\begin{cases} G(\mu, \omega, x) = 1 - \mu[(1 + \omega)x]^2 = (1 + \omega)x, \\ G'_x(\mu, \omega, x) \leq \frac{1+|\omega|}{|1+\omega|}, \end{cases} \tag{5.20}$$

Table 5.1 Surface with 2^r-period-doubling bifurcation

Number of iteration	Fixed curve	Form
2^0iteration	First critical and bifurcating fixed curve on surface	$\mu_0 = \mu_0(\omega)$
2^1 iteration	Second critical and bifurcating fixed curves on surface	$\mu_1 = \mu_1(\omega)$
...	...	...
2^r iteration	2^rth critical and bifurcating fixed curves on surface	$\mu_r = \mu_r(\omega)$
...	...	...

because $G'_x(\mu, \omega, x) = -2\mu(1+\omega)x$. Then, from (5.20), one has

$$\mu_0(\omega) = \frac{1}{4}\left[\left(\frac{1+(1+\omega)^2+|\omega|}{(1+\omega)^2}\right)^2 - 1\right]. \tag{5.21}$$

For the same reason, one can also obtain $\mu_1(\omega)$, but it should be noted that if x is a fixed point of $G(\mu, \omega, x) - (1+\omega)x = 0$, then it must also be a fixed point of

$$G^2(\mu, \omega, x) - (1+\omega)x = 0;$$

therefore, one can obtain

$$\frac{G^2(\mu, \omega, x) - (1+\omega)x}{G(\mu, \omega, x) - (1+\omega)x} = \mu^2\left((1+\omega)x\right)^2 - \mu(1+\omega)x - \mu + 1 = 0,$$

leading to

$$\begin{cases} x_1 = \frac{1-\sqrt{4\mu-3}}{2\mu(1+\omega)}, \\ x_2 = \frac{1+\sqrt{4\mu-3}}{2\mu(1+\omega)}. \end{cases} \tag{5.22}$$

Here, it is noticed that only when $\mu \geq \frac{3}{4}$, solution (5.22) is a real solution. In this case, it represents two fixed curves, so that using the stable boundary condition again to carry out iterations twice yields

$$\left|G'_x(\mu, \omega, x_1) \times G'_x(\mu, \omega, x_2)\right| \leq \frac{1+|\omega|}{|1+\omega|}. \tag{5.23}$$

Combining (5.22) and (5.23), one has

$$\begin{cases} \mu_{1,1}(\omega) = \frac{1}{4}\left(4 - \frac{1+|\omega|}{|1+\omega|}\right), \\ \mu_{1,2}(\omega) = \frac{1}{4}\left(4 + \frac{1+|\omega|}{|1+\omega|}\right). \end{cases} \tag{5.24}$$

Similarly, one can determine the following:

$$\mu_2(\omega), \mu_3(\omega), \ldots, \mu_r(\omega), \ldots.$$

Note, however, that in order to determine the critical and bifurcating fixed curve $\mu_r(\omega)$, $r \geq 3$, one must solve the following fixed-point equation on surface:

$$G^r(\mu, \omega, x) - (1+\omega)x = 0.$$

But $M^0(f(x))$ shows the maximal degree of the polynomial $f(x)$, since $M^0(G^r(\mu, \omega, x) - (1+\omega)x) = 2r$. Observe that

$$\begin{aligned} &G^r(\mu, \omega, x) - (1+\omega)x \\ &= g_{2r}(\omega)x^{2r} + g_{2r-1}(\omega)x^{2r-1} + \cdots + g_1(\omega)x + 1, \end{aligned} \tag{5.25}$$

where the coefficients of every x^i, $i = 2r, 2r-1, \ldots, 1, 0$, are $g_i(\omega)$ with $g_1(\omega) = -\omega$, $g_0(\omega) = 1$. Note that $g_i(\omega)$ is a composite function with respect to the parameter ω, that is, (5.25) is a polynomial of high-order terms with variable coefficients. Therefore, it is very difficult to solve this algebraic equation if $r \geq 3$. But if one takes $\omega = 1$ then it is easy to obtain its critical and bifurcating fixed curve numerically:

$$\begin{aligned} &\mu_3 = 1.368, \mu_4 = 1.394, \mu_5 = 1.396, \\ &\mu_6 = 1.401, \mu_7 = 1.4011, \ldots. \end{aligned} \tag{5.26}$$

It is also noted that when $\omega > 0$, by the stable boundary condition (5.19), one has the following:

$$\left|\prod_{i=1}^{r} G'(\mu, \omega, x_s)\right| \leq \left.\frac{1+|\omega|}{|1+\omega|}\right|_{\omega>0} = 1, \tag{5.27}$$

where x_s is the solution of equation $x_{s+1} = 1 - \mu x_s^2$. Furthermore, one has

$$\mu_1 = \frac{3}{4}, \mu_2 = \frac{5}{4}, \ldots. \tag{5.28}$$

Remark 5.1 As can be seen from the above, the critical and bifurcating fixed curve on surface reduces to a bifurcating fixed point in the 1-D Logistic system, which reflects a fundamental symmetry and a parallelism from the 1-D Logistic system to the 2-D Logistic system for $\omega > 0$.

From (5.21) to (5.24), one can get the critical and bifurcating fixed curve of s-bifurcation on surface, but these points on the critical and bifurcating fixed curve consist of a curve in the intersection between the surface of the r−times iterations and the plane located on the first and the third quadrants. Some typical cases are as follows:

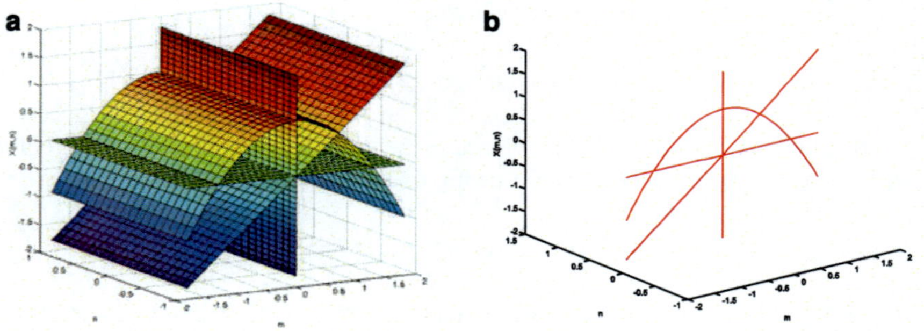

Fig. 5.6 **a** The bifurcating fixed curve of system (5.10) with $\mu = 0.3$, and $\omega = 1$. **b** The section case of system (5.10) with $\mu = 0.3$, and $\omega = 1$

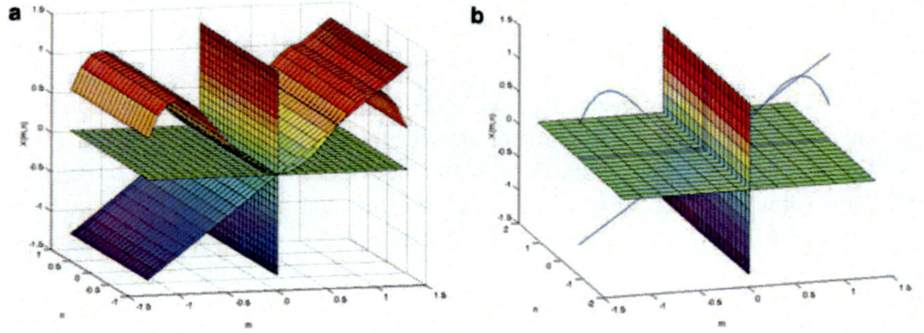

Fig. 5.7 **a** The fixed bifurcation curve of system (5.10) with $\mu = 1$, and $\omega = 1$. **b** The section case of system (5.10) with $\mu = 1$, and $\omega = 1$

(1) When $\mu \in (0, \frac{3}{4})$, the critical and bifurcating fixed curve $\frac{3}{4}$ is just a curve that is the intersection of the spatial surface $G(\mu, \omega, x)$ and the following plane:

$$\text{P(I, III)} = \{z = m \mid (m, z) \in N_r \times N_r\}.$$

Figure 5.6 shows the critical and bifurcating fixed curve for $\mu = 0.3$ and $\omega = 1.0$.

(2) When $\mu \in (\frac{3}{4}, \frac{5}{4})$, system (5.30) has a stable fixed curve of 2-period, and the critical and bifurcating fixed curve $\frac{5}{4}$ is just an intersected curve by the spatial surface $G^2(\mu, \omega, x)$ of the 2-times iteration and P(I, III). Figure 5.7 shows the result with $\mu = 1$ and $\omega = 1$.

(3) When $\mu \in (\frac{5}{4}, 1.37)$, system (5.30) has a stable fixed curve of 2^2-period, and the critical and bifurcating fixed curve at 1.37 is just an intersected curve by the spatial surface $G^2(\mu, \omega, x)$ of the 2^2-times of iterations and P(I, III).

The simulation results shown in Fig. 5.8 are obtained with $\mu = 1.3$ and $\omega = 1$.

With respect to the parameter μ, one has the following:

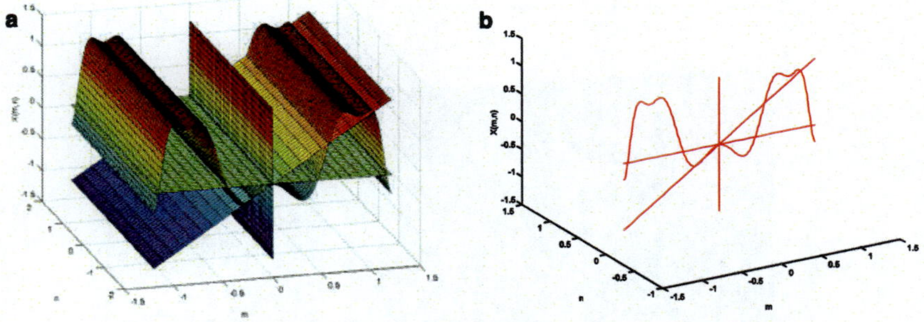

Fig. 5.8 **a** The bifurcating fixed curve of system (5.10) with $\mu = 1.3$, and $\omega = 1$. **b** The section case of system (5.10) with $\mu = 1.3$, and $\omega = 1$

the critical and bifurcating fixed curve of 2^0-stable period

$\rightarrow$ the critical and bifurcating fixed curves of 2^1-stable period

$\rightarrow$ the critical and bifurcating fixed curve of 2^2 -stable period

$\rightarrow \cdots$

$\rightarrow$ the critical and bifurcating fixed curve of 2^r -stable period

$\rightarrow \cdots$

Hence, one obtains the following:

$\rightarrow$ the stable surface of 2^0-period

$\rightarrow$ the stable surface of 2^1-period

$\rightarrow$ the stable surface of 2^2-period

$\rightarrow \cdots$

$\rightarrow$ the stable surface of 2^r-period

$\rightarrow \cdots$

$\rightarrow$ chaos on surface

The surface corresponding to the 2^r-period bifurcation is called the *sequences of the* 2^r*-period-doubling bifurcation on surface*, $r = 0, 1, 2, \ldots$.

In addition, with respect to the parameter μ, one has the s-bifurcation for period 3. More precisely, on a section (or a *segment*) of the plane of the system, i.e., with $\omega = 0,\ n = n_0$ in system (5.30), one has

$$x_{m+1,n_0} = 1 - 1.76x_{mn_0}^2, \tag{5.29}$$

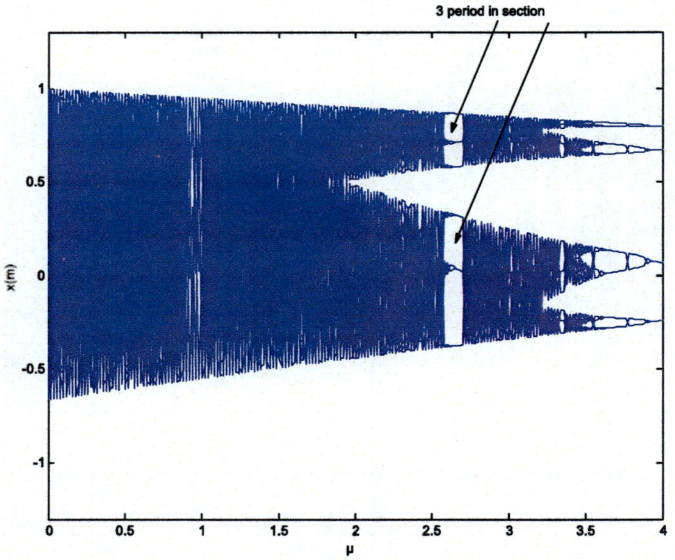

Fig. 5.9 Simulation result on system (5.29)

Table 5.2 Sequence of surface bifurcation

$3 \rhd 5 \rhd 7 \rhd \cdots$	$3.2^r \rhd 5 \cdot 2^r \rhd 7 \cdot 2^r \rhd \cdots$
$3 \cdot 2 \rhd 5 \cdot 2 \rhd 7 \cdot 2 \rhd \cdots$ $3 \cdot 2^2 \rhd 5 \cdot 2^2 \rhd 7 \cdot 2^2 \rhd \cdots$ $\rhd \cdots$	$\rhd \cdots$ $\rhd \cdots \rhd 2^r \rhd \cdots \rhd 2^3 \rhd 2^2 \rhd 2^1 \rhd 2^0$

which is the 1-D Logistic system, with dynamics on the segment shown by Fig. 5.9.

It can be seen from Fig. 5.9 the corresponding 3-period, 6-period, ... bifurcations. On the other hand, with respect to the parameter μ, it also has a nesting sequence of 5-period, 7-period, ... bifurcations. For the spatial system with $\omega \neq 0,\ n = n$, there are also corresponding sequences of 3-period, 6-period, ... bifurcations, as well as nesting 5-period, 7-period, ... bifurcations. In general, one has the following bifurcations on surface:

$$3 \cdot 2^r - period, 5 \cdot 2^r - period, 7 \cdot 2^r - period, \ldots,\ r = 0, 1, 2, \ldots.$$

More precisely, this sequence of surface bifurcation is the following in Table 5.2.

Therefore, the sequence of surface bifurcation is only part of the Sarkorskii sequence. Consequently, by the Sarkorskii Theorem, if system (5.10) has an surface bifurcation of p-period, and if $p \rhd q$, then the system also has an s-bifurcation of q-period.

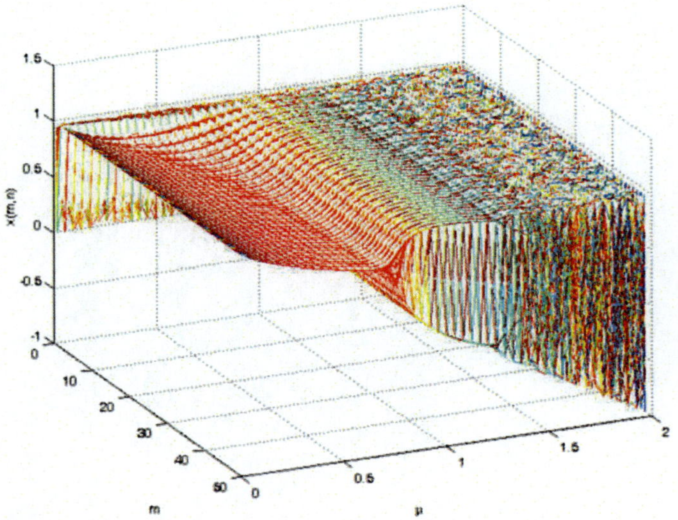

Fig. 5.10 Simulation result of 50-time iteration

5.5.2 Surface Bifurcation and Surface Chaos in the Generalized 2D Logistic System

The generalized 2D Logistic system (5.30):

$$x_{m+1,n} - 0.002x_{m,n+1} = 1 - \mu[0.998x_{m,n}]^2, \tag{5.30}$$

with $\mu \in (0, 2]$ and $\omega = -0.002$, becomes system (4.40). From the definitions and results of surface chaos in above chapters, one knows that this system has bifurcations leading to chaos, and here it has surface bifurcation leading to surface chaos, as visualized by Figs. 5.10 and 5.11.

In addition, one can also take $\mu = \mu_0$ and view the surface bifurcation leading to surface chaos in system (4.40), according to the changes of m and n.

On the other hand, from (5.21) to (5.24), one can obtain the critical and bifurcating fixed curve on surface in system (4.40). For example, the critical and bifurcating fixed curve (here, a spatial straight line) on surface from 2^0-period to 2^1-period is

$$\mu_0 = \frac{1}{4}\left[\left(\frac{1+(1+\omega)^2+|\omega|}{(1+\omega)^2}\right)^2 - 1\right]\Bigg|_{\omega=-0.002} = 0.7540. \tag{5.31}$$

Likewise, the critical and bifurcating fixed curve on surface from 2^1-period to 2^2-period is

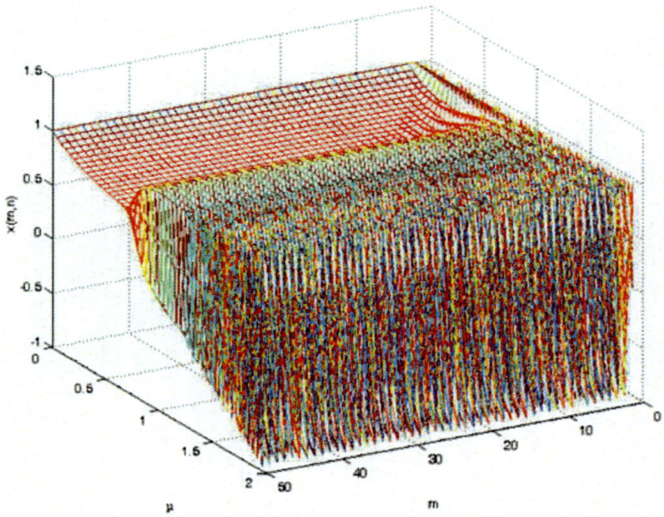

Fig. 5.11 The surface bifurcation leading to surface chao

$$\mu_1 = \frac{1}{4}\left(4 + \frac{1+|\omega|}{|1+\omega|}\right)\bigg|_{\omega=-0.002} = 1.2510.$$

For numerical values of $\mu_0,\ \mu_1$, the system is very close to the 1-D Logistic system, which has $\mu_0' = 0.75,\ \mu_1' = 1.25$.

Similarity, one can obtain $\mu_2,\ \mu_3,\ \mu_4,\ \ldots$.

5.5.3 *Section Analysis for Surface Bifurcation and Surface Chaos in the Generalized 2-D Logistic System*

5.5.3.1 The Case of Fixed *m* and *n* with Changing Parameter μ

Next, section analysis is carried out for surface bifurcation and surface chaos in system (5.10). In the simulations shown in Fig. 5.12a, b, $m = m_0 = 25,\ \mu \in (0, 2]$.

5.5.3.2 The Case of Fixed Parameter μ with Changing *m* and *n*

The Maottor Theorem can be applied to prove that system (5.10) is chaotic, even in the present case on surface when $\mu > 1.55$. Indeed, its Lyapunov exponent $\lambda(R_n^3(x_{0n} + \omega x_{m0})) > 0$, verified by simulations. Moreover, it has surface bifurcations with infinite nesting sequences on surface. Finally, it only has a stable surface of

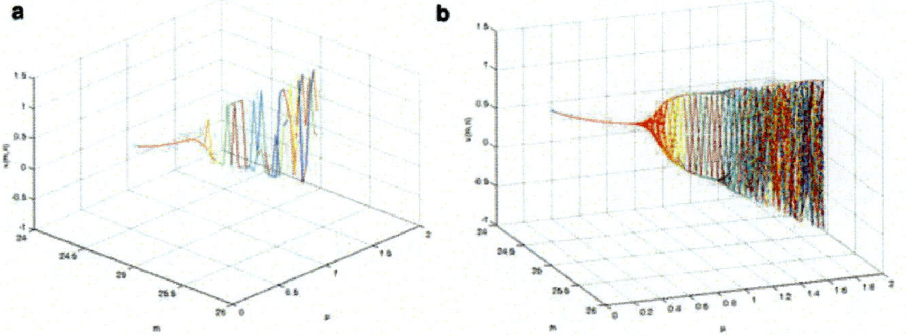

Fig. 5.12 **a** Simulation result for 2-times of iteration. **b** Simulation result for 50-times of iteration

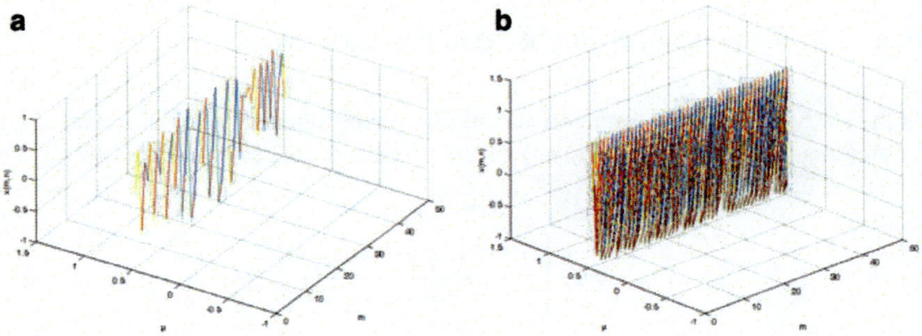

Fig. 5.13 **a** Simulation result of 1-time of iteration. **b** Simulation result of 30-times of iteration

2^0-period from (5.31), that is, the surface has no bifurcation when $\mu \in (0, 0.7540)$. In this case, one may consider two cases as follows:

(i) The case with an infinite bifurcation sequence. For example, take $\mu = \mu_0 = 2$. In this case, there is surface chaos, because $\lambda(R_n^3(x_{0n} + \omega x_{m0})) > 0$. So, the section analysis shows chaos in space, as demonstrated by simulations shown in Fig. 5.13a, b.

(ii) The case of a stable surface. For example, take $\mu = \mu_0 = \frac{1}{5}$. In this case, since $\mu = 0.5 < 0.754$, the system has only one stable surface of 1-period, with Lyapunov exponent $\lambda(R_n^3(x_{0n} + \omega x_{m0})) < 0$. Hence, the section on the spatial surface is also a stable curve (here, it is a spatial straight line), as demonstrated by simulations shown in Fig. 5.14a, b.

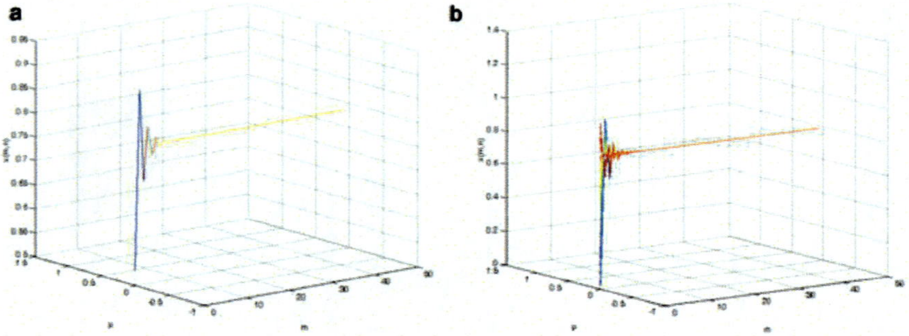

Fig. 5.14 **a** Simulation result of 1-time of iteration. **b** Simulation result of 50-times of iteration

5.6 The Feigenbaum Problem of Surface Chaos

In the study of surface bifurcation and surface chaos, one can obtain the critical and bifurcating fixed curve $\mu_r = \mu_r(\omega)$ of the 2^r-period doubling surface bifurcation, $r = 0, 1, 2, \ldots$; that is,

$$\mu = \mu(\omega) = \frac{1}{4}\left[\left(\frac{1+|\omega|+(1+\omega)^2}{(1+\omega)^2}\right)^2 - 1\right].$$

In specially, when $\omega \in (-\infty, -2] \cup [0, \infty)$, then, by using stable condition on a fixed plane

$$\left|-(1+\omega)(-1 \pm \sqrt{1+4\mu}\right| \leq \frac{1+|\omega|}{|1+\omega|},$$

one has

$$\begin{aligned}
&\left|(-1 \pm \sqrt{1+4\mu}\right| \\
&\leq |-(1+\omega)|\left|(-1 \pm \sqrt{1+4\mu}\right| \\
&= \left|-(1+\omega)(-1 \pm \sqrt{1+4\mu}\right| \\
&\leq \frac{1+|\omega|}{|1+\omega|},
\end{aligned}$$

that is,

$$\left|(-1 \pm \sqrt{1+4\mu}\right| \leq \frac{1+|\omega|}{|1+\omega|},$$

therefore, one obtain

Table 5.3 Relationship between ω and stable condition

$\frac{1+\|\omega\|}{\|1+\omega\|} = -\frac{1-\omega}{1+\omega}$	$\omega \in (-\infty, -1)$
$\frac{1+\|\omega\|}{\|1+\omega\|} = \frac{1-\omega}{1+\omega}$	$\omega \in (-1, 0)$
$\frac{1+\|\omega\|}{\|1+\omega\|} = \frac{1+\omega}{1+\omega} = 1$	$\omega \in [0, \infty)$

$$\mu_0(\omega) = \mu_0(\omega) = \frac{1}{4}\left[\left(\frac{1+\omega+|1+\omega|}{|1+\omega|}\right)^2 - 1\right].$$

Similarly, one has

$$\begin{aligned}
&\mu_1 = \mu_1(\omega) = \tfrac{1}{4}\left(4 + \tfrac{1+|\omega|}{|1+\omega|}\right)\\
&\mu_2 = \mu_2(\omega),\\
&\cdots\\
&\mu_r = \mu_r(\omega),\\
&\cdots.
\end{aligned}$$

For $r \geq 3$, however, since $\mu_r(\omega)$ is involved in a polynomial equation of $2r$-order, it is very difficult to obtain its generic solutions. Notice that

$$\frac{\mu_r(\omega) - \mu_{r-1}(\omega)}{\mu_{r+1}(\omega) - \mu_r(\omega)} \tag{5.32}$$

is a function of parameter ω; therefore if the limit exists in the following form:

$$\lim_{r\to\infty} \frac{\mu_r(\omega) - \mu_{r-1}(\omega)}{\mu_{r+1}(\omega) - \mu_r(\omega)} \stackrel{def}{=} \delta(\omega), \tag{5.33}$$

then the limiting value $\delta(\omega)$ is called the Feigenbaum constant of surface chaos.

On the other hand, since $\omega \in (-\infty, \infty) \setminus \{-1\}$, one has the following (Table 5.3).

Then, every critical and bifurcating fixed curve is a function of the parameter ω for $\omega < 0$ and $\omega \neq -1$. Moreover, when $\omega \in [0, \infty)$, one has $\frac{1+|\omega|}{|1+\omega|} = 1$. So, based on the stable boundary and bifurcating fixed curve, one has

$$\left|\prod_{i=1}^{r} G'_x(\mu, \omega, x_1)\cdots G'_x(\mu, \omega, x_r)\right| \leq \left.\frac{1+|\omega|}{|1+\omega|}\right|_{\omega>0} = 1. \tag{5.34}$$

Condition (5.20) and the condition on the stable boundary of a 1-D discrete dynamic system can be verified, since one can obtain $\mu_0(\omega)$ and $\mu_1(\omega)$. In fact, for $\omega \geqslant 0$, it follows from (5.21) that

$$\mu_0 = \mu_0(\omega \geqslant 0) = \frac{1}{4}\left[\left(\frac{1+|\omega|+|1+\omega|}{|1+\omega|}\right)^2 - 1\right]$$

$$
\begin{aligned}
&= \frac{1}{4}\left[\left(\frac{2+\omega+\omega}{1+\omega}\right)^2 - 1\right] \\
&= \frac{1}{4}\left[\left(\frac{2(1+\omega)}{1+\omega}\right)^2 - 1\right] \\
&= \frac{3}{4}, \\
\mu_1 = \mu_1(\omega \geqslant 0) &= \frac{1}{4}\left(4 + \frac{1+|\omega|}{|1+\omega|}\right) \\
&= \frac{1}{4}\left(4 + \frac{1+\omega}{1+\omega}\right) \\
&= \frac{5}{4}.
\end{aligned}
$$

Thus, using (5.26) again, together with the above discussion, one has the following result:

Theorem 5.3 *(a) For $\omega \geqslant 0$, the Feigenbaum constant of surface chaos is independent of parameter ω, and the Feigenbaum constant for the 1-D Logistic system can be obtained as*

$$
\lim_{r\to\infty} \frac{\mu_r(\omega) - \mu_{r-1}(\omega)}{\mu_{r+1}(\omega) - \mu_r(\omega)} = \lim_{r\to\infty} \frac{\mu_r - \mu_{r-1}}{\mu_{r+1} - \mu_r} = \delta \approx 4.6692.
$$

(b) For $\omega < 0$, the Feigenbaum constant of surface chaos is given by the limit (5.33).

5.7 Other General Surface Bifurcations

The following chaotic behavior of system

$$
x_{m+1,n} + \omega x_{m,n+1} = \sum_{i=1}^{u} p_i(m,n) f_i(\mu_i, (1+\omega) x_{m-\sigma_i, n-\tau_i}),
$$

has not been studied thoroughly because of the complexity in the higher dimensional space. Furthermore, its generalized two order discrete dynamical systems such as:

$$
x_{m+1,n} + a_1 x_{m,n+1} + a_2 x_{m-1,n} + a_3 x_{m,n-1} = \sum_{i=1}^{u} p_i(m,n) f_i\left(\mu_i, (1+\sum_{i=1}^{3} a_i) x_{m-\sigma_i, n-\tau_i}\right), \tag{5.35}
$$

have more complicated behavior clearly illustrated in Fig. 5.15.

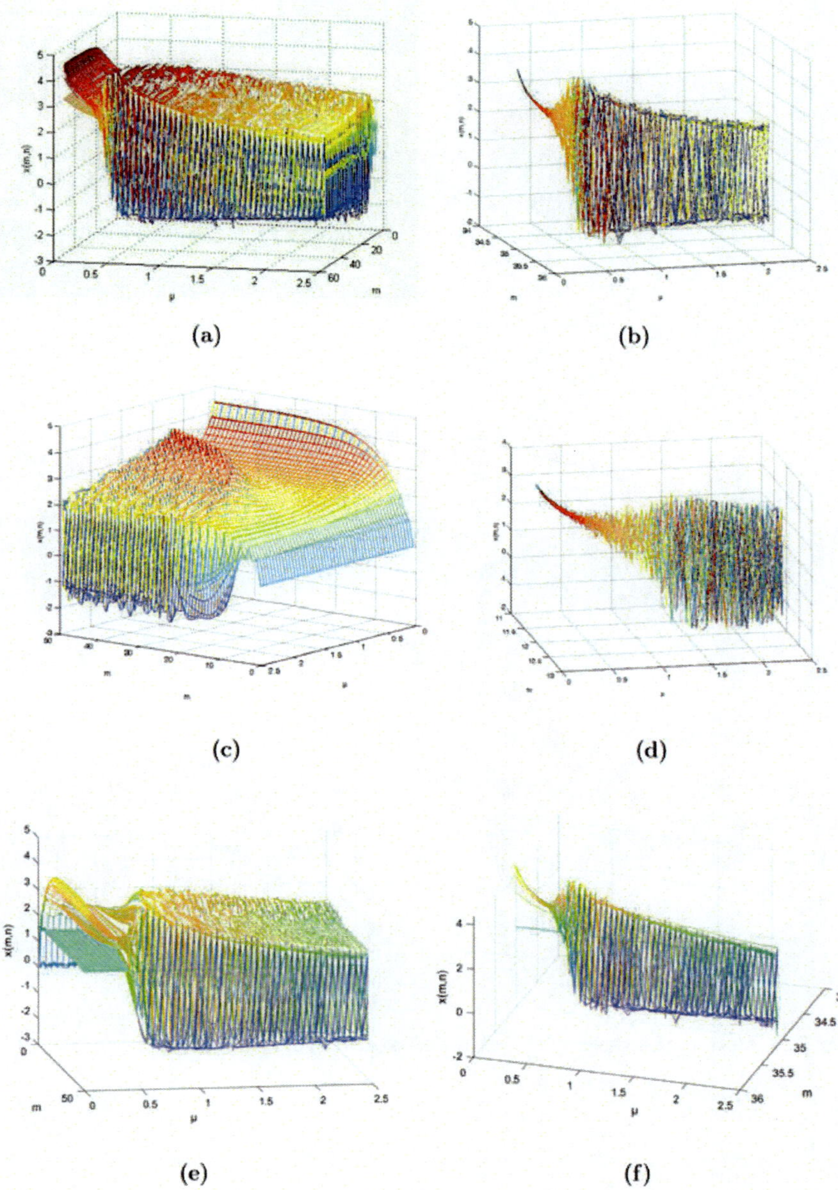

Fig. 5.15 Surface complex behavior of (5.35)

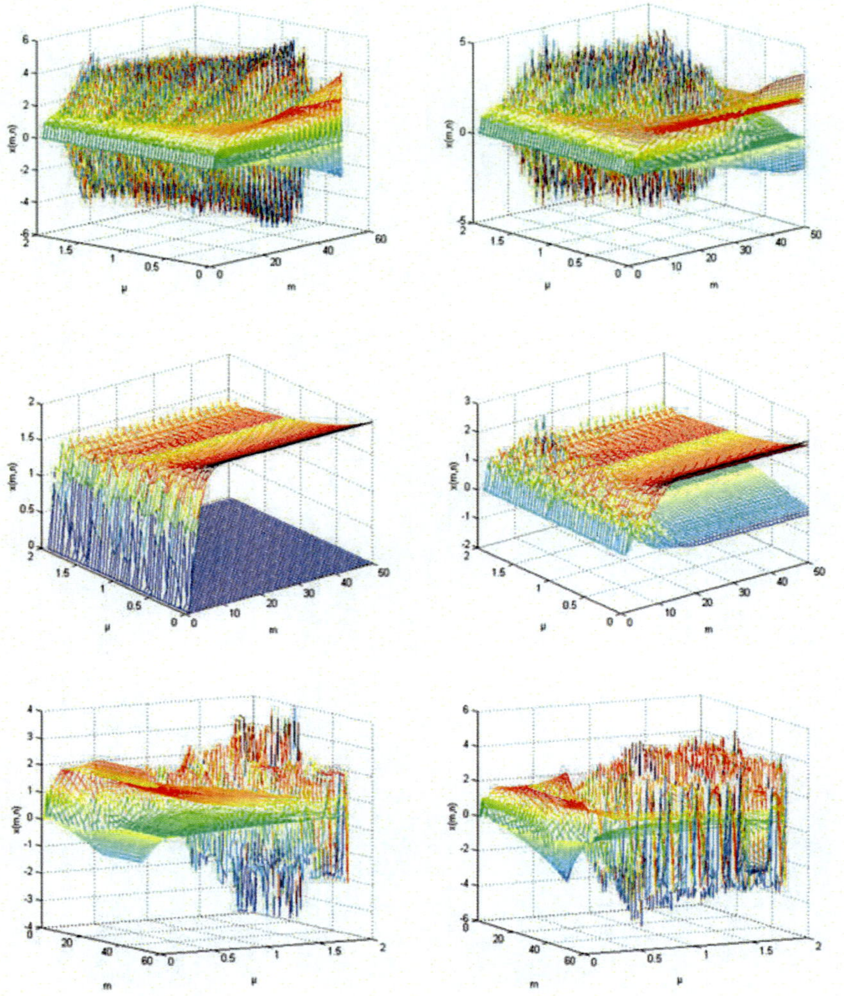

Fig. 5.16 Much more complexity of the bifurcation behavior of system (5.36)

5.8 Some Special Surface Bifurcations

The bifurcation behaviors of system

$$x_{m+1,n} + \omega x_{m,n+1} = \sum_{i=1}^{u} p_i(m, n) f_i(\mu_i, (1+\omega)x_{m-\sigma_i,n-\tau_i}), \tag{5.36}$$

are much more complicated as illustrated in Fig. 5.16 and we need further research on it in the future.

Part II
Control and Synchronization of Surface Chaos

Chapter 6
Prediction-Based Feedback Control of Surface Chaos for Convection System with a Forced Term

In this chapter, we extend the prediction-based feedback control of 1-D discrete dynamical systems, to a class of 2-D discrete dynamical systems. The model is the convection system with a forced term, and control convection system with a forced term. And we focus on how to control the model from the state of spatial chaos onto the fixed plane. The main aim is to study the feedback control for the following 2-D convection system with a forced term

$$x_{m+1,n} + ax_{m,n+1} = f(\mu, (1+a)x_{mn}), \tag{6.1}$$

where f is a nonlinear function called a forced term, $m,\ n \in N_r,\ N_r = \{r, r+1, r+2, ..., r$ is an integer and $r \leq 0\}$, a is a real constant, and μ is a real parameter.

6.1 The Prediction-Based Feedback Control of Surface Chaos

Assume system (6.1) is spatialliy chaotic. Adding the feedback control input $u_{mn} = K(f(\mu, (1+a)x_{mn}) - (1+a)x_{mn})$ to spatially chaotic system (6.1), we obtain the controlled system

$$\begin{aligned} x_{m+1,n} + ax_{m,n+1} &= f(\mu, (1+a)x_{mn}, u_{mn}) \\ &= f(\mu, (1+a)x_{mn}, K(f(\mu, (1+a)x_{mn}) - (1+a)x_{mn})), \end{aligned} \tag{6.2}$$

where K is a gain matrix. Around the fixed plane $x_{mn} = x^* = \frac{f(\mu,(1+a)x^*)}{1+a}$ and $u_{mn} = 0$, system (6.2) can be linearized to be

S. T. Liu and L. Zhang, *Surface Chaos and Its Applications*,
https://doi.org/10.1007/978-981-16-8229-2_6

$$e_{m+1,n} + ae_{m,n+1} = \frac{p + qK(c - (1+a))}{1+a}(1+a)e_{mn}, \tag{6.3}$$

where $p = \partial_x f(\mu, (1+a)x^*, 0) \in \mathfrak{R}$, $q = \partial_u f(\mu, (1+a)x^*, 0) \in \mathfrak{R}$, $e_{m+1,n} + ae_{m,n+1} = x_{m+1,n} + ax_{m,n+1} - (1+a)x^*$, and $e_{mn} + ae_{mn} = x_{mn} + ax_{mn} - (1+a)x^*$.

Let

$$C_{s+1,t+1}(m,n) = x_{s+1,n} + ax_{m,t+1}.$$

Take $s = m, t = n$, which yields

$$C_{m+1,n+1}(m,n) = x_{m+1,n} + ax_{m,n+1},$$

$$C_{s+1,t+1}(m,n) = x_{s+1,n} + ax_{m,t+1}.$$

Let

$$C^* = (1+a)x^*.$$

Then

$$e_{m+1,n+1}(m,n) = C_{m+1,n+1}(m,n) - C^*,$$

and

$$e_{m+1,n+1}(m,n) = \frac{p + qK(c - (1+a))}{1+a} e_{mn}(m,n). \tag{6.4}$$

It can be followed from (6.3) that $e_{mn} = 0$ is a fixed point of controlled system (6.3). When $e_{mn} = 0$ is a stable fixed point, we obtain that $e_{mn} + ae_{mn} \to 0$ as $m, n \to +\infty$, that is, $x_{mn} + ax_{mn} \to (1+a)x^*$ as $m, n \to +\infty$, which shows that the control input u_{mn} can stabilize the fixed plane $x^* = \frac{f(\mu,(1+a)x^*)}{1+a}$.

Using Lyapunov's First Theorem, we obtain that when K satisfies the following inequality:

$$\left| \frac{p + qK(c - (1+a))}{1+a} \right| < 1,$$

the fixed plane is logically asympototically stabilized by the prediction-based feedback control for any a.

6.2 An Illustrative Example

Example 6.1 Consider the following 2-D Logistic system:

$$x_{m+1,n} + ax_{m,n+1} = 1 - \mu((1+a)x_{mn})^2. \tag{6.5}$$

We know that when $\mu = 1.805 > 1.55$, system (6.5) is spatially chaotic in the sense of Marotto. Take $a = -0.051$, $\mu = 1.805 > 1.55$. Then system (6.5) is in the state of spatial chaos, as demonstrated by Fig. 6.1.

The fixed plane of system (6.5) is given by

$$x_{mn} = x^* = \frac{1 - \mu((1+a)x_{mn})^2}{1+a},$$

which is shown in Fig. 6.2 clearly.

By (6.2), we have the following feedback control input:

$$u_{mn} = K(1 - \mu((1+a)x_{mn})^2 - (1+a)x_{mn}).$$

Thus, the controlled system is given by

$$x_{m+1,n} + ax_{m,n+1} = 1 - \mu((1+a)x_{mn})^2 + K(1 - \mu((1+a)x_{mn})^2 - (1+a)x_{mn}).$$

Its linearized system around the fixed plane $x_{mn} = x^* = \frac{1-\mu((1+a)x_{mn})^2}{1+a}$ is described by

$$e_{m+1,n+1}(m,n) = \left((1+K)(1-\sqrt{1+4\mu}) - K\right) e_{mn}(m,n).$$

Take $K = -0.6508$. Then

$$\left|(1+K)(1-\sqrt{1+4\mu}) - K\right|_{\mu=1.8, K=-0.6508} = 0 < 1.$$

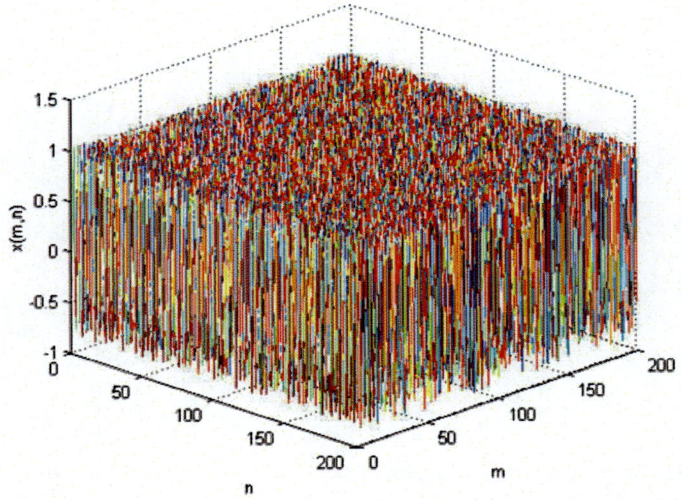

Fig. 6.1 The behavior of spatial chaos for system (6.5)

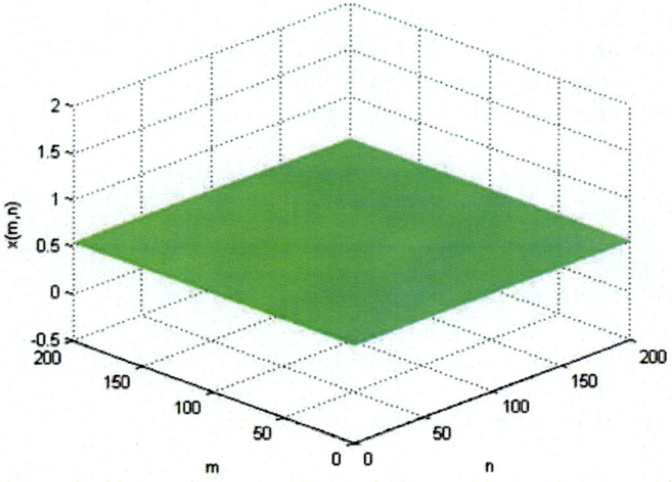

Fig. 6.2 The fixed plane of system (6.5)

Considering that stabilization is guaranteed in a neighborhood of the stabilized orbit in general, we use the following switching control:

$$u = \begin{cases} K(1-\mu((1+a)x_{mn})^2-(1+a)x_{mn}), & \left|x_{m-1,1}+ax_{m-2,2}-(1+a)x_{m-2,1}\right| < \varepsilon \\ 0, \ otherwise \end{cases}$$

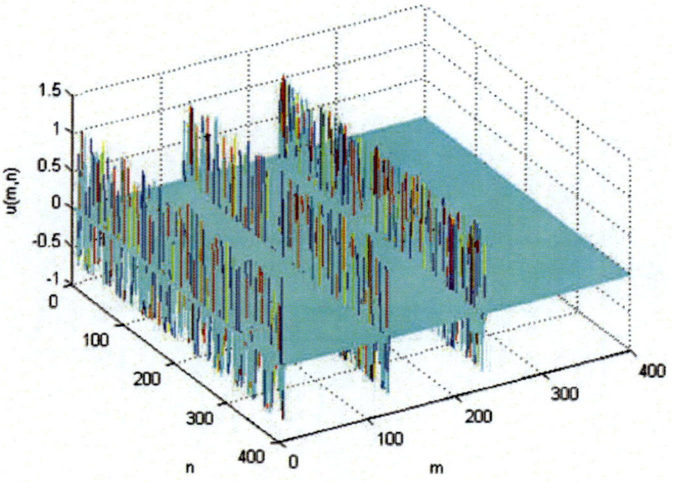

Fig. 6.3 The behavior of controlled system (6.5)

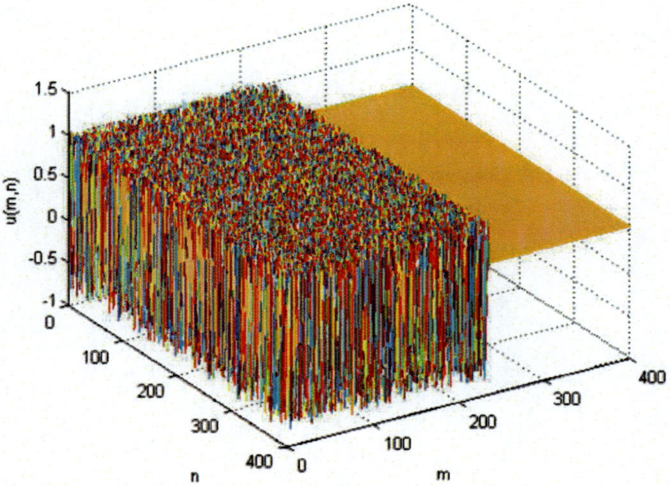

Fig. 6.4 The feedback control input u_{mn} of controlled system (6.5)

where ε is a sufficiently small positive number, $m, n = 1, 2, \cdots, L, L \in [1, +\infty)$ is the number of spatial interate times.

We show the behavior and feedback control input u_{mn} of controlled system (6.5) in Figs. 6.3 and 6.4.

Remark 6.1 It is noted that when $n = n_0$ (constant) and $a = 0$ in system (6.1), system (6.1) becomes

$$x_{m+1,n_0} == f(\mu, x_{m,n_0}), \tag{6.6}$$

which is just the familiar simple case of the 1-D discrete dynamical system. It is clear that system (6.6) is a special case of system (6.1), which shows that 2-D system is a natural extension of 1-D system.

Chapter 7
Spatial Static Bifurcation and Control of 2-D Discrete Dynamical System

In this chapter, we use a unified delayed feedback method to control the spatial static bifurcation of a 2-D discrete dynamical system. By using this control method, we can transfer the existing bifurcation or produce a new transcritical, forked, saddle node bifurcation to realize the discriminant theory and control method of spatial static bifurcation of 2-D discrete dynamic system. Some simulation results verify the effectiveness of the method and the correctness of the theoretical results.

Assume that the central convection equation of 2-D single parameter family is defined as follows:

$$x_{m+1,n} + \omega x_{m,n+1} = F((1+\omega)x_{m,n}, \mu), \tag{7.1}$$

where $f(1+\omega)x_{m,n}, \mu)$ is a nonlinear function, μ means a real number, and $\omega \neq -1$is a constant.

Suppose that $x_{0,0}$ is the fixed point of the Eq. (7.1) (even the fixed point that is not the origin can be translated to be the origin). So we can get

$$\begin{cases} F(0,0) = 0, \\ \frac{\partial F}{\partial x}(0,0) = 1. \end{cases}$$

The following discussion is based on the central manifold which takes the general 2-D single parameter family vector field as an example.

7.1 Spatial Saddle Node Bifurcation and Its Discrimination

There are two characteristics of saddle node bifurcation: only one of the fixed point surfaces $u(x_{m,n})$ passes through the point (0, 0), and satisfies.

S. T. Liu and L. Zhang, *Surface Chaos and Its Applications*,
https://doi.org/10.1007/978-981-16-8229-2_7

(1) It is tangent to the plane $\mu = 0$ in $x_{m,n} = 0$, meaning that

$$\frac{d\mu}{dx}(0,0) = 0. \tag{7.2}$$

(2) The entire surface is located on the side of the plane $\mu = 0$. Locally, this requires the following formula to hold:

$$\frac{d^2\mu}{dx^2}(0,0) \neq 0. \tag{7.3}$$

Let the general 2-D single parameter family vector fields be:

$$x_{m+1,n} + \omega x_{m,n+1} = F((1+\omega)x_{m,n}, \mu). \tag{7.4}$$

Suppose that the fixed point is $(x_{m,n}, \mu) = (0, 0)$, i.e.

$$F(0,0) = 0, \tag{7.5}$$

and the fixed point is not hyperbolic, i.e.

$$\frac{\partial F}{\partial x}(0,0) = 1. \tag{7.6}$$

At the origin, it is given that:

$$h((1+\omega)x_{m,n}, \mu) = F((1+\omega)x_{m,n}, \mu) - x_{m+1,n} - \omega x_{m,n+1}. \tag{7.7}$$

If the transversality condition is satisfied:

$$\frac{\partial h}{\partial \mu}(0,0) = \frac{\partial F}{\partial \mu}(0,0) \neq 0. \tag{7.8}$$

So, from the implicit function theory, for sufficiently small $x_{m,n}$, there is a unique function as

$$\mu = \mu(x_{m,n}), \quad \mu(0) = 0, \tag{7.9}$$

and satisfies

$$\begin{aligned} h((1+\omega)x_{m,n}, \mu(x_{m,n})) &= F((1+\omega)x_{m,n}, \mu(x_{m,n})) && (7.10)\\ &\quad - x_{m+1,n} - \omega x_{m,n+1} && (7.11)\\ &= 0. && (7.12) \end{aligned}$$

In this way, the condition of saddle node bifurcation of (7.4) is formed by (7.2), (7.3), (7.5), (7.6) and (7.8).

Note that (7.3) relationship is not easy to be used because the function between $x_{m,n}$ and μ is implicit. Therefore, it is necessary to improve these two conditions. For this purpose, the implicit derivation of both sides of (7.10) is obtained:

$$\begin{aligned}
&(1+\omega)\frac{\partial h}{\partial x}((1+\omega)x_{m,n}, \mu(x_{m,n})) + \frac{\partial h}{\partial \mu}((1+\omega)x_{m,n}, \mu(x_{m,n})) \cdot \frac{d\mu}{dx} \\
&= (1+\omega)\frac{\partial F}{\partial x}((1+\omega)x_{m,n}, \mu(x_{m,n})) \\
&+ \frac{\partial F}{\partial \mu}((1+\omega)x_{m,n}, \mu(x_{m,n})) \cdot \frac{d\mu}{dx} - 1 - \omega \\
&= 0.
\end{aligned} \tag{7.13}$$

At $(x)_{m,n}, \mu) = (0, 0)$, we consider (7.13) and get

$$\frac{d\mu}{dx}(0,0) = -\frac{(1+\omega)\frac{\partial h}{\partial x}(0,0)}{\frac{\partial h}{\partial \mu}(0,0)} = -\frac{(1+\omega)(\frac{\partial F}{\partial x}(0,0) - 1)}{\frac{\partial F}{\partial \mu}(0,0)} = 0. \tag{7.14}$$

From this we can see that (7.6), (7.8) can ensure that (7.2) is established, that is, at $x_{m,n} = 0$, fixed point surface $x_{m,n}$ and the plane $\mu = 0$ are tangent.

Once again, the implicit derivation of the two sides of the formula (7.13) is obtained:

$$\begin{aligned}
&(1+\omega)^2\frac{\partial^2 h}{\partial x^2}((1+\omega)x_{m,n}, \mu(x_{m,n})) \\
&+ 2(1+\omega)\frac{\partial^2 h}{\partial x \partial \mu}((1+\omega)x_{m,n}, \mu(x_{m,n}))\frac{d\mu}{dx} + \frac{\partial^2 h}{\partial \mu^2}((1+\omega)x_{m,n}, \mu(x_{m,n}))\left(\frac{d\mu}{dx}\right)^2 \\
&+ \frac{\partial h}{\partial \mu}((1+\omega)x_{m,n}, \mu(x_{m,n}))\frac{d^2\mu}{dx^2} \\
&= (1+\omega)^2\frac{\partial^2 F}{\partial x^2}((1+\omega)x_{m,n}, \mu(x_{m,n})) \\
&\quad +2(1+\omega)\frac{\partial^2 F}{\partial x \partial \mu}((1+\omega)x_{m,n}, \mu(x_{m,n}))\frac{d\mu}{dx} \\
&+ \frac{\partial^2 F}{\partial \mu^2}((1+\omega)x_{m,n}, \mu(x_{m,n}))\left(\frac{d\mu}{dx}\right)^2 + \frac{\partial F}{\partial \mu}((1+\omega)x_{m,n}, \mu(x_{m,n}))\frac{d^2\mu}{dx^2} \qquad (7.15) \\
&= 0. \qquad (7.16)
\end{aligned}$$

At $(x_{m,n}, \mu) = (0, 0)$, considering (7.14) and (7.14), the following results can be obtained

$$\frac{d^2\mu}{dx^2}(0,0) = -\frac{(1+\omega)^2\frac{\partial^2 h}{\partial x^2}(0,0)}{\frac{\partial h}{\partial \mu}(0,0)} = -\frac{(1+\omega)^2\frac{\partial^2 F}{\partial x^2}(0,0)}{\frac{\partial F}{\partial \mu}(0,0)}. \tag{7.17}$$

Table 7.1 Stability of every branch in sketch map of the saddle-node bifurcation

$\frac{\partial F}{\partial \mu}(0,0)$	$\frac{\partial^2 F}{\partial x^2}(0,0)$	Stable equilibrium	Unstable equilibrium	No equilibrium
>0	>0	$\mu < 0$ (upper)	$\mu < 0$ (lower)	$\mu > 0$
>0	<0	$\mu > 0$ (upper)	$\mu > 0$ (lower)	$\mu < 0$
<0	>0	$\mu > 0$ (lower)	$\mu > 0$ (upper)	$\mu < 0$
<0	>0	$\mu < 0$ (lower)	$\mu < 0$ (upper)	$\mu > 0$

Therefore, if and only if (7.18) is established, (7.3) holds:

$$\frac{\partial^2 F}{\partial x^2}(0,0) \neq 0. \tag{7.18}$$

So far, the criterion of saddle node bifurcation in (7.4) is obtained as follows:

$$\begin{cases} F(00) = 0, \ \dfrac{\partial F}{\partial x}(0,0) = 1, \\ \dfrac{\partial F}{\partial \mu}(0,0) \neq 0, \ \dfrac{\partial^2 F}{\partial x^2}(0,0) \neq 0. \end{cases} \tag{7.19}$$

The last two formulas are the cross-sectional conditions. The stability of each branch in the bifurcation diagram of the two equations under four different symbol combinations is shown in Table 7.1.

7.1.1 Space Transcritical Bifurcation and Its Discrimination

There are three characteristics of transcritical bifurcation:

(1) Equation (7.1) has two fixed-point surfaces passing through the origin $(x_{m,n}) = (0,0)$.

(2) The fixed point surfaces of (7.2) are distributed on both sides of $(\mu = 0)$.

(3) The stability of the points on any fixed point surface alternates at $\mu = 0$ from stable to unstable or vice versa.

From the characteristic of transcritical bifurcation (7.1), system should satisfy (Otherwise, according to the implicit function theory, only one fixed point surface passes through the origin):

$$\frac{\partial h}{\partial \mu}(0,0) = \frac{\partial F}{\partial \mu}(0,0) = 0. \tag{7.20}$$

If the saddle node bifurcation method discussed above is still used to derive the conditions for the occurrence of transcritical bifurcation, there exists difficulties to be conquered. The reason is that using (7.8) as the condition, we can calculate the fixed point surface passing through the bifurcation point. Then we discuss the vector field on the fixed point surface. Furthermore, the nonlinear properties of fixed point surfaces are discussed by using the derivative information of vector field and implicit function theory. However, this condition does not hold here.

In view that $x_{m,n} = 0$ is a fixed point surface, (7.7) can be written in another form:

$$h((1+\omega)x_{m,n}, \mu) = x_{m,n}H((1+\omega)x_{m,n}, \mu) = x_{m,n}(G((1+\omega)x_{m,n}, \mu) - 1 - \omega). \tag{7.21}$$

Define function:

$$G((1+\omega)x_{m,n}, \mu) = \begin{cases} \dfrac{F((1+\omega)x_{m,n}, \mu)}{x_{m,n}}, & x_{m,n} \neq 0, \\ \dfrac{\partial F}{\partial x}(0, \mu), & x_{m,n} = 0. \end{cases} \tag{7.22}$$

So

$$H((1+\omega)x_{m,n}, \mu) = \begin{cases} \dfrac{h((1+\omega)x_{m,n}, \mu)}{x_{m,n}}, & x_{m,n} \neq 0, \\ \dfrac{\partial h}{\partial x}(0, \mu), & x_{m,n} = 0. \end{cases} \tag{7.23}$$

Suppose $H((1+\omega)x_{m,n}, \mu)$is a pass through origin $(x_{m,n}, \mu) = (0, 0)$.

When the fixed point surface $H((1+\omega)x_{m,n}, \mu)$ is the case with $\mu = (0, 0)$ is distributed on both sides of the surface and the surface is distributed at the same time on both sides of $\mu = 0$, we have

$$\frac{\partial H}{\partial \mu}(0, 0) = \frac{\partial G}{\partial \mu}(0, 0) \neq 0. \tag{7.24}$$

From (7.22) to (7.23), we can get the following results:

$$\frac{\partial^2 F}{\partial x \partial \mu}(0, 0) \neq 0. \tag{7.25}$$

From the implicit function theory, there is a unique function $\mu(x_{m,n})$ for sufficiently small $x_{m,n}$, satisfying:

$$H((1+\omega)x_{m,n}, \mu) = G((1+\omega)x_{m,n}, \mu) - 1 - \omega = 0. \tag{7.26}$$

The implicit derivation on both sides of the formula (7.26) is as follows:

$$\frac{d\mu}{dx}(0, 0) = -\frac{(1+\omega)\frac{\partial H}{\partial x}(0, 0)}{\frac{\partial H}{\partial \mu}(0, 0)} = -\frac{(1+\omega)\frac{\partial G}{\partial x}(0, 0)}{\frac{\partial G}{\partial \mu}(0, 0)}. \tag{7.27}$$

Table 7.2 Stability of every branch in sketch map of the transcritical bifurcation

$\frac{\partial^2 F}{\partial x \partial \mu}\|(0,0) - \frac{\partial^2 F}{\partial x^2}\|(0,0) \cdot \frac{\partial^2 F}{\partial \mu^2}\|_{(0,0)}$,	$\frac{\partial^2 F}{\partial x^2}\|(0,0)$,	Stable equilibrium,	Unstable equilibrium,
$\neq 0$	> 0	Lower	Upper
$\neq 0$	< 0	Upper	Lower

From (7.22) to (7.23), we can get the following results:

$$\frac{d\mu}{dx}(0,0) = -\frac{(1+\omega)\frac{\partial^2 F}{\partial x^2}(0,0)}{\frac{\partial^2 F}{\partial x \partial \mu}(0,0)}. \tag{7.28}$$

So,

$$\frac{d\mu}{dx}(0,0) \neq 0, \tag{7.29}$$

if and only if

$$\frac{\partial^2 F}{\partial x^2}(0,0) \neq 0. \tag{7.30}$$

So far, the criterion for the transcritical bifurcation of (7.4) is obtained as follows:

$$\begin{cases} F(0,0) = 0, \ \frac{\partial F}{\partial x}(0,0) = 1, \ \frac{\partial F}{\partial \mu}(0,0) = 0, \\ \frac{\partial^2 F}{\partial x \partial \mu}(0,0) \neq 0, \ \frac{\partial^2 F}{\partial x^2}(0,0) \neq 0. \end{cases} \tag{7.31}$$

The stability of each branch of transcritical bifurcation is shown in Table 7.2.

7.1.2 *Spatial Fork Bifurcation and Its Discrimination*

Fork bifurcation has three characteristics

(1) There have two fixed point planes passing through the origin point $(x_{m,n}, \mu) = (0,0)$.

(2) The fixed point plane $x_{m,n} = 0$ is distributed on both sides of the plane $\mu = 0$.

(3) Fixed point plane $x_{m,n} = 0$ has different stability on both sides of the plane $\mu = 0$.

It can be seen from the characteristics of fork bifurcation, the occurrence of bifurcation in (7.4) should satisfy the conditions (7.2), (7.3), (7.20), (7.22) and (7.23). Then it can be concluded that the criterion for the fork bifurcation of (7.4) is as follows:

Table 7.3 Stability of every branch in sketch map of the pichfork bifurcation

$\frac{\partial^2 F}{\partial x \partial \mu}(0,0)$,	$\frac{\partial^3 F}{\partial x^3}(0,0)$,	Stable, (1st branch)	Unstable, (1st branch)	Stable, (2nd branch)	Unstable (2nd branch)
>0	>0	$\mu < 0$	$\mu > 0$	–	$\mu < 0$
>0	<0	$\mu < 0$	$\mu > 0$	$\mu > 0$	–
<0	>0	$\mu > 0$	$\mu < 0$	–	$\mu > 0$
<0	>0	$\mu > 0$	$\mu < 0$	$\mu < 0$	–

$$\begin{cases} F(00) = 0, \ \frac{\partial F}{\partial x}(0,0) = 1, \ \frac{\partial F}{\partial \mu}(0,0) = 0, \\ \frac{\partial^2 F}{\partial x \partial \mu}(0,0) \neq 0, \ \frac{\partial^2 F}{\partial x^2}(0,0) \neq 0, \ \frac{\partial^3 F}{\partial x^3}(0,0) \neq 0. \end{cases} \tag{7.32}$$

The stability of each branch of bifurcation is shown in Table 7.3.

7.2 Calculation of Space Static Bifurcation Control

In this section, a 2-D discrete parameterized nonlinear state feedback control system described by (7.32) is taken as an example:

$$x_{m+1,n} + \omega x_{m,n+1} = F((1+\omega)x_{m,n}, \mu) \overset{def}{=} f((1+\omega)x_{m,n}, \mu) + k((1+\omega)x_{m,n}, \mu), \tag{7.33}$$

in which $\mu \in \mathbf{r}$, $k(\cdot)$ are the controllers to be designed, the vector field f is an autonomous continuous vector field (i.e., it does not contain time explicitly).

The space bifurcation analysis and control steps of the control system are as follows:

(1) Structure of controller in (7.1) is initialized for system (7.33) (e.g. linear control or nonlinear control).

(2) Solve the fixed point equation:

$$(1+\omega)x^* = F((1+\omega)x^*, \mu), \tag{7.34}$$

and obtained the solution trajectory $x^*(T, \mu)$.

If the equation has no solution, we can change the structure of controller in step (1), and resolve the fixed point equation until there is a solution.

(3) Given the expected bifurcation value $\mu = \mu^*$ satisfying:

$$\frac{\partial F}{\partial x}|_{x=x^*, \mu=\mu^*} = 1. \tag{7.35}$$

Table 7.4 The table of spatial static bifurcation type determination

$\frac{\partial F}{\partial \mu}(x^*, \mu^*)$	$\frac{\partial^2 F}{\partial x^2}(x^*, \mu^*)$	Bifurcations
$\neq 0$	$\neq 0$	Saddle-node
$= 0$	$\neq 0$	Transcritical
$= 0$	$= 0$	Pitchfork

If (7.35) has no solution, we can change the structure of the controller in step (1), and then return to step (2).

(4) According to Table 7.4, the types of spatial static bifurcations are distinguished.

(5) Based on Tables 7.4, 7.2 and 7.3, the stability of the fixed point corresponding to the bifurcation is judged respectively (Tables 7.1, 7.2 and 7.3 takes the fixed point as an example. When the fixed point is not the origin point, the criterion can be applied to the corresponding fixed point).

7.3 Control of Space Static Bifurcation

Consider 2-D logistic system

$$X_{m+1,n} + \omega X_{m,n+1} = \mu(1+\omega)x_{m,n}[1-(1+\omega)x_{m,n}], \mu > 0. \tag{7.36}$$

The control task is to transfer the original bifurcation point (x^*, μ^*) to a new location (x^0, μ^0), or generate a new bifurcation point (x^0, μ^0).

According to the given system (7.36), the structure of the controller can be determined as follows: from the above steps (2), the results are listed as follows:

$$(1+\omega)x^0 = \mu^0(1+\omega)x^0[1-(1+\omega)x^0] + k|_{x=x^0,\mu=\mu^0}, \tag{7.37}$$

bringing

$$k|_{x=x^0,\mu=\mu^0} = (1+\omega)x^0 - \mu^0(1+\omega)x^0[1-(1+\omega)x^0]. \tag{7.38}$$

Then, from the step (3), we get:

$$\frac{\partial F}{\partial x}|_{x=x^0,\mu=\mu^0} = \mu^0(1+\omega) - 2\mu(1+\omega)^2x^0 + \frac{\partial k}{\partial x}|_{x=x^0,\mu=\mu^0} = 1, \tag{7.39}$$

satisfying

$$\frac{\partial k}{\partial x}|_{x=x^0,\mu=\mu^0} = 1 - \mu^0(1+\omega) - 2\mu^0(1+\omega)^2x^0. \tag{7.40}$$

From the above two systems (7.38) and (7.40), we can determine a simple linear controller as follows:

$$k((1+\omega)x_{m,n},\mu) = [1-(1+\omega)\mu^0 + 2(1+\omega)^2\mu^0 x^0]x_{m,n} \tag{7.41}$$
$$+ \omega x^0 - (1+\omega)^2\mu^0(x^0)^2 \tag{7.42}$$
$$= c_1 x_{m,n} + c_2, \tag{7.43}$$

where $c_1 = 1-(1+\omega)\mu^0 + 2(1+\omega)^2\mu^0 x^0$, $c_2 = \omega x^0 - (1+\omega)^2\mu^0(x^0)^2$.

Based on the controlled 2-D logistic system:

$$x_{m+1,n} + \omega x_{m,n+1} = \mu(1+\omega)x_{m,n}[1-(1+\omega)x_{m,n}] + (c_1 x_{m,n} + c_2), \tag{7.44}$$

we get

$$\frac{\partial F}{\partial \mu}|_{x^0,\mu=\mu^0} = (1+\omega)x^0 - (1+\omega)^2(x^0)^2,$$

$$\frac{\partial^2 F}{\partial x^2}|_{x^0,\mu=\mu^0} = -2(1+\omega)^2(x^0)^2.$$

Therefore, from Table 7.4, there are

(1) when $x^0 = 0$, $\mu^0 \neq 0$, (x^0, μ^0) is a transcritical bifurcation point.

(2) when $x^0 = 0$, $\mu^0 = 0$, the point (x^0, μ^0) is a bifurcation point.

(3) when $x^0 \neq 0$, $\mu^0 \neq 0$, (x^0, μ^0) is the saddle node bifurcation point.

7.3.1 Transcritical Bifurcation Control

It is well known that the uncontrolled 2-D logistic system has transcritical bifurcation $(x^*, \mu^*) = (0, 1)$.

Based on the above results, we choose $x^0 = 0 = x^*$. When $x^0 = 0$, $\omega = -0.002$, the controlled system is simplified as follows:

$$F((1+\omega)x_{m,n},\mu) = [1+(1+\omega)\mu - (1+\omega)\mu^0]x_{m,n} - \mu(1+\omega)^2 x_{m,n}^2. \tag{7.45}$$

Therefore

$$\frac{\partial^2 F}{\partial x \partial \mu}|_{x=x^0,\mu=\mu^0} = (1+\omega) - 2(1+\omega)^2 x^0 = 1+\omega = 0.998 \neq 0. \tag{7.46}$$

According to Table 7.3, the following conclusions are obtained

(*a*) if $\mu^0 < 0$, the lower segment of the fixed point is stable, while the upper segment is unstable.

(*b*) if $\mu^0 > 0$, the upper segment of the fixed point is stable, while the lower segment is unstable.

The simulation is carried out under the condition of $\mu = \frac{3}{2}$, and the original transcritical bifurcation is transferred. Moreover, the original and delayed transcritical bifurcations of the 2-D logistic system are shown in Figs. 7.1 and 7.2, respectively.

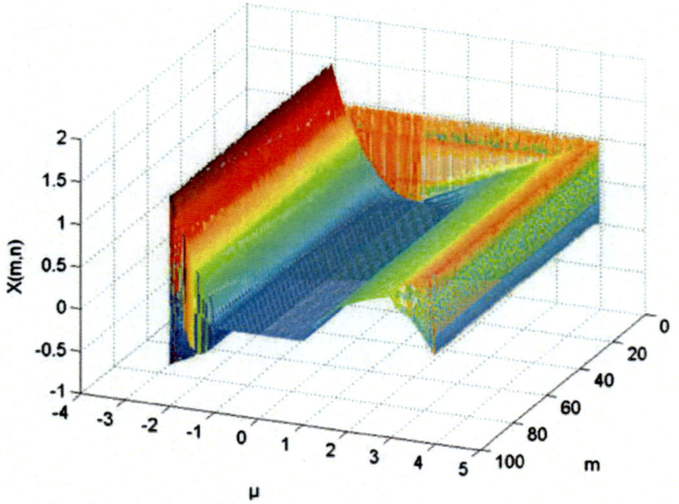

Fig. 7.1 The original transcritical bifurcation of the logistic map

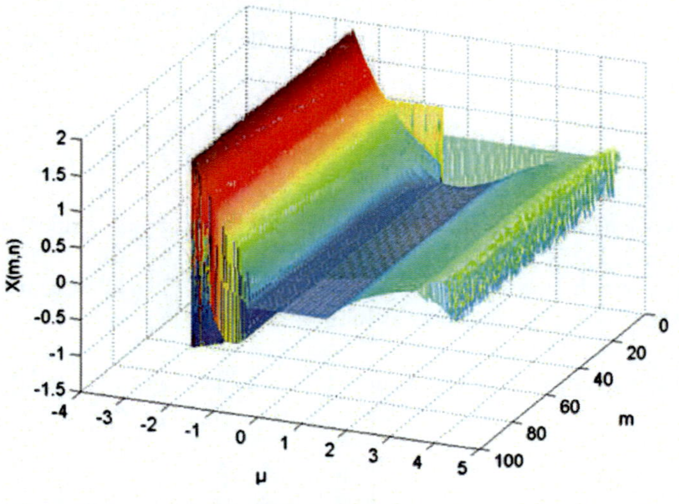

Fig. 7.2 The controlled transcritical bifurcation of the logistic map

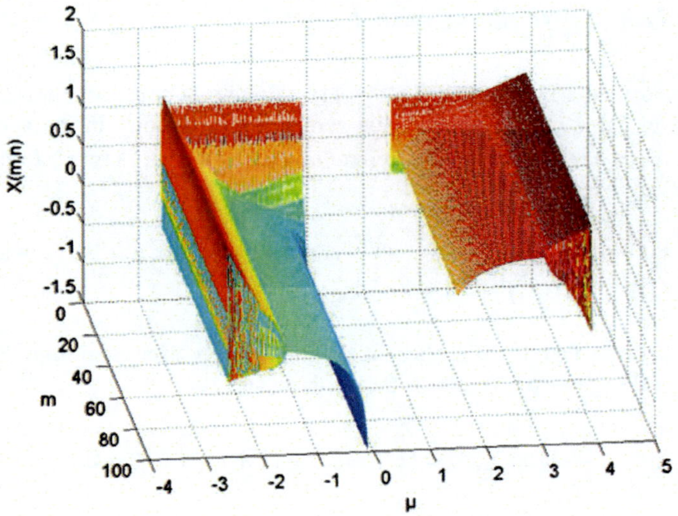

Fig. 7.3 The controlled saddle-node bifurcation of the logistic map

7.3.2 Saddle Node Bifurcation Control

For the controlled 2-D logistic system (7.44), a saddle node bifurcation is constructed according to Table 7.1,

$$\frac{\partial F}{\partial \mu}|_{x=x^0,\mu=\mu^0} = (1+\omega)x^0 - (1+\omega)^2(x^0)^2 = 0.25 > 0,$$

$$\frac{\partial^2 F}{\partial x^2}|_{x=x^0,\mu=\mu^0} = -2(1+\omega)^2\mu^0 = -0.996 > 0,$$

where $(x^0, \mu^0) = (\frac{1}{2}, \frac{3}{2})$, $\omega = -0.002$.

According to Table 7.1, there are saddle node bifurcation points (x^0), (μ^0) and

(a) If $\mu > \mu^0 = \frac{3}{2}$, then the upper branch of the fixed point is stable, while the lower branch is unstable.

(b) There is no fixed point if $\mu < \mu^0 = \frac{3}{2}$.

As shown in Fig. 7.3, the saddle node bifurcation point of the controlled system (7.44) is $(x^0, \mu^0) = (\frac{1}{2}, \frac{3}{2})$.

7.3.3 Fork Bifurcation Control

For the controlled 2-D logistic system (7.44), a fork point bifurcation is constructed. As noted in Table 7.3, there is no non-zero bifurcation point in the system, and $\frac{\partial^3 F}{\partial X^3} \neq 0$. As the previous linear controller (7.44) cannot create a fork bifurcation, we subtract a cubic term, which is:

$$k((1+\omega)x_{m,n}, \mu) = [1-(1+\omega)\mu^0 + 2(1+\omega)^2\mu^0 x^0]x_{m,n} + \omega x^0 \\ -(1+\omega)^2\mu^0(x^0)^2 - (1+\omega)^3(x_{m,n} - x^0)^3.$$

It is easy to verify that the controller satisfies all the necessary conditions. For example, when $(x^0, \mu^0) = (0, 0)$, $\omega = -0.002$, there are

$$\frac{\partial F}{\partial \mu}|_{x=x^0,\mu=\mu^0} = (1+\omega)x^0 - (1+\omega)^2(x^0)^2 = 0,$$

$$\frac{\partial^2 F}{\partial x^2}|_{x=x^0,\mu=\mu^0} = -2(1+\omega)^2\mu^0 = 0,$$

and

$$\frac{\partial^2 F}{\partial x \partial \mu}|_{x=x^0,\mu=\mu^0} = (1+\omega) - 2(1+\omega)^2 x^0 = 0.998 > 0,$$

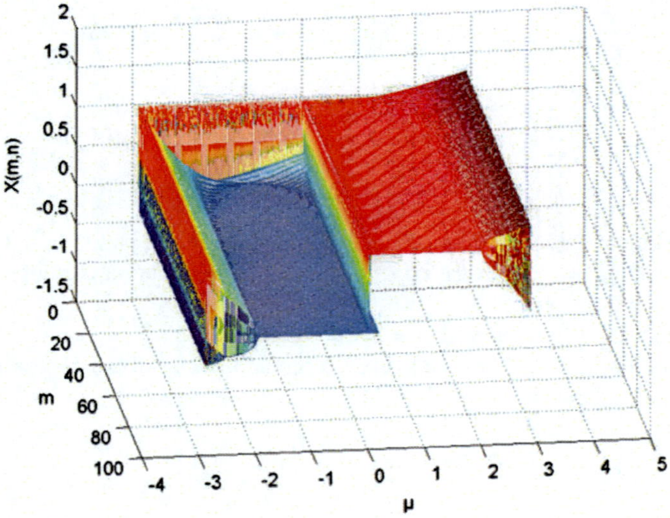

Fig. 7.4 The controlled pichfork bifurcation of the logistic map

$$\frac{\partial^3 F}{\partial x^3}|_{x=x^0,\mu=\mu^0} = -6(1+\omega)^3 = -5.964 < 0.$$

According to Table 7.3, there is a saddle node bifurcation point $(x^0, \mu^0) = (0, 0)$ and
(a) if $\mu < 0 = \mu^0$, then the upper bifurcation on the first fixed point is stable.
(b) if $\mu > 0 = \mu^0$, then the upper branch on the first fixed point is unstable.

Fork bifurcation is composed of $\frac{\partial^2 F}{\partial x^2}|_{x=x^0,\mu=\mu^0}$ shown in Fig. 7.4.

Chapter 8
Holistic Compression Control and Surface Chaos

First we assume that the final spatial chaotic behavior of the period-doubling bifurcation exists. Then for the control of bifurcation on surface chaos, it is either to prevent the structure of bifurcation from coming into being or to produce a new bifurcation (if chaos is beneficial).

We can consider the generalized 2-D spatial Logistic system

$$x_{m+1,n} + \omega x_{m,n+1} = 1 - \mu \left((1+\omega)x_{mn}\right)^2, \tag{8.1}$$

where ω and μ are real parameters, $\mu > 0, \omega \in (-\infty, \infty)/\{-1\}$. And we have discussed its spatial chaotic behavior and behavior of bifurcation on surface chaos. Particularly, detailed analysis about the period-doubling bifurcation of system (8.1) is provided and a set of results on surface bifurcations are obtained.

On the other hand, in fact, system (8.1) is a quite familiar convection equation with a forced term in physics. Therefore, qualitative properties of system (8.1) may lead to some useful information for analyzing this companion partial differential system.

The following mainly investigates the control of bifurcation of surface chaos, and we determine a sufficient condition for the control of surface bifurcation to be stabilized by using the first method of Lyapunov.

8.1 Control Method of Holistic Compression for Bifurcation of Surface Chaos

For the system (8.1), we introduce a term of auxiliary reference feedback, that is

$$x_{m+1,n} + \omega x_{m,n+1} = 1 - \mu \left((1+\omega)x_{mn}\right)^2 + g_{mn}[(1+\omega)x_{mn} - x_{ref}], \tag{8.2}$$

where g_{mn} is a self-turning control gain to be automatically determind at each step, and $[(1+\omega)x_{mn} - x_{ref}]$ is the tracking erroy with respect to an auxiliary constant

S. T. Liu and L. Zhang, *Surface Chaos and Its Applications*,
https://doi.org/10.1007/978-981-16-8229-2_8

amplitude reference signal x_{ref}. The objective of the design is to find a simple and implementable g_{mn} to achieve the goal of automatic control, i.e., $x_{mn} \to \bar{x}$ as $m, n \to \infty$, where $\bar{x}$ is the desired target state which is usually (but not necessarily) an unstable fixed plan (UFP), x^*, of the original map,perhaps of large period. When the control objective is finally realized, $x_{mn} \to \bar{x}$ will be equal to asmall constant (ideally zero) at the time iteration halts. A simple design for the gain is to use a linear combination of the supplied reference sigain and a few available previous system states:

$$g_{mn} = kx_{ref} + \sum_{i=0}^{u}\sum_{j=0}^{v} a_{ij}(1+\omega)x_{m-i,n-j}, \tag{8.3}$$

where m, n are two small integers, independent of the system dimension, and the k and a_{ij}, $i = 0, 1, 2, ...u$, $j = 0, 1, 2, ...v$ are to be determined for stable tracking to the target $\bar{x}$.

Assume that the desired target is $\bar{x}$ and it has been determined, then the target $\bar{x}$ must satisfy (8.2), that is

$$(1+\omega)\bar{x} = 1 - \mu\left((1+\omega)\bar{x}\right)^2 + g_{mn}[(1+\omega)\bar{x} - x_{ref}]. \tag{8.4}$$

To deal with the problem conveniently, we choose the simplest form, that is, take $u = v = 0$, then (8.3) can be changed into

$$g_{mn} = kx_{ref} + a_{00}[(1+\omega)x_{mn}]. \tag{8.5}$$

Substituting (8.5) into (8.4), then we obtain

$$\begin{aligned}(1+\omega)\bar{x} &= 1 - \mu\left((1+\omega)\bar{x}\right)^2 \\ &\quad + \left\{kx_{ref} + a_{00}[(1+\omega)x_{mn}]\right\}[(1+\omega)\bar{x} - x_{ref}].\end{aligned} \tag{8.6}$$

From (8.6), it is easy to see that the $\bar{x}$ is an unstable fixed plan of system (8.1) and also a stable fixed plan of system (8.6). Therefore, the existence of two real roots must be ensured. Note again that, if one take

$$\begin{cases}\alpha = (a_{00} - \mu)(1+\omega)^2, \\ \beta = [(k - a_{00})x_{ref} - 1](1+\omega), \\ \gamma = 1 - kx_{ref}^2,\end{cases} \tag{8.7}$$

then (8.7) reduces to

$$\alpha\bar{x}^2 + \beta\bar{x} + \gamma = 0, \tag{8.8}$$

and so

$$\beta^2 - 4\alpha\gamma > 0,$$

that is

$$\left[\left((k - a_{00})x_{ref} - 1\right)(1 + \omega)\right]^2 - 4(a_{00} - \mu)(1 + \omega)^2(1 - kx_{ref}^2) > 0. \quad (8.9)$$

Here, we just apply (8.9) to confirm the choice confines of the auxiliary reference signal x_{ref}. And for the sake of convenience, we take $k = a_{00}$, $\mu > a_{00}$, then from (8.9) the following form can be directly obtained

$$|x_{ref}| < \sqrt{\frac{1}{k}\left(1 + \frac{1}{4(\mu - k)}\right)} \; (\mu > k)\,. \quad (8.10)$$

Note that, we only desire $x_{ref} > 0$, so we have

$$0 < x_{ref} < \sqrt{\frac{1}{k}\left(1 + \frac{1}{4(\mu - k)}\right)} (\mu > k). \quad (8.11)$$

For example, we can take the $k = a_{00} = \frac{1}{11}$, $\omega = -0.002$. Moreover, to control the spatial bifurcation of s-chaos is a main purpose for us, therefore, the chaotic state of the system must be ensured. Nevertheless, we have proved that the system is chaotic in sense of Maottor when $\mu > 1.55$, hence take $\mu > 1.76$, the following form can be obtained from (8.11):

$$0 < x_{ref} < 4.1337. \quad (8.12)$$

In addition, from (8.9) we can see that the x_{ref} is a variable related to the parameters a_{00}, k, ω, μ, that is x_{ref} is actually a function concerning about the parameters a_{00}, k, ω, μ, which is,

$$x_{ref} = x_{ref}(a_{00}, k, \omega, \mu).$$

Further, if take $a_{00} = k$, $\omega = \omega_0$ to be constant, then the x_{ref} is only a function about the parameter μ.

On the other hand, we can see that the confines of x_{ref} in (8.11) is much larger than those of the one to realize control. But we can make full use of this confine to restrict the choice of the auxiliary reference feedback term, so as to reach the control objective. Next is the further analysis of the whole compression course, which is to realize control. Generally taking $k = a_{00}$, then from (8.5), we have

$$g_{mn} = k\left[x_{ref} + (1 + \omega)x_{mn}\right]. \quad (8.13)$$

Substituting (8.13) into (8.2) gives

$$
\begin{aligned}
& x_{m+1,n} + \omega x_{m,n+1} \\
&= 1 - \mu\left((1+\omega)x_{mn}\right)^2 + \left[k x_{ref} + a_{00}(1+\omega)x_{mn}\right] \times \\
& [(1+\omega)x_{mn} - x_{ref}].
\end{aligned} \tag{8.14}
$$

In view of (8.14), we have

$$
\begin{aligned}
x_{m+1,n} + \omega x_{m,n+1} &= 1 - \mu\left((1+\omega)x_{mn}\right)^2 + (k - a_{00})(1+\omega)x_{mn}x_{ref} \\
&+ a_{00}(1+\omega)^2 x_{ref}^2 - k x_{ref}^2.
\end{aligned} \tag{8.15}
$$

As $k = a_{00}$, therefore,

$$
\begin{aligned}
& x_{m+1,n} + \omega x_{m,n+1} \\
&= 1 - \mu\left((1+\omega)x_{mn}\right)^2 + a_{00}(1+\omega)^2 x_{ref}^2 - k x_{ref}^2.
\end{aligned} \tag{8.16}
$$

Let

$$
\begin{cases}
B_1(x_{mn}) = 1 - \mu\left((1+\omega)x_{mn}\right)^2 + a_{00}(1+\omega)^2 x_{ref}^2 \\
B_2(x_{mn}) = x_{ref}^2,
\end{cases}
$$

then (8.16) reduces to

$$
x_{m+1,n} + \omega x_{m,n+1} = B_1(x_{mn}) - k B_2(x_{mn}). \tag{8.17}
$$

Combine (8.17) and (8.11), then we have

$$
\begin{cases}
x_{m+1,n} + \omega x_{m,n+1} = B_1(x_{mn}) - k B_2(x_{mn}), & (1) \\
x_{ref} \in I = \sqrt{\frac{1}{k}\left(1 + \frac{1}{4(\mu - k)}\right)}, \quad (\mu > k). & (2)
\end{cases} \tag{8.18}
$$

In (1), the $B_1(x_{mn})$ describes the whole restricted state of s-bifurcation of system (8.1) in the area I after the term of control has been added to it. For $kB_2(x_{mn})$, it is increasing in the whole area I with the gradual extending of k. So relative to $B_1(x_{mn})$, $-kB_2(x_{mn})$ provides a negative amount, which says that it provides a compression of the whole structure or decreases the amount. Therefore, $kB_2(x_{mn})$ makes the structure of s-bifurcation entirely compressed in the area I. Thus depending on the feedback information of the variable k, we can reach any desired control objective. We call this method holistic compression control.

On the other hand, if we now insit the $\bar{x} = (1+\omega)x^*$ (unstable fixed plan of (8.1), that is,

$$
x_{mn} \to \bar{x} = (1+\omega)x^*, \text{ as } m, n \to \infty.
$$

Then we must simultaneously satisfy

$$
\mu\left[(1+\omega)x^*\right]^2 + (1+\omega)x^* - 1 = 0, \tag{8.19}
$$

and

$$\alpha x^{*2} + \beta x^{*} + \gamma = 0, \tag{8.20}$$

in which $(1+\omega)x^*$ is an unstable solution of (8.19) but a stable solution of (8.20) as mentioned above. For the quadratic map (8.19),

$$(1+\omega)x^* = \frac{-1+\sqrt{1+4\mu}}{2\mu}. \tag{8.21}$$

Hence, to control the system trajectory to this target, we substitute it into (8.20) and obtain

$$\alpha\left(\frac{-1+\sqrt{1+4\mu}}{2\mu}\right)^2 + \beta\left(\frac{-1+\sqrt{1+4\mu}}{2\mu}\right) + \gamma = 0. \tag{8.22}$$

This simply indicates that the x_{ref} must be chosen to be $(1+\omega)x^*$ in this (substitute $(1+\omega)x^*$ for x_{ref} in (8.22) to recover (8.19), which is consistent with the classical feedback control theory where the target is used in the feedback.

8.2 Sufficient Conditions for Control via Holistic Compression for Bifurcation of Surface Chaos

In this section, we give a sufficient condition for the above compressed control of bifurcation of s-chaos with the first method of Lyapunov.

We determine a sufficient condition for the control of $(1+\omega)x_{mn}$ to the target $(1+\omega)x^*$ for the case where the reference signal $x_{ref} = (1+\omega)x^*$. Substituting $x_{ref} = (1+\omega)x^*$ in (8.14) gives

$$\begin{aligned} &x_{m+1,n} + \omega x_{m,n+1} \\ &= 1 - \mu\left((1+\omega)x_{mn}\right)^2 + \left[k(1+\omega)x^* + a_{00}(1+\omega)x_{mn}\right] \times \\ &\quad [(1+\omega)x_{mn} - (1+\omega)x^*]. \end{aligned} \tag{8.23}$$

Note that $(1+\omega)x^*$satisfies the original system, i.e.,

$$(1+\omega)x^* = 1 - \mu\left((1+\omega)x^*\right)^2. \tag{8.24}$$

A subtraction of these last two equations gives

$$\begin{aligned} &\left(x_{m+1,n} + \omega x_{m,n+1}\right) - (1+\omega)x^* \\ &= \mu(1+\omega)^2(x_{mn} - x^*)\left[(a_{00}-1)(x_{mn} - x^*) + \right. \\ &\quad \left. (a_{00} + k - 2)x^*\right]. \end{aligned} \tag{8.25}$$

Moreover, with the transformation of the spatial structure,

$$C_{s+1,t+1}(m,n) = x_{s+1,n} + \omega x_{m,t+1}, \tag{8.26}$$

we can obtain immediately

$$C_{st}(m,n) = x_{sn} + \omega x_{mt}. \tag{8.27}$$

From (8.26), (8.27), if we take $s = m$, $t = n$, then we have

$$\begin{aligned} C_{m+1,n+1}(m,n) &= x_{m+1,n} + \omega x_{m,n+1}, \\ C_{mn}(m,n) &= x_{mn} + \omega x_{mn} = (1+\omega)x_{mn}, \end{aligned} \tag{8.28}$$

and $C^*_{mn}(m,n) = (1+\omega)x^*$, let again

$$e_{m+1,n+1}(m,n) = C_{m+1,n+1}(m,n) - C^*_{mn}(m,n),$$

then from (8.25), we obtain

$$e_{m+1,n+1}(m,n) = \mu(1+\omega)e_{mn}\left((a_{00}-1)e_{mn} + (k + a_{00} - 2)(1+\omega)x^*\right). \tag{8.29}$$

In view of (8.29), it is easy to see that the control objective can be realized when $e_{mn}(m,n) = 0$ is the stable fixed plan of system (8.29). Now we analyze the Jacobian matrix when (8.29) is at the stable fixed plan

$$J = \left.\frac{\partial(e_{m+1,n+1}(m,n))}{\partial(e_{mn}(m,n))}\right|_{e_{mn}(m,n)} = (k + a_{00} - 2)(1+\omega)x^*.$$

Notice that for this 1-D Jacobian, the eigenvalue of a "matrix" is the scalar itself: $\lambda = (k + a_{00} - 2)(1+\omega)x^*$. Applying the first method of Lyapunov to determine the stability of the controlled system, the sufficient condition is

$$|\lambda| = \left|(k + a_{00} - 2)(1+\omega)x^*\right| < 1,$$

and the stability of system (8.29), which guarantees that $e_{mn}(m,n) \to 0$ as $m, n \to \infty$. Therefore,

$$(1+\omega)x_{mn} \text{ or } C_{m+1,n+1}(m,n) = x_{m+1,n} + \omega x_{m,n+1} \to (1+\omega)x^*, \text{ as } m, n \to \infty.$$

The above analysis indicates that we can control the system orbit either to the unstable fixed point, or one can make use of (8.9) to specify one of a range of possible values for x_{ref}, and tune the control outcome to one of a set of $\bar{x}$ depending on the specific values chosen for the parameters x_{ref}, k, and a_{00}.

Note that while (8.9) do not require explicit knowledge of $(1+\omega)x^*$, one need only make some observations of the system over a short time window to gain parameters.

8.3 Control Process of Holistic Compression for Bifurcation of Surface Chaos

For the spatial dynamic system (8.1), if take $\mu = 1.763$, $\omega = -0.0021$, then system (8.1) reduces to

$$x_{m+1,n} - 0.0021x_{m,n+1} = 1 - 1.763\left(0.9979x_{mn}\right)^2. \tag{8.30}$$

As $\mu = 1.763 > 1.55$, system (8.1) is chaotic in space. Moreover, for the $k = a_{00} = \frac{1}{11}$, we have $0 < x_{ref} < 4$ by (8.11). Because the feedback parameter g_{mn} is restricted by k, a_{00} and $k = a_{00}$. Hence, according to the results of system (8.30), regulate the feedback term g_{mn}, that is the numerical change of k, and we will reach our desired target. The results are simulated as follows in Fig. 8.1.

8.4 Holistic Compression Control for Surface Chaos

This section is mainly about the control of the behavior of surface chaos from the angle of surface chaos. That is, assume a predetermined control target and restrict the behavior of surface chaos to this target (Fig. 8.2).

8.4.1 Chaos Control via Holistic Compression

The bifurcation of surface chaos is controlled with compression. From the angle of surface chaos, this control process is actually the control of the behavior of surface chaos. And the surface chaos is finally controlled to its stable fixed surface, which in fact means it is controlled to its fixed plan as its parameters are determinate. To explain it in details, for system (8.30), which is chaotic, its control system is (8.14). So according to the change of the feedback parameter k, the control system can be controlled to the fixed plan of the original system (8.30), which is

$$x^* = \left.\frac{1+\sqrt{1+4\mu}}{2\mu(1+\omega)}\right|_{\mu=1.763\omega=-0.0021} = 0.52242. \tag{8.31}$$

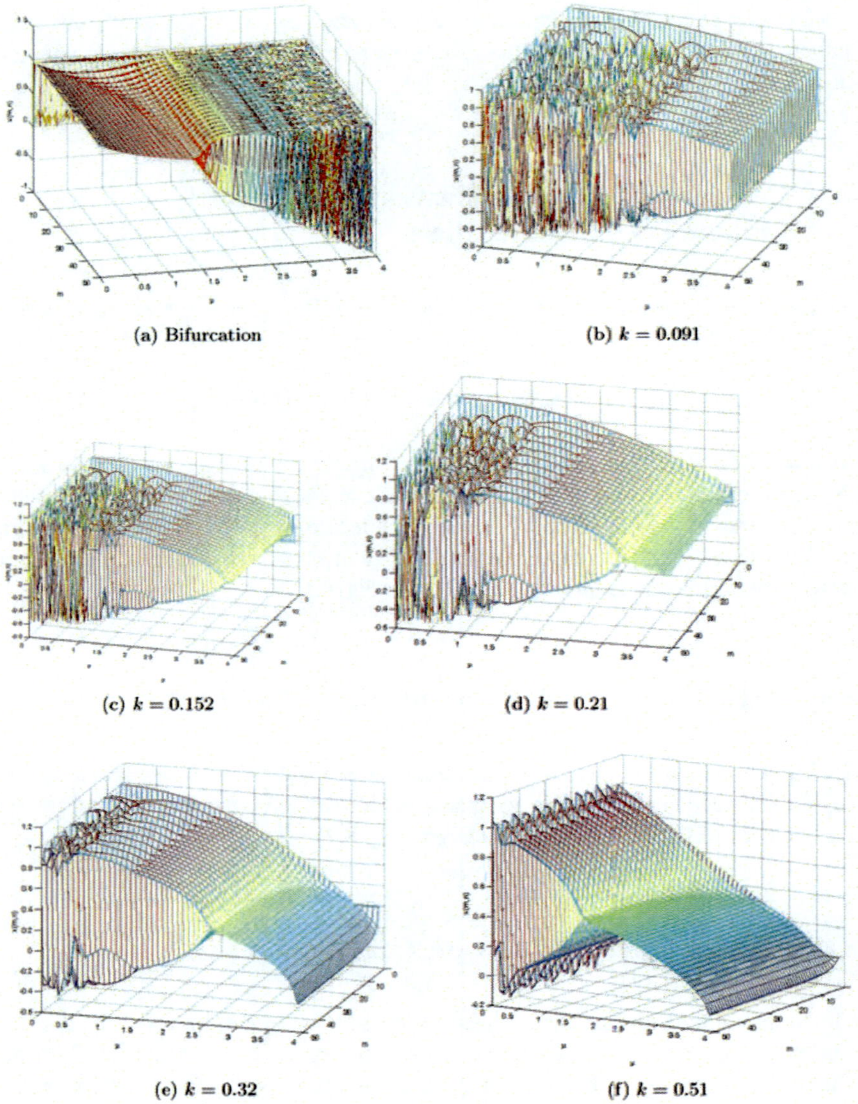

(a) Bifurcation

(b) $k = 0.091$

(c) $k = 0.152$

(d) $k = 0.21$

(e) $k = 0.32$

(f) $k = 0.51$

Fig. 8.1 Variation for the bifurcation of surface chaos in system (8.30)

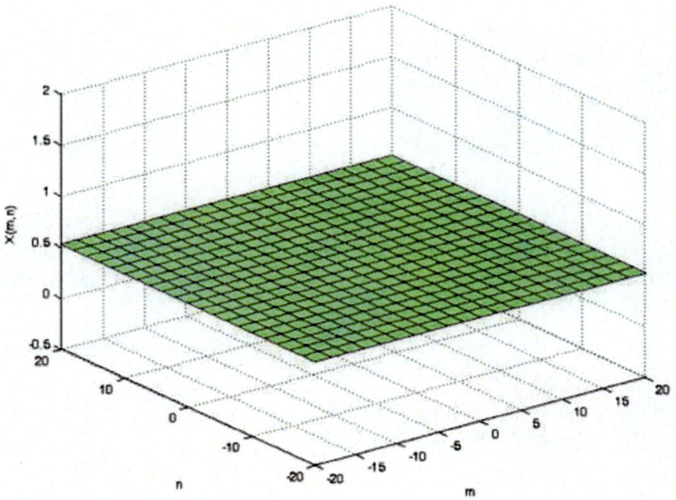

Fig. 8.2 The fixed plan of system (8.30)

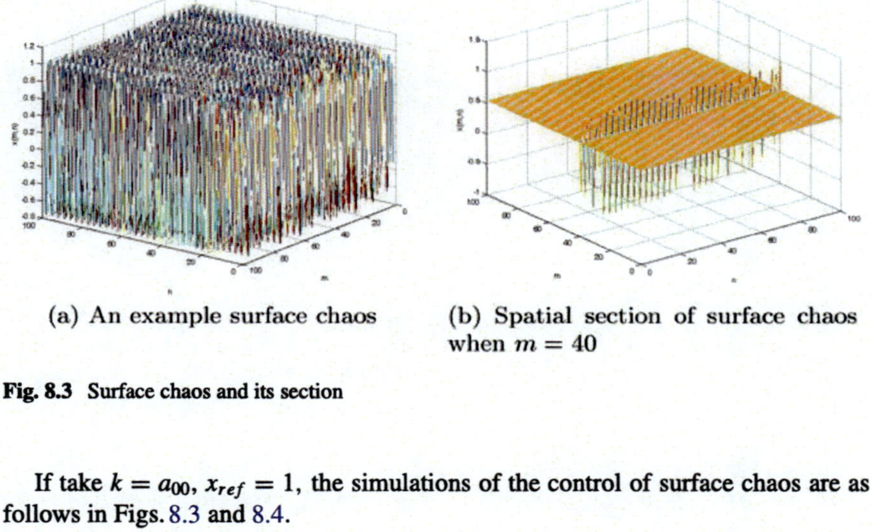

(a) An example surface chaos

(b) Spatial section of surface chaos when $m = 40$

Fig. 8.3 Surface chaos and its section

If take $k = a_{00}$, $x_{ref} = 1$, the simulations of the control of surface chaos are as follows in Figs. 8.3 and 8.4.

8.4.2 *Chaos Control via Local Compression*

For the surface which is in chaotic state, we can control to our desired target from a certain part of it. For example, for system (8.30), $\mu = 1.763$, $\omega = -0.0021$, $x_{ref} = 1$, and the fixed plan of the system is still (8.31). If we require $m = 1, 2, \ldots, 39$,

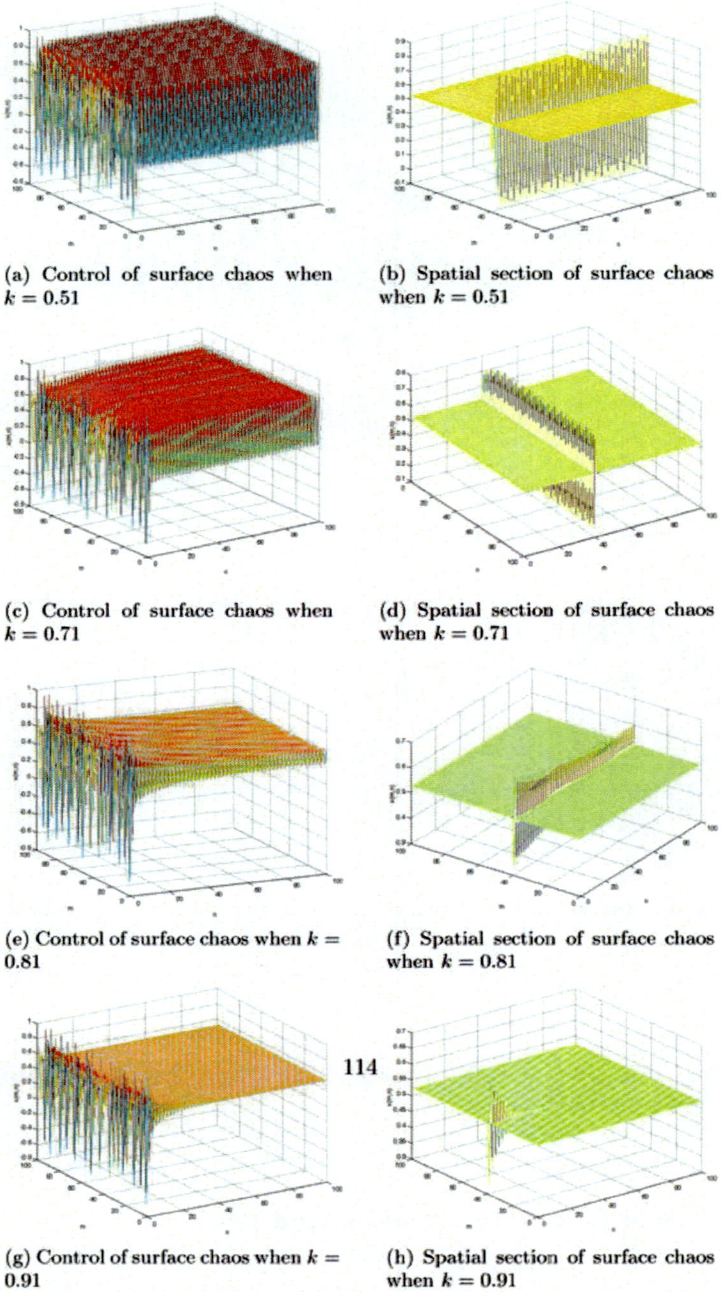

(a) Control of surface chaos when $k = 0.51$

(b) Spatial section of surface chaos when $k = 0.51$

(c) Control of surface chaos when $k = 0.71$

(d) Spatial section of surface chaos when $k = 0.71$

(e) Control of surface chaos when $k = 0.81$

(f) Spatial section of surface chaos when $k = 0.81$

(g) Control of surface chaos when $k = 0.91$

(h) Spatial section of surface chaos when $k = 0.91$

Fig. 8.4 Control of surface chaos and its section

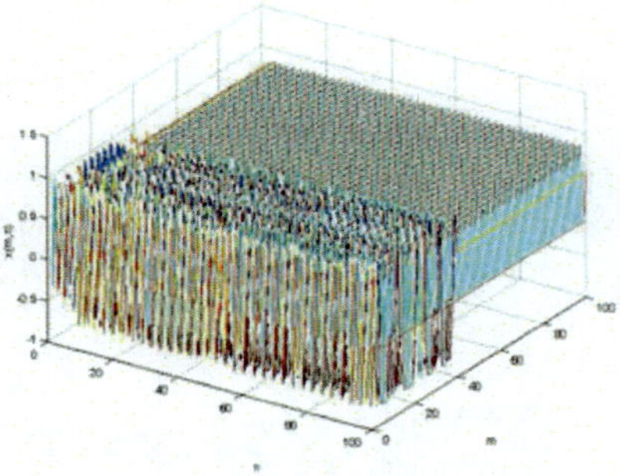

(a) Local control for surface chaos when $k = 0.51$

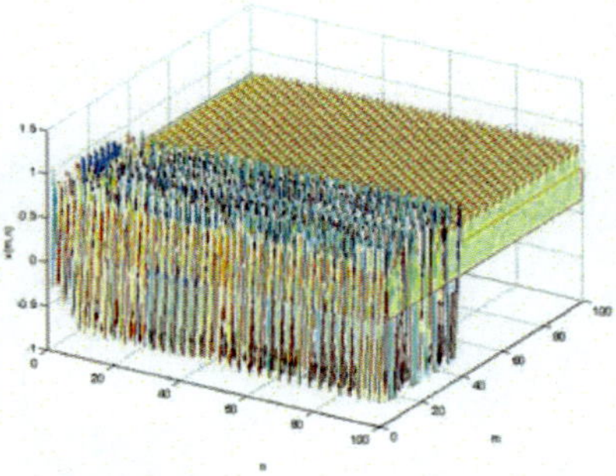

(b) Local control for surface chaos when $k = 0.71$

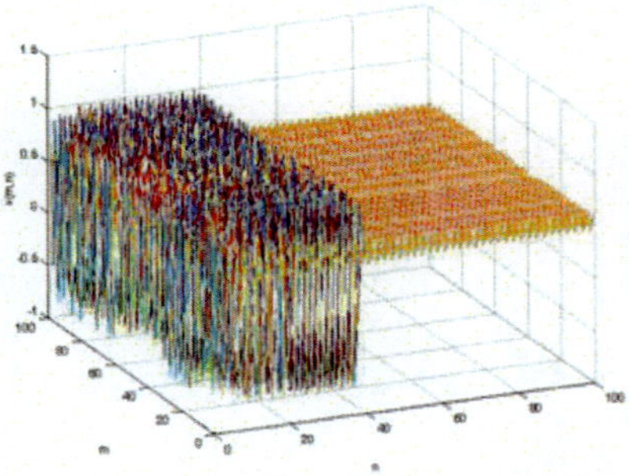

(c) Local control for surface chaos when $k = 0.91$

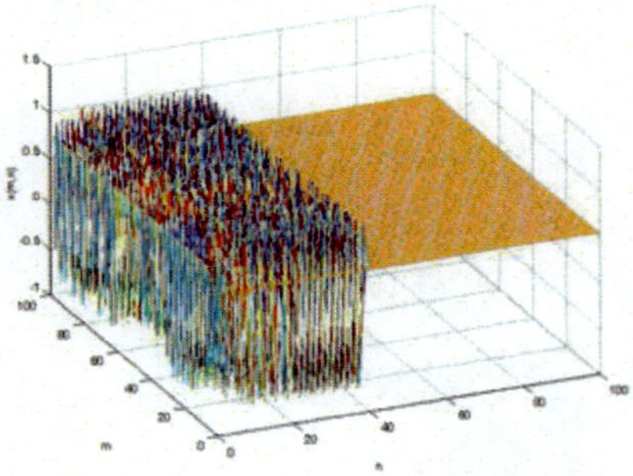

(d) Local control for surface chaos when $k = 1.01$

Fig. 8.5 Local control for surface chaos

to maintain the chaos of the original surface, from $m = 40, 41, \ldots$, $n = 1, 2, \ldots$, we begin to control and control it to its fixed plan, and the simulations are shown in Fig. 8.5.

8.5 The Case of $\omega = 0$

Because $\omega = 0$, then the original system (8.1) reduces to

$$x_{m+1,n} = 1 - \mu x_{mn}^2. \tag{8.32}$$

If take $n = n_0 = 0$, then (8.32) changes to

$$x_{m+1,0} = 1 - \mu x_{m0}^2, \tag{8.33}$$

or

$$x_{m+1} = 1 - \mu x_m^2. \tag{8.34}$$

Take $\mu = 1.763$ again,then the system (8.32) is chaotic. And we can also carry out the control of chaos and bifurcation for one dimension system (8.34). For the sake

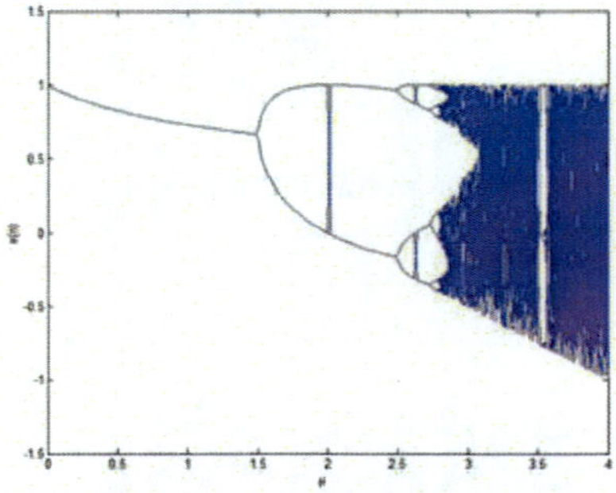

(a) $\mu = 1.763$

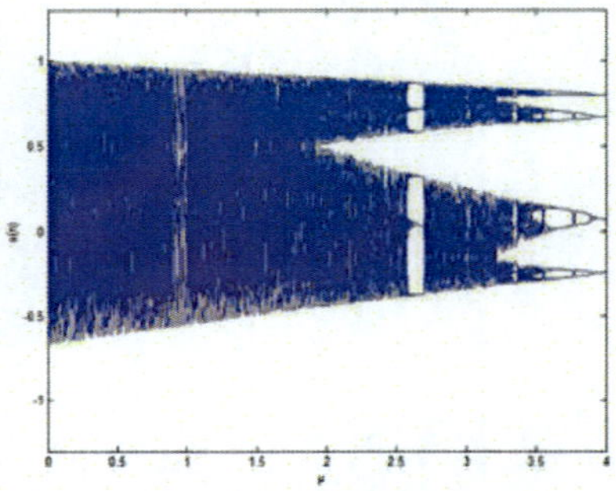

(b) $k = 0.11$

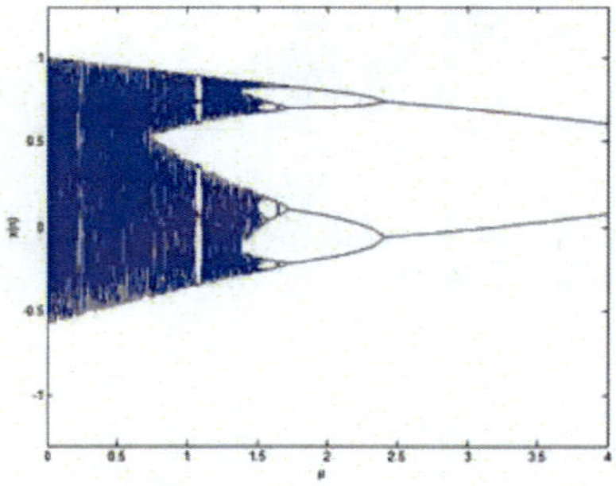

(c) $k = 0.21$

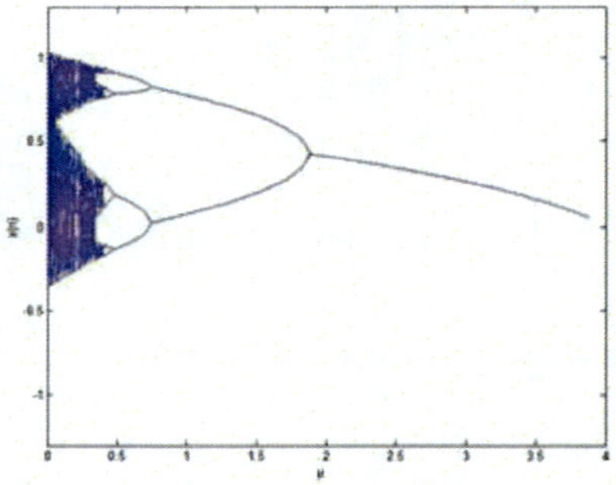

(d) $k = 0.51$

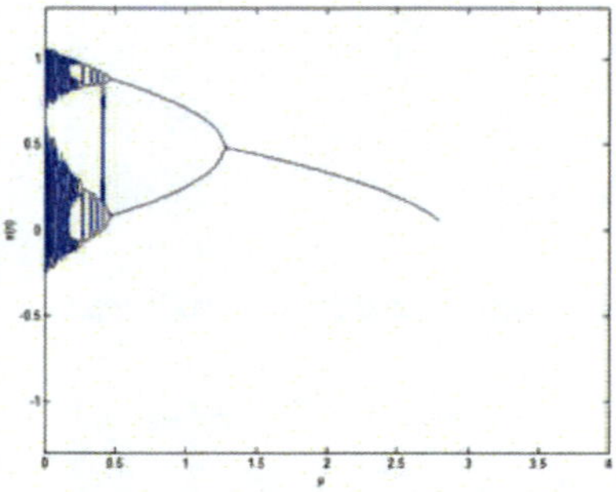

(e) $k = 0.91$

Fig. 8.6 Chaotic bifurcation in system (8.34)

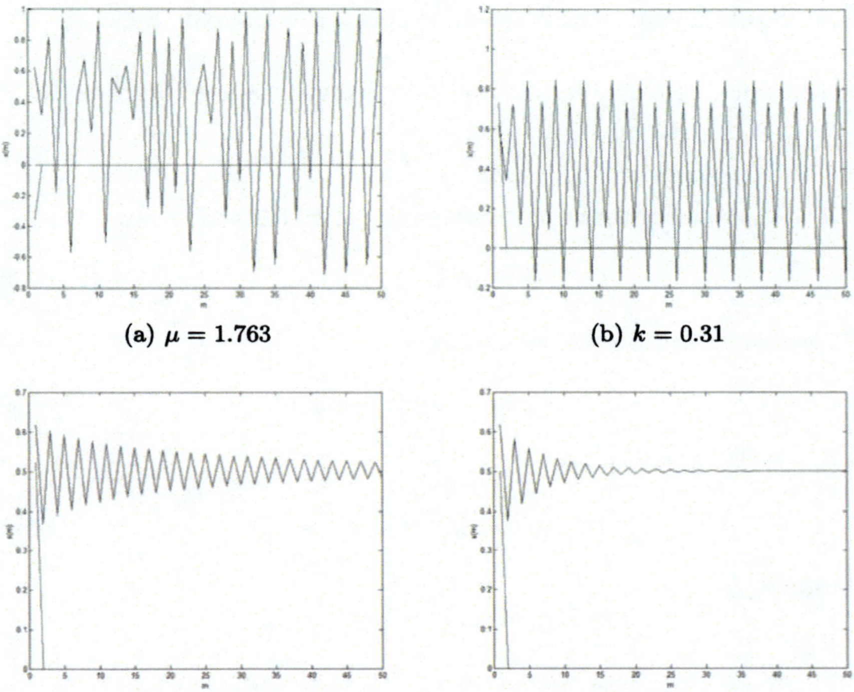

(a) $\mu = 1.763$ (b) $k = 0.31$

(c) $k = 0.81$ (d) $k = 0.91$

Fig. 8.7 The case of chaos in system (8.34)

of convenience, we can devise a simple system on control, for example:

$$x_{m+1} = 1 - \mu x_m^2 + g(x_m - x_{ref}). \tag{8.35}$$

Similar to (8.11), we have

$$0 < x_{ref} < \frac{1}{g}\left(1 + \frac{(g+1)^2}{4\mu}\right). \tag{8.36}$$

Take $g = 0.31$, then we have $0 < x_{ref} < 4.0108$. With the feedback information of g, one can control the bifurcation of one-dimension system (8.34), which can be validated by the following simulations in Fig. 8.6.

Similarly, when $\mu = 1.763$, we can see the chaotic control of x_m with the change of m and it is easy to see that the chaotic system is controlled to its fixed beeline, which can be validated by the following simulations (Fig. 8.7).

8.6 Some Dynamical Properties of Spatial Logistic System

In the section, we mainly analyze some basic dynamic characteristics of system

$$x_{m+1,n} + \omega x_{m,n+1} = f(\mu, x_{mn} + \omega x_{mn}), \tag{8.37}$$

and also apply the mathematical transformation of the following, that is

$$C_{s+1,t+1}(m,n) = x_{s+1,n} + \omega x_{m,t+1}. \tag{8.38}$$

Then from (8.38), we obtain immediately

$$C_{st}(m,n) = x_{sn} + \omega x_{mt}. \tag{8.39}$$

Using (8.38), (8.39), then get

$$C_{s+1,t+1}(m,n) = f(\mu, C_{st}(m,n)). \tag{8.40}$$

Generally, one has

$$C_{s+2,t+2}(m,n) = f^2(\mu, C_{st}(m,n)),$$
$$\cdots\cdots$$
$$C_{s+p,t+p}(m,n) = f^p(\mu, C_{st}(m,n)).$$

Specially, if we take $f(\mu, C_{st}(m,n)) = 1 - \mu C_{st}^2(m,n)$, then system (8.40) reduces to

$$C_{s+1,t+1}(m,n) = 1 - \mu C_{st}^2(m,n). \tag{8.41}$$

In addition, from (8.38), (8.39), if we take $s = m$, $t = n$, then we have

$$\begin{aligned} C_{m+1,n+1}(m,n) &= x_{m+1,n} + \omega x_{m,n+1}, \\ C_{mn}(m,n) &= x_{mn} + \omega x_{mn} = (1+\omega)x_{mn}, \end{aligned} \tag{8.42}$$

hence we have

$$C_{m+1,n+1}(m,n) = 1 - \mu C_{mn}^2(m,n), \tag{8.43}$$

also, the system (8.43) reduces to system (8.1), that is,

$$x_{m+1,n} + \omega x_{m,n+1} = 1 - \mu[(1+\omega)x_{mn}]^2,$$

which is called a spatially generalized Logistic system.

In fact, note that system (8.40) is a system of Euclid space of four-dimension, therefore, we can discuss the properties of the system in 4-D space. But to exhibit the dynamical characteristics of system (8.1) perfectly, we still consider system (8.1) literally in 2-D space.

We directly view the duality function of the following

$$G(\mu, \omega, x, y) = 1 - \mu[(1+\omega)x]^2, \tag{8.44}$$

we can see it is a particular duality function because it has no variable y on the right side of (8.44).

Moreover, Robert L. Devaney has given the defintions of the"critical point", "non-degenerate", "attracting fixed point (an attractor)" and "repelling fixed point (a repellor)". Hence, we have the following dynamical characteristics on topology of system (8.1).

Property I *The zero is a critical point of $G(\mu, \omega, x, y)$, and for any x, $(x, y) \in R^2$, it is non-degenerate of $G(\mu, \omega, x, y)$.*

Proof Since $G'_x(\mu, \omega, x, y) = -2\mu(1+\omega)x$, thus for any $y \in R$, we always have

$$G'_x(\mu, \omega, 0, y) = -2\mu(1+\omega)x|_{x=0} = 0,$$

and also $G''_x(\mu, \omega, x, y) = -2\mu(1+\omega)$, $\mu > 0$, so for any $(x, y) \in R^2$, we always have $G''_x(\mu, \omega, x, y) = -2\mu(1+\omega) \neq 0$. With the definition of critical point, we know zero is the critical point of system (8.1), that is to say any $(x, y) \in R^2$ is non-degenerate of $G(\mu, \omega, x, y)$.

Property II *The system (8.1) has an attracting fixed point (an attractor) or a sink, and also a repelling fixed point (a repellor) or source.*

Proof Consider the fixed surface of system (8.1), that is,

$$1 - \mu((1+\omega)x)^2 = (1+\omega)x,$$

then we can determine two pieces of fixed surface, that is,

$$P_+(\mu, \omega) = \frac{-1+\sqrt{1+4\mu}}{2\mu(1+\omega)}, \quad P_-(\mu, \omega) = \frac{-1-\sqrt{1+4\mu}}{2\mu(1+\omega)},$$

because

$$G'_x(\mu, \omega, P_+(\mu, \omega), y) = 1 - \sqrt{1+4\mu} < 0$$

and

$$G'_x(\mu, \omega, P_-(\mu, \omega), y) = 1 + \sqrt{1+4\mu} > 0,$$

hence $P_+(\mu, \omega)$ is an attracting fixed point (an attractor) or a sink, and $P_-(\mu, \omega)$ is a repelling fixed point (a repellor) or source.

Property III *For the $G(\mu, \omega, x, y)$, there is at most one attracting periodic orbit for each μ.*

Proof We have shown that

$$S(G(\mu, \omega, x, y)) = \frac{G''}{G'} - \frac{3}{2}\left(\frac{G''}{G'}\right)^2$$
$$= -\frac{3}{2}\left(\frac{1}{x}\right)^2 < 0,$$

and if $|x|$ is sufficiently large, it is easy to see $\left|G_x^n(\mu, \omega, x, y)\right| \to \infty$, (where $G_x^n(\mu, \omega, x, y) = \underbrace{G_x(G_x ... G_x}_{n \text{ times}}(\mu, \omega, x, y)...))$. Hence there is no attracting periodic orbit with infinite stable sets.

Lemma 8.1 *Suppose $S(G(\mu, \omega, x, y)) < 0\,(S(G(\mu, \omega, x, y)) = 0$ is allowed). Suppose $G(\mu, \omega, x, y)$ has n Critical points. Then $G(\mu, \omega, x, y)$ has at most $n + 2$ attracting periodic orbits.*

Property IV *For the $G(\mu, \omega, x, y)$, there are at most three attracting periodic orbits for each μ.*

Proof Because $G'(\mu, \omega, 0, y) = 0$, zero is a critical point, there are at most three attracting periodic orbits for each μ.

Chapter 9
Linear Generalized Synchronization of Surface Chaos

9.1 Preliminaries

In 1976, the French mathematician and astronomer Michel Hénon studied the following famous two-dimensional Hénon map [358]

$$\begin{cases} x_{m+1} = 1 + \beta y_m - \alpha x_m^2, & \alpha = 1.4, \\ y_{m+1} = x_m, & \beta = 0.3, \end{cases} \tag{9.1}$$

which is a typical example of plane chaos.

It is worth mentioning that the considered system of these aforementioned papers can be turned into the discrete form of one-dimensional convection equation with a forcing term:

$$x_{m+1,n} + \omega x_{m,n+1} = f(\mu, (1+\omega)x_{m,n}),$$

where ω is a real constant, $f(\cdot)$ is a nonlinear function, $m, n \in N_r = \{r, r+1, r+2, \cdots \mid r \text{ is an integer and } r \leq 0\}$ and μ is a real parameter.

However, to the best of our knowledge, there are few results dealing with the following high dimensional spatial model

$$\begin{cases} x_{m+1,n} + \omega x_{m,n+1} = \\ \quad f(\omega, \alpha, \beta, (1+\omega)x_{m,n}, (1+\omega)y_{m,n}), \\ y_{m+1,n} + \omega y_{m,n+1} = \\ \quad g(\omega, \alpha, \beta, (1+\omega)x_{m,n}, (1+\omega)y_{m,n}), \end{cases} \tag{9.2}$$

where ω is a real constant, α, β are positive parameters and $m, n \in N_r = \{r, r+1, r+2, \cdots \mid r \text{ is an integer and } r \leq 0\}$, $f(\cdot), g(\cdot)$ are nonlinear functions. Obviously, system (9.2) is a general form of system (9.1). And thus it is important and valuable to study the qualitative analysis and control of this high dimensional spatial model

S. T. Liu and L. Zhang, *Surface Chaos and Its Applications*,
https://doi.org/10.1007/978-981-16-8229-2_9

(9.2). In fact, compared with low dimensional system, the high dimensional system is more universal, and its application is more extensive in practical engineering areas including secure communication, power converters and information processing.

Motivated by the discussions above, this chapter focuses on generalized synchronization of the following two spatially generalized 2-D Hénon systems

$$\begin{cases} x_{m+1,n} + \omega x_{m,n+1} = 1 + \beta_1(1+\omega)y_{m,n} - \\ \qquad\qquad\qquad \alpha_1\left[(1+\omega)x_{m,n}\right]^2, \\ y_{m+1,n} + \omega y_{m,n+1} = (1+\omega)x_{m,n}, \end{cases} \tag{9.3}$$

$$\begin{cases} z_{m+1,n} + \omega z_{m,n+1} = 1 + \beta_2(1+\omega)w_{m,n} - \\ \qquad\qquad\qquad \alpha_2\left[(1+\omega)z_{m,n}\right]^2, \\ w_{m+1,n} + \omega w_{m,n+1} = (1+\omega)z_{m,n}, \end{cases} \tag{9.4}$$

where ω is a real constant, α_i, β_i, $(i = 1, 2)$ are positive parameters.

9.2 Spatially Fixed Plane and Its Stability

In fact, system (9.2) has very complex dynamical properties. In this section, we studies spatial fixed plane and its stability of system (9.2).

On the other hand, when $\omega = 0$ and $n = 0$, system (9.2) reduces to

$$\begin{cases} x_{m+1} = f(\alpha, \beta, x_m, y_m), \\ y_{m+1} = g(\alpha, \beta, x_m, y_m), \end{cases} \tag{9.5}$$

and this is what we are very familiar with about the case of two-dimensional discrete dynamic systems. Further, when $f(\alpha, \beta, x_m, y_m)=1 + \beta y_m - \alpha x_m^2$, $g(\alpha, \beta, x_m, y_m)$ $= x_n$, system (9.5) becomes the following two-dimensional Hénon map

$$\begin{cases} x_{m+1} = 1 + \beta y_m - \alpha x_m^2, \\ y_{m+1} = x_m. \end{cases} \tag{9.6}$$

As is known, a zero point (x^*, y^*) of equations

$$\begin{cases} x - f(\alpha, \beta, x, y) = 0, \\ y - g(\alpha, \beta, x, y) = 0, \end{cases}$$

is said to be a fixed point of (9.5). The fixed points of the two-dimensional Hénon map (9.6) are given as follows

$$\begin{cases} x^* = \dfrac{(\beta - 1) \pm \sqrt{(1-\beta)^2 - 4\alpha}}{2\alpha}, \\ y^* = x^*. \end{cases}$$

Definition 9.1 Let $(x_{m,n}, y_{m,n}) = (x^*, y^*)$ for all $m, n \in N_r$. Then, (x^*, y^*) is called a fixed plane of the spatial 2-D discrete dynamical system (9.2), if (x^*, y^*) is a zero solution of equations

$$\begin{cases} (1+\omega)x^* - f(\omega, \alpha, \beta, (1+\omega)x^*, (1+\omega)y^*) = 0, \\ (1+\omega)y^* - g(\omega, \alpha, \beta, (1+\omega)x^*, (1+\omega)y^*) = 0. \end{cases}$$

A fixed plane is denoted by L(2D).

Definition 9.2 Any $(x^*, y^*) \in L(2D)$ is said to be a fixed point of the spatial 2-D discrete dynamical system (9.2).

When $f(\omega, \alpha, \beta, (1+\omega)x_{m,n}, (1+\omega)y_{m,n}) = 1 + \beta(1+\omega)y_{m,n} - \alpha\left[(1+\omega)x_{m,n}\right]^2$, $g(\omega, \alpha, \beta, (1+\omega)x_{m,n}, (1+\omega)y_{m,n}) = (1+\omega)x_{m,n}$, the spatial 2-D discrete dynamical system (9.2) becomes

$$\begin{cases} x_{m+1,n} + \omega x_{m,n+1} = 1 + \beta(1+\omega)y_{m,n} - \\ \qquad\qquad \alpha\left[(1+\omega)x_{m,n}\right]^2, \\ y_{m+1,n} + \omega y_{m,n+1} = (1+\omega)x_{m,n}, \end{cases} \tag{9.7}$$

and it is called a spatially generalized 2D Hénon system. There are the basic characteristics of system (9.7), such as the spatial Lyapunov exponents, spatial chaos, spatial strange attractors and so on.

In order to obtain the following conclusions, hereafter, we assume that

$$\omega \in (-\infty, +\infty)\backslash\{-1\}. \tag{9.8}$$

Theorem 9.1 *For all $m, n \in N_r$, a fixed plane of the spatially generalized 2-D Henon system (9.7) is given by*

$$(x_{m,n}, y_{m,n}) = (x^*, y^*) = \left(\frac{(\beta-1)\pm\sqrt{(\beta-1)^2+4\alpha}}{2\alpha(1+\omega)}, \frac{(\beta-1)\pm\sqrt{(\beta-1)^2+4\alpha}}{2\alpha(1+\omega)}\right). \tag{9.9}$$

Theorem 9.2 *Assume that (9.8) holds, a stability condition for the fixed plane (x^*, y^*) of the spatial 2-D system (9.2) is*

$$|\lambda| \leq \begin{cases} \dfrac{\omega-1}{1+\omega}, & \omega \in (-\infty,\ -1), \\ \dfrac{1-\omega}{1+\omega}, & \omega \in (-1,\ 0), \\ 1, & \omega \in (0,\ \infty). \end{cases} \tag{9.10}$$

Proof For any $m, n \in N_r$, take $(x^*, y^*) \in L(2D)$, and start the orbit from a point with a small distance away from (x^*, y^*). Define small quantities $(\varepsilon_{m,n}, \eta_{m,n})$ by

$$\begin{cases} x_{m,n} = x^* + \varepsilon_{m,n}, \\ y_{m,n} = y^* + \eta_{m,n}, \end{cases} \tag{9.11}$$

and, similarly,

$$\begin{cases} x_{m+1,n} = x^* + \varepsilon_{m+1,n}, \\ x_{m,n+1} = x^* + \varepsilon_{m,n+1}, \\ y_{m+1,n} = y^* + \eta_{m+1,n}, \\ y_{m,n+1} = y^* + \eta_{m,n+1}, \end{cases} \tag{9.12}$$

then

$$\begin{cases} x_{m+1,n} + \omega x_{m,n+1} = (1+\omega)x^* + \\ \qquad (\varepsilon_{m+1,n} + \omega\varepsilon_{m,n+1}), \\ y_{m+1,n} + \omega y_{m,n+1} = (1+\omega)y^* + \\ \qquad (\eta_{m+1,n} + \omega\eta_{m,n+1}). \end{cases} \tag{9.13}$$

Substituting (9.13) into system (9.2), we obtain

$$\begin{cases} (1+\omega)x^* + (\varepsilon_{m+1,n} + \omega\varepsilon_{m,n+1}) = \\ \qquad f(\omega, \alpha, \beta, (1+\omega)(x^* + \varepsilon_{m,n}), \\ \qquad (1+\omega)(y^* + \eta_{m,n})), \\ (1+\omega)y^* + (\eta_{m+1,n} + \omega\eta_{m,n+1}) = \\ \qquad g(\omega, \alpha, \beta, (1+\omega)(x^* + \varepsilon_{m,n}), \\ \qquad (1+\omega)(y^* + \eta_{m,n})). \end{cases} \tag{9.14}$$

In the case of spatial 2-D system (9.2), stability simply means that when approaching the fixed point $(x^*,\ y^*)$, the difference between the kth iterate $(x_{m,n},\ y_{m,n})$ and $(x^*,\ y^*)$ keeps decreasing. In other words, it follows from (9.11) and (9.12) that

$$\left|\frac{\varepsilon_{m+1,n}}{\varepsilon_{m,n}}\right| < 1, \quad \left|\frac{\varepsilon_{m,n+1}}{\varepsilon_{m,n}}\right| < 1, \tag{9.15}$$

and

$$\left|\frac{\eta_{m+1,n}}{\eta_{m,n}}\right| < 1, \quad \left|\frac{\eta_{m,n+1}}{\eta_{m,n}}\right| < 1. \tag{9.16}$$

Therefore, when m and n are sufficiently large, applying (9.15) and (9.16) yields

$$\begin{cases} \left|\dfrac{\varepsilon_{m+1,n}}{(1+\omega)\varepsilon_{m,n}}\right| = \dfrac{1}{|1+\omega|}\left|\dfrac{\varepsilon_{m+1,n}}{\varepsilon_{m,n}}\right| < \dfrac{1}{|1+\omega|}, \\ \left|\dfrac{\varepsilon_{m,n+1}}{(1+\omega)\varepsilon_{m,n}}\right| = \dfrac{1}{|1+\omega|}\left|\dfrac{\varepsilon_{m,n+1}}{\varepsilon_{m,n}}\right| < \dfrac{1}{|1+\omega|}, \end{cases} \tag{9.17}$$

and

$$\begin{cases} \left|\dfrac{\eta_{m+1,n}}{(1+\omega)\eta_{m,n}}\right| = \dfrac{1}{|1+\omega|}\left|\dfrac{\eta_{m+1,n}}{\eta_{m,n}}\right| < \dfrac{1}{|1+\omega|}, \\ \left|\dfrac{\eta_{m,n+1}}{(1+\omega)\eta_{m,n}}\right| = \dfrac{1}{|1+\omega|}\left|\dfrac{\eta_{m,n+1}}{\eta_{m,n}}\right| < \dfrac{1}{|1+\omega|}. \end{cases} \tag{9.18}$$

Applying (9.17) and (9.18), and using (9.15) and (9.16), we obtain

$$\begin{aligned} &\left|\frac{\varepsilon_{m+1,n}+\omega\varepsilon_{m,n+1}}{(1+\omega)\varepsilon_{m,n}}\right| \\ &\leq \left|\frac{\varepsilon_{m+1,n}}{(1+\omega)\varepsilon_{m,n}}\right| + |\omega|\left|\frac{\varepsilon_{m,n+1}}{(1+\omega)\varepsilon_{m,n}}\right| \\ &< \frac{1}{|1+\omega|} + \frac{|\omega|}{|1+\omega|} \\ &= \frac{1+|\omega|}{|1+\omega|} \\ &= \begin{cases} \dfrac{\omega-1}{1+\omega}, & \omega \in (-\infty,\ -1), \\ \dfrac{1-\omega}{1+\omega}, & \omega \in (-1,\ 0), \\ 1, & \omega \in (0,\ \infty), \end{cases} \end{aligned}$$

and

$$
\begin{aligned}
&\left|\frac{\eta_{m+1,n}+\omega\eta_{m,n+1}}{(1+\omega)\eta_{m,n}}\right| \\
&\le \left|\frac{\eta_{m+1,n}}{(1+\omega)\eta_{m,n}}\right| + |\omega|\left|\frac{\eta_{m,n+1}}{(1+\omega)\eta_{m,n}}\right| \\
&< \frac{1}{|1+\omega|}+\frac{|\omega|}{|1+\omega|} \\
&= \frac{1+|\omega|}{|1+\omega|} \\
&= \begin{cases} \dfrac{\omega-1}{1+\omega}, & \omega\in(-\infty,\ -1), \\ \dfrac{1-\omega}{1+\omega}, & \omega\in(-1,\ 0), \\ 1, & \omega\in(0,\ \infty). \end{cases}
\end{aligned}
$$

Next, we expand system (9.2) to its Taylor expansion and only retain the linear part, then we have

$$
\begin{cases}
\varepsilon_{m+1,n}+\omega\varepsilon_{m,n+1} = \\
\qquad \left.\dfrac{\partial f}{\partial x}\right|_{(x^*,y^*)}(1+\omega)\varepsilon_{m,n}+ \\
\qquad \left.\dfrac{\partial f}{\partial y}\right|_{(x^*,y^*)}(1+\omega)\eta_{m,n}, \\
\eta_{m+1,n}+\omega\eta_{m,n+1} = \\
\qquad \left.\dfrac{\partial g}{\partial x}\right|_{(x^*,y^*)}(1+\omega)\varepsilon_{m,n}+ \\
\qquad \left.\dfrac{\partial g}{\partial y}\right|_{(x^*,y^*)}(1+\omega)\eta_{m,n},
\end{cases}
\tag{9.19}
$$

where

$$
\begin{cases}
\left.\dfrac{\partial f}{\partial x}\right|_{(x^*,y^*)} = \dfrac{\partial f}{\partial x}(\omega,\alpha,\beta,(1+\omega)x^*,(1+\omega)y^*), \\
\left.\dfrac{\partial f}{\partial y}\right|_{(x^*,y^*)} = \dfrac{\partial f}{\partial y}(\omega,\alpha,\beta,(1+\omega)x^*,(1+\omega)y^*), \\
\left.\dfrac{\partial g}{\partial x}\right|_{(x^*,y^*)} = \dfrac{\partial g}{\partial x}(\omega,\alpha,\beta,(1+\omega)x^*,(1+\omega)y^*), \\
\left.\dfrac{\partial g}{\partial y}\right|_{(x^*,y^*)} = \dfrac{\partial g}{\partial y}(\omega,\alpha,\beta,(1+\omega)x^*,(1+\omega)y^*).
\end{cases}
$$

As the linear difference equations are with constant coefficients, the structure of the solution is as follows

$$\begin{cases} \varepsilon_{m+1,n} + \omega\varepsilon_{m,n+1} = \lambda(1+\omega)\varepsilon_{m,n}, \\ \eta_{m+1,n} + \omega\eta_{m,n+1} = \lambda(1+\omega)\eta_{m,n}, \end{cases} \tag{9.20}$$

substituting (9.20) into (9.19), we write it into the matrix form

$$\begin{bmatrix} \frac{\partial f}{\partial x} - \lambda & \frac{\partial f}{\partial y} \\ \frac{\partial g}{\partial x} & \frac{\partial g}{\partial y} - \lambda \end{bmatrix}_{(x^*,y^*)} \begin{bmatrix} (1+\omega)\varepsilon_{m,n} \\ (1+\omega)\eta_{m,n} \end{bmatrix} = 0. \tag{9.21}$$

For linear homogeneous equations (9.21), there are non-zero solution conditions

$$\begin{vmatrix} \frac{\partial f}{\partial x} - \lambda & \frac{\partial f}{\partial y} \\ \frac{\partial g}{\partial x} & \frac{\partial g}{\partial y} - \lambda \end{vmatrix}_{(x^*,y^*)} = 0,$$

which leads to the stability condition for the fixed point (x^*, y^*) as

$$\begin{aligned} |\lambda| &= \left|\frac{\varepsilon_{m+1,n} + \omega\varepsilon_{m,n+1}}{(1+\omega)\varepsilon_{m,n}}\right| = \left|\frac{\eta_{m+1,n} + \omega\eta_{m,n+1}}{(1+\omega)\eta_{m,n}}\right| \\ &< \frac{1+|\omega|}{|1+\omega|} \\ &= \begin{cases} \frac{\omega-1}{1+\omega}, & \omega \in (-\infty, -1), \\ \frac{1-\omega}{1+\omega}, & \omega \in (-1, 0), \\ 1, & \omega \in (0, \infty). \end{cases} \end{aligned}$$

Corollary 9.1 *In particular, when $\omega = 0$ and $n = 0$, the stability condition for system (9.2) is*

$$|\lambda| < 1.$$

This is just a simple case of the familiar two-dimensional system, and it shows that spatial 2-D system (9.2) is a natural generalization of the two-dimensional map (9.5).

Theorem 9.3 *A stability condition for the fixed point (x^*, y^*) of the 2-D spatially generalized Hénon system (9.7) is*

$$\Pi = \left\{ (\alpha, \beta) \middle| \Delta_1 < \alpha < \Delta_2, \ |\beta| < 1 \right\}, \tag{9.22}$$

where

$$\Delta_1 = \frac{1}{4}\left[\left(\frac{1}{1+|\omega|}\beta - \frac{1+|\omega|}{(1+\omega)^2}\right)^2 - 2(\beta-1)\left(\frac{1}{1+|\omega|}\beta - \frac{1+|\omega|}{(1+\omega)^2}\right)\right],$$

$$\Delta_2 = \frac{1}{4}\left[\left(\frac{1}{1+|\omega|}\beta - \frac{1+|\omega|}{(1+\omega)^2}\right)^2 + 2(\beta-1)\left(\frac{1}{1+|\omega|}\beta - \frac{1+|\omega|}{(1+\omega)^2}\right)\right].$$

Proof For the system (9.7), by Theorem (9.1), its fixed plane

$$(x^*,\ y^*) = \left(\frac{(\beta-1)\pm\sqrt{(\beta-1)^2+4\alpha}}{2\alpha(1+\omega)}, \frac{(\beta-1)\pm\sqrt{(\beta-1)^2+4\alpha}}{2\alpha(1+\omega)}\right),$$

then, the Jacobi matrix of the fixed plane at $(x^*,\ y^*)$ is

$$J = \begin{bmatrix} A & B \\ C & D \end{bmatrix},$$

where

$$A = -2\alpha(1+\omega)^2 x^*,\ B = \beta(1+\omega),\ C = 1,\ D = 0.$$

From $|\lambda I - J| = 0$, the characteristic equation is obtained

$$\lambda^2 - (A+D)\lambda + AD - BC = 0,$$

the eigenvalue are

$$\lambda_{1,2} = \frac{(A+D)\pm\sqrt{(A+D)^2-4(AD-BC)}}{2},$$

and by Theorem 9.2, one has

$$|\lambda_{1,2}| < \frac{1+|\omega|}{|1+\omega|}. \tag{9.23}$$

Solving (9.7) for α yields (9.22).

Corollary 9.2 *In particular, when $\omega = 0$ and $n = 0$, the stability condition for the two-dimensional Hénon map (9.6) is*

$$\Pi = \left\{ \begin{array}{c} (\alpha,\ \beta) \left| -\frac{1}{4}(\beta - 1)^2 < \alpha < \right. \\ \frac{1}{4}(\beta - 1)^2, |\beta| < 1 \end{array} \right\}. \tag{9.24}$$

Remark 9.1 Note that (9.24) is just the familiar stability condition for the two-dimensional Hénon map, which shows again that the spatial 2-D system is a natural extension of the two-dimensional map.

9.3 Generalized Synchronization of Surface Chaos

Then spatial 2-D Hénon systems (9.3) and (9.4) have a transversal homoclinic orbit for all $|\beta_i| < \varepsilon\ (i = 1, 2)$ for some $\varepsilon > 0$. In particular, for $|\beta_i| < \varepsilon\ (i = 1, 2)$ systems (9.3) and (9.4) are chaotic. Since the generalized synchronization is studied, assume that $\alpha_1 \neq \alpha_2$, $\beta_1 \neq \beta_2$, without loss of generality, let

$$\alpha_1 > \alpha_2,\ \beta_1 < \beta_2.$$

Next, apply a linear coupling term to (9.3) and consider

$$\begin{cases} x_{m+1,n} + \omega x_{m,n+1} = 1 + \beta_1(1+\omega)y_{m,n} - \\ \quad \alpha_1\left[(1+\omega)x_{m,n}\right]^2 + \xi_1\left(h_1(x_{m,n}, z_{m,n})\right)^2 + \\ \quad \xi_2\left(h_2(y_{m,n}, w_{m,n})\right), \\ y_{m+1,n} + \omega y_{m,n+1} = (1+\omega)x_{m,n}, \end{cases} \tag{9.25}$$

$$\begin{cases} z_{m+1,n} + \omega z_{m,n+1} = 1 + \beta_2(1+\omega)w_{m,n} - \\ \quad \alpha_2\left[(1+\omega)z_{m,n}\right]^2, \\ w_{m+1,n} + \omega w_{m,n+1} = (1+\omega)z_{m,n}, \end{cases} \tag{9.26}$$

where $\xi_i\ (i = 1, 2)$ are the coupling constant and $h_1(x_{m,n}, z_{m,n})$, $h_2(y_{m,n}, w_{m,n})$ is determined. It is clear that system (9.25) is driven by system (9.26). Next, let

$$\begin{cases} h_1(x_{m,n}, z_{m,n}) = \alpha_1 x_{m,n} - \alpha_2 z_{m,n}, \\ h_2(y_{m,n}, w_{m,n}) = \beta_1 y_{m,n} - \beta_2 w_{m,n}, \end{cases} \tag{9.27}$$

and introduce

$$\begin{cases} e_{m,n}(1) = H_1(x_{m,n}, z_{m,n}), \\ e_{m,n}(2) = H_2(y_{m,n}, w_{m,n}), \end{cases}$$

and hope that four functions $H_1(x_{m,n}, z_{m,n})$, $H_2(y_{m,n}, w_{m,n})$, and $h_1(x_{m,n}, z_{m,n})$, $h_2(y_{m,n}, w_{m,n})$ can be found such that $e_{m+1,n}(i)$, $e_{m,n+1}(i)$, and $e_{m,n}(i)$ $(i = 1, 2)$, satisfy the spatially generalized 2-D Hénon system

$$\begin{cases} e_{m+1,n}(1) + \omega e_{m,n+1}(1) = \\ \qquad 1 + \beta(1+\omega)e_{m,n}(2) - \\ \qquad \alpha\left[(1+\omega)e_{m,n}(1)\right]^2, \\ e_{m+1,n}(2) + \omega e_{m,n+1}(2) = (1+\omega)e_{m,n}(1). \end{cases} \tag{9.28}$$

In general, there exists a functional relation between α, β and ω, α_i, β_i $(i = 1, 2)$, that is, $\alpha = \alpha(\alpha_1, \alpha_2, \omega)$, $\beta = \beta(\beta_1, \beta_2, \omega)$.

If two appropriate values of ξ_1, ξ_2 can be found such that

$$\Pi = \left\{ (\alpha,\ \beta) \middle| \Delta_1 < \alpha < \Delta_2,\ |\beta| < 1 \right\},$$

where Δ_1, Δ_2 are defined in Theorem 9.3, then $(e_{m,n}(1), e_{m,n}(2))$ will tend to a stable fixed point $(e(1), e(2))$. Hence, $H_1(x_{m,n}, z_{m,n}) = constant$, $H_2(y_{m,n}, w_{m,n}) = constant$, namely, $(x_{m,n}, y_{m,n})$ and $(z_{m,n}, w_{m,n})$ satisfy a deterministic functional relation, so that the generalized synchronization of chaos is realized.

Substituting (9.27) into (9.25), we obtain

$$\begin{cases} x_{m+1,n} + \omega x_{m,n+1} = \\ \quad 1 + \beta_1(1+\omega)y_{m,n} - \alpha_1\left[(1+\omega)x_{m,n}\right]^2 + \\ \quad \xi_1\left(\alpha_1 x_{m,n} - \alpha_2 z_{m,n}\right)^2 + \\ \quad \xi_2\left(\beta_1 y_{m,n} - \beta_2 w_{m,n}\right), \\ y_{m+1,n} + \omega y_{m,n+1} = (1+\omega)x_{m,n}. \end{cases} \tag{9.29}$$

On the other hand, choosing

$$\begin{cases} H_1(x_{m,n}, z_{m,n}) = a_1 x_{m,n} + b_1 z_{m,n}, \\ H_2(y_{m,n}, w_{m,n}) = a_2 y_{m,n} + b_2 w_{m,n}, \end{cases}$$

and taking

$$\begin{cases} e_{m,n}(1) = a_1 x_{m,n} + b_1 z_{m,n}, \\ e_{m,n}(2) = a_2 y_{m,n} + b_2 w_{m,n}, \end{cases}$$

we have

$$\begin{cases} e_{m+1,n}(1) = a_1 x_{m+1,n} + b_1 z_{m+1,n}, \\ e_{m,n+1}(1) = a_1 x_{m,n+1} + b_1 z_{m,n+1}, \\ e_{m+1,n}(2) = a_2 y_{m+1,n} + b_2 w_{m+1,n}, \\ e_{m,n+1}(2) = a_2 y_{m,n+1} + b_2 w_{m,n+1}. \end{cases}$$

Therefore,

$$\begin{cases} e_{m+1,n}(1) + \omega e_{m,n+1}(1) = a_1 (x_{m+1,n} + \\ \quad \omega x_{m,n+1}) + b_1 (z_{m+1,n} + \omega z_{m,n+1}), \\ e_{m+1,n}(2) + \omega e_{m,n+1}(2) = a_2 (y_{m+1,n} + \\ \quad \omega y_{m,n+1}) + b_2 (w_{m+1,n} + \omega w_{m,n+1}). \end{cases} \tag{9.30}$$

Substituting (9.26) and (9.29) into (9.30) we have

$$\begin{cases} e_{m+1,n}(1) + \omega e_{m,n+1}(1) = a_1 + a_1 \beta_1 (1 + \\ \quad \omega) y_{m,n} - a_1 \alpha_1 \left[(1+\omega) x_{m,n}\right]^2 + \\ \quad a_1 \xi_1 \left(\alpha_1 x_{m,n} - \alpha_2 z_{m,n}\right)^2 + \\ \quad a_1 \xi_2 \left(\beta_1 y_{m,n} - \beta_2 w_{m,n}\right) + \\ \quad b_1 + b_1 \beta_2 (1+\omega) w_{m,n} - \\ \quad b_1 \alpha_2 \left[(1+\omega) z_{m,n}\right]^2, \\ e_{m+1,n}(2) + \omega e_{m,n+1}(2) = a_2 (1+\omega) x_{m,n} + \\ \quad b_2 (1+\omega) z_{m,n}. \end{cases} \tag{9.31}$$

We also get the following equation from (9.28)

$$\begin{cases} e_{m+1,n}(1) + \omega e_{m,n+1}(1) = \\ \quad 1 + \beta (1+\omega)(a_2 y_{m,n} + b_2 w_{m,n}) - \\ \quad \alpha \left[(1+\omega)(a_1 x_{m,n} + b_1 z_{m,n})\right]^2 \\ e_{m+1,n}(2) + \omega e_{m,n+1}(2) = \\ \quad a_1 (1+\omega) x_{m,n} + b_1 (1+\omega) z_{m,n}. \end{cases} \tag{9.32}$$

From (9.31) and (9.32), we get

$$\begin{cases} a_1 + b_1 = 1, \\ a_1\xi_1\alpha_1^2 - a_1\alpha_1(1+\omega)^2 = -\alpha a_1^2(1+\omega)^2, \\ a_1\xi_1\alpha_2^2 - b_1\alpha_2(1+\omega)^2 = -\alpha b_1^2(1+\omega)^2, \\ -2a_1\xi_1\alpha_1\alpha_2 = -2\alpha a_1 b_1(1+\omega)^2, \\ a_1 = a_2, \\ b_1 = b_2, \\ a_1\beta_1(1+\omega) + a_1\beta_1\xi_2 = \beta a_2(1+\omega), \\ b_1\beta_2(1+\omega) - a_1\beta_2\xi_2 = \beta b_2(1+\omega). \end{cases} \tag{9.33}$$

Solving the algebraic equations in (9.33) gives

$$\begin{cases} a_1 = a_2 = \dfrac{\xi_1\alpha_1 - (1+\omega)^2}{\xi_1(\alpha_1-\alpha_2) - (1+\omega)^2}, \\ b_1 = b_2 = \dfrac{\xi_1\alpha_2}{(1+\omega)^2 - \xi_1(\alpha_1-\alpha_2)}, \\ \alpha = \dfrac{\alpha_1}{(1+\omega)^2}\left[(1+\omega)^2 - \xi_1(\alpha_1-\alpha_2)\right], \\ \beta = \dfrac{\beta_1(1+\omega) + \beta_1\xi_2}{(1+\omega)}. \end{cases} \tag{9.34}$$

It follows from system (9.32) that $(e_{m,n}(1), e_{m,n}(2))$ satisfies the following spatially generalized 2-D Hénon system

$$\begin{cases} e_{m+1,n}(1) + \omega e_{m,n+1}(1) = \\ \qquad 1 + \beta(1+\omega)e_{m,n}(2) - \\ \qquad \alpha\left[(1+\omega)e_{m,n}(1)\right]^2 \\ e_{m+1,n}(2) + \omega e_{m,n+1}(2) = (1+\omega)e_{m,n}(1). \end{cases} \tag{9.35}$$

On the other hand, to ensure the stability of the fixed points for $(e_{m,n}(1), e_{m,n}(2))$, it also needs to satisfy (9.22). Consequently, we obtain a range of the stable coupling strength ξ_1, ξ_2, that is,

$$\begin{cases} \dfrac{(1+\omega)^2}{\alpha_1-\alpha_2}\left(1 - \dfrac{\Delta_2}{\alpha_1}\right) < \xi_1 < \\ \qquad \dfrac{(1+\omega)^2}{\alpha_1-\alpha_2}\left(1 - \dfrac{\Delta_1}{\alpha_1}\right), \\ -\dfrac{1}{\beta_1}(1+\omega)(1-\beta_1) < \xi_2 < \\ \qquad \dfrac{1}{\beta_1}(1+\omega)(1-\beta_1), \end{cases} \tag{9.36}$$

where Δ_1, Δ_2 are defined in Theorem 9.3.

9.4 An Illustrative Example

In systems (9.3) and (9.4), take $\omega = -0.02$, $\alpha_1 = 1.94$, $\alpha_2 = 1.65$, Then, it follows from Lemma 9.1 that systems (9.3) and (9.4) has transversal homoclinic orbits, when $\beta_i (i = 1, 2)$ are sufficiently small (e.g. $\beta_1 = 0.03$, $\beta_2 = 0.05$). In particular, systems (9.3) and (9.4) are surface chaotic, as demonstrated in Figs. 9.1 and 9.2.

Now, in view of condition (9.36), let $\xi_1 = 2.8$, $\xi_2 = 0.1$. Then, $(x_{m,n}, y_{m,n})$ and $(z_{m,n}, w_{m,n})$ satisfy the stable condition, so that they both tend to a plane, as demonstrated in Fig. 9.3. To achieve generalized synchronization of surface chaos, according to formula (9.34), one needs to use

$$\begin{cases} \alpha = \dfrac{\alpha_1}{(1+\omega)^2}\left[(1+\omega)^2 - \xi_1(\alpha_1 - \alpha_2)\right] \approx 0.2998, \\ \beta = \dfrac{\beta_1(1+\omega) + \beta_1\xi_2}{(1+\omega)} \approx 0.0331, \end{cases}$$

which yields $\alpha(1+\omega)^2 = 0.28792792$, $\beta(1+\omega) = 0.032438$. And so system (9.35) reduces to

$$\begin{cases} e_{m+1,n}(1) - 0.02e_{m,n+1}(1) = \\ \qquad 1 + 0.032438e_{m,n}(2) - \\ \qquad 0.28792792\left(e_{m,n}(1)\right)^2 \\ e_{m+1,n}(2) - 0.02e_{m,n+1}(2) = 0.98e_{m,n}(1). \end{cases} \tag{9.37}$$

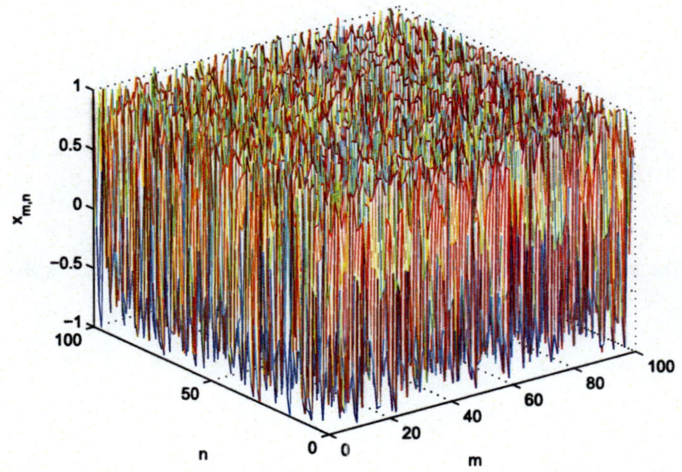

Fig. 9.1 Spatial chaos behavior of system (9.3) with $\omega = -0.02$, $\alpha_1 = 1.94$, $\beta_1 = 0.03$

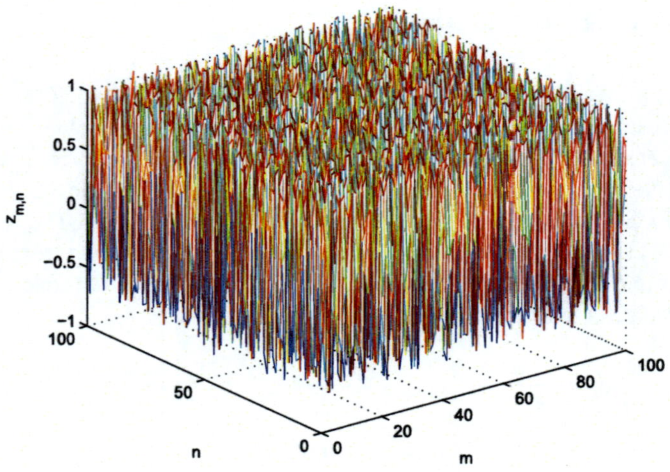

Fig. 9.2 Surface chaos behavior of system (9.3) with $\omega = -0.02$, $\alpha_1 = 1.65$, $\beta_1 = 0.03$

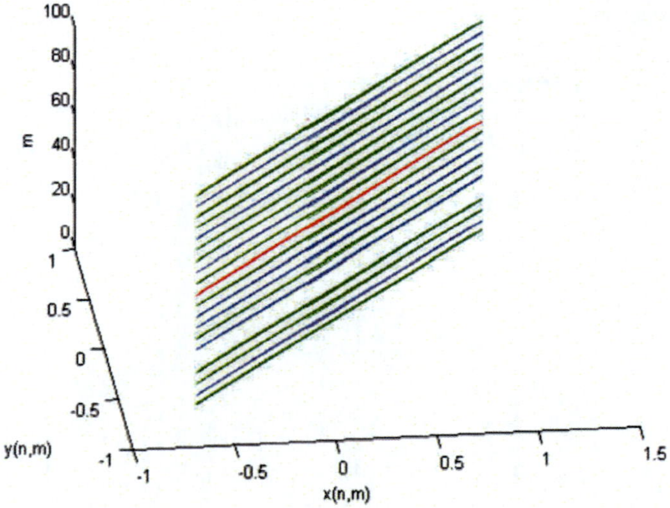

Fig. 9.3 The states $(x_{m,n}, y_{m,n})$ and $(z_{m,n}, w_{m,n})$ satisfy the stable condition with $\omega = -0.02$, $\alpha_1 = 1.94$, $\alpha_2 = 1.65$, $\beta = 0.04$ and $\xi_1 = 2.8$, $\xi_2 = 0.1$

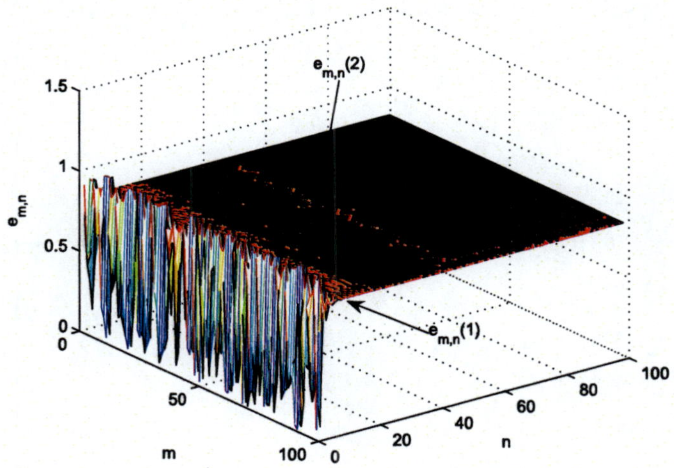

Fig. 9.4 The surface chaotic states $(x_{m,n}, y_{m,n})$ and $(z_{m,n}, w_{m,n})$ achieve generalized synchronization

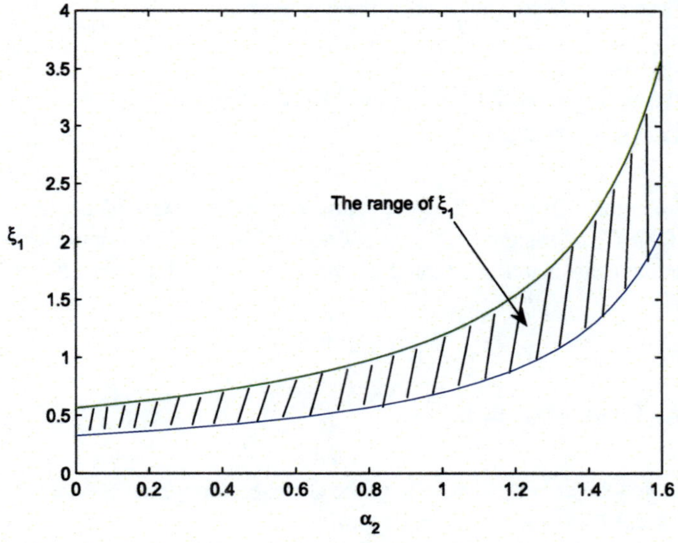

Fig. 9.5 The range of ξ_1 that ensures chaos synchronization

Then, $(x_{m,n}, z_{m,n})$ and $(y_{m,n}, w_{m,n})$ achieve generalized synchronization, as demonstrated by Fig. 9.4. The range of the ξ_1, ξ_2 values that ensure surface chaos synchronization is shown by the shadowed part of Figs. 9.5 and 9.6.

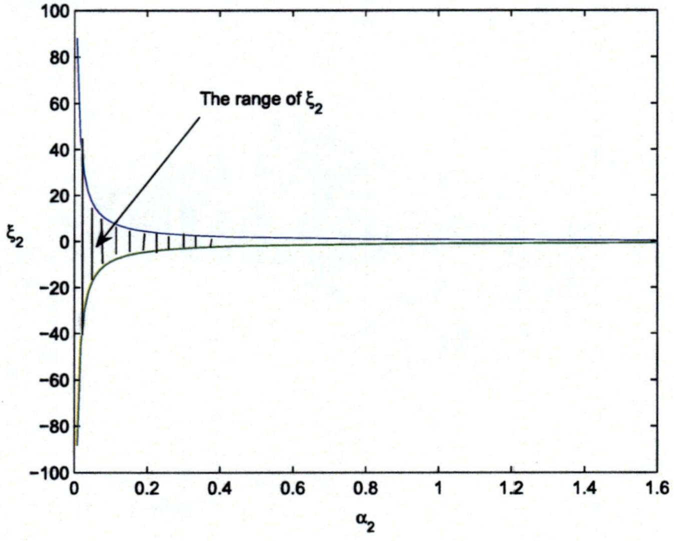

Fig. 9.6 The range of ξ_2 that ensures chaos synchronization

9.5 The Relationship Between Fixed Plane and Synchronization

Using Theorems 9.1 and 9.2, we obtains the stable fixed plane $(e(1), e(2)) = (0.8407, 0.8407)$ of system (9.37), as shown by Fig. 9.7. The controllable result of generalized synchronization attained on the fixed plane of systems (9.37) is shown by Fig. 9.7.

9.6 The Case of $\omega = 0$

Specially, we take $\omega = 0,\ n = 0$, then the corresponding spatially generalized 2-D Hénon systems (9.3) and (9.4) can be written

$$\begin{cases} x_{m+1} = 1 + \beta_1 y_m - \alpha_1 x_m^2, \\ y_{m+1} = x_m, \end{cases} \tag{9.38}$$

$$\begin{cases} z_{m+1} = 1 + \beta_2 w_m - \alpha_2 z_m^2, \\ w_{m+1} = z_m. \end{cases} \tag{9.39}$$

Its coupled systems (9.25) and (9.26) becomes

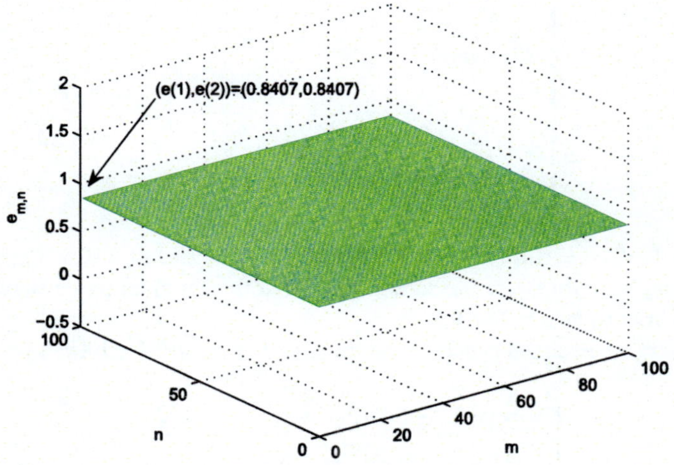

Fig. 9.7 Stable fixed plane of system (9.37)

$$
\begin{cases} x_{m+1} = 1 + \beta_1 y_m - \alpha_1 x_m^2 + \\ \qquad \xi_1 \left(h_1(x_m, z_m)\right)^2 + \xi_2 \left(h_2(y_m, w_m)\right), \\ y_{m+1} = x_m, \end{cases} \tag{9.40}
$$

$$
\begin{cases} z_{m+1} = 1 + \beta_2 w_m - \alpha_2 z_m^2, \\ w_{m+1} = z_m. \end{cases} \tag{9.41}
$$

Similarly, let its corresponding form

$$
\begin{cases} h_1(x_m, z_m) = \alpha_1 x_m - \alpha_2 z_m, \\ h_2(y_m, w_m) = \beta_1 y_m - \beta_2 w_m, \end{cases}
$$

we obtain

$$
\begin{cases} a_1 = a_2 = \dfrac{\xi_1 \alpha_1 - 1}{\xi_1(\alpha_1 - \alpha_2) - 1}, \\ b_1 = b_2 = \dfrac{\xi_1 \alpha_2}{1 - \xi_1(\alpha_1 - \alpha_2)}, \\ \alpha = \alpha_1 \left[1 - \xi_1(\alpha_1 - \alpha_2)\right], \\ \beta = \beta_1 + \beta_1 \xi_2, \end{cases}
$$

and the stable range of the coupling strength ξ_1, ξ_2, as

$$\begin{cases} \dfrac{1}{\alpha_1 - \alpha_2}\left(1 - \dfrac{(\beta-1)^2}{4\alpha_1}\right) < \xi_1 < \\ \qquad\qquad \dfrac{1}{\alpha_1 - \alpha_2}\left(1 + \dfrac{(\beta-1)^2}{4\alpha_1}\right), \\ -\dfrac{1}{\beta_1}(1-\beta_1) < \xi_2 < \dfrac{1}{\beta_1}(1-\beta_1). \end{cases}$$

Let $\alpha_1 = 1.94, \alpha_2 = 1.65, \beta_1 = 0.03, \beta_2 = 0.05, \xi_1 = 2.8,\ \xi_2 = 0.1$. Then (x_m, y_m) and (z_m, w_m) satisfy the stable condition, so that they both tend to a stable straight line, as demonstrated by Fig. 9.8.

Similarly, generalized synchronization of surface chaos is achieved with

$$\begin{cases} \alpha = \alpha_1 \left[1 - \xi_1(\alpha_1 - \alpha_2)\right] = 0.36472, \\ \beta = \beta_1\beta_1\xi_2 = 0.033, \end{cases}$$

so that system (9.35) reduces to

$$\begin{cases} e_{m+1}(1) = 1 + 0.033 e_m(2) - 0.36472\,(e_m(1))^2\,, \\ e_{m+1}(2) = e_m(1). \end{cases}$$

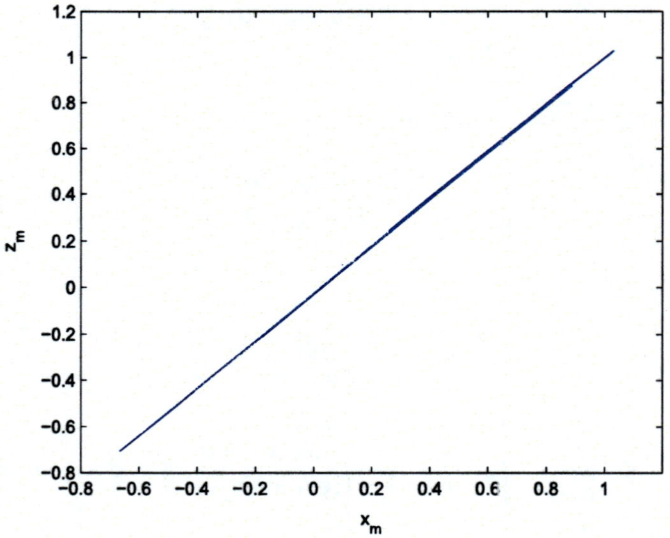

Fig. 9.8 The chaotic states (x_m, y_m) and (z_m, w_m) satisfy the stable condition with $\alpha_1 = 1.94$, $\alpha_2 = 1.65$, $\beta_1 = 0.03$, $\beta_2 = 0.04$ and $\xi_1 = 2.8$, $\xi_2 = 0.1$

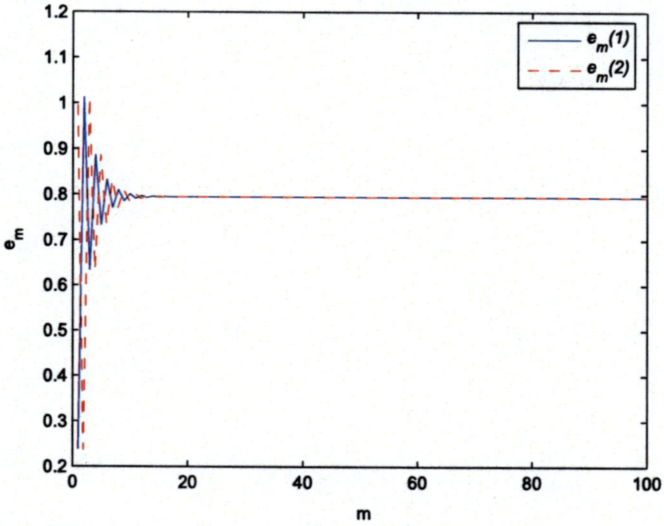

Fig. 9.9 The chaotic states (x_m, y_m) and (z_m, w_m) reach generalized synchronization

Then, the chaotic states (x_m, y_m) and (z_m, w_m) reach generalized synchronization, as demonstrated by Fig. 9.9.

Chapter 10
Generalized Feedback Synchronization of Surface Chaos

10.1 Introduction

In this chapter, the following two spatially generalized Logistic systems which can be deduced from Eq. (3.1) are studied:

$$\begin{cases} x_{m+1,n} + \omega x_{m,n+1} = 1 - \mu_1 \left[(1+\omega)x_{mn}\right]^2, \\ y_{m+1,n} + \omega y_{m,n+1} = 1 - \mu_2 \left[(1+\omega)y_{mn}\right]^2, \end{cases} \tag{10.1}$$

where ω is a real constant and $\mu_i, i = 1, 2$, are positive parameters. Notably, this kind of generalized Logistic systems can be surface chaotic in the (generalized) sense of Li and Yorke.

The following definition, given in [360], will be needed. Consider the coupled system

$$\begin{cases} \dot{x} = f(x, y), \\ \dot{y} = g(x, y), \end{cases} \tag{10.2}$$

where $x, y \in R^s$, and $f, g : \Omega \subset R^s \times R^s \to R^s$ are C^1 maps. Suppose that there exists a functional relation $y = H(x) : R^s \to R^s$, and the manifold $y = H(x)$, $x \in R^s$, is denoted by M_H below. Also, let $x(t; x_0)$ and $y(t; y_0)$ be a solution to system (10.2). If

$$\lim_{t\to\infty} |y(t, y_0) - H(x(t, x_0))| = 0, \tag{10.3}$$

holds for all initial values x_0 and y_0 in the vicinity of M_H, then it is said that $y(t, y_0)$ and $x(t, x_0)$ are in generalized synchronization.

Similarly, the following definition is introduced.

S. T. Liu and L. Zhang, *Surface Chaos and Its Applications*,
https://doi.org/10.1007/978-981-16-8229-2_10

Definition 10.1 Consider two spatially chaotic systems:

$$\begin{cases} x_{m+1,n} + \omega x_{m,n+1} = f(x_{mn}; \mu_1, \omega), \\ y_{m+1,n} + \omega y_{m,n+1} = g(y_{mn}; \mu_2, \omega), \end{cases} \tag{10.4}$$

which may be surface chaotic. They are said to be in *generalized synchronization*, if there is a well-defined functional relation $y_{mn} = \varphi(x_{mn})$ for all $x_{mn} \in R^s$, $y_{mn} \in R^s$, and for all large enough values of n, m.

In this chapter, only the case of $s = 3$ is discussed.

10.2 A Feedback Control Method for Nonlinear Generalized Synchronization

First, let us review the following concepts.

Lemma 10.1 *A stability condition for the fixed point x^* of the 2-D spatially generalized system (10.1) is*

$$\mu_i < \frac{1}{4}\left[\left(\frac{1 + |\omega| + (\omega + 1)^2}{(\omega + 1)^2}\right)^2 - 1\right], \quad i = 1, 2.$$

Lemma 10.2 *A fixed plane of the spatially generalized Logistic system (10.1) is given by*

$$x_{mn} = x^* = \frac{-1 \pm \sqrt{1 + 4\mu}}{2(1 + \omega)\mu}$$

for all $m, n \in N_r$.

Then, we recall the concept of discrete chaos in the sense of Li and Yorke for 2-D discrete systems Definition 3.2.

The following result can be proved easily by Lemma 4.6.

Lemma 10.3 *Assume that $\mu_i \geq 1.55$, $i = 1, 2$. Then, the spatially generalized 2-D Logistic system (10.1) is surface chaotic in the sense of Li and Yorke.*

Since generalized chaos synchronization is studied in this chapter, it will be assumed that in system (10.1), $\mu_1 \neq \mu_2$, and without loss of generality, let

$$\mu_1 > \mu_2 \geq 1.55. \tag{10.5}$$

Now, we apply a nonlinear feedback control term to system (10.1), thereby considering the controlled system

$$\begin{cases} x_{m+1,n} + \omega x_{m,n+1} = 1 - \mu_1 \left[(1+\omega)x_{mn}\right]^2 + \xi \left(h(x_{mn}, y_{mn})\right)^2, \\ y_{m+1,n} + \omega y_{m,n+1} = 1 - \mu_2 \left[(1+\omega)y_{mn}\right]^2, \end{cases} \tag{10.6}$$

where ξ is a real constant (coupling strength), it is clear that the first subsystem of system (10.6) is driven by the second one.

Next, we introduce a new notation,

$$e_{mn} = H(x_{mn}, y_{mn}), \tag{10.7}$$

and hope that two functions $H(x_{mn}, y_{mn})$ and $h(x_{mn}, y_{mn})$ can be precisely found such that $e_{m+1,n}$, $e_{m,n+1}$, and e_{mn} satisfy the following spatially generalized Logistic system:

$$e_{m+1,n} + \omega e_{m,n+1} = 1 - \mu \left[(1+\omega)e_{mn}\right]^2. \tag{10.8}$$

In general, there exists a functional relation between μ and the other parameters: μ_1, μ_2, ξ, ω; that is, $\mu = \mu(\mu_1, \mu_2, \xi, \omega)$. Here the objective is to find and specify this nonlinear function μ, so as to achieve a generalized synchronization of system (10.1).

According to Lemma (10.1), if an appropriate value of ξ can be found such that

$$0 < \mu \leq \frac{1}{4}\left[\left(\frac{1 + |\omega| + (\omega+1)^2}{(\omega+1)^2}\right)^2 - 1\right],$$

then e_{mn} will tend to a stable fixed point e^*. Lemma (10.3) then implies that

$$e^* = \frac{-1 \pm \sqrt{1+4\mu}}{2(1+\omega)\mu}$$

for all $m, n \in N_r$. Hence, $H(x_{mn}, y_{mn})$ = constant, namely, x_{mn} and y_{mn} satisfy a deterministic functional relation, so that the intended generalized synchronization of surface chaos is achieved.

10.3 Generalized Synchronization of Surface Chaos

For system (10.1), each Logistic subsystem can be chaotic if one index is fixed to be constant. However, with two variable indexes, its chaotic behavior is more complex in the 2-D space and needs to be investigated. If, moreover, two such 2-D subsystems are coupled together, then the surface chaotic behavior and synchronization become non-trivial. In this section, stability analysis is applied to guarantee the intended generalized synchronization of surface chaos. To do so, let

$$h(x_{mn}, y_{mn}) = \mu_1 x_{mn} + \mu_2^2 y_{mn}^2 - \mu_2.$$

Then, the coupled systems (10.6) can be rewritten as

$$\begin{cases} x_{m+1,n} + \omega x_{m,n+1} = 1 - \mu_1 \left[(1+\omega)x_{mn}\right]^2 + \xi \left(\mu_1 x_{mn} + \mu_2^2 y_{mn} - \mu_2\right)^2, \\ y_{m+1,n} + \omega y_{m,n+1} = 1 - \mu_2 \left[(1+\omega)y_{mn}\right]^2. \end{cases} \tag{10.9}$$

On the other hand, we choose

$$e_{mn} = H(x_{mn}, y_{mn}) := \alpha x_{mn} + \beta y_{mn}^2 + \gamma, \tag{10.10}$$

where α, β and γ will be to determined later. In addition, to guarantee (10.8) to hold, we consider the following multi-variable transform:

$$e_{st} = \alpha x_{st} + \beta \omega^{t-n} y_{st}^2 + \left(2\beta\omega y_{m+1,n} y_{m,n+1}\right)^{s-m} + \gamma, \tag{10.11}$$

where $s, t, m, n \in N_0 = \{0, 1, 2, \dots\}$, with $s \geq m$ and $t \geq n$. Here, as usual, $0^0 := 1$. This transform (10.11) leads to

$$\begin{cases} e_{m+1,n} = \alpha x_{m+1,n} + \beta y_{m+1,n}^2 + 2\beta\omega y_{m+1,n} y_{m,n+1} + \gamma, \\ e_{m,n+1} = \alpha x_{m,n+1} + \beta\omega y_{m+1,n}^2 + \gamma + 1, \\ e_{mn} = \alpha x_{mn} + \beta y_{mn}^2 + \gamma + 1. \end{cases} \tag{10.12}$$

It follows from the first two equations in (10.12) that

$$\begin{aligned} & e_{m+1,n} + \omega e_{m,n+1} \\ &= \alpha(x_{m+1,n} + \omega x_{m,n+1}) + \beta(y_{m+1,n}^2 + 2\beta\omega y_{m+1,n} y_{m,n+1} + \omega^2 y_{m+1,n}^2) \\ &\quad + \gamma(1+\omega) + \omega \\ &= \alpha(x_{m+1,n} + \omega x_{m,n+1}) + \beta(y_{m+1,n} + \omega y_{m+1,n})^2 + \gamma(1+\omega) + \omega. \end{aligned} \tag{10.13}$$

Substituting the third equation of (10.12) and (10.13) into (10.8) yields

$$\begin{aligned} & \alpha(x_{m+1,n} + \omega x_{m,n+1}) + \beta(y_{m+1,n} + \omega y_{m+1,n})^2 + \gamma(1+\omega) + \omega \\ &= 1 - \mu\left[(1+\omega)\left(\alpha x_{mn} + \beta y_{mn}^2 + (\gamma+1)\right)\right]^2. \end{aligned} \tag{10.14}$$

Then, substituting the coupled systems (10.9) into (10.14) yields

$$\begin{cases} \alpha + \alpha\xi\mu_2^2 + \beta + \gamma(1+\omega) + \omega = 1 - \mu(1+\omega)^2(1+\gamma)^2, \\ \alpha\xi\mu_1^2 - \alpha\mu_1(1+\omega)^2 = -\mu\alpha^2(1+\omega)^2, \\ -2\alpha\xi\mu_1\mu_2 = -2\alpha\mu(1+\gamma)(1+\omega)^2, \\ 2\alpha\xi\mu_1\mu_2^2 = -2\alpha\beta\mu(1+\omega), \\ -\left[2\alpha\xi\mu_2^3 + 2\beta\mu_2(1+\omega)^2\right] = -2\beta\mu(1+\gamma)(1+\omega)^2, \\ \alpha\xi\mu_2^4 + \beta\mu_2^2(1+\omega)^4 = -\beta^2\mu(1+\omega)^2. \end{cases} \tag{10.15}$$

To this end, solving the algebraic equations in (10.15) gives

$$\begin{aligned} \alpha &= \frac{2\mu_1 P}{Q + \xi\mu_1\mu_2^2}, \\ \beta &= \frac{2\xi\mu_1\mu_2^2}{Q + \xi\mu_1\mu_2^2}, \\ \gamma &= \frac{2\mu_2 P}{\xi\mu_1\mu_2^2 - Q} - 1, \\ \mu &= \frac{\left[(1+\omega)^2 - \xi\mu_1\right]\left(Q + \xi\mu_1\mu_2^2\right)}{2(1+\omega)^2 P}, \end{aligned} \tag{10.16}$$

where

$$\begin{aligned} P &= \left[\xi\mu_1 - (1+\omega)^2\right], \\ Q &= \left[\mu_1 - \mu_2(1+\omega)\left(1 - \mu_2(1+\omega)\right)\right]\left[\xi\mu_1 - (1+\omega)^2\right]. \end{aligned}$$

It then follows from the third equation of (10.12), and (10.13) as well as (10.14) that e_{mn} satisfies the following spatially generalized Logistic system:

$$e_{m+1,n} + \omega e_{m,n+1} = 1 - \mu\left[(1+\omega)e_{mn}\right]^2. \tag{10.17}$$

To further ensure the stability of the fixed points for e_{mn} in (10.17), it is required that

$$0 < \mu \le \frac{1}{4}\left[\left(\frac{1 + |\omega| + (\omega+1)^2}{(\omega+1)^2}\right)^2 - 1\right].$$

Then, letting

$$R = \left[\mu_1 - \mu_2(1+\omega)\left(1 - \mu_2(1+\omega)\right)\right],$$

and

$$S = \frac{1}{4}\left[\left(\frac{1 + |\omega| + (\omega+1)^2}{(\omega+1)^2}\right)^2 - 1\right]$$

yields, from (10.16), that

$$\mu = \frac{\left[(1+\omega)^2 - \xi\mu_1\right]\left\{R\left[\xi\mu_1 - (1+\omega)^2\right] + \xi\mu_1\mu_2^2\right\}}{2(1+\omega)^2\left[\xi\mu_1 - (1+\omega)^2\right]}. \tag{10.18}$$

Furthermore, applying (10.18) with $\mu > 0$, we obtain

$$\xi < \frac{R(1+\omega)^2}{\mu_1(R+\mu_2^2)}. \tag{10.19}$$

Now, take

$$\Omega_1 = \left\{\xi \;\middle|\; \xi < \frac{R(1+\omega)^2}{\mu_1(R+\mu_2^2)}\right\},$$

and let $\mu < S$, we have

$$\mu_1^2(R+\mu_2^2)\xi^2 - (1-2S)\mu_1(1+\omega)^2\xi - 2S(1+\omega)^4 \geq 0. \tag{10.20}$$

On the other hand, take

$$\Omega_2 = \left\{\xi \;\middle|\; \mu_1^2(R+\mu_2^2)\xi^2 - (1-2S)\mu_1(1+\omega)^2\xi - 2S(1+\omega)^4 \geq 0\right\}.$$

Thus, the range of the stable coupling strength ξ is given by

$$\Omega = \Omega_1 \cap \Omega_2.$$

10.4 Illustrative Examples

Example 10.1 In systems (10.1), take $\mu_1 = 1.78$, $\mu_2 = 1.75$, and $\omega = -0.05$. Then, it follows from Lemma 10.3 that the two subsystems in (10.1) are both surface chaotic, as demonstrated in Figs. 10.1 and 10.2.

Example 10.2 In view of (10.19) and (10.20), let $\xi = 0.24$. Then, x_{mn} and y_{mn} satisfy the stability condition, so that they both tend to a fixed plane (a curve in this case), as demonstrated in Fig. 10.3.

Example 10.3 To achieve generalized synchronization of chaos, according to formula (10.16), we use

$$\mu = \frac{\left[(1+\omega)^2 - \xi\mu_1\right]\left\{R\left[\xi\mu_1 - (1+\omega)^2\right] + \xi\mu_1\mu_2^2\right\}}{2(1+\omega)^2\left[\xi\mu_1 - (1+\omega)^2\right]} \approx 0.27321\,,$$

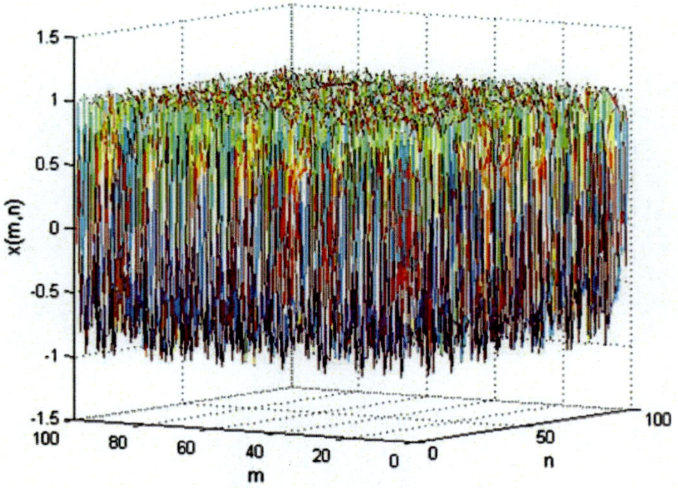

Fig. 10.1 Surface chaotic behavior of system (10.1) with $\omega = -0.05$, $\mu = 1.78$

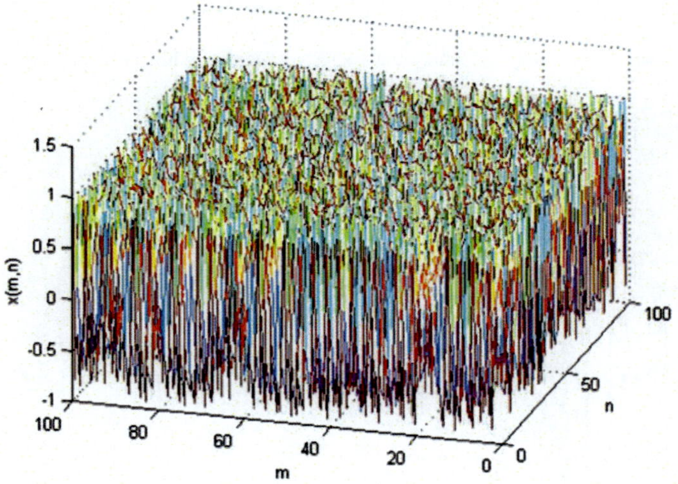

Fig. 10.2 Chaotic behavior of system (10.1) with $\omega = -0.05$, $\mu = 1.75$

which yields $\mu(1+\omega)^2 = 0.246575$, so that system (10.17) becomes

$$e_{m+1,n} - 0.05e_{m,n+1} = 1 - \mu\,[(1+\omega)e_{mn}]^2 = 1 - 0.246575e_{mn}^2. \tag{10.21}$$

As a result, x_{mn} and y_{mn} achieve generalized synchronization for all large enough values of m, n, as demonstrated in Fig. 10.4.

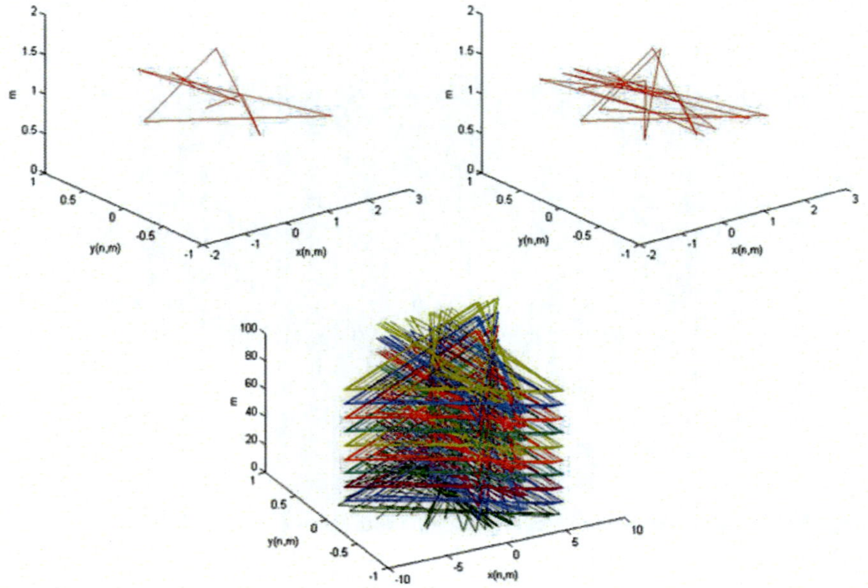

Fig. 10.3 States $x_{m,n}$ and $y_{m,n}$ satisfy the stability condition with $\mu_1 = 1.78$, $\mu_2 = 1.75$ and $\xi = 0.24$

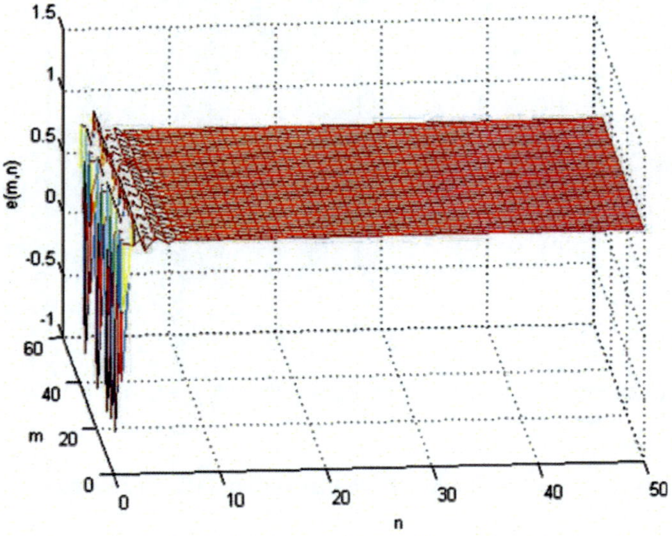

Fig. 10.4 Chaotic states $x_{m,n}$ and $y_{m,n}$ achieve generalized synchronization with respect to $n(a)$ and with respect to $m(b)$

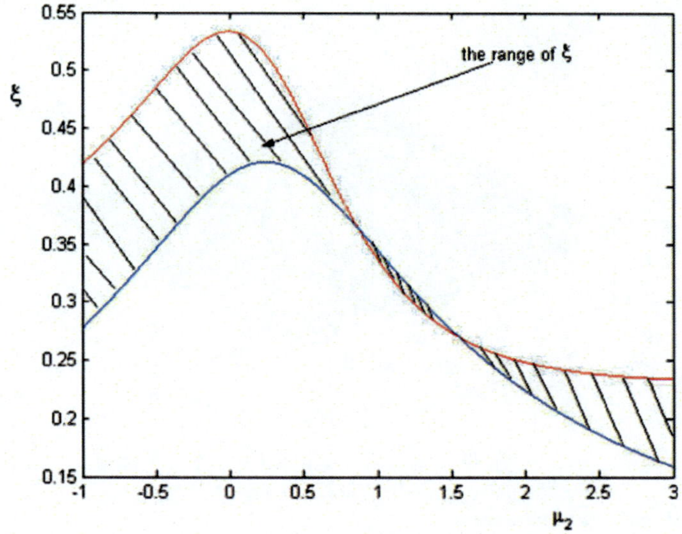

Fig. 10.5 The range of ξ that ensures generalized surface chaos synchronization

Example 10.4 Applying (10.19) and (10.20) again, with $\mu_1 = 1.78$, yields

$$\begin{cases} \xi \le \frac{0.9025(1.78-0.95\mu_2+0.9025\mu_2^2)}{1.78(1.78-0.95\mu_2+1.9025\mu_2^2)}, \\ 3.17(1.78 - 0.95\mu_2 + 1.9025\mu_2^2)\xi^2 + 1.35\xi - 1.5 \ge 0. \end{cases} \tag{10.22}$$

The range of the ξ values that ensure surface chaos synchronization is determined by (10.22), which is shown in the shadowed part of Fig. 10.5.

10.5 Further Study of the Synchronized System

Using Lemma 10.3, we obtain the fixed plane of system (10.23) as

$$x_{mn} = x^* = \frac{-1 + \sqrt{1+4\mu}}{2(1+\omega)\mu} \approx 0.86,$$

for all $m, n \in N_r$.

The controlled result of generalized surface chaos synchronization for all large enough values of m, n, attained on the fixed plane of system (10.21), is shown in Figs. 10.6 and 10.7.

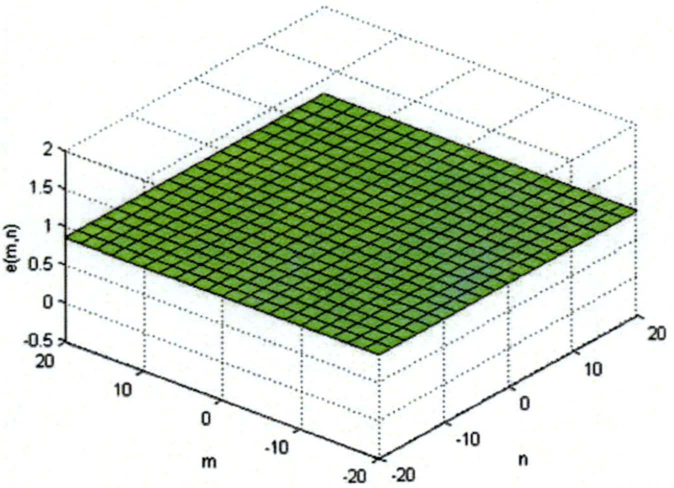

Fig. 10.6 Fixed plane of the system

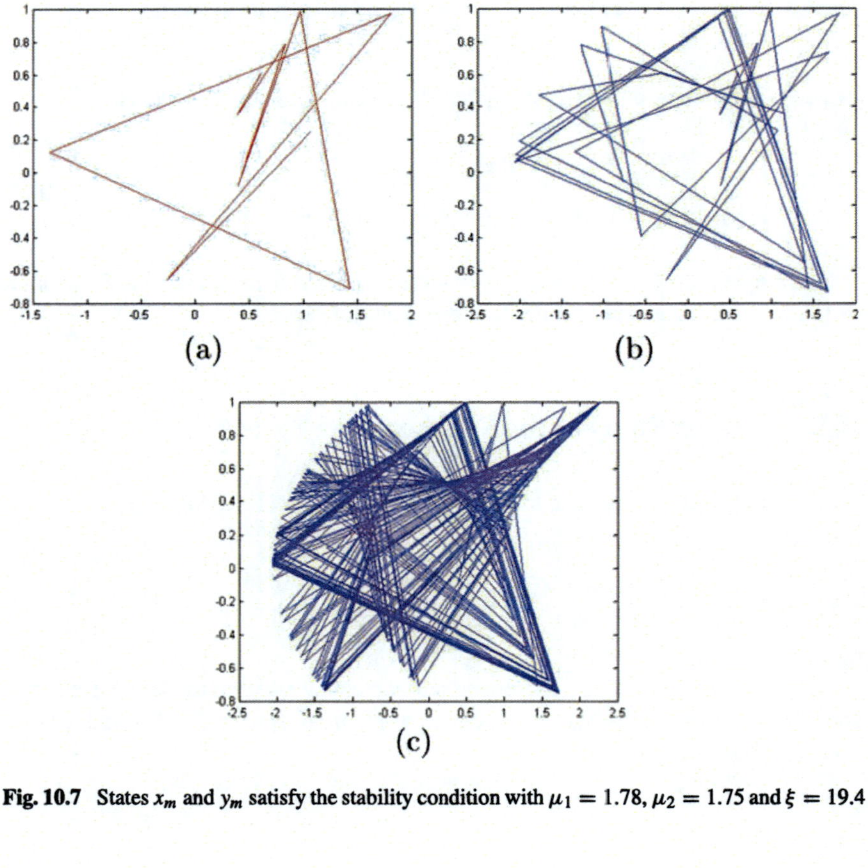

Fig. 10.7 States x_m and y_m satisfy the stability condition with $\mu_1 = 1.78$, $\mu_2 = 1.75$ and $\xi = 19.4$

The case of $\omega = 0$ and $n = n_0$ is studied here in details.

Since $\omega = 0$ and $n = n_0$, we may take $n_0 = 0$ for simplicity. In this section, let $x_{m,n_0} = x_m$ for simplicity. Then, the corresponding spatially generalized Logistic system (10.1) can be rewritten as

$$\begin{cases} x_{m+1} = 1 - \mu_1 x_m^2, \\ y_{m+1} = 1 - \mu_2 y_m^2, \end{cases} \tag{10.23}$$

where $\mu_i, i = 1, 2$, are positive parameters. Their coupled systems (10.9) become

$$\begin{cases} x_{m+1} = 1 - \mu_1 x_m^2 + \xi \left(\mu_1 x_m + \mu_2^2 y_m - \mu_2\right)^2, \\ y_{m+1} = 1 - \mu_2 y_{mn}^2. \end{cases} \tag{10.24}$$

Let $H(x_{mn}, y_{mn})$ be in the form (10.11). Similarly, we can obtain

$$\begin{aligned} \alpha &= \frac{\mu_1 \xi - 1}{\xi(\mu_1 - \mu_2) - 1}, \\ \beta &= \frac{\xi \mu_2^2}{\xi(\mu_1 - \mu_2) - 1}, \\ \gamma &= \frac{-\xi \mu_2}{\xi(\mu_1 - \mu_2) - 1}, \\ \mu &= \mu_1 \left[1 - \xi(\mu_1 - \mu_2)\right]. \end{aligned}$$

We can also obtain the stability range for the coupling strength ξ, as

$$\frac{1}{(\mu_1 - \mu_2)} \left(1 - \frac{3}{4\mu_1}\right) \leq \xi < \frac{1}{\mu_1 - \mu_2}.$$

Now, let $\mu_1 = 1.78$, $\mu_2 = 1.75$, $n = 0$, and $\xi = 19.4$. Then, x_m and y_m satisfy the stability condition, so that both x_m and y_m tend to a fixed plane (a curve in this case), as demonstrated in Fig. 10.7.

Similarly, the controlled generalized surface chaos synchronization is achieved, with

$$\mu = \mu_1 \left[1 - \xi(\mu_1 - \mu_2)\right] = 0.6186,$$

and thus system (10.18) reduces to

$$e_{m+1} = 1 - \mu e_m^2 = 1 - 0.6186 e_m^2.$$

Finally, the chaotic states x_{mn} and y_{mn} reach generalized synchronization, as demonstrated in Fig. 10.8.

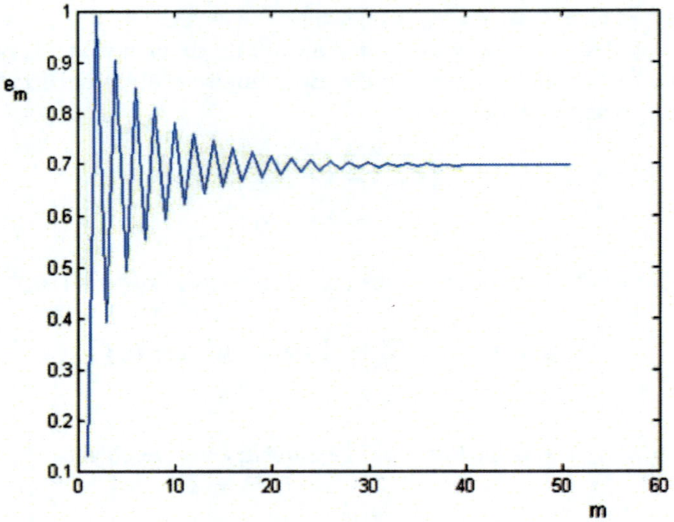

Fig. 10.8 Surface chaotic states x_m and y_m reach generalized synchronization

Part III
Wave Behavior in the Process of Tending to Surface Chaos

Chapter 11
Surface Determining Wave Behavior of a Delay 2-D Discrete System

11.1 General Wave Delay 2-D System

In this chapter, we consider the general delay 2-D discrete system:

$$\mu u_{m+1,n} + \xi u_{m,n+1} - p u_{m,n} + \sum_{i}^{k} q(i) u_{m-\sigma(i),n-\tau(i)} = 0, \tag{11.1}$$

where $p, q(i)$ are real numbers, m, n are nonnegative integers, μ, ξ are any parameters, and $\sigma(i), \tau(i)$ are nonzero integers. System (11.1) can be used to model many problems such as population dynamics with spatial migrations, heat diffusion and random walk in the real world [361–364]. Moreover, system (11.1) can be viewed as a discrete analogue of the delay functional partial differential equation:

$$\frac{\partial u(x, y)}{\partial x} + \frac{\partial u(x, y)}{\partial y} + (2 - p)u(x, y) + qu(x - \sigma, y - \tau) = 0, \tag{11.2}$$

where p, q are the same as in (11.1) and σ, τ are delays such that σ, τ are nonnegative real numbers. In fact, (11.2) is a convection system with a source term and two delays in physics. Therefore, the study of (11.1) may lead to some useful information for analyzing the continuous system (11.2).

In the following part, we focus on the oscillatory behavior of system (11.1). With the conditions that $p > 0$ and $q \geq 0$, the oscillatory behavior of system (11.1) has been investigated by using reduction to absurdity and fixed points theorem [363, 364]. With no conditions attached on constant coefficients p and q, we introduce a new technique to establish some necessary and sufficient conditions for the oscillation of system (11.1) by means of the envelope theory of a family of straight lines. To the best of our knowledge, no research has been done on the analysis of the oscillatory behavior of delay 2-D discrete systems from the perspective of the envelope theory of a family of straight lines [248, 364–369].

S. T. Liu and L. Zhang, *Surface Chaos and Its Applications*,
https://doi.org/10.1007/978-981-16-8229-2_11

11.2 The Case as $\mu = \xi = 1, i = 1, \sigma(1) = k, \tau(i) = l$

11.2.1 Preliminaries

Definition 11.1 ([363]) A solution of system (11.1) as $\mu = \xi = 1, i = 1, \sigma(1) = k,$ $\tau(i) = l$ is a real double sequence $\{u_{m,n}\}$ which is defined for $m \geq -k, n \geq -l$ and satisfies (11.1) for $m \geq 0$ and $n \geq 0$.

Definition 11.2 ([363]) A solution $\{u_{m,n}\}$ of system (11.1) as $\mu = \xi = 1$, $i = 1$, $\sigma(1) = k$, $\tau(i) = l$ is said to be eventually positive (or negative) if $u_{m,n} > 0$ (or $u_{m,n} < 0$) for large numbers m and n. It is said to be oscillatory if it is neither eventually positive nor eventually negative. System (11.1) is called oscillatory if all of its nontrivial solutions are oscillatory.

Lemma 11.1 ([363]) *The following statements are equivalent:*

(1) *Every solution of system (11.1) as $\mu = \xi = 1$, $i = 1$, $\sigma(1) = k$, $\tau(i) = l$ is oscillatory.*

(2) *The characteristic equation of system (11.1) as $\mu = \xi = 1$, $i = 1$, $\sigma(1) = k$, $\tau(i) = l$*

$$\lambda + \mu - p + q\lambda^{-k}\mu^{-l} = 0$$

has no positive roots.

Lemma 11.2 *Suppose that $f(x, y), g(x, y)$ and $h(x, y)$ are differentiable on $(-\infty, +\infty) \times (-\infty, +\infty)$. Let Γ be a two-parameter family of straight lines defined by the equation*

$$f(\lambda, \mu)x + g(\lambda, \mu)y = h(\lambda, \mu)$$

where λ and μ are parameters. Let Σ be the envelope of the family Γ. Then the equation

$$f(\lambda, \mu)a + g(\lambda, \mu)b = h(\lambda, \mu)$$

has no real roots if and only if there is no tangent of Σ passing through the point (a, b) in $xy-plane$.

Proof By the theory of envelopes, we can see that the set Γ is just the set of tangents of Σ.

For necessity. Suppose to the contrary that there exists a straight line $l_{(\lambda^*,\mu^*)}$ in Γ which passes through the point (a, b) in the plane, that is,

$$f(\lambda^*, \mu^*)a + g(\lambda^*, \mu^*)b = h(\lambda^*, \mu^*).$$

This implies that the equation

$$f(\lambda, \mu)a + g(\lambda, \mu)b = h(\lambda, \mu)$$

has a real solution $(\lambda^*, \mu^*) \in (-\infty, +\infty) \times (-\infty, +\infty)$, which is a contradiction.

For sufficiency. Suppose to the contrary that there exists a pair (λ^*, μ^*) of real numbers such that

$$f(\lambda, \mu)a + g(\lambda, \mu)b = h(\lambda, \mu).$$

That is,

$$f(\lambda^*, \mu^*)a + g(\lambda^*, \mu^*)b = h(\lambda^*, \mu^*).$$

This implies that there is a straight line $l_{(\lambda^*,\mu^*)}$ in Γ which passes through the point (a, b) in $xy - plane$, which is also a contradiction. This completes the proof.

Lemma 11.3 ([368]) *Suppose that $f(x)$ is differentiable on $(0, +\infty)$ such that $f(x)$ is not identically zero on $(0, +\infty)$ and $\lim_{x\to+\infty} f(x) > 0$ or $\lim_{x\to 0^+} f(x) > 0$. Then*

$$F(x, y) = y + f(x) = 0$$

has no positive roots on $(0, +\infty) \times (0, +\infty)$ if and only if $f(x) = 0$ has no positive roots on $(0, +\infty)$.

Lemma 11.4 ([370]) *Let*

$$f(x, p, q) = x^{\sigma+1} - px^{\sigma} + q,$$

where σ is a positive integer, p and q are real parameters. Then the equation $f(x, p, q) = 0$ has no positive roots if and only if $p \le 0$ and $q \ge 0$ or $p > 0$ and

$$q > \frac{\sigma^{\sigma}}{(\sigma + 1)^{\sigma+1}} p^{\sigma+1}.$$

Lemma 11.5 ([361]) *Every solution of system*

$$u_{m+1,n} + u_{m,n+1} - pu_{m,n} = 0$$

oscillates if and only if $p \le 0$.

11.2.2 Surface Wave Criteria

In this section, we establish criteria to determine the surface oscillation behavior of (11.1) as $\mu = \xi = 1, i = 1, \sigma(1) = k, \tau(i) = l$. For convenience of discussion, it is necessary to classify (11.1) into four mutually exclusive cases: (i) $k > 0$ and $l > 0$, (ii) $k = 0$ and $l > 0$, (iii) $k > 0$ and $l = 0$, (iv) $k = 0$ and $l = 0$.

Theorem 11.1 *Assume that $k > 0$ and $l > 0$. Then every solution of (11.1) as $\mu = \xi = 1$ $i = 1$, $\sigma(1) = k$, $\tau(i) = l$ surface oscillates if and only if $p \leq 0$ and $q \geq 0$ or $p > 0$ and*

$$q > \frac{k^k l^l}{(k+l+1)^{k+l+1}} p^{k+l+1}.$$

Proof When $k > 0$ and $l > 0$, the characteristic equation of (11.1) as $\mu = \xi = 1$ $i = 1, \sigma(1) = k, \tau(i) = l$ is

$$\phi(p, q, \lambda, \mu) = \lambda + \mu - p + q\lambda^{-k}\mu^{-l} = 0. \tag{11.3}$$

Let

$$F(p, q, \lambda, \mu) = \lambda^k \mu^l \phi(p, q, \lambda, \mu) = \lambda^{k+1}\mu^l + \lambda^k \mu^{l+1} - p\lambda^k \mu^l + q = 0. \tag{11.4}$$

It is clear that (11.3) has no positive roots if and only if (11.4) has no positive roots. Since we focus on oscillatory solutions of (11.1) as $\mu = \xi = 1$ $i = 1, \sigma(1) = k$, $\tau(i) = l$, by Lemma 11.1, we only need to consider (11.4) for $\lambda > 0$ and $\mu > 0$. In fact, $F(x, y, \lambda, \mu) = 0$ may be considered as an equation describing a two-parameter family of straight lines in $xy - plane$, where x and y are the coordinates of points of the straight lines in the plane and λ, μ are two parameters. Since the pair (p, q) of real numbers can be viewed as a point in $xy - plane$, we try to find the exact regions containing points (p, q) in $xy - plane$ such that (11.4) has no positive roots.

The points of the envelope of the two-parameter family of straight lines defined by (11.4) satisfy the system of equations:

$$\begin{cases} F(x, y, \lambda, \mu) = 0, \\ F_\lambda(x, y, \lambda, \mu) = (k+1)\lambda^k \mu^l + k\lambda^{k-1}\mu^{l+1} - xk\lambda^{k-1}\mu^l = 0, \\ F_\mu(x, y, \lambda, \mu) = (l+1)\lambda^k \mu^l + l\lambda^{k+1}\mu^{l-1} - xl\lambda^k \mu^{l-1} = 0, \end{cases} \tag{11.5}$$

where $\lambda > 0$ and $\mu > 0$. By the theory of envelopes, eliminating the two parameters λ and μ from (11.5) gives the equation of the envelope:

$$y(x) = \frac{k^k l^l}{(k+l+1)^{k+l+1}} x^{k+l+1}, \tag{11.6}$$

where $x > 0$. From (11.6), we have

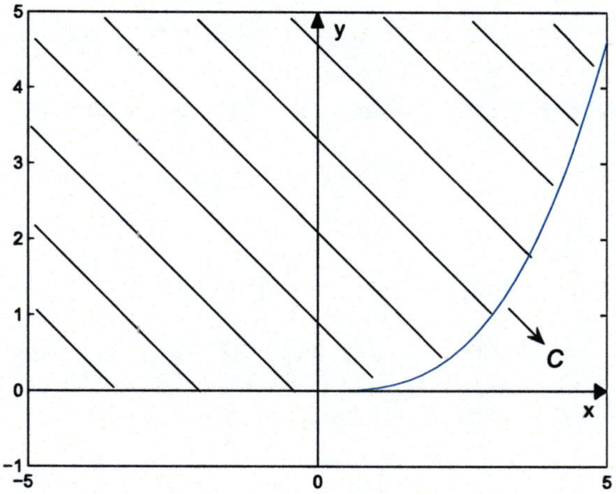

Fig. 11.1 Envelope curve for $k = 1$ and $l = 1$

$$y'(x) = \frac{k^k l^l}{(k+l+1)^{k+l}} x^{k+l},$$
$$y''(x) = \frac{(k+l)k^k l^l}{(k+l+1)^{k+l}} x^{k+l-1},$$

where $x > 0$. Since $y'(x) > 0$ and $y''(x) > 0$ for $x \in (0, +\infty)$, $y(x)$ is a strictly increasing and strictly convex function on $(0, +\infty)$. Furthermore, we have $\lim_{x\to+\infty} y(x) = \lim_{x\to+\infty} k^k l^l x^{k+l+1}/(k+l+1)^{k+l+1} = +\infty$, $\lim_{x\to0^+} y(x) = \lim_{x\to0^+} k^k l^l x^{k+l+1}/(k+l+1)^{k+l+1} = 0$ and $y(x) > 0$ for $x \in (0, +\infty)$. The envelope described by (11.6) is a strictly increasing and strictly convex curve C over $(0, +\infty)$ as shown in Fig. 11.1. In view of Fig. 11.1 and the obtained information of $y(x)$, it is easily seen that when the point (p, q) is in the closed upper left plane, that is, $p \le 0$ and $q \ge 0$, there does not be any tangent of the envelope C which also passes through this point, and when the point (p, q) is vertically above the envelope C, that is, $p > 0$ and $q > k^k l^l p^{k+l+1}/(k+l+1)^{k+l+1}$, there cannot exist any tangent of the envelope C which also passes through this point, otherwise, there always exists a tangent of the envelope C which passes through this point. It follows from Lemma 11.2 that Eq. (11.4) does not have any positive roots if and only if $p \le 0$ and $q \ge 0$ or $p > 0$ and

$$q > \frac{k^k l^l}{(k+l+1)^{k+l+1}} p^{k+l+1}.$$

Note that conditions for the existence of positive solutions of the characteristic Eq. (11.3) are the same as for (11.4). By Lemma 11.1, the statement of this theorem holds. The proof is thus completed.

Theorem 11.2 *Assume that* $k = 0$ *and* $l > 0$. *Then every solution of system (11.1) surface oscillates if and only if* $p \le 0$ *and* $q \ge 0$ *or* $p > 0$ *and*

$$q > \frac{l^l}{(l+1)^{l+1}} p^{l+1}.$$

Proof When $k = 0$ and $l > 0$, then the characteristic equation of system (11.1) is

$$\phi(p, q, \lambda, \mu) = \lambda + \mu - p + q\mu^{-l} = 0. \tag{11.7}$$

Let

$$f(p, q, \mu) = \mu - p + q\mu^{-l}.$$

Since $\lim_{\mu \to +\infty} f(p, q, \mu) > 0$ and $f(p, q, \mu)$ is differentiable with respect to μ on $(0, +\infty)$, Lemma 11.3 implies that (11.7) has no positive roots if and only if $f(p, q, \mu) = 0$ has no positive roots. Let

$$F(p, q, \mu) = \mu^l f(p, q, \mu) = \mu^{l+1} - p\mu^l + q = 0. \tag{11.8}$$

It is clear that (11.8) has no positive roots if and only if $f(p, q, \mu) = 0$ has no positive roots. It follows from Lemma 11.4 that (11.8) does not have any positive roots if and only if $p \le 0$ and $q \ge 0$ or $p > 0$ and

$$q > \frac{l^l}{(l+1)^{l+1}} p^{l+1}.$$

Note that (11.7) has no positive roots if and only if (11.8) has no positive roots. By Lemma 11.1, the statement of this theorem holds. The proof is thus completed.

Remark 11.1 If we set $k = 0$ in Theorem 11.1 and define $0^0 = 1$, then Theorem 11.1 implies Theorem 11.2. This means that Theorem 11.2 is a special case of Theorem 11.1.

Theorem 11.3 *Assume that $k > 0$ and $l = 0$. Then every solution of system (11.1) as $\mu = \xi = 1, i = 1, \sigma(1) = k, \tau(i) = l$ as $i = 1, \sigma(1) = k, \tau(i) = l$ surface oscillates if and only if $p \le 0$ and $q \ge 0$ or $p > 0$ and*

$$q > \frac{k^k}{(k+1)^{k+1}} p^{k+1}.$$

Proof When $k > 0$ and $l = 0$, the characteristic equation of system (11.1) as $\mu = \xi = 1, i = 1, \sigma(1) = k, \tau(i) = l$ is

$$\phi(p, q, \lambda, \mu) = \lambda + \mu - p + q\lambda^{-k} = 0. \tag{11.9}$$

Let

$$f(p, q, \lambda) = \lambda - p + q\lambda^{-k}.$$

Since $\lim_{\lambda \to +\infty} f(p, q, \lambda) > 0$ and $f(p, q, \lambda)$ is differentiable with respect to λ on $(0, +\infty)$, Lemma 11.3 implies that (11.9) has no positive roots if and only if $f(p, q, \lambda) = 0$ has no positive roots. Let

$$F(p, q, \lambda) = \lambda^k f(p, q, \lambda) = \lambda^{k+1} - p\lambda^k + q = 0. \tag{11.10}$$

It is clear that (11.10) has no positive roots if and only if $f(p, q, \lambda) = 0$ has no positive roots. It follows from Lemma 11.4 that (11.10) does not have any positive roots if and only if $p \le 0$ and $q \ge 0$ or $p > 0$ and

$$q > \frac{k^k}{(k+1)^{k+1}} p^{k+1}.$$

Note that (11.9) has no positive roots if and only if (11.10) has no positive roots. By Lemma 11.1, the statement of this theorem holds. The proof is thus completed.

Remark 11.2 If we set $l = 0$ in Theorem 11.1 and define $0^0 = 1$, then Theorem 11.1 implies Theorem 11.3. This means that Theorem 11.3 is a special case of Theorem 11.1.

Theorem 11.4 *Assume that $k = 0$ and $l = 0$. Then every solution of system (11.1) as $i = 1$, $\sigma(1) = k$, $\tau(i) = l$ surface oscillates if and only if $q \ge p$.*

Proof When $k = 0$ and $l = 0$, one can rewrite (11.1) as $\mu = \xi = 1, i = 1, \sigma(1) = k$, $\tau(i) = l$ as

$$u_{m+1,n} + u_{m,n+1} + (q - p)u_{m,n} = 0.$$

From Lemma 11.5, we can see that every solution of system (11.1) oscillates if and only if $q \ge p$. This completes the proof.

11.2.3 Illustrative Examples

In this part, we illustrate the obtained results above by some examples.

Example 11.1 Consider the delay 2-D discrete system

$$u_{m+1,n} + u_{m,n+1} - 0.06u_{m,n} + 0.06u_{m-1,n-1} = 0. \tag{11.11}$$

From (11.11), we have $k = 1, l = 1$, $p = 0.06 > 0, q = 0.06 > 0$ and

$$q = 0.06 > \frac{1}{27}(0.06)^3 = \frac{k^k l^l}{(k+l+1)^{k+l+1}} p^{k+l+1}.$$

Since the conditions of Theorem 11.1 are satisfied, every solution of (11.11) is surface oscillatory. The surface oscillatory behavior of (11.11) is shown in Fig. 11.2.

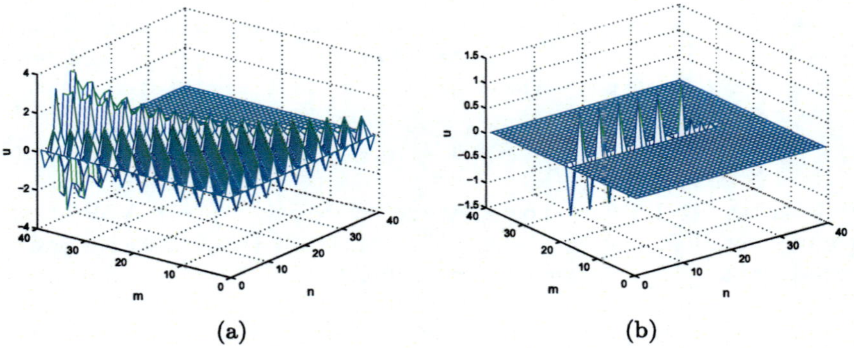

Fig. 11.2 **a** Surface oscillatory behavior of (11.11). **b** Surface oscillatory behavior of (11.11) with m = 20

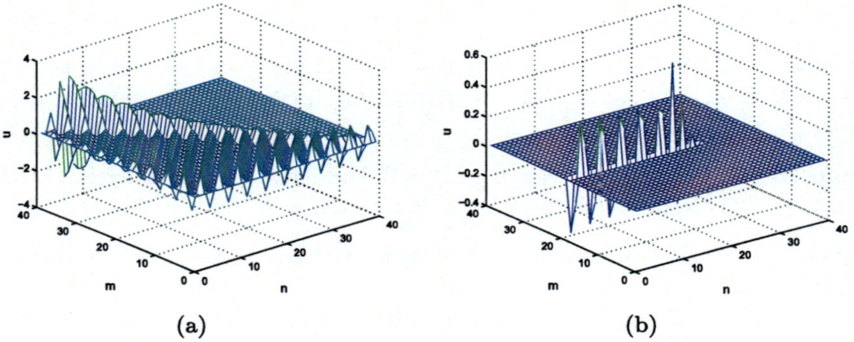

Fig. 11.3 **a** Oscillatory behavior of (11.12). **b** Oscillatory behavior of (11.12) with m = 20

Example 11.2 Consider the delay 2-D discrete system

$$u_{m+1,n} + u_{m,n+1} - 0.07u_{m,n} + 0.05u_{m,n-1} = 0. \tag{11.12}$$

From (11.12), we have ($k = 0, l = 1, p = 0.07 > 0, q = 0.05 > 0$) and

$$q = 0.05 > \frac{1}{4}(0.07)^2 = \frac{l^l}{(l+1)^{l+1}} p^{l+1}.$$

Since the conditions of Theorem 11.2 are satisfied, every solution of (11.12) is surface oscillatory. The surface oscillatory behavior of (11.12) is shown in Fig. 11.3.

Example 11.3 Consider the delay 2-D discrete system

$$u_{m+1,n} + u_{m,n+1} - 0.03u_{m,n} + 0.1u_{m-2,n} = 0. \tag{11.13}$$

From (11.13), we have ($k = 2, l = 0, p = 0.03 > 0, q = 0.1 > 0$) and

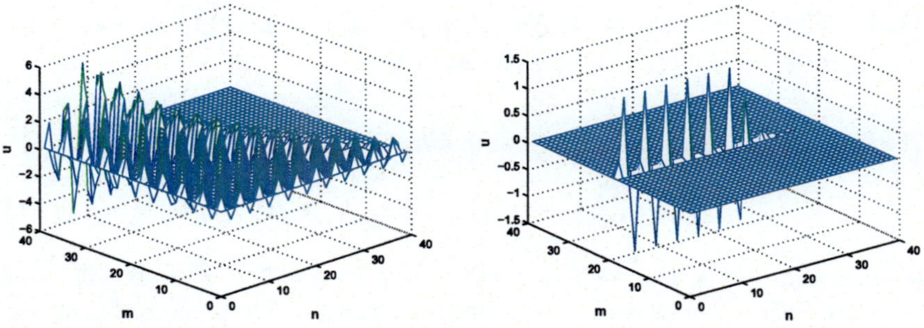

(a) (b)

Fig. 11.4 **a** Surface oscillatory behavior of (11.13). **b** Surface oscillatory behavior of (11.13) with m = 20

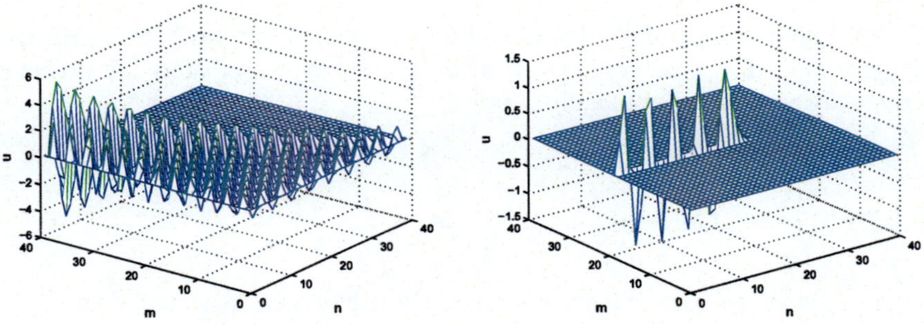

(a) (b)

Fig. 11.5 **a** Surface oscillatory behavior of (11.14). **b** Surface oscillatory behavior of (11.14) with m = 20

$$q = 0.1 > \frac{4}{27}(0.03)^3 = \frac{k^k}{(k+1)^{k+1}} p^{k+1}.$$

Since the conditions of Theorem 11.3 are satisfied, every solution of (11.13) is surface oscillatory. The surface oscillatory behavior of (11.13) is shown in Fig. 11.4.

Example 11.4 Consider the delay 2-D discrete system

$$u_{m+1,n} + u_{m,n+1} + 0.09u_{m,n} = 0. \tag{11.14}$$

From (11.14), we have $k = 0, l = 0$ and $q - p = 0.09 > 0$. Since the conditions of Theorem 11.4 are satisfied, every solution of (11.14) is surface oscillatory. The surface oscillatory behavior of (11.14) is shown in Fig. 11.5.

11.3 The Case as $\mu = a, \xi = b, p = -c\ i = 2, \sigma(1) = -2, \tau(1) = 0, \sigma(2) = 0, \tau(2) = -2$

The system (11.1) becomes the following form

$$u_{n+2,m} + u_{n,m+2} + au_{n+1,m} + bu_{n,m+1} + cu_{n,m} = 0, \tag{11.15}$$

where $\mu = a,\ \xi = b,\ p = -c\ i = 2,\ \sigma(1) = -2,\ \tau(1) = 0,\ \sigma(2) = 0,\ \tau(2) = -2$, a, b, c are real numbers and $m,\ n$ are nonnegative integers. The purpose of this part is to develop a new method for the analysis of the surface oscillation of (11.15) based on the theory of envelopes.

By a solution of (11.15), we mean a nontrivial double sequence $\{A_{m,n}\}$ of real numbers which is defined for $m \geq 0$ and $n \geq 0$, and satisfies (11.15) for $m \geq 0$ and $n \geq 0$.

A solution $\{A_{m,n}\}$ of (11.15) is said to be eventually positive (or negative) if $A_{m,n} \geq 0$ (or $A_{m,n} \leq 0$) for large m and n. It is said to be oscillatory if it is neither eventually positive nor eventually negative.

11.3.1 Some Lemmas

In this section, we give some preliminaries that will be used in next section.

From results in Ref. [363], we can easily obtain the following lemma.

Lemma 11.6 *The following statements are equivalent:*

(a) Every solution of (11.15) is oscillatory.

(b) The characteristic equation of (11.15)

$$\lambda^2 + \mu^2 + a\lambda + b\mu + c = 0$$

has no positive roots.

Lemma 11.7 *Suppose that $f(x, y)$, $g(x, y)$, $h(x, y)$ and $v(x, y)$ are differentiable on $(-\infty, +\infty) \times (-\infty, +\infty)$. Let Γ be a two-parameter family of planes defined by the equation*

$$f(\lambda, \mu)x + g(\lambda, \mu)y + h(\lambda, \mu)z = v(\lambda, \mu),$$

where λ and μ are parameters. Let Σ be the envelope of the family Γ. Then the equation

$$f(\lambda, \mu)a + g(\lambda, \mu)b + h(\lambda, \mu)c = v(\lambda, \mu)$$

has no real roots if and only if there is no tangent plane of Σ passing through the point (a, b, c) in Euclidean space $\mathbb{R}^3$.

Lemma 11.8 *For the linear homogeneous difference equation*

$$u_{n+k} + a_1 u_{n+k-1} + \cdots + a_k u_n = 0, \tag{11.16}$$

where n is nonnegative integer, k is positive integer and $a_1, a_2, \cdots, a_k$ *are real numbers, the following statements are equivalent:*
(a) Every solution of 11.16 is surface oscillatory.
(b) The characteristic equation of 11.16

$$\lambda^k + a_1\lambda^{k-1} + \cdots + a_k = 0$$

has no positive roots.

11.3.2 Main Results

In this section, we establish the necessary and sufficient condition for oscillations of all solutions of (11.15) by the envelope theory.

Theorem 11.5 *Every solution of 11.16 surface oscillates if and only if* $a \geq 0$, $b \geq 0$ *and* $c \geq 0$ *or* $(a < 0)$, $b < 0$ *and*

$$c > \frac{a^2 + b^2}{4}.$$

Proof The characteristic equation of (11.15) is

$$\lambda^2 + \mu^2 + a\lambda + b\mu + c = 0.$$

Let

$$f(a, b, c, \lambda, \mu) = \lambda^2 + \mu^2 + a\lambda + b\mu + c = 0. \tag{11.17}$$

Since we mainly discuss the oscillations of (11.15), by Lemma 11.6, attention will be restricted to the case where $(\lambda > 0)$ and $(\mu > 0)$. To this end, we will think of (a, b, c) as a point in Euclidean space R^3 and try to find the the exact regions containing points (a, b, c) in Euclidean space R^3 such that (11.17) has no positive roots. In fact, $f(x, y, z, \lambda, \mu) = 0$ can be regarded as an equation describing a two-parameter family of planes in Euclidean space R^3, where x, y and z are the coordinates of points of the planes in Euclidean space R^3 and λ, μ are two parameters.

By the envelope theory, the points of the envelope of the two-parameter family of planes described by (11.17) satisfy the system of equations:

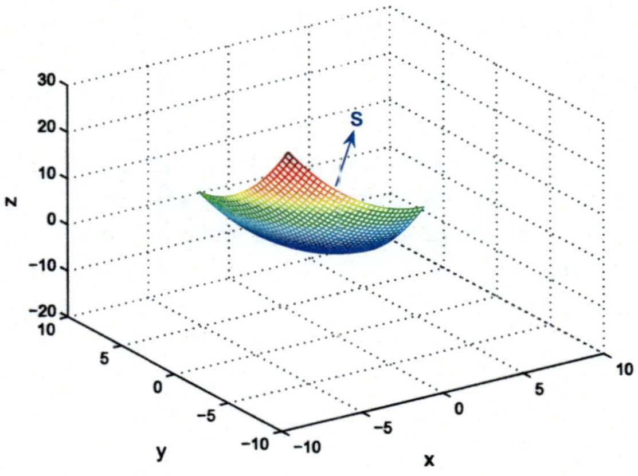

Fig. 11.6 Envelope surface for $(z(x, y) = (x^2 + y^2)/4(x < 0, y < 0))$.

$$\begin{cases} f(x, y, z, \lambda, \mu) = \lambda x + \mu y + z + \lambda^2 + \mu^2 = 0, \\ f_\lambda(x, y, z, \lambda, \mu) = x + 2\lambda = 0, \\ f_\mu(x, y, z, \lambda, \mu) = y + 2\mu = 0, \end{cases} \tag{11.18}$$

where $\lambda > 0$ and $\mu > 0$. Eliminating the two parameters λ and μ from (11.18) leads to the equation of the envelope:

$$z(x, y) = \frac{x^2 + y^2}{4}, \tag{11.19}$$

where $x < 0$ and $y < 0$. From (11.19), we have $z(x, y) > 0$ for $x < 0$ and $y < 0$, $\partial^2 z/\partial x^2 = 1/2 > 0$, $\partial^2 z/\partial y^2 = 1/2 > 0$ and $\partial^2 z/\partial x^2 \cdot \partial^2 z/\partial y^2 - (\partial^2 z/\partial x \partial y)^2 = 1/4 > 0$. Hence, $z(x, y)$ is a positive and strictly convex function on $(-\infty, 0) \times (-\infty, 0)$. Furthermore, the envelope defined by (11.19) is a convex surface S over $(-\infty, 0) \times (-\infty, 0)$ as depicted in Fig. 11.6. In view of Fig. 11.6 and the obtained information of $z(x, y)$, it is easily seen that when the point (a, b, c) is in the first closed octant, that is, $a \geq 0, b \geq 0$ and $c \geq 0$, there cannot be any tangent plane of the envelope S which also passes through this point, and when the point (a, b, c) is vertically above the envelope S, that is, $a < 0, b < 0$ and $c > (a^2 + b^2)/4$, there cannot exist any tangent plane of the envelope S which also passes through this point, while the point (a, b, c) is situated elsewhere, there always exists a tangent plane of the envelope S which also passes through this point. By Lemma 11.7, the characteristic Eq. (11.17) does not have any positive roots if and only if $a \geq 0, b \geq 0$ and $c \geq 0$ or $a < 0, b < 0$ and

$$c > \frac{a^2 + b^2}{4}.$$

Lemma 11.6 implies that the statement of this theorem is true. The proof is thus completed.

If we let $\mu = 0$ in (11.17), then (11.17) reduces to the form

$$\lambda^2 + a\lambda + c = 0. \tag{11.20}$$

The corresponding ordinary difference equation of (11.20) takes the form

$$u_{n+2} + au_{n+1} + cu_n = 0. \tag{11.21}$$

On the basis of Theorem 11.5, we can obtain the following results.

Corollary 11.1 *Equation (11.20) has no positive roots if and only if $c \geq 0$ and $a \geq 0$ or $a < 0$ and $c > a^2/4$.*

Remark 11.3 In fact, Corollary 11.1 is just Theorem 11.5 of [370].

Corollary 11.2 *Every solution of (11.21) surface oscillates if and only if $c \geq 0$ and $a \geq 0$ or $a < 0$ and $c > a^2/4$.*

Proof Combining the results of Corollary 11.1 and Lemma 11.8 yields the proof.

Remark 11.4 Corollary 11.2 shows that Theorem 11.5 is a natural generalization for the corresponding result of the ordinary difference Eq. (11.21) and the ordinary difference Eq. (11.21) is a special case of our study.

11.3.3 Illustrative Examples

Example 11.5 Consider the partial difference equation

$$u_{n+2,m} + u_{n,m+2} - 0.3u_{n+1,m} - 0.1u_{n,m+1} + 0.28u_{n,m} = 0, \tag{11.22}$$

where m, n are nonnegative integers.

From (11.22), we have $a = -0.3 < 0$, $b = -0.1 < 0$ and $(a^2 + b^2)/4 = ((-0.3)^2 + (-0.1)^2)/4 = 0.025 < 0.28 = c$. By Theorem 11.5, every solution of (11.22) is surface oscillatory. The surface oscillatory behavior of (11.22) is demonstrated by Fig. 11.7.

Example 11.6 Consider the partial difference equation

$$u_{n+2,m} + u_{n,m+2} + 0.02u_{n+1,m} + 0.12u_{n,m+1} + 0.12u_{n,m} = 0, \tag{11.23}$$

where m, n are nonnegative integers.

From (11.23), we have $a = 0.02 > 0$, $b = 0.12 > 0$ and $c = 0.12 > 0$. It follows from Theorem 11.5 that every solution of (11.23) is surface oscillatory. The surface oscillatory behavior of (11.23) is demonstrated by Fig. 11.8.

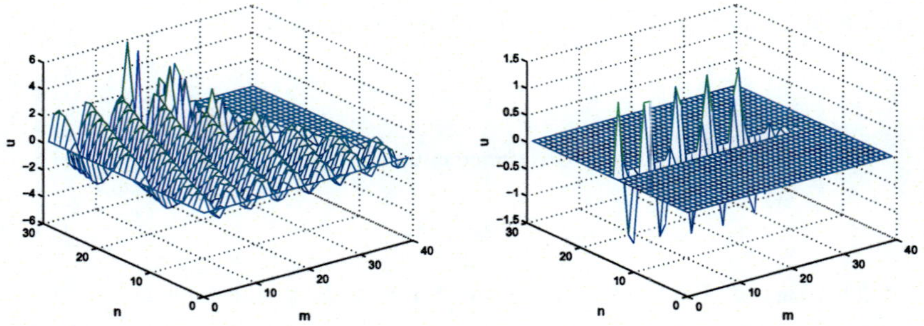

(a) Surface oscillatory behavior of System (11.22).

(b) Surface oscillatory behavior of System (11.22) with n=15.

Fig. 11.7 Surface oscillatory behavior of system (11.22)

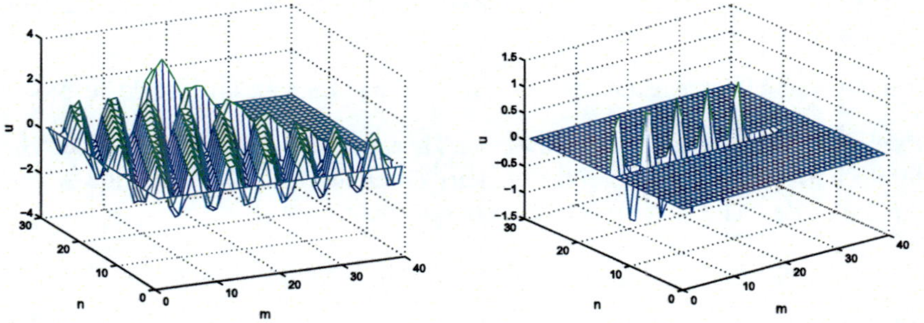

(a) Surface oscillatory behavior of System (11.23).

(b) Surface oscillatory behavior of System (11.23) with n=15.

Fig. 11.8 Surface oscillatory behavior of system (11.23)

11.4 The Case as $\mu = 1, \xi = \hat{p}, p = -q, i = 1, q(1) = r$

Taken the the parameters as $\mu = 1, \xi = \hat{p}, p = -q, i = 1, q(1) = r$ in system (11.1), it follows that

$$u_{m+1,n} + pu_{m,n+1} + qu_{m,n} + ru_{m-\sigma,n-\tau} = 0, \tag{11.24}$$

where $\hat{p}, q, r$ are real numbers with $\hat{p}^2 + q^2 + r^2 \neq 0$ and m, n, σ, τ are nonnegative integers. The corresponding continuous form of system (11.24) is

$$\frac{\partial u(x,y)}{\partial x} + \hat{p}\frac{\partial u(x,y)}{\partial y} + (1 + \hat{p} + q)u(x,y) + ru(x - \xi, y - \eta) = 0, \tag{11.25}$$

where $\hat{p}, q, r$ are the same as in system (11.24) and ξ, η are delays such that ξ, η are nonnegative real numbers. In fact, system (11.25) is a convection system with a source term and two delays. It is well known that it is a formidable task to derive exact analytical solutions of system (11.25) in practice. Therefore, the study of system (11.24) may provide some useful information for analyzing the continuous system (11.25).

In this paper, we focus on surface oscillation behavior of system (11.24). The main objective of this paper is to employ a new method, based on the envelope theory of the family of planes, to derive necessary and sufficient conditions for the delay 2-D discrete system (11.24) to be surface oscillatory.

Definition 11.3 A solution of (11.24) is a real double sequence $\{u_{m,n}\}$ which is defined for $m \geq -\sigma, n \geq -\tau$ and satisfies (11.24) for $m \geq 0$ and $n \geq 0$.

Definition 11.4 A solution $\{u_{m,n}\}$ of (11.24) is said to be eventually positive (or negative) if $u_{m,n} > 0$ (or $u_{m,n} < 0$) for large numbers m and n. It is said to be oscillatory if it is neither eventually positive nor eventually negative. System (11.24) is called oscillatory if all of its nontrivial solutions are oscillatory.

It should be pointed out that by taking $u_{m,n} = u^*$ for $m \in \{-\sigma, -\sigma + 1, \ldots\}$ and $n \in \{-\tau, -\tau + 1, \ldots\}$ in (11.24), one has

$$(p + q + r + 1)u^* = 0. \tag{11.26}$$

Equation (11.26) is called the fixed plane equation of system (11.24). Therefore, the fixed plane equation of system (11.24) is $u^* = 0$. Hence, the surface oscillation behavior of system (11.24) is that the solutions of system (11.24) surface oscillate about the fixed plane with arbitrarily large zeros.

11.4.1 Preliminaries

In this section, some lemmas are presented for the proofs of the main results in the next section.

Lemma 11.9 ([366]) *The following statements are equivalent:*

(1) *Every solution of system (11.24) is surface oscillatory.*

(2) *The characteristic equation of system (11.24)*

$$\lambda + \hat{p}\mu + q + r\lambda^{-\sigma}\mu^{-\tau} = 0$$

has no positive roots.

Lemma 11.10 ([370]) *Let*

$$f(x, q, r) = x^{\sigma+1} - qx^{\sigma} + r,$$

where σ is a positive integer, q and r are real parameters. Then the equation $f(x, q, r) = 0$ has no positive roots if and only if $q \leq 0$ and $r \geq 0$; or $q > 0$ and

$$r > \frac{\sigma^\sigma}{(\sigma+1)^{\sigma+1}} q^{\sigma+1}.$$

Lemma 11.11 *Suppose that $f(x, y)$, $g(x, y)$, $h(x, y)$ and $v(x, y)$ are differentiable on $(-\infty, +\infty) \times (-\infty, +\infty)$. Let Γ be a two-parameter family of planes defined by the equation*

$$f(\lambda, \mu)x + g(\lambda, \mu)y + h(\lambda, \mu)z = v(\lambda, \mu),$$

where λ and μ are parameters. Let Σ be the envelope of the family Γ. Then the equation

$$f(\lambda, \mu)a + g(\lambda, \mu)b + h(\lambda, \mu)c = v(\lambda, \mu)$$

has no real roots if and only if there is no tangent plane of Σ passing through the point (a, b, c) in $xyz - space$.

Proof By the theory of envelopes, we can see that the set Γ is just the set of tangent planes of Σ.

For necessity. Suppose to the contrary that there exists a tangent plane $\hat{p}_{(\lambda^*,\mu^*)}$ in Γ which passes through the point (a, b, c) in $xyz - space$. Thus, one has

$$f(\lambda^*, \mu^*)a + g(\lambda^*, \mu^*)b + h(\lambda^*, \mu^*)c = v(\lambda^*, \mu^*).$$

This implies that the equation

$$f(\lambda, \mu)a + g(\lambda, \mu)b + h(\lambda, \mu)c = v(\lambda, \mu)$$

has a real solution $(\lambda^*, \mu^*) \in (-\infty, +\infty) \times (-\infty, +\infty)$. This is a contradiction.

For sufficiency. Suppose to the contrary that there exists a pair (λ^*, μ^*) of real numbers such that

$$f(\lambda, \mu)a + g(\lambda, \mu)b + h(\lambda, \mu)c = v(\lambda, \mu).$$

Therefore, we have

$$f(\lambda^*, \mu^*)a + g(\lambda^*, \mu^*)b + h(\lambda^*, \mu^*)c = v(\lambda^*, \mu^*).$$

This implies that there is a tangent plane $\hat{p}_{(\lambda^*,\mu^*)}$ in Γ which passes through the point (a, b, c) in $xyz - space$. This is also a contradiction. This completes the proof.

Lemma 11.12 *Suppose that $f(x)$, $g(x)$, $h(x)$ and $v(x)$ are differentiable on $(-\infty, +\infty)$. Let Γ be the one-parameter family of planes defined by the equation*

$$f(\lambda)x + g(\lambda)y + h(\lambda)z = v(\lambda),$$

where λ is a parameter. Let Σ be the envelope of the family Γ. Then the equation

$$f(\lambda)a + g(\lambda)b + h(\lambda)c = v(\lambda)$$

has no real roots if and only if there is no tangent plane of Σ passing through the point (a, b, c) in $xyz - space$.

Proof In view of the theory of envelopes, we can see that the set Γ is just the set of tangent planes of Σ.

For necessity. Suppose to the contrary that there exists a tangent plane p_{λ^*} in Γ which passes through the point (a, b, c) in $xyz - space$. Thus, we obtain

$$f(\lambda^*)a + g(\lambda^*)b + h(\lambda^*)c = v(\lambda^*),$$

which implies that the equation

$$f(\lambda)a + g(\lambda)b + h(\lambda)c = v(\lambda)$$

has a real solution $\lambda^* \in (-\infty, +\infty)$. This is a contradiction.

For sufficiency. Suppose to the contrary that there is a real number λ^* such that

$$f(\lambda)a + g(\lambda)b + h(\lambda)c = v(\lambda).$$

Hence, we get

$$f(\lambda^*)a + g(\lambda^*)b + h(\lambda^*)c = v(\lambda^*),$$

which implies that there exists a tangent plane $\hat{p}_{\lambda^*}$ in Γ which passes through the point (a, b, c) in $xyz - space$. This is a contradiction. This completes the proof.

Lemma 11.13 *Suppose that $f(x)$ is differentiable on $(0, +\infty)$ such that $f(x)$ is not identically zero on $(0, +\infty)$ and $\lim_{x\to+\infty} f(x) > 0$ or $\lim_{x\to0^+} f(x) > 0$. Then*

$$F(x, y) = y + f(x)$$

has no positive roots on $(0, +\infty) \times (0, +\infty)$ if and only if $f(x)$ has no positive roots on $(0, +\infty)$.

Proof Sufficiency. Suppose that $f(x)$ has no positive roots on $(0, +\infty)$. Since $f(x)$ is not identically zero on $(0, +\infty)$ and $\lim_{x\to+\infty} f(x) > 0$ or $\lim_{x\to0^+} f(x) > 0$, it follows that $f(x) \geq 0$ for any $x \in (0, +\infty)$. Hence, for any $(x, y) \in (0, +\infty) \times (0, +\infty)$, we obtain

$$F(x, y) = y + f(x) > 0.$$

Therefore,

$$F(x, y) = y + f(x)$$

has no positive roots on $(0, +\infty) \times (0, +\infty)$.

Necessity. Suppose to the contrary that there exists $x_0 \in (0, +\infty)$ such that $f(x_0) = 0$. Then we have

$$F(x_0, 0) = 0 + f(x_0) = 0.$$

In view of the implicit function theorem and that $f(x)$ is not identically zero on $(0, +\infty)$, there exists a point $(x^*, y^*) \in (0, +\infty) \times (0, +\infty)$ such that $F(x^*, y^*) = 0$. This is a contradiction. Therefore, this means that $f(x)$ has no positive roots on $(0, +\infty)$. This completes the proof.

11.4.2 Main Results

In this section, some necessary and sufficient conditions for surface oscillations of all solutions of system (11.24) are established.

Since σ, τ are nonnegative integers, to facilitate discussions, we can divide system (11.24) into four mutually exclusive cases: (i) $\sigma \geq 1$ and $\tau \geq 1$, (ii) $\sigma \geq 1$ and $\tau = 0$, (iii) $\sigma = 0$ and $\tau \geq 1$, (iv) $\sigma = 0$ and $\tau = 0$.

Theorem 11.6 *Assume that $\sigma \geq 1$ and $\tau \geq 1$. Then every solution of system (11.24) surface oscillates if and only if $\hat{p} \geq 0$, $q \geq 0$ and $r \geq 0$ or $\hat{p} > 0$, $q < 0$ and*

$$r > (-1)^{\sigma+\tau+1} \frac{\sigma^\sigma \tau^\tau q^{\sigma+\tau+1}}{(\sigma+\tau+1)^{\sigma+\tau+1} \hat{p}^\tau}.$$

Proof When $\sigma \geq 1$ and $\tau \geq 1$, the characteristic equation of system (11.24) is

$$\phi(\hat{p}, q, r, \lambda, \mu) = \lambda + \hat{p}\mu + q + r\lambda^{-\sigma}\mu^{-\tau} = 0. \tag{11.27}$$

Let

$$F(\hat{p}, q, r, \lambda, \mu) = \lambda^\sigma \mu^\tau \phi(\hat{p}, q, r, \lambda, \mu) = \lambda^{\sigma+1}\mu^\tau + \hat{p}\lambda^\sigma \mu^{\tau+1} + q\lambda^\sigma \mu^\tau + r = 0. \tag{11.28}$$

It is clear that (11.27) has no positive roots if and only if (11.28) has no positive roots. Since we are concerned with surface oscillatory solutions of (11.24), we only need to consider the case where (11.28) has no positive roots, that is, $\lambda > 0$ and $\mu > 0$. We will treat the triple (p, q, r) as a point in $xyz - space$, and try to find the exact regions containing points $(\hat{p}, q, r)$ in $xyz - space$ such that (11.28) has no positive roots. Note that $F(x, y, z, \lambda, \mu) = 0$ can be regarded as an equation describing a two-parameter family of planes in $xyz - space$, where x, y and z are the coordinates of point of the planes in $xyz - space$ and λ, μ are parameters. By giving the pair (λ, μ) a specific value in $(0, +\infty) \times (0, +\infty)$, we obtain the equation of one of the planes of the two-parameter family of planes.

In view of the envelop theory, the points of the envelope of the two-parameter family of planes defined by (11.28) satisfy the system of equations:

$$\begin{cases} F(x, y, z, \lambda, \mu) = 0, \\ F_\lambda(x, y, z, \lambda, \mu) = (\sigma + 1)\lambda^\sigma \mu^\tau + \sigma\lambda^{\sigma-1}\mu^{\tau+1}x + \sigma\lambda^{\sigma-1}\mu^\tau y = 0, \\ F_\mu(x, y, z, \lambda, \mu) = \tau\lambda^{\sigma+1}\mu^{\tau-1} + (\tau + 1)\lambda^\sigma\mu^\tau x + \tau\lambda^\sigma\mu^{\tau-1}y = 0, \end{cases} \quad (11.29)$$

where $\lambda > 0$ and $\mu > 0$. Eliminating the two parameters λ and μ from (11.29), we get the equation of the envelope:

$$z(x, y) = (-1)^{\sigma+\tau+1}\frac{\sigma^\sigma\tau^\tau y^{\sigma+\tau+1}}{(\sigma + \tau + 1)^{\sigma+\tau+1}x^\tau}, \quad (11.30)$$

where $x > 0$ and $y < 0$. From (11.30), we get

$$\frac{\partial z}{\partial x} = (-1)^{\sigma+\tau+2}\frac{\sigma^\sigma\tau^{\tau+1}y^{\sigma+\tau+1}}{(\sigma + \tau + 1)^{\sigma+\tau+1}x^{\tau+1}},$$

$$\frac{\partial z}{\partial y} = (-1)^{\sigma+\tau+1}\frac{\sigma^\sigma\tau^\tau y^{\sigma+\tau}}{(\sigma + \tau + 1)^{\sigma+\tau}x^\tau},$$

$$\frac{\partial^2 z}{\partial x^2} = (-1)^{\sigma+\tau+3}\frac{(\tau + 1)\sigma^\sigma\tau^{\tau+1}y^{\sigma+\tau+1}}{(\sigma + \tau + 1)^{\sigma+\tau+1}x^{\tau+2}},$$

$$\frac{\partial^2 z}{\partial y^2} = (-1)^{\sigma+\tau+1}\frac{(\sigma + \tau)\sigma^\sigma\tau^\tau y^{\sigma+\tau-1}}{(\sigma + \tau + 1)^{\sigma+\tau}x^\tau},$$

$$\frac{\partial^2 z}{\partial x\partial y} = (-1)^{\sigma+\tau+2}\frac{\sigma^\sigma\tau^{\tau+1}y^{\sigma+\tau}}{(\sigma + \tau + 1)^{\sigma+\tau}x^{\tau+1}}.$$

Notice that when $x > 0$ and $y < 0$, we have $\partial^2 z/\partial x^2 > 0$, $\partial^2 z/\partial y^2 > 0$, $\partial^2 z/\partial x^2 \cdot \partial^2 z/\partial y^2 - (\partial^2 z/\partial x\partial y)^2 = \sigma^{2\sigma+1}\tau^{2\tau+1}y^{2\sigma+2\tau}x^{-2\tau-2}/(\sigma + \tau + 1)^{2\sigma+2\tau+1} > 0$ and $z(x, y) > 0$. Thus, $z(x, y)$ is a positive and strictly convex function on $(0, +\infty) \times (-\infty, 0)$. Furthermore, the envelope described by (11.30) is a strictly convex surface S over $(0, +\infty) \times (-\infty, 0)$ as depicted in Fig. 11.9. In view of Fig. 11.9 and the obtained information of $z(x, y)$, it is easily seen that when the point (p, q, r) is in the first closed octant, that is, $p \geq 0$, $q \geq 0$ and $r \geq 0$, there cannot be any tangent plane of the envelope S which also passes through this point, and when the point (p, q, r) is vertically above the envelope S, that is, $p > 0, q < 0$ and $r > (-1)^{\sigma+\tau+1}\sigma^\sigma\tau^\tau q^{\sigma+\tau+1}/(\sigma + \tau + 1)^{\sigma+\tau+1}p^\tau$, there cannot exist any tangent plane of the envelope S which also passes through this point, while the point (p, q, r) is situated elsewhere, such a tangent plane can be drawn. By Lemma 11.11, (11.28) does not have any positive roots if and only if $p \geq 0$, $q \geq 0$ and $r \geq 0$ or $p > 0, q < 0$ and

$$r > (-1)^{\sigma+\tau+1}\frac{\sigma^\sigma\tau^\tau q^{\sigma+\tau+1}}{(\sigma + \tau + 1)^{\sigma+\tau+1}\hat{p}^\tau}.$$

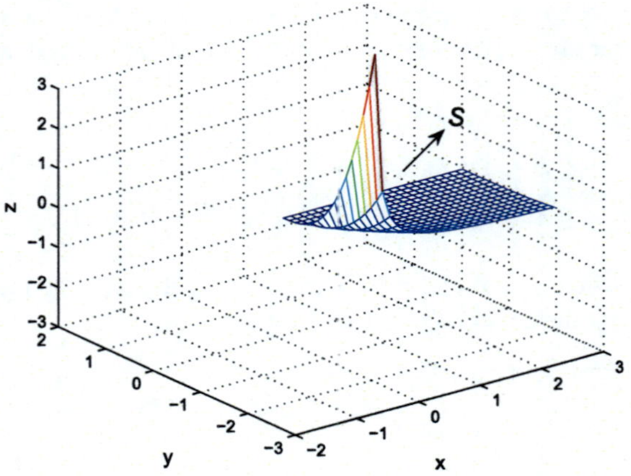

Fig. 11.9 Envelope surface for $\sigma = 1$ and $\tau = 1$

Since (11.27) is identical with (11.28) for the existence of positive solutions, Lemma 11.9 implies the statement of this theorem. The proof is thus completed.

Theorem 11.7 *Assume that* $\sigma \geq 1$ *and* $\tau = 0$. *Then every solution of system (11.24) surface oscillates if and only if* $\hat{p} \geq 0$, $q \geq 0$ *and* $r \geq 0$ *or* $\hat{p} \geq 0$, $q < 0$ *and*

$$r > (-1)^{\sigma+1} \frac{\sigma^\sigma}{(\sigma+1)^{\sigma+1}} q^{\sigma+1}.$$

Proof When $\sigma \geq 1$ and $\tau = 0$, then the characteristic equation of system (11.24) is

$$\phi(\hat{p}, q, r, \lambda, \mu) = \lambda + \hat{p}\mu + q + r\lambda^{-\sigma} = 0. \tag{11.31}$$

When $p < 0$, it is clear that (11.31) has positive solutions. We only need to consider two cases: (1) $\hat{p} = 0$ and (2) $\hat{p} > 0$.

Case (1) $\hat{p} = 0$. In this case, (11.31) can be written as

$$\phi(q, r, \lambda) = \lambda + q + r\lambda^{-\sigma} = 0. \tag{11.32}$$

Let

$$f(q, r, \lambda) = \lambda^\sigma \phi(q, r, \lambda) = \lambda^{\sigma+1} + q\lambda^\sigma + r = 0. \tag{11.33}$$

Since (11.33) is identical with (11.32) for the existence of positive solutions, it follows from Lemma 11.10 that (11.32) has no positive roots if and only if $q \geq 0$ and $r \geq 0$ or $q < 0$ and

$$r > (-1)^{\sigma+1} \frac{\sigma^\sigma}{(\sigma+1)^{\sigma+1}} q^{\sigma+1}.$$

Case (2) $\hat{p} > 0$. In this case, (11.31) can be rewritten as

$$\phi(\hat{p}, q, r, \lambda, \mu) = \hat{p}\left(\frac{1}{\hat{p}}\lambda + \mu + \frac{q}{\hat{p}} + \frac{r}{\hat{p}}\lambda^{-\sigma}\right) = 0.$$

Let

$$f\left(\frac{1}{\hat{p}}, \frac{q}{\hat{p}}, \frac{r}{\hat{p}}, \lambda\right) = \frac{1}{\hat{p}}\lambda + \frac{q}{\hat{p}} + \frac{r}{\hat{p}}\lambda^{-\sigma} = 0. \tag{11.34}$$

Since $\lim_{\lambda \to +\infty} f(1/\hat{p}, q/\hat{p}, r/\hat{p}, \lambda) > 0$ for $\hat{p} > 0$ and $f(1/\hat{p}, q/\hat{p}, r/\hat{p}, \lambda)$ is differentiable with respect to λ on $(0, +\infty)$, Lemma 11.13 implies that (11.31) has no positive roots if and only if (11.34) has no positive roots. Let

$$F\left(\frac{1}{\hat{p}}, \frac{q}{\hat{p}}, \frac{r}{\hat{p}}, \lambda\right) = \lambda^{\sigma} f\left(\frac{1}{\hat{p}}, \frac{q}{\hat{p}}, \frac{r}{\hat{p}}, \lambda\right) = \frac{1}{\hat{p}}\lambda^{\sigma+1} + \frac{q}{\hat{p}}\lambda^{\sigma} + \frac{r}{\hat{p}} = 0. \tag{11.35}$$

It is clear that (11.34) has no positive roots if and only if (11.35) has no positive roots. Since we are concerned with surface oscillatory solutions of (11.24), we only need to consider the case where (11.35) has no positive roots, that is, $\lambda > 0$. We will treat the triple $(1/\hat{p}, q/\hat{p}, r/\hat{p})$ as a point in $xyz - space$ and look for the exact regions containing points$(1/\hat{p}, q/\hat{p}, r/\hat{p})$ in $xyz - space$ such that (11.35) has no positive roots. In fact, $F(x, y, z, \lambda) = 0$ can be regarded as an equation describing a one-parameter family of planes in $xyz - space$, where x, y and z are the coordinates of point of the plane in $xyz - space$ and λ is a parameter. By giving λ a specific number value in $(0, +\infty)$, we obtain the equation of one of the planes of the one-parameter family of planes.

From the theory of envelopes, the points of the envelope of the one-parameter family of planes described by (11.35) satisfy the system of equations:

$$\begin{cases} F(x, y, z, \lambda) = \lambda^{\sigma+1}x + \lambda^{\sigma}y + z = 0, \\ F_{\lambda}(x, y, z, \lambda) = (\sigma+1)\lambda^{\sigma}x + \sigma\lambda^{\sigma-1}y = 0, \end{cases} \tag{11.36}$$

where $\lambda > 0$. Eliminating the parameter $\lambda > 0$ from (11.36), we obtain the equation of the envelope:

$$z(x, y) = (-1)^{\sigma+1}\frac{\sigma^{\sigma}y^{\sigma+1}}{(\sigma+1)^{\sigma+1}x^{\sigma}}, \tag{11.37}$$

where $x > 0$ and $y < 0$. From (11.37), we have

$$\begin{aligned}
\frac{\partial z}{\partial x} &= (-1)^{\sigma+2}\frac{\sigma^{\sigma+1}y^{\sigma+1}}{(\sigma+1)^{\sigma+1}x^{\sigma+1}}, \\
\frac{\partial z}{\partial y} &= (-1)^{\sigma+1}\frac{\sigma^{\sigma}y^{\sigma}}{(\sigma+1)^{\sigma}x^{\sigma}}, \\
\frac{\partial^2 z}{\partial x^2} &= (-1)^{\sigma+3}\frac{\sigma^{\sigma+1}y^{\sigma+1}}{(\sigma+1)^{\sigma}x^{\sigma+2}},
\end{aligned}$$

$$\frac{\partial^2 z}{\partial y^2} = (-1)^{\sigma+1} \frac{\sigma^{\sigma+1} y^{\sigma-1}}{(\sigma+1)^{\sigma} x^{\sigma}},$$
$$\frac{\partial^2 z}{\partial x \partial y} = (-1)^{\sigma+2} \frac{\sigma^{\sigma+1} y^{\sigma}}{(\sigma+1)^{\sigma} x^{\sigma+1}}.$$

Note that when $x > 0$ and $y < 0$, we have $z(x, y) > 0, \partial^2 z/\partial x^2 > 0, \partial^2 z/\partial y^2 > 0$ and $\partial^2 z/\partial x^2 \cdot \partial^2 z/\partial y^2 - (\partial^2 z/\partial x \partial y)^2 = 0$, Therefore, $z(x, y)$ is a positive and convex function on $(0, +\infty) \times (-\infty, 0)$. Furthermore, the envelope defined by (11.37) is a convex surface S over $(0, +\infty) \times (-\infty, 0)$ as depicted in Fig. 11.10. In view of Fig. 11.10 and the obtained information of $z(x, y)$, it is easy to see that when the point $(1/\hat{p}, q/\hat{p}, r/\hat{p})$ is in the first closed octant except on the positive coordinate axis $x = 0$, that is, $1/\hat{p} > 0$, $q/\hat{p} \geq 0$ and $r/\hat{p} \geq 0$, which are equivalent to $\hat{p} > 0$, $q \geq 0$ and $r \geq 0$, there cannot be any tangent plane of the envelope S which also passes through this point, and when the point $(1/\hat{p}, q/\hat{p}, r/\hat{p})$ is vertically above the envelope S in the forth octant, that is, $1/\hat{p} > 0$, $q/\hat{p} < 0$ and $r/\hat{p} > (-1)^{\sigma+1}\sigma^{\sigma}(q/\hat{p})^{\sigma+1}/(\sigma+1)^{\sigma+1}(1/\hat{p})^{\sigma}$, which are equivalent to $\hat{p} > 0$, $q < 0$ and $r > (-1)^{\sigma+1}\sigma^{\sigma}q^{\sigma+1}/(\sigma+1)^{\sigma+1}$, there cannot exist any tangent plane of the envelope S which also passes through this point, while the point $(1/\hat{p}, q/\hat{p}, r/\hat{p})$ is situated elsewhere, such a tangent plane can be drawn. Since (11.31) is identical with (11.35) for the existence of positive solutions, it follows from Lemma 11.12 that (11.31) does not have any positive roots if and only if $\hat{p} > 0$, $q \geq 0$ and $r \geq 0$ or $\hat{p} > 0$, $q < 0$ and

$$r > (-1)^{\sigma+1} \frac{\sigma^{\sigma}}{(\sigma+1)^{\sigma+1}} q^{\sigma+1}.$$

Combining case (1) with case (2), one can see that (11.31) does not have any positive roots if and only if $\hat{p} \geq 0$, $q \geq 0$ and $r \geq 0$ or $\hat{p} \geq 0$, $q < 0$ and

Fig. 11.10 Envelope surface for $\sigma = 1$ and $\tau = 0$

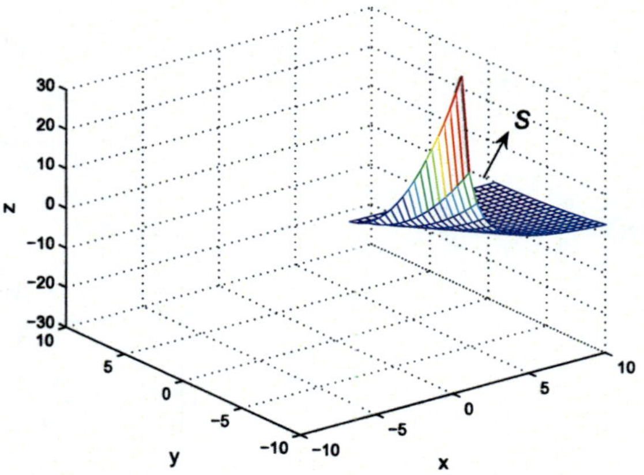

$$r > (-1)^{\sigma+1}\frac{\sigma^\sigma}{(\sigma+1)^{\sigma+1}}q^{\sigma+1}.$$

Lemma 11.9 implies the statement of this theorem. This completes the proof.

Theorem 11.8 *Assume that $\sigma = 0$ and $\tau \geq 1$. Then every solution of system (11.24) surface oscillates if and only if $\hat{p} \geq 0$, $q \geq 0$ and $r \geq 0$ or $\hat{p} > 0$, $q < 0$ and*

$$r > (-1)^{\tau+1}\frac{\tau^\tau q^{\tau+1}}{(\tau+1)^{\tau+1}\hat{p}^\tau}.$$

Proof When $\sigma = 0$ and $\tau \geq 1$, then the characteristic equation of system (11.24) is

$$\phi(\hat{p}, q, r, \lambda, \mu) = \lambda + \hat{p}\mu + q + r\mu^{-\tau} = 0. \tag{11.38}$$

When $\hat{p} < 0$, it is clear that (11.38) has positive solutions. We only need to consider two cases: (1) $\hat{p} = 0$ and (2) $\hat{p} > 0$.

Case (1) $\hat{p} = 0$. In this case, it is easy to verify that (11.38) has no positive roots if and only if $q \geq 0$ and $r \geq 0$.

Case (2) $\hat{p} > 0$. In this case, let

$$f(\hat{p}, q, r, \mu) = \hat{p}\mu + q + r\mu^{-\tau} = 0. \tag{11.39}$$

Since $\lim_{\mu\to+\infty} f(\hat{p}, \hat{p}, r, \mu) > 0$ for $\hat{p} > 0$ and $f(\hat{p}, \hat{p}, r, \mu)$ is differentiable with respect to μ on $(0, +\infty)$, Lemma 11.13 implies that (11.38) has no positive roots if and only if (11.39) has no positive roots. Let

$$F(\hat{p}, q, r, \mu) = \mu^\tau f(\hat{p}, q, r, \mu) = \hat{p}\mu^{\tau+1} + q\mu^\tau + r = 0. \tag{11.40}$$

It is clear that (11.39) has no positive roots if and only if (11.40) has no positive roots. Since we are concerned with positive roots of (11.40), we will restrict our attention to the case where $\mu > 0$. We will treat the triple $(\hat{p}, q, r)$ as a point in $xyz - space$ and look for the exact regions containing points $(\hat{p}, q, r)$ in $xyz - space$ such that (11.40) has no positive roots. In fact, $F(x, y, z, \mu) = 0$ can be regarded as an equation describing a one-parameter family of planes in $xyz - space$, where x, y and z are the coordinates of point of the plane in $xyz - space$ and μ is a parameter. By giving μ a specific number value in $(0, +\infty)$, we obtain the equation of one of the planes of the one-parameter family of planes.

From the theory of envelopes, the points of the envelope of the one-parameter family of planes described by (11.40) satisfy the system of equations:

$$\begin{cases} F(x, y, z, \mu) = \mu^{\tau+1}x + \mu^\tau y + z = 0, \\ F_\mu(x, y, z, \mu) = (\tau+1)\mu^\tau x + \tau\mu^{\tau-1}y = 0, \end{cases} \tag{11.41}$$

where $\mu > 0$. Eliminating the parameter μ from (11.41), we obtain the equation of the envelope:

$$z(x, y) = (-1)^{\tau+1} \frac{\tau^{\tau} y^{\tau+1}}{(\tau+1)^{\tau+1} x^{\tau}}, \tag{11.42}$$

where $x > 0$ and $y < 0$. From (11.42), we have

$$\begin{aligned} \frac{\partial z}{\partial x} &= (-1)^{\tau+2} \frac{\tau^{\tau+1} y^{\tau+1}}{(\tau+1)^{\tau+1} x^{\tau+1}}, \\ \frac{\partial z}{\partial y} &= (-1)^{\tau+1} \frac{\tau^{\tau} y^{\tau}}{(\tau+1)^{\tau} x^{\tau}}, \\ \frac{\partial^2 z}{\partial x^2} &= (-1)^{\tau+3} \frac{\tau^{\tau+1} y^{\tau+1}}{(\tau+1)^{\tau} x^{\tau+2}}, \\ \frac{\partial^2 z}{\partial y^2} &= (-1)^{\tau+1} \frac{\tau^{\tau+1} y^{\tau-1}}{(\tau+1)^{\tau} x^{\tau}}, \\ \frac{\partial^2 z}{\partial x \partial y} &= (-1)^{\tau+2} \frac{\tau^{\tau+1} y^{\tau}}{(\tau+1)^{\tau} x^{\tau+1}}. \end{aligned}$$

Note that when $x > 0$ and $y < 0$, we have $z(x, y) > 0$, $\partial^2 z/\partial x^2 > 0$, $\partial^2 z/\partial y^2 > 0$ and $\partial^2 z/\partial x^2 \cdot \partial^2 z/\partial y^2 - (\partial^2 z/\partial x \partial y)^2 = 0$. Therefore, $z(x, y)$ is a positive and convex function on $(0, +\infty) \times (-\infty, 0)$. Furthermore, the envelope defined by (11.42) is a convex surface S over $(0, +\infty) \times (-\infty, 0)$ as depicted in Fig. 11.11. In view of Fig. 11.11 and the obtained information of $z(x, y)$, it is easy to see that when the point $(\hat{p}, q, r)$ is in the first closed octant except on the positive coordinate axis $x = 0$, that is, $\hat{p} > 0$, $q \geq 0$ and $r \geq 0$, there cannot be any tangent plane of the envelope S which also passes through this point, and when the point $(\hat{p}, q, r)$ is vertically above the envelope S in the forth octant, that is, $\hat{p} > 0$, $q < 0$

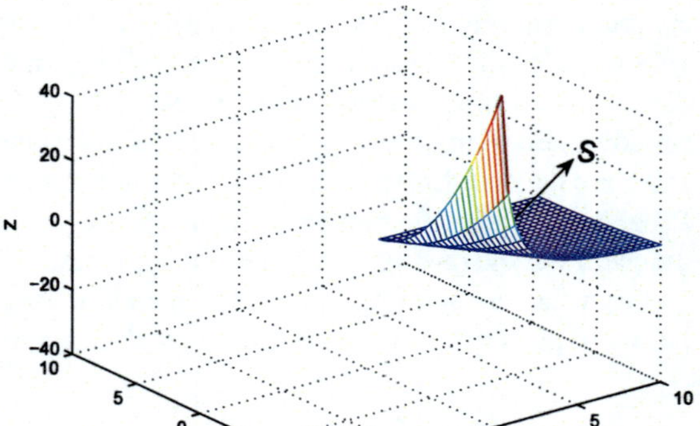

Fig. 11.11 Envelope surface for $\sigma = 0$ and $\tau = 1$

and $r > (-1)^{\tau+1}\tau^{\tau}q^{\tau+1}/(\tau+1)^{\tau+1}\hat{p}^{\tau}$, there cannot exist any tangent plane of the envelope S which also passes through this point, while the point $(\hat{p}, q, r)$ is situated elsewhere, such a tangent plane can be drawn. Since (11.38) is identical with (11.40) for the existence of positive solutions, it follows from Lemma 11.12 that (11.38) does not have any positive roots if and only if $\hat{p} > 0$, $q \geq 0$ and $r \geq 0$ or $\hat{p} > 0$, $q < 0$ and

$$r > (-1)^{\tau+1}\frac{\tau^{\tau}q^{\tau+1}}{(\tau+1)^{\tau+1}p^{\tau}}.$$

Combining case (1) with case (2) implies that the characteristic equation (11.38) does not have any positive roots if and only if $\hat{p} \geq 0$, $q \geq 0$ and $r \geq 0$ or $\hat{p} > 0$, $q < 0$ and

$$r > (-1)^{\tau+1}\frac{\tau^{\tau}q^{\tau+1}}{(\tau+1)^{\tau+1}\hat{p}^{\tau}}.$$

Lemma 11.9 implies the statement of this theorem. This completes the proof.

Remark 11.5 If we set $\sigma = 0$ in Theorem 11.6 and define $0^0 = 1$, then the result of Theorem 11.6 gives the result of Theorem 11.8. This means that Theorem 11.8 is a special case of Theorem 11.6.

Theorem 11.9 *Assume that $\sigma = 0$ and $\tau = 0$. Then every solution of system (11.24) surface oscillates if and only if $p \geq 0$ and $q + r \geq 0$.*

Proof When $\sigma = 0$ and $\tau = 0$, one can rewrite system (11.24) as

$$u_{m+1,n} + \hat{p}u_{m,n+1} + (q+r)u_{m,n} = 0. \tag{11.43}$$

The characteristic equation of system (11.43) is

$$\lambda + \hat{p}\mu + (q+r) = 0. \tag{11.44}$$

It is easy to see that (11.44) does not have any positive roots if and only if $\hat{p} \geq 0$ and $q + r \geq 0$. From Lemma 11.9, we can see that every solution of system (11.24) surface oscillates if and only if $\hat{p} \geq 0$ and $q + r \geq 0$. This completes the proof.

11.4.3 Illustrative Examples

Example 11.7 Consider the delay discrete convection system

$$u_{m+1,n} + 0.91u_{m,n+1} - 0.1u_{m,n} + 0.13u_{m-1,n-1} = 0. \tag{11.45}$$

From system (11.45), we have $\sigma = 1, \tau = 1, \hat{p} = 0.91, q = -0.1$ and $r = 0.13$. Since $\hat{p} = 0.91 > 0, q = -0.1 < 0$ and

$$r = 0.13 > \frac{0.1^3}{27 \times 0.91} = (-1)^{\sigma+\tau+1} \frac{\sigma^\sigma \tau^\tau q^{\sigma+\tau+1}}{(\sigma+\tau+1)^{\sigma+\tau+1} p^\tau},$$

according to Theorem 11.6, every solution of system (11.45) is surface oscillatory. The surface oscillatory behavior of system (11.45) is demonstrated by Fig. 11.12.

Example 11.8 Consider the delay discrete convection system

$$u_{m+1,n} + 0.7u_{m,n+1} + 0.26u_{m,n} + 0.2u_{m-1,n-1} = 0. \tag{11.46}$$

From system (11.46), we have $\sigma = 1, \tau = 1, \hat{p} = 0.7, q = 0.26$ and $r = 0.2$. Since $\hat{p} = 0.7 > 0, q = 0.26 > 0$ and $r = 0.2 > 0$, according to Theorem 11.6, every solution of system (11.46) is surface oscillatory. The surface oscillatory behavior of system (11.46) is demonstrated by Fig. 11.13.

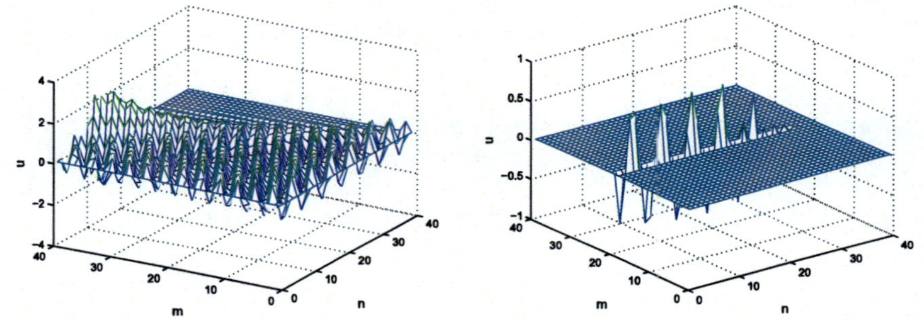

(a) (b)

Fig. 11.12 **a** Surface oscillatory behavior of system (11.45). **b** Surface oscillatory behavior of system (11.45) with m = 20

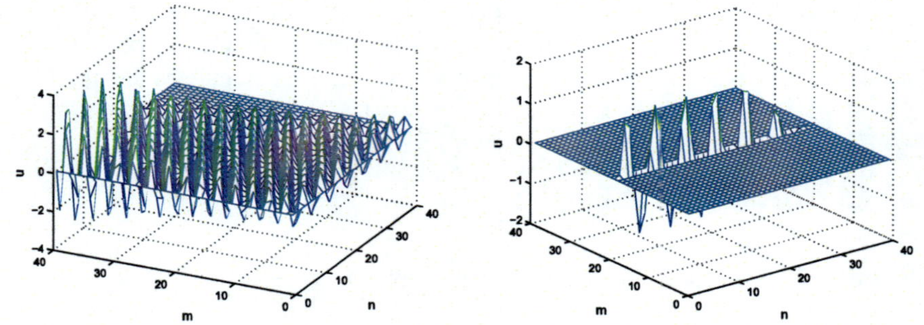

(a) (b)

Fig. 11.13 **a** Surface oscillatory behavior of system (11.46). **b** Surface oscillatory behavior of system (11.46) with m = 20

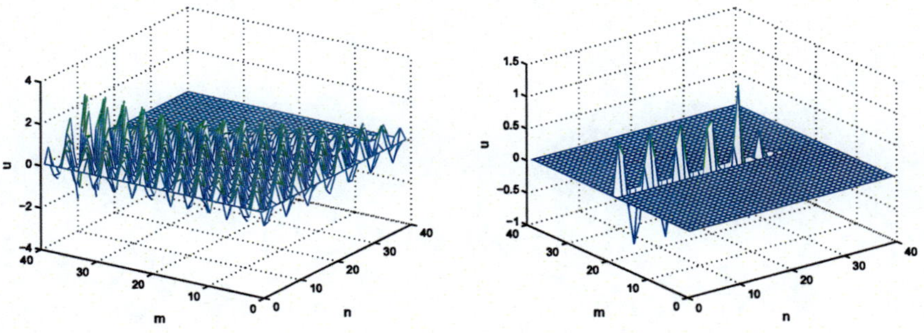

Fig. 11.14 **a** Surface oscillatory behavior of system (11.47). **b** Surface oscillatory behavior of system (11.47) with m = 20

Example 11.9 Consider the delay discrete convection system

$$u_{m+1,n} + 0.9u_{m,n+1} - 0.1_{m,n} + 0.17u_{m-1,n} = 0. \tag{11.47}$$

From system (11.47), we have $\sigma = 1, \tau = 0, \hat{p} = 0.9, q = -0.1$ and $r = 0.17$. Since $\hat{p} = 0.9 > 0, q = -0.1 < 0$ and

$$r = 0.17 > \frac{1}{4}(-0.1)^2 = (-1)^{\sigma+1}\frac{\sigma^\sigma}{(\sigma+1)^{\sigma+1}}q^{\sigma+1},$$

in view of Theorem 11.7, every solution of system (11.47) is surface oscillatory. The surface oscillatory behavior of system (11.47) is demonstrated by Fig. 11.14.

Example 11.10 Consider the delay discrete convection system

$$u_{m+1,n} + u_{m,n+1} - 0.06u_{m,n} + 0.07u_{m,n-1} = 0. \tag{11.48}$$

From system (11.48), we have $\sigma = 0, \tau = 1, \hat{p} = 1, q = -0.06$ and $r = 0.07$. Since $\hat{p} = 1 > 0, q = -0.06 < 0$ and

$$r = 0.07 > \frac{1}{4}(-0.06)^2 = (-1)^{\tau+1}\frac{\tau^\tau q^{\tau+1}}{(\tau+1)^{\tau+1}p^\tau},$$

according to Theorem 11.8, every solution of system (11.48) is surface oscillatory. The surface oscillatory behavior of system (11.48) is demonstrated by Fig. 11.15.

Example 11.11 Consider the delay discrete convection system

$$u_{m+1,n} + 0.9u_{m,n+1} + (0.1 + 0.08)u_{m,n} = 0. \tag{11.49}$$

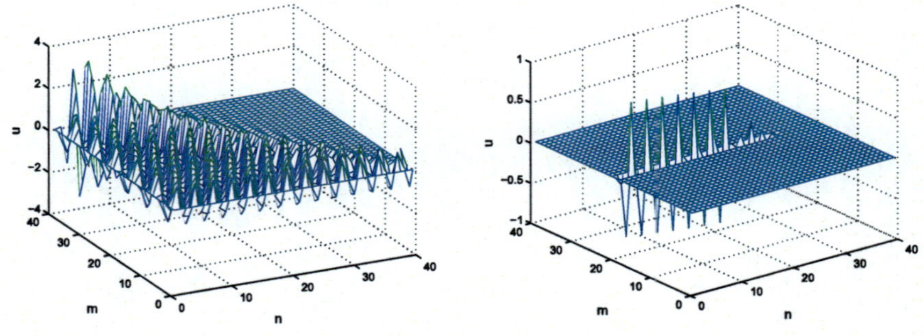

(a) (b)

Fig. 11.15 **a** Surface oscillatory behavior of system (11.48). **b** Surface oscillatory behavior of system (11.48) with m = 20

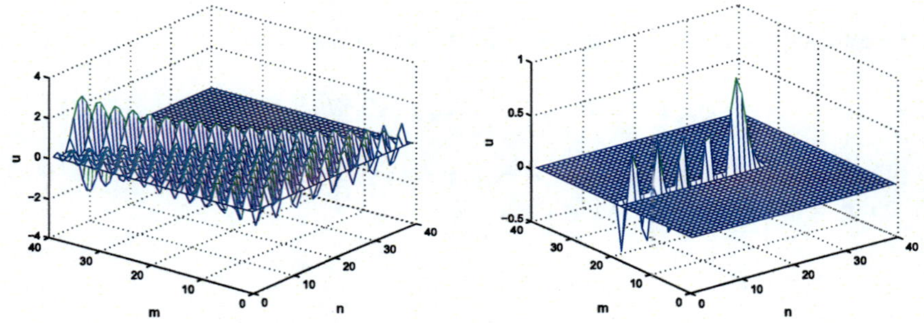

(a) (b)

Fig. 11.16 **a** Surface oscillatory behavior of system (11.49). **b** Surface oscillatory behavior of system (11.49) with m = 20

From system (11.49), we have $\sigma = 0, \tau = 0$, $\hat{p} = 0.9, q = 0.1$ and $r = 0.08$. Since $\hat{p} = 0.9 > 0$ and $q + r = 0.18 > 0$, according to Theorem 11.9, every solution of system (11.49) is surface oscillatory. The surface oscillatory behavior of system (11.49) is demonstrated by Fig. 11.16.

11.5 The Case as $\mu = \hat{p}, \xi = q, p = 1, i = 1, q(1) = r$

As $\mu = \hat{p}$, $\xi = q$, $p = 1$, $i = 1$, $q(1) = r$, we consider the following mixed 2-D discrete convection system from system (11.1):

$$\hat{p}u_{m+1,n} + qu_{m,n+1} - u_{m,n} + ru_{m-\sigma,n-\tau} = 0, \tag{11.50}$$

where $\hat{p}, q, r$ are real numbers with $\hat{p}^2 + q^2 + r^2 \neq 0, m, n$ are nonnegative integers, and σ, τ are nonzero integers with $\sigma\tau < 0$.

In fact, system (11.50) can be regarded as a discrete analogue of the following continuous convection system with a source term:

$$\hat{p}\frac{\partial u(x, y)}{\partial x} + q\frac{\partial u(x, y)}{\partial y} + (\hat{p} + q - 1)u(x, y) + ru(x - \xi, y - \eta) = 0, \quad (11.51)$$

where $\hat{p}, q, r$ are the same as in system (11.50), and ξ, η are real numbers with $\xi\eta < 0$. Nevertheless, in general it is difficult to derive exact analytical solutions of system (11.51) in practice. Therefore, the study of system (11.50) will be helpful for understanding the dynamical behaviors of system (11.51).

The purpose of this chapter is to apply a new method, based on the envelope theory of the family of planes, to propose some effective criteria to determine the surface oscillation behavior of the mixed discrete system (11.50).

11.5.1 Preliminaries

Definition 11.5 System (11.50) is said to be a system of mixed type if $\sigma\tau < 0$.

Definition 11.6 A solution of (11.50) is a real double sequence $\{u_{m,n}\}$ which is defined for $m \geq -max\{0, \sigma\}$, $n \geq -max\{0, \tau\}$, and which satisfies (11.50) for $m \geq 0$ and $n \geq 0$.

Definition 11.7 A solution $\{u_{m,n}\}$ of (11.50) is said to be eventually positive (or negative) if $u_{m,n} > 0$ (or $u_{m,n} < 0$) for large numbers m and n. It is said to be oscillatory if it is neither eventually positive nor eventually negative. System (11.50) is called oscillatory if all of its nontrivial solutions are oscillatory.

It is clear that the fixed plane equation of system (11.50) is $u^* = 0$. Thus, the surface oscillation behavior of system (11.50) is that the solutions of system (11.50) oscillate about the fixed plane with an infinite number of zeros.

Lemma 11.14 *The following statements are equivalent:*

(a) Every solution of system (11.50) is oscillatory.

(b) The characteristic equation of system (11.50)

$$\hat{p}\lambda + q\mu - 1 + r\lambda^{-\sigma}\mu^{-\tau} = 0$$

has no positive roots.

Lemma 11.15 *Suppose that $f(x, y)$, $g(x, y)$, $h(x, y)$, and $v(x, y)$ are differentiable on $(-\infty, +\infty) \times (-\infty, +\infty)$. Let Γ be a two-parameter family of planes defined by the equation*

$$f(\lambda, \mu)x + g(\lambda, \mu)y + h(\lambda, \mu)z = v(\lambda, \mu),$$

where λ and μ are parameters. Let Σ be the envelope of the family Γ. Then the equation

$$f(\lambda, \mu)a + g(\lambda, \mu)b + h(\lambda, \mu)c = v(\lambda, \mu)$$

has no real roots if and only if there is no tangent plane of Σ passing through the point (a, b, c) in $xyz - space$.

Lemma 11.16 *Suppose that $f(x), g(x), h(x)$, and $v(x)$ are differentiable on $(-\infty, +\infty)$. Let Γ be the one-parameter family of planes defined by the equation*

$$f(\lambda)x + g(\lambda)y + h(\lambda)z = v(\lambda),$$

where λ is a parameter. Let Σ be the envelope of the family Γ. Then the equation

$$f(\lambda)a + g(\lambda)b + h(\lambda)c = v(\lambda)$$

has no real roots if and only if there is no tangent plane of Σ passing through the point (a, b, c) in $xyz - space$.

11.5.2 Main Results

In this section, we will establish the necessary and sufficient conditions for surface oscillations of all solutions of system (11.50).

Theorem 11.10 *Assume that $\sigma < 0$, $\tau > 0$, and $\sigma < -\tau - 1$. Then every solution of system (11.50) surface oscillates if and only if $\hat{p} \le 0$, $q \le 0$, and $r \le 0$; or $\hat{p} > 0$, $q < 0$, and*

$$r < \frac{\sigma^\sigma \tau^\tau}{(\sigma + \tau + 1)^{\sigma+\tau+1} p^\sigma q^\tau}.$$

Proof When $\sigma < 0$, $\tau > 0$, and $\sigma < -\tau - 1$, the characteristic equation of system (11.50) is

$$\phi(\hat{p}, q, r, \lambda, \mu) = \hat{p}\lambda + q\mu - 1 + r\lambda^{-\sigma}\mu^{-\tau} = 0. \tag{11.52}$$

From Lemma 11.14, we only need to consider positive roots of (11.52), that is $\lambda > 0$ and $\mu > 0$. We will treat the triple $(\hat{p}, q, r)$ as a point in $xyz - space$ and try to find the exact regions containing points $(\hat{p}, q, r)$ in $xyz - space$ such that (11.52) has no positive roots. In fact, $\phi(x, y, z, \lambda, \mu) = 0$ can be regarded as an equation describing a two-parameter family of planes in $xyz - space$, where x, y, and z are the coordinates of points of the planes in $xyz - space$ and λ, μ are parameters. By giving the pair (λ, μ) a specific value in $(0, +\infty) \times (0, +\infty)$, we obtain the equation of one of the planes of the two-parameter family of planes.

In view of the envelope theory, the points of the envelope of the two-parameter family of planes defined by (11.52) satisfy the system of equations:

$$\begin{cases} \phi(x, y, z, \lambda, \mu) = 0, \\ \phi_\lambda(x, y, z, \lambda, \mu) = x - \sigma\lambda^{-\sigma-1}\mu^{-\tau}z = 0, \\ \phi_\mu(x, y, z, \lambda, \mu) = y - \tau\lambda^{-\sigma}\mu^{-\tau-1}z = 0, \end{cases} \tag{11.53}$$

where $\lambda > 0$ and $\mu > 0$. Eliminating the two parameters λ and μ from (11.53), we get the equation of the envelope:

$$z(x, y) = \frac{\sigma^\sigma\tau^\tau}{(\sigma + \tau + 1)^{\sigma+\tau+1}}x^{-\sigma}y^{-\tau}, \tag{11.54}$$

where $x > 0$ and $y < 0$. From (11.54), we get

$$\begin{aligned} \frac{\partial z}{\partial x} &= -\frac{\sigma^{\sigma+1}\tau^\tau}{(\sigma + \tau + 1)^{\sigma+\tau+1}}x^{-\sigma-1}y^{-\tau}, \\ \frac{\partial z}{\partial y} &= -\frac{\sigma^\sigma\tau^{\tau+1}}{(\sigma + \tau + 1)^{\sigma+\tau+1}}x^{-\sigma}y^{-\tau-1}, \\ \frac{\partial^2 z}{\partial x^2} &= \frac{(\sigma + 1)\sigma^\sigma\tau^\tau}{(\sigma + \tau + 1)^{\sigma+\tau+1}}x^{-\sigma-2}y^{-\tau}, \\ \frac{\partial^2 z}{\partial y^2} &= \frac{(\tau + 1)\sigma^\sigma\tau^{\tau+1}}{(\sigma + \tau + 1)^{\sigma+\tau+1}}x^{-\sigma}y^{-\tau-2}, \\ \frac{\partial^2 z}{\partial x\partial y} &= \frac{\sigma^{\sigma+1}\tau^{\tau+1}}{(\sigma + \tau + 1)^{\sigma+\tau+1}}x^{-\sigma-1}y^{-\tau-1}, \end{aligned} \tag{11.55}$$

where $x > 0$ and $y < 0$. It follows from (11.54) and (11.55) that $\partial^2 z/\partial x^2 < 0$, $\partial^2 z/\partial y^2 < 0$, $\partial^2 z/\partial x^2 \cdot \partial^2 z/\partial y^2 - (\partial^2 z/\partial x\partial y)^2 = \sigma^{2\sigma+1}\tau^{2\tau+1}/x^{2\sigma+2}y^{2\tau+2}$ $(\sigma + \tau + 1)^{2\sigma+2\tau+1} > 0$, and $z(x, y) < 0$ all hold for $x > 0$ and $y < 0$. Thus, $z(x, y)$ is negative and concave on $(0, +\infty) \times (-\infty, 0)$. Furthermore, the envelope described by (11.54) is a concave surface S over $(0, +\infty) \times (-\infty, 0)$ as shown in Fig. 11.17. In view of Fig. 11.17 and the obtained information of $z(x, y)$, it is easily seen that when the point $(\hat{p}, q, r)$ is in the seventh closed octant, that is, $\hat{p} \leq 0, q \leq 0$, and $r \leq 0$, there cannot be any tangent plane of the envelope S which also passes through this point, and when the point $(\hat{p}, q, r)$ is vertically below the envelope S in the eighth octant, that is, $\hat{p} > 0$, $q < 0$, and $r < \sigma^\sigma\tau^\tau/(\sigma + \tau + 1)^{\sigma+\tau+1}p^\sigma q^\tau$, there cannot exist any tangent plane of the envelope S which also passes through this point, while the point $(\hat{p}, q, r)$ is situated elsewhere, such a tangent plane can be drawn. By Lemma 11.15, we can see that (11.52) does not have any positive roots if and only if $\hat{p} \leq 0$, $q \leq 0$, and $r \leq 0$; or $\hat{p} > 0$, $q < 0$, and

$$r < \frac{\sigma^\sigma\tau^\tau}{(\sigma + \tau + 1)^{\sigma+\tau+1}p^\sigma q^\tau}.$$

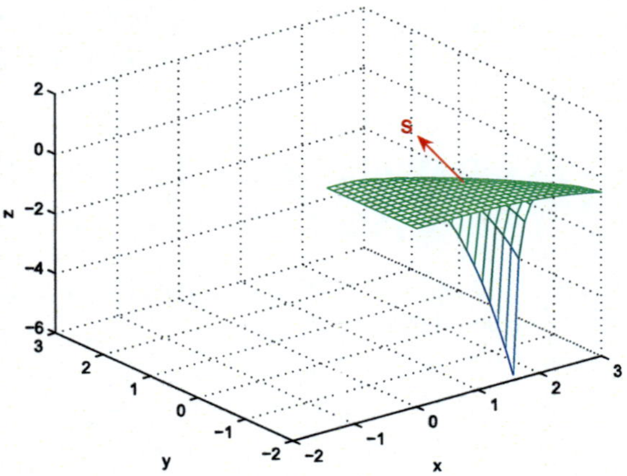

Fig. 11.17 Envelope surface for $\sigma = -3$ and $\tau = 1$

Lemma 11.14 implies the statement of this theorem. The proof is thus completed.

Theorem 11.11 *Assume that $\sigma < 0$, $\tau > 0$, and $\sigma = -\tau - 1$. Then every solution of system (11.50) surface oscillates if and only if $\hat{p} \leq 0$, $q \leq 0$, and $r \leq 0$; or $\hat{p} > 0$, $q < 0$, and*

$$r < (-1)^{\tau+1} \frac{\tau^{\tau} \hat{p}^{\tau+1}}{(\tau+1)^{\tau+1} q^{\tau}}.$$

Proof When $\sigma < 0$, $\tau > 0$, and $\sigma = -\tau - 1$, the characteristic equation of system (11.50) is

$$\phi(\hat{p}, q, r, \lambda, \mu) = \hat{p}\lambda + q\mu - 1 + r\lambda^{\tau+1}\mu^{-\tau} = 0. \tag{11.56}$$

Substituting $\mu = c\lambda (c > 0)$ into (11.56) yields

$$\phi(\hat{p}, q, r, \lambda) = (\hat{p} + qc + rc^{-\tau})\lambda - 1 = 0. \tag{11.57}$$

It is clear that (11.56) has no positive roots if and only if (11.57) has no positive roots for any $c \in (0, +\infty)$, that is, $qc^{\tau+1} + \hat{p}c^{\tau} + r < 0$ always holds for any $c \in (0, +\infty)$. Let

$$f(\hat{p}, q, r, c) = qc^{\tau+1} + \hat{p}c^{\tau} + r = 0. \tag{11.58}$$

When $q > 0$, since $\lim_{c \to +\infty} f(\hat{p}, q, r, c) = +\infty$ and $f(\hat{p}, q, r, c)$ is continuous on $(0, +\infty)$, there exists $c^* \in (0, +\infty)$ such that $f(\hat{p}, q, r, c^*) > 0$. Thus, We only need to consider two cases:(a) $q = 0$ and (b) $q < 0$.

Case (a): $q = 0$. In this case, it is clear that $f(\hat{p}, q, r, c) < 0$ always holds for any $c \in (0, +\infty)$ if and only if $\hat{p} \leq 0$ and $r \leq 0$. Hence, (11.56) has no positive roots if and only if $p \leq 0$ and $r \leq 0$.

Case (b): $q < 0$. In this case, it is clear that $f(\hat{p}, q, r, c) < 0$ always holds for any $c \in (0, +\infty)$ if and only if (11.58) has no positive roots. Since we are concerned with positive roots of (11.58), we will restrict our attention to the case where $c > 0$. We will treat the triple $(\hat{p}, q, r)$ of real numbers as a point in $xyz - space$ and determine the exact regions containing points $(\hat{p}, q, r)$ in $xyz - space$ in which (11.58) has no positive roots. In fact, $f(x, y, z, c) = 0$ can be regarded as an equation describing a one-parameter family of planes in $xyz - space$, where x, y, and z are the coordinates of points of the plane in $xyz - space$ and c is a parameter. By giving c a specific value in $(0, +\infty)$, we obtain the equation of one of the planes of the one-parameter family of planes.

From the theory of envelopes, the points of the envelope of the one-parameter family of planes described by (11.58) satisfy the system of equations:

$$\begin{cases} f(x, y, z, c) = c^{\tau}x + c^{\tau+1}y + z = 0, \\ f_c(x, y, z, \lambda) = \tau c^{\tau-1}x + (\tau+1)c^{\tau}y = 0, \end{cases} \tag{11.59}$$

where $c > 0$. Eliminating the parameter $c > 0$ from (11.59), we obtain the equation of the envelope:

$$z(x, y) = (-1)^{\tau+1}\frac{\tau^{\tau}x^{\tau+1}}{(\tau+1)^{\tau+1}y^{\tau}}, \tag{11.60}$$

where $x > 0$ and $y < 0$. From (11.60), we have

$$\begin{aligned} \frac{\partial z}{\partial x} &= (-1)^{\tau+1}\frac{\tau^{\tau}x^{\tau}}{(\tau+1)^{\tau}y^{\tau}}, \\ \frac{\partial z}{\partial y} &= (-1)^{\tau+2}\frac{\tau^{\tau+1}x^{\tau+1}}{(\tau+1)^{\tau+1}y^{\tau+1}}, \\ \frac{\partial^2 z}{\partial x^2} &= (-1)^{\tau+1}\frac{\tau^{\tau+1}x^{\tau-1}}{(\tau+1)^{\tau}y^{\tau}}, \\ \frac{\partial^2 z}{\partial y^2} &= (-1)^{\tau+3}\frac{\tau^{\tau+1}x^{\tau+1}}{(\tau+1)^{\tau}y^{\tau+2}}, \\ \frac{\partial^2 z}{\partial x \partial y} &= (-1)^{\tau+2}\frac{\tau^{\tau+1}x^{\tau}}{(\tau+1)^{\tau}y^{\tau+1}}, \end{aligned} \tag{11.61}$$

where $x > 0$ and $y < 0$. It follows from (11.60) and (11.61) that $z(x, y) < 0$, $\partial^2 z/\partial x^2 > 0$, $\partial^2 z/\partial y^2 > 0$, and $\partial^2 z/\partial x^2 \cdot \partial^2 z/\partial y^2 - (\partial^2 z/\partial x \partial y)^2 = 0$ all hold for $x > 0$ and $y < 0$. Thus, $z(x, y)$ is negative and concave on $(0, +\infty) \times (-\infty, 0)$. Furthermore, the envelope described by (11.60) is a concave surface S over $(0, +\infty) \times (-\infty, 0)$ as shown in Fig. 11.18. In view of Fig. 11.18 and the obtained information of $z(x, y)$, it is easy to see that when the point $(\hat{p}, q, r)$ is in the seventh closed octant, that is, $\hat{p} \le 0$, $q \le 0$, and $r \le 0$, there cannot be any tangent plane of the envelope S which also passes through this point, and when the point $(\hat{p}, q, r)$ is vertically below the envelope S in the eighth octant, that is, $\hat{p} > 0$, $q < 0$, and $r < (-1)^{\tau+1}\tau^{\tau}\hat{p}^{\tau+1}/(\tau+1)^{\tau+1}q^{\tau}$, there cannot exist any tangent plane of the envelope S which also passes through this point, while the point $(\hat{p}, q, r)$ is situated

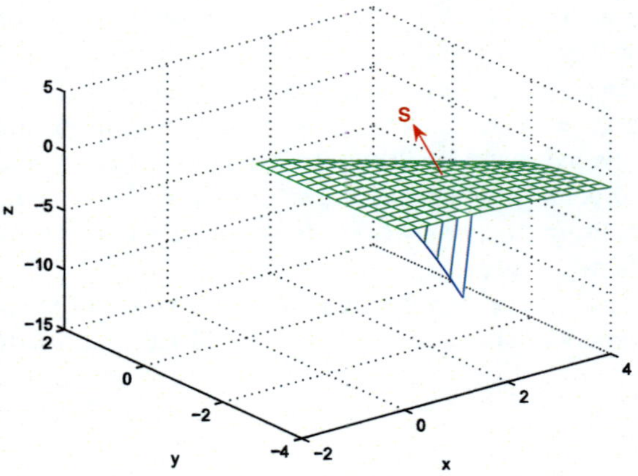

Fig. 11.18 Envelope surface for $\sigma = -2$ and $\tau = 1$

elsewhere, such a tangent plane can be drawn. Notice that (11.56) is identical with (11.58) for the existence of positive solutions and $q < 0$. It follows from Lemma 11.16 that (11.56) does not have any positive roots if and only if $\hat{p} \le 0$, $q < 0$, and $r \le 0$; or $\hat{p} > 0$, $q < 0$, and

$$r < (-1)^{\tau+1} \frac{\tau^{\tau} \hat{p}^{\tau+1}}{(\tau+1)^{\tau+1} q^{\tau}}.$$

Combining case (*a*) with case (*b*) implies that the characteristic Eq. (11.56) does not have any positive roots if and only if $\hat{p} \le 0$, $q \le 0$, and $r \le 0$; or $\hat{p} > 0$, $q < 0$, and

$$r < (-1)^{\tau+1} \frac{\tau^{\tau} \hat{p}^{\tau+1}}{(\tau+1)^{\tau+1} q^{\tau}}.$$

Lemma 11.14 implies the statement of this theorem. This completes the proof.

Remark 11.6 If we set $\sigma = -\tau - 1$ in Theorem 11.10 and define $0^0 = 1$, then the result of Theorem 11.10 gives the result of Theorem 11.11, which means that Theorem 11.11 is a special case of Theorem 11.10.

Theorem 11.12 *Assume that $\sigma < 0$, $\tau > 0$, and $\sigma > -\tau - 1$. Then every solution of system (11.50) surface oscillates if and only if $\hat{p} \le 0$, $q \le 0$, and $r \le 0$.*

Proof The proof is similar to that of Theorem 11.10 and hence will be sketched. When $\sigma < 0$, $\tau > 0$, and $\sigma > -\tau - 1$, the characteristic equation of system (11.50) is given by (11.52), where $\lambda > 0$ and $\mu > 0$. As in the proof of Theorem 11.10, $\phi(x, y, z, \lambda, \mu) = 0$ can be regarded as an equation describing a two-parameter family of planes in $xyz - space$, where x, y, and z are the coordinates of points of the planes in $xyz - space$ and λ, μ are parameters. We will restrict our attention to the case where $\lambda > 0$ and $\mu > 0$. The envelope of the two-parameter family of planes is determined by (11.53). The equation of this envelope is given by (11.54)

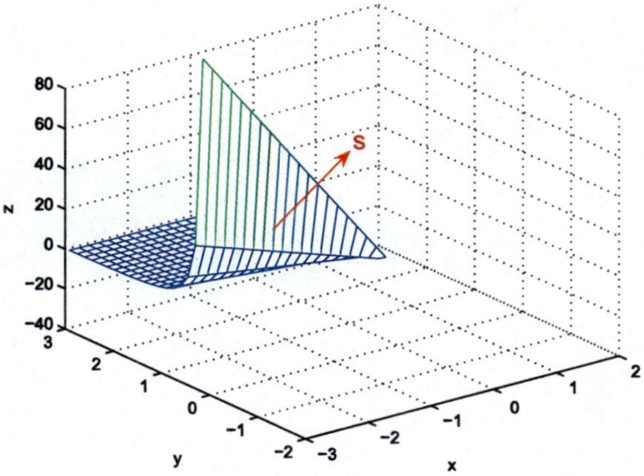

Fig. 11.19 Envelope surface for $\sigma = -1$ and $\tau = 2$

for $x < 0$ and $y > 0$, and the partial derivatives $\partial z/\partial x$, $\partial z/\partial y$, $\partial^2 z/\partial x^2$, $\partial^2 z/\partial y^2$, and $\partial^2 z/\partial x \partial y$ are given by (11.55), respectively. However, notice that when $x < 0$ and $y > 0$, we have $\partial^2 z/\partial x^2 < 0$, $\partial^2 z/\partial y^2 < 0$, and $\partial^2 z/\partial x^2 \cdot \partial^2 z/\partial y^2 - (\partial^2 z/\partial x \partial y)^2 = \sigma^{2\sigma+1}\tau^{2\tau+1}/x^{2\sigma+2}y^{2\tau+2}(\sigma+\tau+1)^{2\sigma+2\tau+1} < 0$. Thus, $z(x, y)$ is neither concave nor convex on $(-\infty, 0) \times (0, +\infty)$. Furthermore, $z(x, y) > 0$ for $x < 0$ and $y > 0$. The corresponding envelope S is neither concave nor convex surface over $(-\infty, 0) \times (0, +\infty)$ as shown in Fig. 11.19. In view of Fig. 11.19 and the obtained information of $z(x, y)$, it is easily seen that when the point (p, q, r) is in the seventh closed octant, that is, $p \le 0$, $q \le 0$, and $r \le 0$, there cannot be any tangent plane of the envelope S which also passes through this point, while the point (p, q, r) is situated elsewhere, such a tangent plane can be drawn. By Lemma 11.15, (11.52) does not have any positive roots if and only if $\hat{p} \le 0$, $q \le 0$, and $r \le 0$. From Lemma 11.14, we can see that every solution of system (11.50) oscillates if and only if $\hat{p} \le 0$, $q \le 0$, and $r \le 0$. This completes the proof.

Theorem 11.13 *Assume that $\sigma > 0$, $\tau < 0$, and $\sigma < -\tau - 1$. Then every solution of system (11.50) surface oscillates if and only if $\hat{p} \le 0$, $q \le 0$, and $r \le 0$; or $\hat{p} < 0$, $q > 0$, and*

$$r < \frac{\sigma^\sigma \tau^\tau}{(\sigma+\tau+1)^{\sigma+\tau+1}\hat{p}^\sigma q^\tau}.$$

Proof The proof is similar to that of Theorem 11.10 and hence will be sketched. When $\sigma > 0$, $\tau < 0$, and $\sigma < -\tau - 1$, the characteristic equation of system (11.50) is given by (11.52), where $\lambda > 0$ and $\mu > 0$. As in the proof of Theorem 11.10, $\phi(x, y, z, \lambda, \mu) = 0$ can be regarded as an equation describing a two-parameter family of planes in $xyz - space$, where x, y, and z are the coordinates of points of the planes in $xyz - space$ and λ, μ are parameters. We will restrict our attention to the case where $\lambda > 0$ and $\mu > 0$. The envelope of the two-parameter family of planes is determined by (11.53). The equation of this envelope is given by (11.54) for

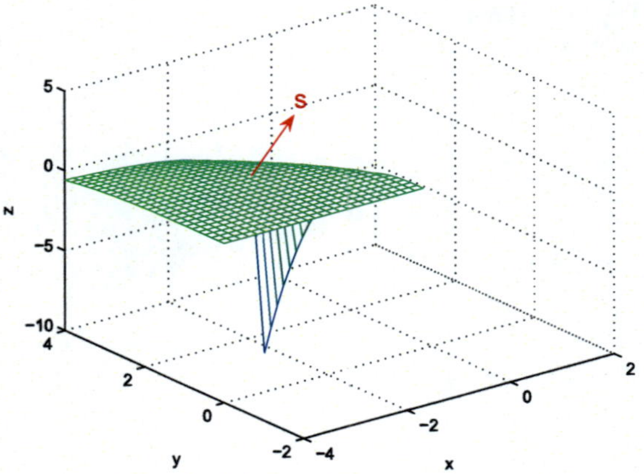

Fig. 11.20 Envelope surface for $\sigma = 1$ and $\tau = -3$

$x < 0$ and $y > 0$, and the partial derivatives $\partial z/\partial x$, $\partial z/\partial y$, $\partial^2 z/\partial x^2$, $\partial^2 z/\partial y^2$, and $\partial^2 z/\partial x \partial y$ are given by (11.55), respectively. However, note that when $x < 0$ and $y > 0$, we have $z(x, y) < 0$, $\partial^2 z/\partial x^2 < 0$, $\partial^2 z/\partial y^2 < 0$, and $\partial^2 z/\partial x^2 \cdot \partial^2 z/\partial y^2 - (\partial^2 z/\partial x \partial y)^2 = \sigma^{2\sigma+1}\tau^{2\tau+1}/x^{2\sigma+2}y^{2\tau+2}(\sigma+\tau+1)^{2\sigma+2\tau+1} > 0$. Thus, $z(x, y)$ is negative and concave on $(-\infty, 0) \times (0, +\infty)$. Furthermore, the corresponding envelope S is concave surface over $(-\infty, 0) \times (0, +\infty)$ as shown in Fig. 11.20. In view of Figure 11.20 and the obtained information of $z(x, y)$, it is easily seen that when the point $(\hat{p}, q, r)$ is in the seventh closed octant, that is, $\hat{p} \le 0$, $q \le 0$, and $r \le 0$, there cannot be any tangent plane of the envelope S which also passes through this point, and when the point $(\hat{p}, q, r)$ is vertically below the envelope S in the sixth octant, that is, $\hat{p} < 0$, $q > 0$, and $r < \sigma^\sigma \tau^\tau/(\sigma+\tau+1)^{\sigma+\tau+1}\hat{p}^\sigma q^\tau$, there cannot exist any tangent plane of the envelope S which also passes through this point, while the point $(\hat{p}, q, r)$ is situated elsewhere, such a tangent plane can be drawn. By Lemma 11.15, (11.52) does not have any positive roots if and only if $\hat{p} \le 0$, $q \le 0$, and $r \le 0$; or $\hat{p} < 0$, $q > 0$, and

$$r < \frac{\sigma^\sigma \tau^\tau}{(\sigma+\tau+1)^{\sigma+\tau+1}\hat{p}^\sigma q^\tau}.$$

Lemma 11.14 implies the statement of this theorem. The proof is thus completed.

Theorem 11.14 *Assume that $\sigma > 0$, $\tau < 0$, and $\sigma = -\tau - 1$. Then every solution of system (11.50) surface oscillates if and only if $\hat{p} \le 0$, $q \le 0$, and $r \le 0$; or $\hat{p} < 0$, $q > 0$, and*

$$r < (-1)^{\sigma+1}\frac{\sigma^\sigma q^{\sigma+1}}{(\sigma+1)^{\sigma+1}\hat{p}^\sigma}.$$

Proof When $\sigma > 0$, $\tau < 0$, and $\sigma = -\tau - 1$, the characteristic equation of system (11.50) is

$$\phi(\hat{p}, q, r, \lambda, \mu) = \hat{p}\lambda + q\mu - 1 + r\lambda^{-\sigma}\mu^{\sigma+1} = 0. \tag{11.62}$$

Substituting $\lambda = c\mu (c > 0)$ into (11.62) yields

$$\phi(\hat{p}, q, r, \mu) = (q + \hat{p}c + rc^{-\sigma})\mu - 1 = 0. \tag{11.63}$$

It is clear that (11.62) has no positive roots if and only if (11.63) has no positive roots for any $c \in (0, +\infty)$, that is, $\hat{p}c^{\sigma+1} + qc^{\sigma} + r < 0$ always holds for any $c \in (0, +\infty)$. Let

$$f(\hat{p}, q, r, c) = \hat{p}c^{\sigma+1} + qc^{\sigma} + r = 0. \tag{11.64}$$

When $\hat{p} > 0$, since $\lim_{c\to+\infty} f(\hat{p}, q, r, c) = +\infty$ and $f(\hat{p}, q, r, c)$ is continuous on $(0, +\infty)$, there exists $c^* \in (0, +\infty)$ such that $f(\hat{p}, q, r, c^*) > 0$. Hence, We only need to consider two cases: (a) $\hat{p} = 0$ and (b) $p < 0$.

Case (a): $\hat{p} = 0$. In this case, it is clear that $f(\hat{p}, q, r, c) < 0$ always holds for any $c \in (0, +\infty)$ if and only if $q \leq 0$ and $r \leq 0$. Hence, (11.62) has no positive roots if and only if $q \leq 0$ and $r \leq 0$.

Case (b): $\hat{p} < 0$. In this case, it is clear that $f(\hat{p}, q, r, c) < 0$ always holds for any $c \in (0, +\infty)$ if and only if $f(\hat{p}, q, r, c) = 0$ has no positive roots. Since we are concerned with positive roots of (11.64), our attention will be restricted to the case where $c > 0$. We will treat the triple $(\hat{p}, q, r)$ of real numbers as a point in $xyz-space$ and look for the exact regions containing points $(\hat{p}, q, r)$ in $xyz-space$ such that (11.64) has no positive roots. In fact, $f(x, y, z, c) = 0$ can be regarded as an equation describing a one-parameter family of planes in $xyz-space$, where x, y, and z are the coordinates of points of the plane in $xyz-space$ and c is a parameter. By giving c a specific value in $(0, +\infty)$, we obtain the equation of one of the planes of the one-parameter family of planes.

From the theory of envelopes, the points of the envelope of the one-parameter family of planes described by (11.64) satisfy the system of equations:

$$\begin{cases} f(x, y, z, c) = c^{\tau+1}x + c^{\tau}y + z = 0, \\ f_c(x, y, z, \lambda) = (\tau + 1)c^{\tau}x + \tau c^{\tau-1}y = 0, \end{cases} \tag{11.65}$$

where $c > 0$. Eliminating the parameter $c > 0$ from (11.65), we obtain the equation of the envelope:

$$z(x, y) = (-1)^{\sigma+1}\frac{\sigma^{\sigma}y^{\sigma+1}}{(\sigma + 1)^{\sigma+1}x^{\sigma}}, \tag{11.66}$$

where $x < 0$ and $y > 0$. From (11.66), we have

$$\begin{aligned}
\frac{\partial z}{\partial x} &= (-1)^{\sigma+2}\frac{\sigma^{\sigma+1}y^{\sigma+1}}{(\sigma+1)^{\sigma+1}x^{\sigma+1}},\\
\frac{\partial z}{\partial y} &= (-1)^{\sigma+1}\frac{\sigma^{\sigma}y^{\sigma}}{(\sigma+1)^{\sigma}x^{\sigma}},\\
\frac{\partial^2 z}{\partial x^2} &= (-1)^{\sigma+3}\frac{\sigma^{\sigma+1}y^{\sigma+1}}{(\sigma+1)^{\sigma}x^{\sigma+2}},\\
\frac{\partial^2 z}{\partial y^2} &= (-1)^{\sigma+1}\frac{\sigma^{\sigma+1}y^{\sigma-1}}{(\sigma+1)^{\sigma}x^{\sigma}},\\
\frac{\partial^2 z}{\partial x\partial y} &= (-1)^{\sigma+2}\frac{\sigma^{\sigma+1}y^{\sigma}}{(\sigma+1)^{\sigma}x^{\sigma+1}},
\end{aligned} \tag{11.67}$$

where $x < 0$ and $y > 0$. It follows from (11.66) and (11.67) that $z(x, y) < 0$, $\partial^2 z/\partial x^2 < 0$, $\partial^2 z/\partial y^2 < 0$, and $\partial^2 z/\partial x^2 \cdot \partial^2 z/\partial y^2 - (\partial^2 z/\partial x\partial y)^2 = 0$ all hold for $x < 0$ and $y > 0$. Thus, $z(x, y)$ is negative and concave on $(-\infty, 0) \times (0, +\infty)$. Furthermore, the envelope described by (11.66) is a concave surface S over $(-\infty, 0) \times (0, +\infty)$ as shown in Fig. 11.21. In view of Fig. 11.21 and the obtained information of $z(x, y)$, it is easy to see that when the point $(\hat{p}, q, r)$ is in the seventh closed octant, that is, $\hat{p} \le 0$, $q \le 0$, and $r \le 0$, there cannot be any tangent plane of the envelope S which also passes through this point, and when the point $(\hat{p}, q, r)$ is vertically below the envelope S in the sixth octant, that is, $\hat{p} < 0$, $q > 0$, and $r < (-1)^{\sigma+1}\sigma^{\sigma}q^{\sigma+1}/(\sigma+1)^{\sigma+1}\hat{p}^{\sigma}$, there cannot exist any tangent plane of the envelope S which also passes through this point, while the point $(\hat{p}, q, r)$ is situated elsewhere, such a tangent plane can be drawn. Note that (11.62) is identical with (11.64) for the existence of positive solutions and $\hat{p} < 0$. It follows from Lemma 11.16 that (11.62) does not have any positive roots if and only if $\hat{p} < 0$, $q \le 0$, and $r \le 0$; or $\hat{p} < 0$, $q > 0$, and

$$r < (-1)^{\sigma+1}\frac{\sigma^{\sigma}q^{\sigma+1}}{(\sigma+1)^{\sigma+1}\hat{p}^{\sigma}}.$$

Combining case (*a*) with case (*b*) implies that the characteristic Eq. (11.62) has no positive roots if and only if $\hat{p} \le 0$, $q \le 0$, and $r \le 0$; or $\hat{p} < 0$, $q > 0$, and

$$r < (-1)^{\sigma+1}\frac{\sigma^{\sigma}q^{\sigma+1}}{(\sigma+1)^{\sigma+1}\hat{p}^{\sigma}}.$$

Lemma 11.14 implies the statement of this theorem. This completes the proof.

Remark 11.7 If we set $\sigma = -\tau - 1$ in Theorem 11.13 and define $0^0 = 1$, then the result of Theorem 11.13 gives the result of Theorem 11.14, which means that Theorem 11.14 is a special case of Theorem 11.13.

Theorem 11.15 *Assume that* $\sigma > 0$, $\tau < 0$, *and* $\sigma > -\tau - 1$. *Then every solution of system (11.50) surface oscillates if and only if* $\hat{p} \le 0$, $q \le 0$, *and* $r \le 0$.

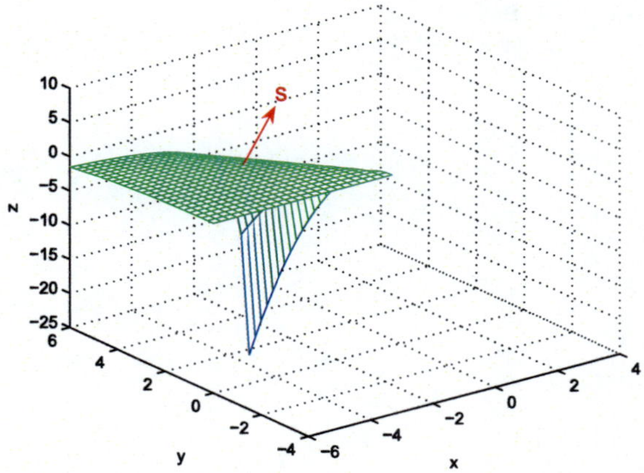

Fig. 11.21 Envelope surface for $\sigma = 1$ and $\tau = -2$

Proof The proof is similar to that of Theorem 11.10 and hence will be sketched. When $\sigma > 0, \tau < 0$, and $\sigma > -\tau - 1$, the characteristic equation of system (11.50) is given by (11.52), where $\lambda > 0$ and $\mu > 0$. As in the proof of Theorem 11.10, $\phi(x, y, z, \lambda, \mu) = 0$ can be regarded as an equation describing a two-parameter family of planes in $xyz - space$, where x, y, and z are the coordinates of points of the planes in $xyz - space$ and λ, μ are parameters. We will restrict our attention to the case where $\lambda > 0$ and $\mu > 0$. The envelope of the two-parameter family of planes is determined by (11.53). The equation of this envelope is given by (11.54) for $x > 0$ and $y < 0$, and the partial derivatives $\partial z/\partial x$, $\partial z/\partial y$, $\partial^2 z/\partial x^2$, $\partial^2 z/\partial y^2$, and $\partial^2 z/\partial x \partial y$ are given by (11.55), respectively. However, note that when $x > 0$ and $y < 0$, we have $\partial^2 z/\partial x^2 < 0$, $\partial^2 z/\partial y^2 < 0$, and $\partial^2 z/\partial x^2 \cdot \partial^2 z/\partial y^2 - (\partial^2 z/\partial x \partial y)^2 = \sigma^{2\sigma+1}\tau^{2\tau+1}/x^{2\sigma+2}y^{2\tau+2}(\sigma + \tau + 1)^{2\sigma+2\tau+1} < 0$. Thus, $z(x, y)$ is neither concave nor convex on $(0, +\infty) \times (-\infty, 0)$. Furthermore, $z(x, y) < 0$ for $x > 0$ and $y < 0$. The corresponding envelope S is neither concave nor convex surface over $(0, +\infty) \times (-\infty, 0)$ as shown in Fig. 11.22. In view of Fig. 11.22 and the obtained information of $z(x, y)$, it is easily seen that when the point $(\hat{p}, q, r)$ is in the seventh closed octant, that is, $\hat{p} \le 0$, $q \le 0$, and $r \le 0$, there cannot be any tangent plane of the envelope S which also passes through this point, while the point $(\hat{p}, q, r)$ is situated elsewhere, such a tangent plane can be drawn. Lemma 11.15 implies that (11.52) does not have any positive roots if and only if $\hat{p} \le 0, q \le 0$, and $r \le 0$. From Lemma 11.14, we can see that every solution of system (11.50) oscillates if and only if $\hat{p} \le 0, q \le 0$, and $r \le 0$. This completes the proof.

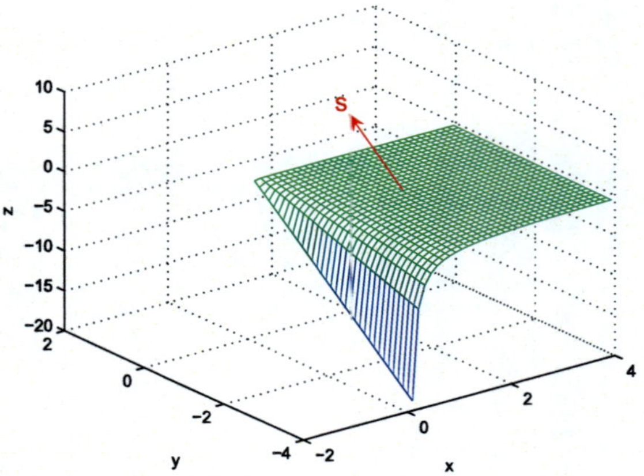

Fig. 11.22 Envelope surface for $\sigma = 2$ and $\tau = -1$

11.5.3 Illustrative Examples

Example 11.12 Consider the mixed discrete system

$$0.24u_{m+1,n} - 0.16u_{m,n+1} - u_{m,n} - u_{m+3,n-1} = 0. \tag{11.68}$$

From system (11.68), we have $\sigma = -3, \tau = 1, \hat{p} = 0.24, q = -0.16$, and $r = -1$. Since $\hat{p} = 0.24 > 0, q = -0.16 < 0$, and

$$r = -1 < \frac{0.24^3}{3^3(-0.16)} = \frac{\sigma^\sigma \tau^\tau}{(\sigma+\tau+1)^{\sigma+\tau+1} p^\sigma q^\tau},$$

according to Theorem 11.10, every solution of system (11.68) is surface oscillatory. The surface oscillatory behavior of system (11.68) is demonstrated by Fig. 11.23.

Example 11.13 Consider the mixed discrete convection system

$$0.1u_{m+1,n} - 0.1u_{m,n+1} - u_{m,n} - u_{m+2,n-1} = 0. \tag{11.69}$$

From system (11.69), we have $\sigma = -2, \tau = 1, \hat{p} = 0.1, q = -0.1$, and $r = -1$. Since $\hat{p} = 0.1 > 0, q = -0.1 < 0$, and

$$r = -1 < -\frac{0.1^2}{2^2(-0.1)} = (-1)^{\sigma+1} \frac{\tau^\tau p^\sigma}{(\tau+1)^{\tau+1} q^\tau},$$

according to Theorem 11.11, every solution of system (11.69) is surface oscillatory. The surface oscillatory behavior of system (11.69) is demonstrated by Fig. 11.24.

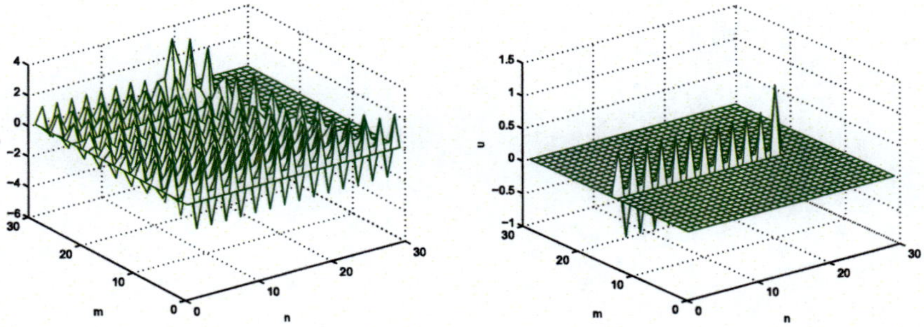

(a) (b)

Fig. 11.23 **a** Surface oscillatory behavior of system (11.68). **b** Surface oscillatory behavior of system (11.68) with m = 15

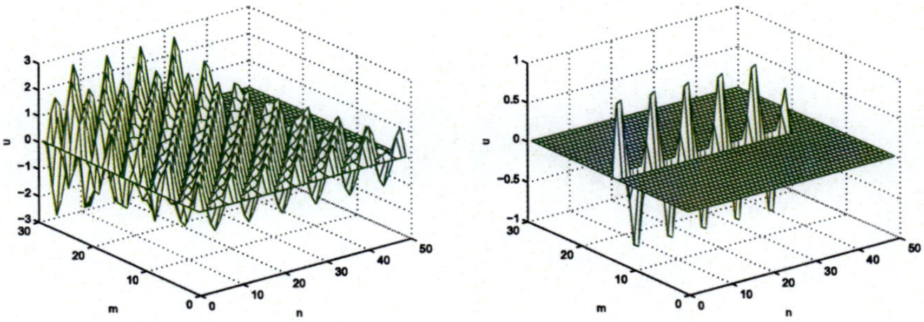

(a) (b)

Fig. 11.24 **a** Surface oscillatory behavior of system (11.69). **b** Surface oscillatory behavior of system (11.69) with m = 15

Example 11.14 Consider the mixed discrete convection system

$$- u_{m+1,n} - 0.1u_{m,n+1} - u_{m,n} - 0.02u_{m+1,n-2} = 0. \tag{11.70}$$

From system (11.70), we have $\sigma = -1, \tau = 2, \hat{p} = -1, q = -0.1$, and $r = -0.02$. Since $\hat{p} = -1 < 0, q = -0.1 < 0$, and $r = -0.02 < 0$, in view of Theorem 11.12, every solution of system (11.70) is surface oscillatory. The surface oscillatory behavior of system (11.70) is demonstrated by Fig. 11.25.

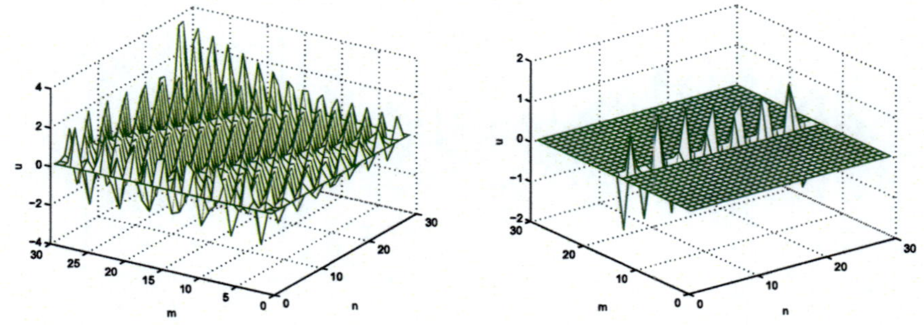

(a) (b)

Fig. 11.25 **a** Surface oscillatory behavior of system (11.70). **b** Surface oscillatory behavior of system (11.70) with m = 15

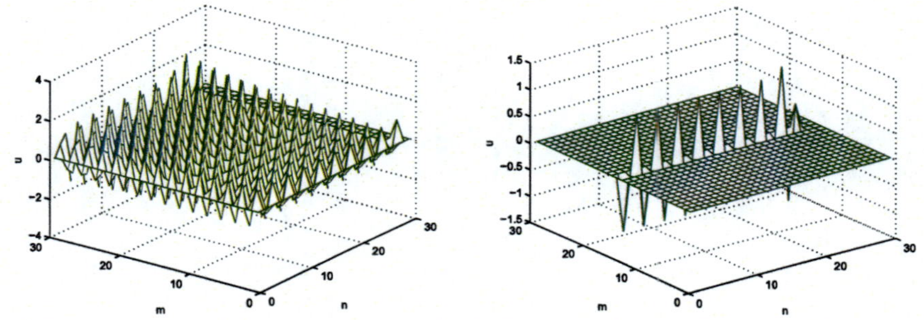

(a) (b)

Fig. 11.26 **a** Surface oscillatory behavior of system (11.71). **b** Surface oscillatory behavior of system (11.71) with m = 15

Example 11.15 Consider the mixed discrete system

$$-u_{m+1,n} + 0.05u_{m,n+1} - u_{m,n} - 0.01u_{m-1,n+3} = 0. \tag{11.71}$$

From system (11.71), we have $\sigma = 1, \tau = -3, \hat{p} = -1, q = 0.05$, and $r = -0.01$. Since $\hat{p} = -1 < 0, q = 0.05 > 0$, and

$$r = -0.01 < -\frac{0.05^3}{3^3} = \frac{\sigma^\sigma \tau^\tau}{(\sigma + \tau + 1)^{\sigma+\tau+1} \hat{p}^\sigma q^\tau},$$

according to Theorem 11.13, every solution of system (11.71) is surface oscillatory. The surface oscillatory behavior of system (11.71) is demonstrated by Fig. 11.26.

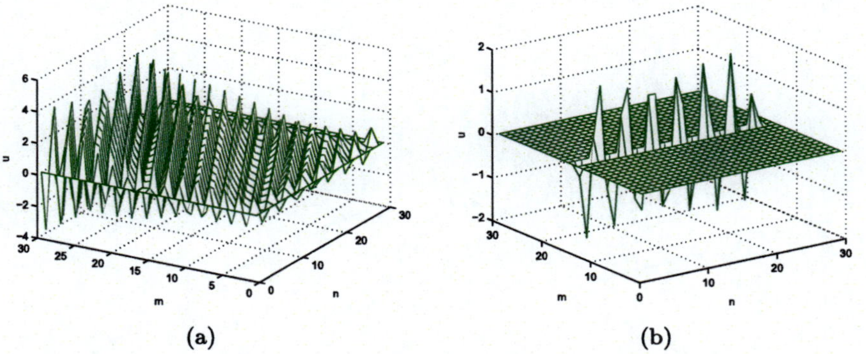

Fig. 11.27 **a** Surface oscillatory behavior of system (11.72). **b** Surface oscillatory behavior of system (11.72) with m = 15

Example 11.16 Consider the mixed discrete system

$$-u_{m+1,n} + 0.12u_{m,n+1} - u_{m,n} - 0.04u_{m-1,n+2} = 0. \tag{11.72}$$

From system (11.72), we have $\sigma = 1, \tau = -2, \hat{p} = -1, q = 0.12$, and $r = -0.04$. Since $\hat{p} = -1 < 0, q = 0.12 > 0$, and

$$r = -0.04 < -\frac{0.12^2}{2^2} = (-1)^{\sigma+1}\frac{\sigma^\sigma q^{\sigma+1}}{(\sigma+1)^{\sigma+1}\hat{p}^\sigma},$$

according to Theorem 11.14, every solution of system (11.72) is surface oscillatory. The surface oscillatory behavior of system (11.72) is demonstrated by Fig. 11.27.

Example 11.17 Consider the mixed discrete system

$$-u_{m+1,n} - 0.08u_{m,n+1} - u_{m,n} - 0.01u_{m-1,n+1} = 0. \tag{11.73}$$

From system (11.73), we have $\sigma = 1, \tau = -1, \hat{p} = -1, q = -0.08$, and $r = -0.01$. Since $\hat{p} = -1 < 0$ $q = -0.08 < 0$, and $r = -0.01 < 0$, according to Theorem 11.15, every solution of system (11.73) is surface oscillatory. The surface oscillatory behavior of system (11.73) is demonstrated by Fig. 11.28.

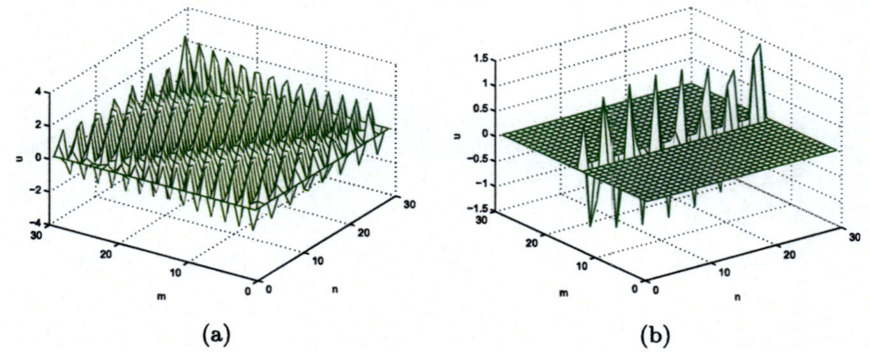

Fig. 11.28 **a** Surface oscillatory behavior of system (11.73). **b** Surface oscillatory behavior of system (11.73) with m = 15

11.6 The Case as $\mu = \hat{p}, \xi = q, p = 1, i = 1, \sigma(1) = -\sigma, \tau(1) = -\tau$

In this part,we consider the case as $\mu = \hat{p}, \xi = q, p = 1, i = 1, \sigma(1) = -\sigma, \tau(1) = -\tau$ for system (11.1):

$$\hat{p}u_{m+1,n} + qu_{m,n+1} - u_{m,n} + ru_{m+\sigma,n+\tau} = 0 \tag{11.74}$$

where $\hat{p}, q, r$ are real numbers with $\hat{p}^2 + q^2 + r^2 \neq 0$, and m, n, σ, τ are nonnegative integers.

In fact, system (11.74) may be viewed as a discrete form of the following continuous convection system with a source term and two advances:

$$\hat{p}\frac{\partial u(x, y)}{\partial x} + q\frac{\partial u(x, y)}{\partial y} + (\hat{p} + q - 1)u(x, y) + ru(x + \xi, y + \eta) = 0, \tag{11.75}$$

where $\hat{p}, q, r$ are the same as in system (11.74), and ξ, η are advances such that ξ, η are nonnegative real numbers. Nevertheless, in general it is a formidable task to derive exact solutions of system (11.75) directly in practice. Therefore, the study of system (11.74) may offer some useful information for analyzing the continuous system (11.75).

In recent years, some issues of 2-D discrete systems such as stability, oscillation, chaos, generalized synchronization of spatial chaos, have been investigated extensively (see [366, 370–373] and the references therein). All these have laid a good theoretical foundation for those theoretical studies and practical applications of 2-D discrete systems.

In this chapter, we will be mainly concerned with surface oscillation behavior of system (11.74). we try to employ a new method, based on the envelope theory of the family of planes, to derive some effective criteria to determine the surface oscillation behavior of the advanced discrete system (11.74).

11.6.1 Preliminaries

Definition 11.8 System (11.74) is said to be a system of advanced type if $\sigma \geq 0, \tau \geq 0$, and $\sigma^2 + \tau^2 \neq 0$.

Definition 11.9 A solution of (11.74) is a real double sequence $\{u_{m,n}\}$ which is defined for $m \geq 0, n \geq 0$, and which satisfies (11.74) for $m \geq 0$ and $n \geq 0$.

Definition 11.10 A solution $\{u_{m,n}\}$ of (11.74) is said to be eventually positive (or negative) if $u_{m,n} > 0$ (or $u_{m,n} < 0$) for large numbers m and n. It is said to be oscillatory if it is neither eventually positive nor eventually negative. System (11.74) is called oscillatory if all of its nontrivial solutions are oscillatory.

It is worth noting that the fixed plane equation of system (11.74) is $u^* = 0$. Therefore, the oscillation behavior of system (11.74) is that the solutions of system (11.74) oscillate about the fixed plane with an infinite number of zeros.

Lemma 11.17 ([366]) *The following statements are equivalent:*
(1) *Every solution of system (11.74) is oscillatory.*
(2) *The characteristic equation of system (11.74)*

$$\hat{p}\lambda + q\mu - 1 + r\lambda^{\sigma}\mu^{\tau} = 0$$

has no positive roots.

Lemma 11.18 ([366]) *Let*

$$f(x, \hat{p}, r) = x - \hat{p} + x^{\sigma} r,$$

where σ is a positive integer greater than or equal to 2, $\hat{p}$ and r are real parameters. Then the equation $f(x, \hat{p}, r) = 0$ has no positive roots if and only if $\hat{p} \leq 0$ and $r \geq 0$; or $\hat{p} > 0$ and

$$r < -\frac{(\sigma - 1)^{\sigma-1}}{\sigma^{\sigma}} \hat{p}^{1-\sigma}.$$

Lemma 11.19 ([370]) *Suppose that $f(x, y)$, $g(x, y)$, $h(x, y)$, and $v(x, y)$ are differentiable on $(-\infty, +\infty) \times (-\infty, +\infty)$. Let Γ be a two-parameter family of planes defined by the equation*

$$f(\lambda, \mu)x + g(\lambda, \mu)y + h(\lambda, \mu)z = v(\lambda, \mu),$$

where λ and μ are parameters. Let Σ be the envelope of the family Γ. Then the equation

$$f(\lambda, \mu)a + g(\lambda, \mu)b + h(\lambda, \mu)c = v(\lambda, \mu)$$

has no real roots if and only if there is no tangent plane of Σ passing through the point (a, b, c) in $xyz - space$.

Lemma 11.20 ([370]) *Suppose that $f(x)$, $g(x)$, $h(x)$, and $v(x)$ are differentiable on $(-\infty, +\infty)$. Let Γ be the one-parameter family of planes defined by the equation*

$$f(\lambda)x + g(\lambda)y + h(\lambda)z = v(\lambda),$$

where λ is a parameter. Let Σ be the envelope of the family Γ. Then the equation

$$f(\lambda)a + g(\lambda)b + h(\lambda)c = v(\lambda)$$

has no real roots if and only if there is no tangent plane of Σ passing through the point (a, b, c) in $xyz - space$.

Lemma 11.21 ([370]) *Suppose that $f(x)$ is differentiable on $(0, +\infty)$ such that $f(x)$ is not identically zero on $(0, +\infty)$ and $\lim_{x\to+\infty} f(x) > 0$ or $\lim_{x\to 0^+} f(x) > 0$. Then*

$$F(x, y) = y + f(x)$$

has no positive roots on $(0, +\infty) \times (0, +\infty)$ if and only if $f(x)$ has no positive roots on $(0, +\infty)$.

11.6.2 Main Results

In this section, we will establish the necessary and sufficient conditions for oscillations of all solutions of system (11.74).

Theorem 11.16 *Assume that $\sigma > 0$ and $\tau > 0$. Then every solution of system (11.74) surface oscillates if and only if $\hat{p} \leq 0$, $q \leq 0$, and $r \leq 0$.*

Proof When $\sigma > 0$ and $\tau > 0$, then the characteristic equation of system (11.74) is

$$\phi(\hat{p}, q, r, \lambda, \mu) = \hat{p}\lambda + q\mu - 1 + r\lambda^{\sigma}\mu^{\tau} = 0. \tag{11.76}$$

Since we are concerned with positive roots of (11.76), we only need to consider the case where $\lambda > 0$ and $\mu > 0$. We will treat the triple $(\hat{p}, q, r)$ as a point in $xyz - space$, and try to find the exact regions containing points $(\hat{p}, q, r)$ in $xyz - space$ such that (11.76) has no positive roots. In fact, $\phi(x, y, z, \lambda, \mu) = 0$ can be regarded as an equation describing a two-parameter family of planes in $xyz - space$, where

x, y, and z are the coordinates of points of the planes in $xyz - space$ and λ, μ are parameters. By giving the pair (λ, μ) a specific value in $(0, +\infty) \times (0, +\infty)$, we obtain the equation of one of the planes of the two-parameter family of planes.

In view of the envelop theory, the points of the envelope of the two-parameter family of planes defined by (11.76) satisfy the system of equations:

$$\begin{cases} \phi(x, y, z, \lambda, \mu) = 0, \\ \phi_\lambda(x, y, z, \lambda, \mu) = x + \sigma\lambda^{\sigma-1}\mu^\tau z = 0, \\ \phi_\mu(x, y, z, \lambda, \mu) = y + \tau\lambda^\sigma\mu^{\tau-1}z = 0, \end{cases} \tag{11.77}$$

where $\lambda > 0$ and $\mu > 0$. Eliminating the two parameters λ and μ from (11.77), we get the equation of the envelope:

$$z(x, y) = -\frac{(\sigma + \tau - 1)^{\sigma+\tau-1}}{\sigma^\sigma\tau^\tau}x^\sigma y^\tau, \tag{11.78}$$

where $x > 0$ and $y > 0$. From (11.78), we get

$$\begin{aligned}
\frac{\partial z}{\partial x} &= -\frac{(\sigma + \tau - 1)^{\sigma+\tau-1}}{\sigma^{\sigma-1}\tau^\tau}x^{\sigma-1}y^\tau, \\
\frac{\partial z}{\partial y} &= -\frac{(\sigma + \tau - 1)^{\sigma+\tau-1}}{\sigma^\sigma\tau^{\tau-1}}x^\sigma y^{\tau-1}, \\
\frac{\partial^2 z}{\partial x^2} &= -\frac{(\sigma - 1)(\sigma + \tau - 1)^{\sigma+\tau-1}}{\sigma^{\sigma-1}\tau^\tau}x^{\sigma-2}y^\tau, \\
\frac{\partial^2 z}{\partial y^2} &= -\frac{(\tau - 1)(\sigma + \tau - 1)^{\sigma+\tau-1}}{\sigma^\sigma\tau^{\tau-1}}x^\sigma y^{\tau-2}, \\
\frac{\partial^2 z}{\partial x\partial y} &= -\frac{(\sigma + \tau - 1)^{\sigma+\tau-1}}{\sigma^{\sigma-1}\tau^{\tau-1}}x^{\sigma-1}y^{\tau-1}.
\end{aligned}$$

Notice that when $x > 0$ and $y > 0$, we have $\partial^2 z/\partial x^2 < 0$, $\partial^2 z/\partial y^2 < 0$ and $\partial^2 z/\partial x^2 \cdot \partial^2 z/\partial y^2 - (\partial^2 z/\partial x\partial y)^2 = -x^{2\sigma-2}y^{2\tau-2}(\sigma + \tau - 1)^{2\sigma+2\tau-1}/\sigma^{2\sigma-1}\tau^{2\tau-1} < 0$. Thus, $z(x, y)$ is neither concave nor convex on $(0, +\infty) \times (0, +\infty)$. Furthermore, $z(x, y) < 0$ for $x > 0$ and $y > 0$, $\lim_{(x,y)\to(+\infty,+\infty)} z(x, y) = -\infty$ and $\lim_{(x,y)\to(0^+,0^+)} z(x, y) = 0$. The envelope described by (11.78) is a neither concave nor convex surface S over $(0, +\infty) \times (0, +\infty)$ as depicted in Fig. 11.29. In view of Fig. 11.29 and the obtained information of $z(x, y)$, it is easily seen that when the point $(\hat{p}, q, r)$ is in the seventh closed octant, that is, $\hat{p} \leq 0, q \leq 0$, and $r \leq 0$, there cannot be any tangent plane of the envelope S which also passes through this point, while the point (p, q, r) is situated elsewhere, such a tangent plane can be drawn. By Lemmas in this section, (11.76) does not have any positive roots if and only if $\hat{p} \leq 0, q \leq 0$, and $r \leq 0$. The proof is thus completed.

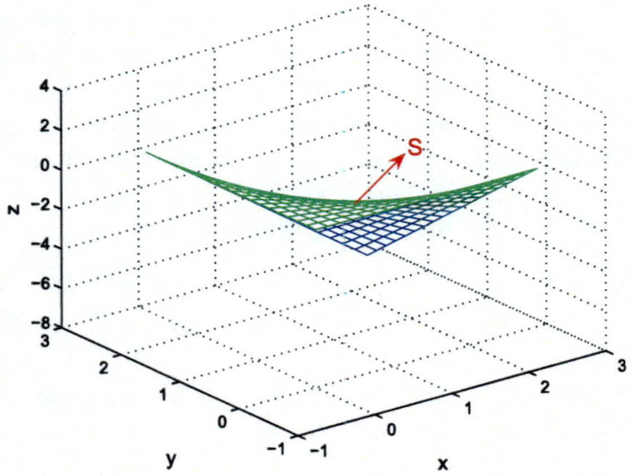

Fig. 11.29 Envelope surface for $\sigma = 1$ and $\tau = 1$

Theorem 11.17 *Assume that $\sigma > 1$ and $\tau = 0$. Then every solution of system (11.74) surface oscillates if and only if $\hat{p} \leq 0$, $q \leq 0$, and $r \leq 0$; or $\hat{p} > 0$, $q \leq 0$, and*

$$r < -\frac{(\sigma-1)^{\sigma-1}}{\sigma^\sigma}\hat{p}^\sigma.$$

Proof When $\sigma > 1$ and $\tau = 0$, then the characteristic equation of system (11.74) is

$$\phi(\hat{p}, q, r, \lambda, \mu) = \hat{p}\lambda + q\mu - 1 + r\lambda^\sigma = 0. \tag{11.79}$$

When $q > 0$, it is clear that (11.79) has positive solutions. We only need to consider two cases: (1) $q = 0$ and (2) $q < 0$.

Case (1) $q = 0$. In this case, (11.79) can be written as

$$\phi(\hat{p}, r, \lambda) = \hat{p}\lambda - 1 + r\lambda^\sigma = 0.$$

It easily follows from Lemma 11.18 that $\phi(p, r, \lambda) = 0$ has no positive roots if and only if $\hat{p} \leq 0$ and $r \leq 0$; or $\hat{p} > 0$ and

$$r < -\frac{(\sigma-1)^{\sigma-1}}{\sigma^\sigma}p^\sigma.$$

Case (2) $q < 0$. In this case, (11.79) can be rewritten as

$$\phi(\hat{p}, q, r, \lambda, \mu) = q\left(\mu + \frac{\hat{p}}{q}\lambda - \frac{1}{q} + \frac{r}{q}\lambda^\sigma\right) = 0.$$

Let

$$F\left(\frac{1}{q}, \frac{\hat{p}}{q}, \frac{r}{q}, \lambda\right) = \frac{\hat{p}}{q}\lambda - \frac{1}{q} + \frac{r}{q}\lambda^\sigma = 0. \tag{11.80}$$

Since $\lim_{\lambda \to 0^+} F(1/q, \hat{p}/q, r/q, \lambda) = -1/q > 0$ for $q < 0$, Lemma 11.21 implies that (11.79) has no positive roots if and only if (11.80) has no positive roots. Since we are concerned with positive roots of (11.80), we will restrict our discussion to the case where $\lambda > 0$. We will treat the triple $(1/q, \hat{p}/q, r/q)$ as a point in $xyz-space$ and look for the exact regions containing points $(1/q, \hat{p}/q, r/q)$ in $xyz-space$ such that (11.80) has no positive roots. In fact, $F(x, y, z, \lambda) = 0$ can be regarded as an equation describing a one-parameter family of planes in $xyz-space$, where x, y, and z are the coordinates of points of the plane in $xyz-space$ and λ is a parameter. By giving λ a specific value in $(0, +\infty)$, we obtain the equation of one of the planes of the one-parameter family of planes.

From the theory of envelopes, the points of the envelope of the one-parameter family of planes described by (11.80) satisfy the system of equations:

$$\begin{cases} F(x, y, z, \lambda) = -x + \lambda y + \lambda^\sigma z = 0, \\ F_\lambda(x, y, z, \lambda) = y + \sigma\lambda^{\sigma-1} z = 0, \end{cases} \tag{11.81}$$

where $\lambda > 0$. Eliminating the parameter $\lambda > 0$ from (11.81), we obtain the equation of the envelope:

$$z(x, y) = -\frac{(\sigma-1)^{\sigma-1} y^\sigma}{\sigma^\sigma x^{\sigma-1}}, \tag{11.82}$$

where $x < 0$ and $y < 0$. From (11.82), we have

$$\begin{aligned} \frac{\partial z}{\partial x} &= \frac{(\sigma-1)^\sigma y^\sigma}{\sigma^\sigma x^\sigma}, \\ \frac{\partial z}{\partial y} &= -\frac{(\sigma-1)^{\sigma-1} y^{\sigma-1}}{\sigma^{\sigma-1} x^{\sigma-1}}, \\ \frac{\partial^2 z}{\partial x^2} &= -\frac{(\sigma-1)^\sigma y^\sigma}{\sigma^{\sigma-1} x^{\sigma+1}}, \\ \frac{\partial^2 z}{\partial y^2} &= -\frac{(\sigma-1)^\sigma y^{\sigma-2}}{\sigma^{\sigma-1} x^{\sigma-1}}, \\ \frac{\partial^2 z}{\partial x \partial y} &= \frac{(\sigma-1)^\sigma y^{\sigma-1}}{\sigma^{\sigma-1} x^\sigma}. \end{aligned}$$

Notice that when $x < 0$ and $y < 0$, we have $z(x, y) > 0, \partial^2 z/\partial x^2 > 0, \partial^2 z/\partial y^2 > 0$, and $\partial^2 z/\partial x^2 \cdot \partial^2 z/\partial y^2 - (\partial^2 z/\partial x \partial y)^2 = 0$. Thus, $z(x, y)$ is a positive and convex function on $(-\infty, 0) \times (-\infty, 0)$. Furthermore, the envelope defined by (11.82) is a convex surface S over $(-\infty, 0) \times (-\infty, 0)$ as depicted in Fig. 11.30. In view of Fig. 11.30 and the obtained information of $z(x, y)$, it is easy to see that when the point $(1/q, \hat{p}/q, r/q)$ is in the second closed octant except on the negative coordinate axis $x = 0$, that is, $1/q < 0$, $\hat{p}/q \geq 0$, and $r/q \geq 0$, which reduce to $\hat{p} \leq 0$, $q < 0$, and $r \leq 0$, there cannot be any tangent plane of the envelope S which also passes through this point, and when the point $(1/q, \hat{p}/q, r/q)$ is vertically above the envelope S in the

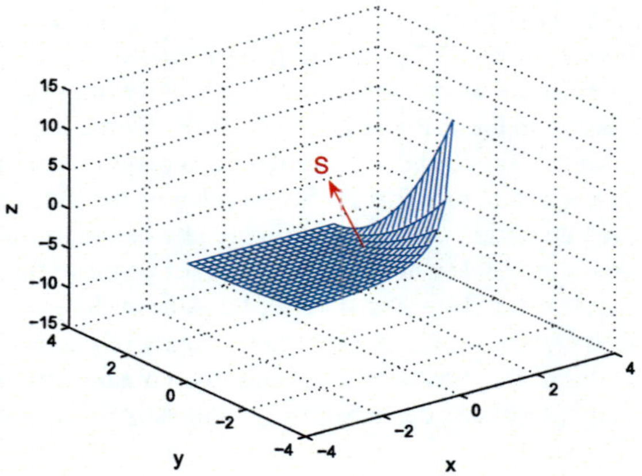

Fig. 11.30 Envelope surface for $\sigma = 0$ and $\tau = 1$

third octant, that is, $1/q < 0, \hat{p}/q < 0$, and $r/q > -(\hat{p}/q)^{\sigma}(\sigma-1)^{\sigma-1}/(1/q)^{\sigma-1}\sigma^{\sigma}$, which reduce to $\hat{p} > 0, q < 0$, and $r < -(\sigma-1)^{\sigma-1}\hat{p}^{\sigma}/\sigma^{\sigma}$, there cannot exist any tangent plane of the envelope S which also passes through this point, while the point$(1/q, \hat{p}/q, r/q)$ is situated elsewhere, such a tangent plane can be drawn. Since (11.79) is identical with (11.80) for the existence of positive solutions, Lemma 11.20 implies that (11.79) does not have any positive roots if and only if $\hat{p} \le 0, q < 0$, and $r \le 0$; or $\hat{p} > 0, q < 0$, and

$$r < -\frac{(\sigma-1)^{\sigma-1}}{\sigma^{\sigma}}p^{\sigma}.$$

Combining case (1) with case (2) implies that (11.79) does not have any positive roots if and only if $\hat{p} \le 0, q \le 0$, and $r \le 0$; or $\hat{p} > 0, q \le 0$, and

$$r < -\frac{(\sigma-1)^{\sigma-1}}{\sigma^{\sigma}}p^{\sigma}.$$

Lemma 11.17 implies the statement of this theorem. This completes the proof.

Theorem 11.18 *Assume that $\sigma = 0$ and $\tau > 1$. Then every solution of system (11.74) surface oscillates if and only if $\hat{p} \le 0$, $q \le 0$, and $r \le 0$; or $\hat{p} \le 0$, $q > 0$, and*

$$r < -\frac{(\tau-1)^{\tau-1}}{\tau^{\tau}}q^{\tau}.$$

Proof When $\sigma = 0$ and $\tau > 1$, then the characteristic equation of system (11.74) is

$$\phi(\hat{p}, q, r, \lambda, \mu) = \hat{p}\lambda + q\mu - 1 + r\mu^{\tau} = 0. \tag{11.83}$$

When $\hat{p} > 0$, it is clear that (11.83) has positive solutions. we only need to consider two cases: (1) $\hat{p} = 0$ and (2) $\hat{p} < 0$.

Case (1) $\hat{p} = 0$. In this case, (11.83) can be written as

$$\phi(q, r, \mu) = q\mu - 1 + r\mu^{\tau} = 0.$$

It easily follows from Lemma 11.18 that $\phi(q, r, \mu) = 0$ has no positive roots if and only if $q \leq 0$ and $r \leq 0$, or $q > 0$ and

$$r < -\frac{(\tau - 1)^{\tau-1}}{\tau^{\tau}} q^{\tau}.$$

Case (2) $p < 0$. In this case, (11.83) can be rewritten as

$$\phi(\hat{p}, q, r, \lambda, \mu) = \hat{p}\left(\lambda + \frac{q}{\hat{p}}\mu - \frac{1}{\hat{p}} + \frac{r}{\hat{p}}\mu^{\tau}\right) = 0.$$

Let

$$F\left(\frac{1}{\hat{p}}, \frac{q}{\hat{p}}, \frac{r}{\hat{p}}, \lambda\right) = \frac{q}{\hat{p}}\mu - \frac{1}{\hat{p}} + \frac{r}{\hat{p}}\mu^{\tau} = 0. \tag{11.84}$$

Since $\lim_{\mu \to 0^+} F(1/\hat{p}, q/\hat{p}, r/\hat{p}, \mu) = -1/\hat{p} > 0$ for $\hat{p} < 0$, Lemma 11.21 implies that (11.83) has no positive roots if and only if (11.84) has no positive roots. Since we are concerned with positive roots of (11.84), we will restrict our discussion to the case where $\mu > 0$. We will treat the triple $(1/\hat{p}, q/\hat{p}, r/\hat{p})$ as a point in $xyz - space$ and look for the exact regions containing points $(1/p, q/\hat{p}, r/\hat{p})$ in $xyz - space$ such that (11.84) has no positive roots. In fact, $F(x, y, z, \mu) = 0$ can be regarded as an equation describing a one-parameter family of planes in $xyz - space$, where x, y, and z are the coordinates of points of the plane in $xyz - space$ and μ is a parameter. By giving μ a specific value in $(0, +\infty)$, we obtain the equation of one of the planes of the one-parameter family of planes.

From the theory of envelopes, the points of the envelope of the one-parameter family of planes described by (11.84) satisfy the system of equations:

$$\begin{cases} F(x, y, z, \mu) = -x + \mu y + \mu^{\tau} z = 0, \\ F_{\mu}(x, y, z, \mu) = y + \tau\mu^{\tau-1} z = 0, \end{cases} \tag{11.85}$$

where $\mu > 0$. Eliminating the parameter μ from (11.85), we obtain the equation of the envelope:

$$z(x, y) = -\frac{(\tau - 1)^{\tau-1} y^{\tau}}{\tau^{\tau} x^{\tau-1}}, \tag{11.86}$$

where $x < 0$ and $y < 0$. From (11.86), we have

$$\frac{\partial z}{\partial x} = \frac{(\tau - 1)^{\tau} y^{\tau}}{\tau^{\tau} x^{\tau}},$$

$$\frac{\partial z}{\partial y} = -\frac{(\tau-1)^{\tau-1}y^{\tau-1}}{\tau^{\tau-1}x^{\tau-1}},$$
$$\frac{\partial^2 z}{\partial x^2} = -\frac{(\tau-1)^{\tau}y^{\tau}}{\tau^{\tau-1}x^{\tau+1}},$$
$$\frac{\partial^2 z}{\partial y^2} = -\frac{(\tau-1)^{\tau}y^{\tau-2}}{\tau^{\tau-1}x^{\tau-1}},$$
$$\frac{\partial^2 z}{\partial x\partial y} = \frac{(\tau-1)^{\tau}y^{\tau-1}}{\tau^{\tau-1}x^{\tau}}.$$

Notice that when $x < 0$ and $y < 0$, we have $z(x, y) > 0$, $\partial^2 z/\partial x^2 > 0$, $\partial^2 z/\partial y^2 > 0$, and $\partial^2 z/\partial x^2 \cdot \partial^2 z/\partial y^2 - (\partial^2 z/\partial x\partial y)^2 = 0$. Thus, $z(x, y)$ is a positive and convex function on $(-\infty, 0) \times (-\infty, 0)$. Furthermore, the envelope defined by (11.86) is a convex surface S over $(x, y) \in (-\infty, 0) \times (-\infty, 0)$ as depicted in Fig. 11.31. In view of Fig. 11.31 and the obtained information of $z(x, y)$, it is easy to see that when the point $(1/\hat{p}, q/\hat{p}, r/\hat{p})$ is in the second closed octant except on the negative coordinate axis $x = 0$, that is, $1/\hat{p} < 0$, $q/\hat{p} \geq 0$, and $r/\hat{p} \geq 0$, which reduce to $\hat{p} < 0$, $q \leq 0$, and $r \leq 0$, there cannot be any tangent plane of the envelope S which also passes through this point, and when the point $(1/\hat{p}, q/\hat{p}, r/\hat{p})$ is vertically above the envelope S in the third octant, that is, $1/\hat{p} < 0$, $q/\hat{p} < 0$, and $r/\hat{p} > -(q/\hat{p})^{\tau}(\tau-1)^{\tau-1}/(1/\hat{p})^{\tau-1}\tau^{\tau}$, which reduce to $\hat{p} < 0$, $q > 0$, and $r < -(\tau-1)^{\tau-1}q^{\tau}/\tau^{\tau}$, there cannot exist any tangent plane of the envelope S which also passes through this point, while the point $(1/\hat{p}, q/\hat{p}, r/\hat{p})$ is situated elsewhere, such a tangent plane can be drawn. Since (11.83) is identical with (11.84) for the existence of positive solutions, Lemma 11.20 implies that (11.83) does not have any positive roots if and only if $\hat{p} < 0$, $q \leq 0$, and $r \leq 0$; or $\hat{p} < 0$, $q > 0$, and

$$r < -\frac{(\tau-1)^{\tau-1}}{\tau^{\tau}}q^{\tau}.$$

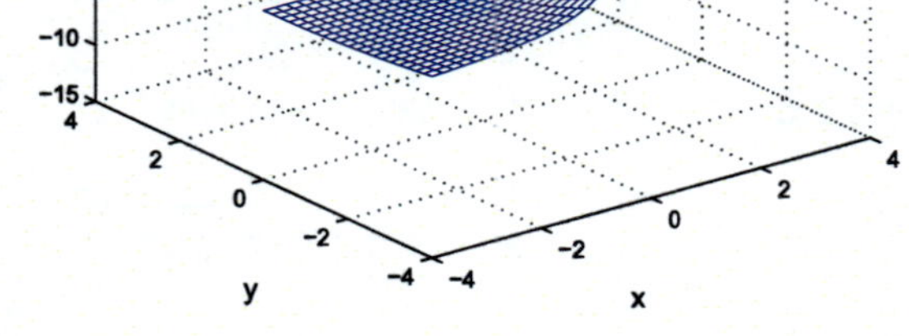

Fig. 11.31 Envelope surface for $\sigma = 0$ and $\tau = 2$

Combining case (1) with case (2) implies that (11.83) does not have any positive roots if and only if $\hat{p} \leq 0, q \leq 0$, and $r \leq 0$; or $\hat{p} \leq 0, q > 0$, and

$$r < -\frac{(\tau-1)^{\tau-1}}{\tau^\tau} q^\tau.$$

Lemma 11.17 implies the statement of this theorem. This completes the proof.

Theorem 11.19 *Assume that $\sigma = 0$ and $\tau = 1$. Then every solution of system (11.74) surface oscillates if and only if $\hat{p} \leq 0$ and $(q + r) \leq 0$.*

Proof When $\sigma = 0$ and $\tau = 1$, one can rewrite system (11.74) as

$$\hat{p}u_{m+1,n} + (q + r)u_{m,n+1} - u_{m,n} = 0. \tag{11.87}$$

The characteristic equation of system (11.87) is

$$\hat{p}\lambda + (q + r)\mu - 1 = 0. \tag{11.88}$$

It is clear that (11.88) does not have any positive roots if and only if $\hat{p} \leq 0$ and $(q + r) \leq 0$. Lemma 11.17 implies the statement of this theorem. This completes the proof.

Theorem 11.20 *Assume that $\sigma = 1$ and $\tau = 0$. Then every solution of system (11.74) surface oscillates if and only if $q \leq 0$ and $(\hat{p} + r) \leq 0$.*

Proof The proof is similar to the proof of theorem 11.19 and thus is omitted.

11.6.3 Illustrative Examples

In this section, some examples are given to illustrate the results above.

Example 11.18 Consider the advanced discrete system

$$-0.21u_{m+1,n} - 0.1u_{m,n+1} - u_{m,n} - u_{m+1,n+1} = 0. \tag{11.89}$$

From system (11.89), we have $\sigma = 1, \tau = 1, p = -0.21, q = -0.1$, and $r = -1$. Since $\hat{p} = -0.21 < 0, q = -0.1 < 0$, and $r = -1 < 0$, according to Theorem 11.16, every solution of system (11.89) is surface oscillatory. The surface oscillatory behavior of system (11.89) is demonstrated by Fig. 11.32.

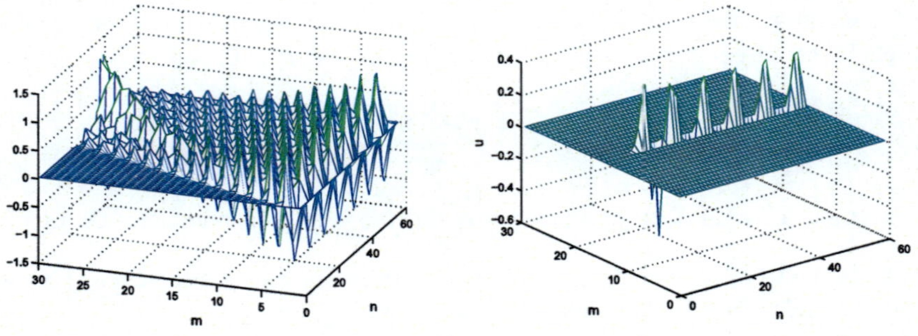

(a) (b)

Fig. 11.32 **a** Fixed plane of system (11.89). **b** Surface oscillatory behavior of system (11.89). **c** Surface oscillatory behavior of system (11.89) along its fixed plane. **d** Surface oscillatory behavior of system (11.89) with m = 15

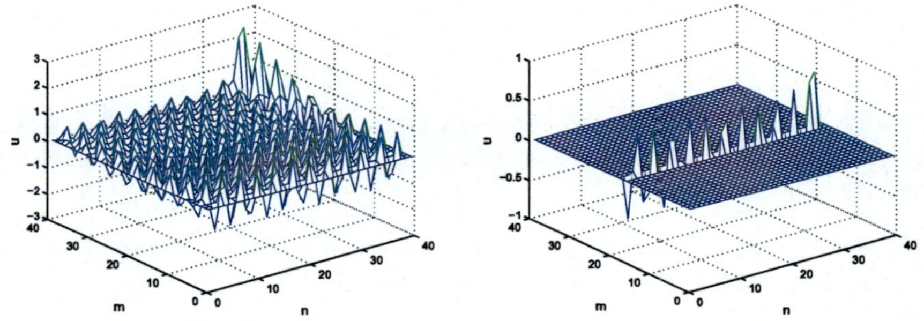

(a) (b)

Fig. 11.33 **a** Fixed plane of system (11.90). **b** Surface oscillatory behavior of system (11.90). **c** Surface oscillatory behavior of system (11.90) along its fixed plane. **d** Surface oscillatory behavior of system (11.90) with m = 18

Example 11.19 Consider the advanced discrete convection system

$$0.1u_{m+1,n} - 0.17u_{m,n+1} - u_{m,n} - u_{m+2,n} = 0. \tag{11.90}$$

From system (11.90), we have $\sigma = 2, \tau = 0, p = 0.1, q = -0.17$, and $r = -1$. Since $\hat{p} = 0.1 > 0, q = -0.17 < 0$, and

$$r = -1 < -\frac{1}{2^2}(0.1)^2 = -\frac{(\sigma - 1)^{\sigma-1}}{\sigma^\sigma}p^\sigma,$$

according to Theorem 11.16, every solution of system (11.90) is surface oscillatory. The surface oscillatory behavior of system (11.90) is demonstrated by Fig. 11.33.

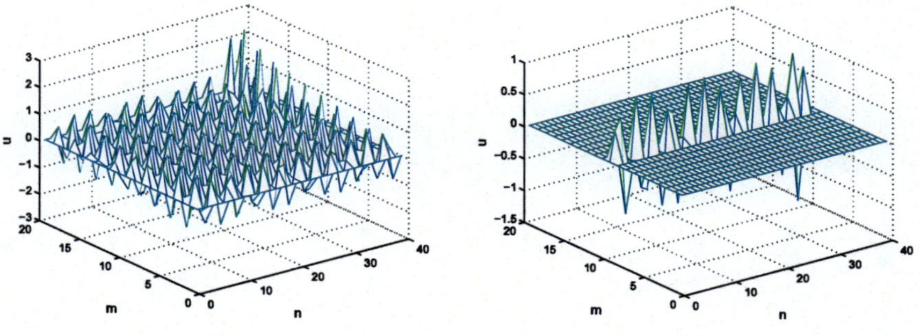

(a) (b)

Fig. 11.34 **a** Fixed plane of system (11.91). **b** Surface oscillatory behavior of system (11.91). **c** Surface oscillatory behavior of system (11.91) along its fixed plane. **d** Surface oscillatory behavior of system (11.91) with m = 10

Example 11.20 Consider the advanced discrete convection system

$$-u_{m+1,n} + 0.1u_{m,n+1} - u_{m,n} - 0.06u_{m,n+2} = 0. \tag{11.91}$$

From system (11.91), we have $\sigma = 0, \tau = 2, \hat{p} = -1, q = 0.1$, and $r = -0.06$. Since $\hat{p} = -1 < 0, q = 0.1 > 0$, and

$$r = -0.06 < -\frac{1}{2^2}(0.1)^2 = -\frac{(\tau - 1)^{\tau-1}}{\tau^\tau} q^\tau,$$

in view of Theorem 11.17, every solution of system (11.91) is surface oscillatory. The surface oscillatory behavior of system (11.91) is demonstrated by Fig. 11.34.

Example 11.21 Consider the advanced discrete system

$$-u_{m+1,n} - 0.1u_{m,n+1} - u_{m,n} = 0. \tag{11.92}$$

From system (11.92), we have $\sigma = 0, \tau = 1, \hat{p} = -1$, and $q + r = -0.1$. Since $\hat{p} = -1 < 0$ and $q + r = -0.1 < 0$, according to Theorem 11.19, every solution of system (11.92) is surface oscillatory. The surface oscillatory behavior of system (11.92) is demonstrated by Fig. 11.35.

Example 11.22 Consider the advanced discrete system

$$-u_{m+1,n} - 0.08u_{m,n+1} - u_{m,n} = 0. \tag{11.93}$$

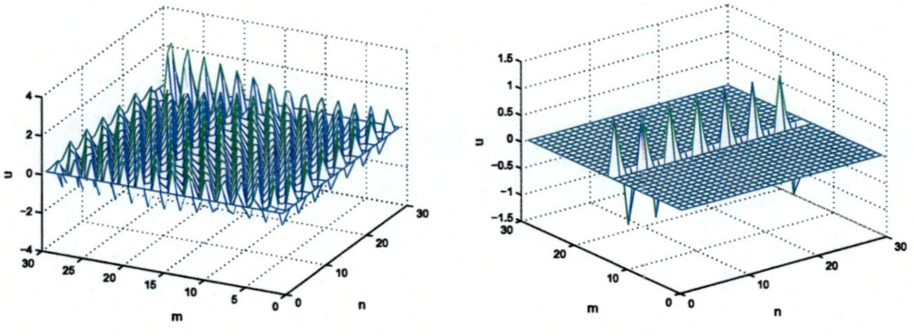

(a) (b)

Fig. 11.35 **a** Fixed plane of system (11.92). **b** Surface oscillatory behavior of system (11.92). **c** Surface oscillatory behavior of system (11.92) along its fixed plane. **d** Surface oscillatory behavior of system (11.92) with m = 15

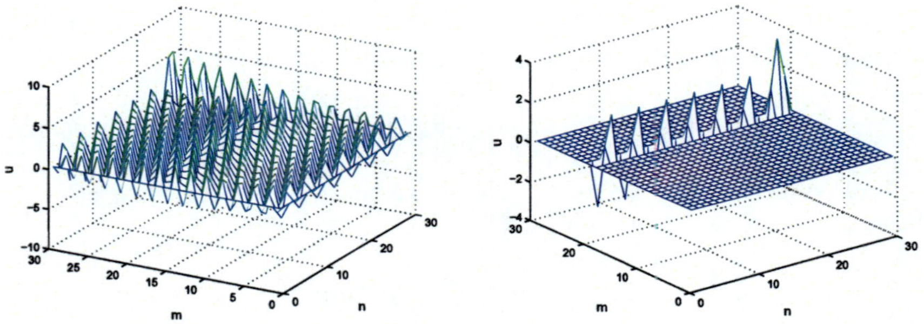

(a) (b)

Fig. 11.36 **a** Fixed plane of system (11.93). **b** Surface oscillatory behavior of system (11.93). **c** Surface oscillatory behavior of system (11.93) along its fixed plane. **d** Surface oscillatory behavior of system (11.93) with m = 20

From system (11.93), we have $\sigma = 1$, $\tau = 0$, $q = -0.08$, and $\hat{p} + r = -1$. Since $q = -0.08 < 0$ and $\hat{p} + r = -1 < 0$, according to Theorem 11.20, every solution of system (11.93) is surface oscillatory. The surface oscillatory behavior of system (11.93) is demonstrated by Fig. 11.36.

Chapter 12
Nonlinear Analysis of the Process from the Wave to Surface Chaos

This chapter is of particular interest to introduce the oscillatory behaviors which is called as wave behaviors in this book for second-order nonlinear partial difference equations, because they are discrete analogues of second-order nonlinear partial differential equations and have many physical applications (see, for example, [382–384]). Recently, there has been an increasing interest in the study of the asymptotic behaviors of solutions, especially oscillatory behaviors, of delay partial difference equations (see, for example, [385–391]).

It is an interesting question to ask if one can extend oscillatory criteria for second-order nonlinear partial differential equations to its discrete counterpart–nonlinear two-dimensional partial difference systems, and, if so, how. This chapter attempts to provide some answers to this question.

More precisely, general nonlinear two-dimensional partial difference systems are considered in this part, which also include half-linear and quasilinear partial difference equations as special cases.

Consider the following nonlinear two-dimensional partial difference systems:

$$\begin{cases} \Delta_1(x_{mn}) - b_{mn}g(y_{mn}) = 0, & m, n \in N_i = \{i, i+1, \ldots, \}, \\ T(\Delta_1, \Delta_2)(y_{mn}) + a_{mn}f(x_{mn}) = 0, & i = 0, 1, 2, \ldots, \end{cases} \tag{12.1}$$

where $T(\Delta_1, \Delta_2) = \Delta_1 + \Delta_2 + I$, $\Delta_1 y_{mn} = y_{m+1,n} - y_{mn}$, $\Delta_2 y_{mn} = y_{m,n+1} - y_{mn}$, $I_{mn} y_{mn} = y_{mn}$, $\{a_{mn}\}$ and $\{b_{mn}\}$ are real sequences, for $(m, n) \in N_0$, f, g are continuous real value functions on R with the following sign property:

$$uf(u) > 0 \text{ and } ug(u) > 0 \text{ for all } u \neq 0. \tag{12.2}$$

Its continuous form is listed as below:

$$\begin{cases} \frac{\partial X(x,y)}{\partial x} - b(x, y)g(Y(x, y)) = 0, \\ \frac{\partial Y(x,y)}{\partial x} + \frac{\partial Y(x,y)}{\partial y} - Y(x, y) + a(x, y)f(X(x, y)) = 0. \end{cases} \tag{12.3}$$

S. T. Liu and L. Zhang, *Surface Chaos and Its Applications*,
https://doi.org/10.1007/978-981-16-8229-2_12

The generalized system can be considered as follows:

$$\begin{cases} T(\nabla_1, \nabla_2)(x_{mn}) + b_{mn}g(y_{mn}) = 0, \\ T(\Delta_1, \Delta_2)(y_{mn}) + a_{mn}f(x_{mn}) = 0, \end{cases} \tag{12.4}$$

which implies:

$$\begin{cases} x_{m+1,n} + x_{m,n+1} - x_{m,n} + b_{m,n}g(y_{m,n}) = 0, \\ y_{m-1,n} + y_{m,n-1} - y_{m,n}y_{m,n}) + a_{m,n}f(x_{m,n}) = 0, \end{cases} \tag{12.5}$$

where, $m, n \in N_0 = \{0, 1, 2, \ldots, \}$, $T(\nabla_1, \nabla_2) = \nabla_1 + \nabla_2 + I$, $\nabla_1 y_{mn} = y_{m-1,n} - y_{mn}$, $\nabla_2 y_{mn} = y_{m,n-1} - y_{mn}$. And its continuous PDE form is listed as follows:

$$\begin{cases} \frac{\partial X}{\partial x} + \frac{\partial X}{\partial y} + b(x, y)g(Y(x, y)) = 0, \\ \frac{\partial Y}{\partial x} + \frac{\partial Y}{\partial y} + a(x, y)f(X(x, y)) = 0. \end{cases} \tag{12.6}$$

As usual, a real sequence defined on N_0 is said to be oscillatory if it is eventually neither positive nor negative, and it is said to be nonoscillatory otherwise. A solution, $(\{x_{mn}\}, \{y_{mn}\})$, of system (12.1) or (12.4) is said to be oscillatory if both of its components are oscillatory, and it is said to be nonoscillatory otherwise. System (12.1) or (12.4) is said to be oscillatory if all its solutions are oscillatory.

Note that if $b_{mn} \geq 0$ and $b_{mn} \neq 0$ for infinitely many values of m, n, then it follows easily from the first equation in (12.1) or (12.4) that for any solution $(\{x_{mn}\}, \{y_{mn}\})$ of the system, the oscillation of $\{x_{mn}\}$ implies the oscillation of $\{y_{mn}\}$ as wall. Thus, if $(\{x_{mn}\}, \{y_{mn}\})$ is a nonoscillatory solution of (12.1) or (12.4), then $\{x_{mn}\}$ is always nonoscillatory. Furthermore, if $a_{mn} \geq 0$ and $a_{mn} \neq 0$ for infinitely many m, n, then it follows from the second equation in (12.1) or (12.4) that $\{y_{mn}\}$ is an eventually one-sign sequence.

In the case where $b_{mn} > 0$ for all $m, n \in N_0$, and $g(u) = u$ with $u \in R$, the difference system (12.1) and (12.4) reduces to the following second-order nonlinear difference equations:

$$T(\Delta_1, \Delta_2)\left\{\frac{1}{b_{mn}}[\varDelta_1(x_{mn})]\right\} + a_{mn}f(x_{mn}) = 0, \tag{12.7}$$

and

$$T(\Delta_1, \Delta_2)\left\{-\frac{1}{b_{mn}}\left[T(\nabla_1, \nabla_2)(x_{mn})\right]\right\} + a_{mn}f(x_{mn}) = 0, \tag{12.8}$$

respectively.

Furthermore, if $b_{mn} = 1$ for $m \geq m_0, n \geq n_0$, and $f(u) = |u|^{\lambda}\text{sgn}\, u$, these equations become

$$T(\Delta_1, \Delta_2)[\varDelta_1(x_{mn})] + a_{mn}\,|x_{mn}|^{\lambda}\,\text{sgn}\, x_{mn} = 0, \tag{12.9}$$

and

$$T(\Delta_1, \Delta_2)\left[-T(\nabla_1, \nabla_2)(x_{mn})\right] + a_{mn}\left|x_{mn}\right|^{\lambda} sgn x_{mn} = 0. \tag{12.10}$$

Also, if $g(u) = u^{\alpha}$, where α is a ratio of odd positive integers, then systems (12.1) and (12.4) reduce to the following quasilinear difference equations:

$$T(\Delta_1, \Delta_2)\left\{\frac{1}{b_{mn}^{1/\alpha}}\left[\Delta_1(x_{mn})\right]^{1/\alpha}\right\} + a_{mn} f(x_{mn}) = 0, \tag{12.11}$$

and

$$T(\Delta_1, \Delta_2)\left\{-\frac{1}{b_{mn}^{1/\alpha}}\left[T(\nabla_1, \nabla_2)(x_{mn})\right]^{1/\alpha}\right\} + a_{mn} f(x_{mn}) = 0, \tag{12.12}$$

respectively. Regarding oscillation criteria for (12.7) and (12.8). It follows from (12.7) to (12.12) that system (12.1) and (12.4) have rich dynamics, respectively. Therefore, a careful study on these systems may in turn yield useful information for their companion partial differential equations.

Throughout this chapter, we focus on not only the solutions of the coupled systems (12.1) and (12.4) that exist for $m, n \in N_0$, but also the solutions for the uncoupled systems.

12.1 Some Preliminary Lemmas

In this section, some useful preliminary results are first established, which will be needed for the proofs of the main results given in the next section.

The following notation will be used:

$$\sum_{i=v}^{u} *_i = 0 \quad \text{for } u < v. \tag{12.13}$$

Lemma 12.1

$$\sum_{i=m-k}^{m}\sum_{j=n-l}^{n}(A_{i+1,j} + A_{i,j+1} - A_{ij})$$
$$= \sum_{i=m+1-k}^{m+1}\sum_{j=n+1-l}^{n} A_{ij} + \sum_{i=m-k}^{m} A_{i,n+1} - A_{m-k,n-l} + A_{m+1,n-l}.$$

Lemma 12.2 *Suppose that* $w_{mn} \geq 0$. *Then,*

$$\sum_{s=m_0}^{m-1}\sum_{t=n_0}^{n-1} B_{st}\,\Delta_1(w_{st}) \ge B_{m-1,n-1}w_{m,n-1} - \sum_{s=m_0}^{m-1}\sum_{t=n_0}^{n-1} b_{st}w_{s+1,t}, \tag{12.14}$$

where

$$B_{m,n} = \sum_{i=m_0}^{m-1}\sum_{j=n_0}^{n-1} b_{ij}.$$

Proof First, note that

$$\sum_{s=m_0}^{m-1}\sum_{t=n_0}^{n-1} B_{st}\,\Delta_1(w_{st}) = \sum_{s=m_0}^{m-1}\sum_{t=n_0}^{n-1}\sum_{i=m_0}^{s-1}\sum_{j=n_0}^{t-1} b_{ij}(w_{s+1,t}-w_{st})$$

$$= \sum_{s=m_0}^{m-1}\sum_{t=n_0}^{n-1}\left[\left(\sum_{i=m_0}^{s}\sum_{j=n_0}^{t} b_{ij} - \sum_{i=m_0}^{s-1} b_{it} - \sum_{j=n_0}^{t-1} b_{sj} - b_{st}\right)\left(w_{s+1,t}-w_{st}\right)\right]. \tag{12.15}$$

Then, note that

$$\sum_{i=m_0}^{s}\sum_{j=n_0}^{t} b_{ij} = \sum_{i=m_0}^{s-1}\sum_{j=n_0}^{t-1} b_{ij} + \sum_{j=n_0}^{t-1} b_{sj} + \sum_{i=m_0}^{s-1} b_{it} + b_{st}$$

$$\ge \sum_{i=m_0}^{s-1}\sum_{j=n_0}^{t-1} b_{ij} + \sum_{j=n_0}^{t-1} b_{sj} + \sum_{i=m_0}^{s-1} b_{it}. \tag{12.16}$$

Substituting (12.16) into (12.15), and using (12.13), one obtains

$$\sum_{s=m_0}^{m-1}\sum_{t=n_0}^{n-1} B_{st}\,\Delta_1(w_{st})$$

$$\ge \sum_{s=m_0}^{m-1}\sum_{t=n_0}^{n-1}\left(\sum_{i=m_0}^{s-1}\sum_{j=n_0}^{t-1} b_{ij} - b_{st}\right)(w_{s+1,t}-w_{st})$$

$$= \sum_{s=m_0}^{m-1}\sum_{t=n_0}^{n-1}\left(\sum_{i=m_0}^{s-1}\sum_{j=n_0}^{t-1} b_{ij}w_{s+1,t} - \sum_{i=m_0}^{s-1}\sum_{j=n_0}^{t-1} b_{ij}w_{st} - b_{st}w_{s+1,t} + b_{st}w_{st}\right)$$

$$\ge \sum_{s=m_0}^{m-1}\sum_{t=n_0}^{n-1}\left(\sum_{i=m_0}^{s-1}\sum_{j=n_0}^{t-1} b_{ij}w_{s+1,t} - \sum_{i=m_0}^{s-1}\sum_{j=n_0}^{t-1} b_{ij}w_{st} - b_{st}w_{s+1,t}\right)$$

$$= \sum_{s=m_0}^{m-1}\left(\sum_{t=n_0}^{n-1}\sum_{i=m_0}^{s-1}\sum_{j=n_0}^{t-1} b_{ij}w_{s+1,t} - \sum_{t=n_0}^{n-1}\sum_{i=m_0}^{s-1}\sum_{j=n_0}^{t-1} b_{ij}w_{st}\right) - \sum_{s=m_0}^{m-1}\sum_{t=n_0}^{n-1} b_{st}w_{s+1,t}$$

$$\geq \sum_{s=m_0}^{m-1}\sum_{t=n_0}^{n-1} b_{ij} w_{m,n-1} - \sum_{s=m_0}^{m-1}\sum_{t=n_0}^{n-1} b_{st} w_{s+1,t}$$

$$= B_{m-1,n-1} w_{m,n-1} - \sum_{s=m_0}^{m-1}\sum_{t=n_0}^{n-1} b_{st} w_{s+1,t}. \tag{12.17}$$

This implies that

$$\sum_{s=m_0}^{m-1}\sum_{t=n_0}^{n-1} B_{st}\, \Delta_1\, (w_{st}) \geq B_{m-1,n-1} w_{m,n-1} - \sum_{s=m_0}^{m-1}\sum_{t=n_0}^{n-1} b_{st} w_{s+1,t}, \tag{12.18}$$

so that (12.14) is obtained.

Lemma 12.3

$$\sum_{i=m_1}^{m}\sum_{j=n_1}^{n} (A_{i-1,j} + A_{i,j-1} - A_{ij})$$

$$= \sum_{i=m_1}^{m-1}\sum_{j=n_1}^{n} A_{i,j-1} + \sum_{j=n_1}^{n} A_{m_1-1,j} + A_{m,n_1-1} - A_{m,n}.$$

Lemma 12.4 *Suppose that* $w_{mn} \geq 0$. *Then*

$$\sum_{s=m_0}^{m-1}\sum_{t=n_0}^{n-1} B_{st} T(\Delta_1, \Delta_2)(w_{st}) \tag{12.19}$$

$$\geq B_{m-1,n-1}(w_{m,n-1} + w_{m-1,n}) - \sum_{s=m_0}^{m-1}\sum_{t=n_0}^{n-1} b_{st}(w_{s+1,t} + w_{s,t+1}),$$

where,

$$B_{m,n} = \sum_{i=m_0}^{m-1}\sum_{j=n_0}^{n-1} b_{ij}.$$

Proof First, note that

$$\sum_{s=m_0}^{m-1}\sum_{t=n_0}^{n-1} B_{st} T(\Delta_1, \Delta_2)(w_{st}) \tag{12.20}$$

$$= \sum_{s=m_0}^{m-1}\sum_{t=n_0}^{n-1} \left[\sum_{i=m_0}^{s-1}\sum_{j=n_0}^{t-1} b_{ij}(w_{s+1,t} + w_{s,t+1} - w_{st}) \right]$$

$$
\begin{aligned}
&\geq \sum_{s=m_0}^{m-1}\sum_{t=n_0}^{n-1}\left[\sum_{i=m_0}^{s-1}\sum_{j=n_0}^{t-1} b_{ij}(w_{s+1,t}+w_{s,t+1}-2w_{st})\right] \qquad (12.21)\\
&= \sum_{s=m_0}^{m-1}\sum_{t=n_0}^{n-1}\sum_{i=m_0}^{s-1}\sum_{j=n_0}^{t-1} b_{ij}\left((w_{s+1,t}-w_{st})+(w_{s,t+1}-w_{st})\right)\\
&= \sum_{s=m_0}^{m-1}\sum_{t=n_0}^{n-1}\sum_{i=m_0}^{s-1}\sum_{j=n_0}^{t-1} b_{ij}(w_{s+1,t}-w_{st})\\
&+ \sum_{s=m_0}^{m-1}\sum_{t=n_0}^{n-1}\sum_{i=m_0}^{s-1}\sum_{j=n_0}^{t-1} b_{ij}(w_{s,t+1}-w_{st}). \qquad (12.22)
\end{aligned}
$$

Then, note that

$$
\begin{aligned}
&\sum_{s=m_0}^{m-1}\sum_{t=n_0}^{n-1}\sum_{i=m_0}^{s-1}\sum_{j=n_0}^{t-1} b_{ij}(w_{s+1,t}-w_{st})\\
&= \sum_{s=m_0}^{m-1}\sum_{t=n_0}^{n-1}\left[\left(\sum_{i=m_0}^{s}\sum_{j=n_0}^{t} b_{ij}-\sum_{i=m_0}^{s-1} b_{it}-\right.\right.\\
&\qquad\left.\left.\sum_{j=n_0}^{t-1} b_{sj}-b_{st}\right)(w_{s+1,t}-w_{st})\right] \qquad (12.23)
\end{aligned}
$$

and

$$
\begin{aligned}
\sum_{i=m_0}^{s}\sum_{j=n_0}^{t} b_{ij} &= \sum_{i=m_0}^{s-1}\sum_{j=n_0}^{t-1} b_{ij}+\sum_{j=n_0}^{t-1} b_{sj}+\sum_{i=m_0}^{s-1} b_{it}+b_{st}\\
&\geq \sum_{i=m_0}^{s-1}\sum_{j=n_0}^{t-1} b_{ij}+\sum_{j=n_0}^{t-1} b_{sj}+\sum_{i=m_0}^{s-1} b_{it}. \qquad (12.24)
\end{aligned}
$$

Substituting (12.24) into (12.23), one obtains

$$
\begin{aligned}
&\sum_{s=m_0}^{m-1}\sum_{t=n_0}^{n-1}\sum_{i=m_0}^{s-1}\sum_{j=n_0}^{t-1} b_{ij}(w_{s+1,t}-w_{st})\\
&\geq \sum_{s=m_0}^{m-1}\sum_{t=n_0}^{n-1}\left(\sum_{i=m_0}^{s-1}\sum_{j=n_0}^{t-1} b_{ij}-b_{st}\right)(w_{s+1,t}-w_{st})\\
&= \sum_{s=m_0}^{m-1}\sum_{t=n_0}^{n-1}\left(\sum_{i=m_0}^{s-1}\sum_{j=n_0}^{t-1} b_{ij}w_{s+1,t}-\sum_{i=m_0}^{s-1}\sum_{j=n_0}^{t-1} b_{ij}w_{st}-b_{st}w_{s+1,t}+b_{st}w_{st}\right)
\end{aligned}
$$

$$\geq \sum_{s=m_0}^{m-1} \sum_{t=n_0}^{n-1} \left(\sum_{i=m_0}^{s-1} \sum_{j=n_0}^{t-1} b_{ij} w_{s+1,t} - \sum_{i=m_0}^{s-1} \sum_{j=n_0}^{t-1} b_{ij} w_{st} - b_{st} w_{s+1,t} \right)$$

$$= \sum_{s=m_0}^{m-1} \left(\sum_{t=n_0}^{n-1} \sum_{i=m_0}^{s-1} \sum_{j=n_0}^{t-1} b_{ij} w_{s+1,t} - \sum_{t=n_0}^{n-1} \sum_{i=m_0}^{s-1} \sum_{j=n_0}^{t-1} b_{ij} w_{st} \right) - \sum_{s=m_0}^{m-1} \sum_{t=n_0}^{n-1} b_{st} w_{s+1,t}$$

$$\geq \sum_{s=m_0}^{m-1} \sum_{t=n_0}^{n-1} b_{ij} w_{m,n-1} - \sum_{s=m_0}^{m-1} \sum_{t=n_0}^{n-1} b_{st} w_{s+1,t}$$

$$= B_{m-1,n-1} w_{m,n-1} - \sum_{s=m_0}^{m-1} \sum_{t=n_0}^{n-1} b_{st} w_{s+1,t},$$

that is

$$\sum_{s=m_0}^{m-1} \sum_{t=n_0}^{n-1} \sum_{i=m_0}^{s-1} \sum_{j=n_0}^{t-1} b_{ij} (w_{s+1,t} - w_{st}) \geq B_{m-1,n-1} w_{m,n-1} - \sum_{s=m_0}^{m-1} \sum_{t=n_0}^{n-1} b_{st} w_{s+1,t}. \tag{12.25}$$

Similarly, one also has

$$\sum_{s=m_0}^{m-1} \sum_{t=n_0}^{n-1} \sum_{i=m_0}^{s-1} \sum_{j=n_0}^{t-1} b_{ij} (w_{s,t+1} - w_{st}) \geq B_{m-1,n-1} w_{m-1,n} - \sum_{s=m_0}^{m-1} \sum_{t=n_0}^{n-1} b_{st} w_{s,t+1}. \tag{12.26}$$

Combining (12.21), (12.25) and (12.26) gives (12.19).

12.2 Main Results for the Coupled System

In this section, the above preliminary results are applied and generalized to system (12.1). We assume and utilize the following conditions:

(c_1) $a_{mn} \geq 0$ and $b_{mn} \geq 0$ for all $m, n \in N_0$, and neither sequence vanishes identically for $m, n \in N_0$.

(c_2)

$$\lim_{m,n \to \infty} B_{mn} = \infty,$$

where

$$B_{mn} = \sum_{s=m_0}^{m-1} \sum_{t=n_0}^{n-1} b_{st}.$$

(c_3) $f(uv) \geq f(u) f(v)$ for all $u \geq 0$, $v \geq 0$.

(c_4) $\int_0^{\pm\alpha} \frac{du}{f(g(u))} < \infty$ for all $\alpha > 0$.

(c_5)

$$\sum_{m=m_0}^{\infty}\sum_{n=n_0}^{\infty} f\left(\frac{1}{n-n_0}\right) f(B_{mn})a_{mn} = \infty.$$

(c_6) f and g are nondecreasing.
(c_7) Taking boundary conditions $x_{m_0 n} = c_n \geq 0$ and $x_{mn_0} = d_m \geq 0$ such that

$$\sum_{n=n_0}^{\infty} c_n < \infty$$

and

$$\sum_{m=m_0}^{\infty} d_m < \infty.$$

(c_8) There exist nonnegative functions g_1 and f_1 such that $f(u) - f(v) = (u - v)f_1(u, v)$, $g(u) - g(v) = (u - v)g_1(u, v)$ and $g_1(u, v) \geq \xi > 0$, for $u, v \neq 0$.
(c_9)

$$\int_{\pm\alpha}^{\pm\infty} \frac{du}{f(u)} < \infty$$

for every $\alpha > 0$.
(c_{10})

$$\sum_{m=m_0}^{\infty}\sum_{n=n_0}^{\infty} B_{mn}a_{mn} = \infty.$$

(c_{11}) $\frac{u}{f(u)}$ is bounded.

Theorem 12.1 *Suppose that conditions* (c_1)–(c_7) *and* (12.2) *are satisfied. Then system* (12.1) *is oscillatory.*

Proof Suppose that system (12.1) has a nonoscillatory solution $(\{x_{mn}\}, \{y_{mn}\})$, for $m, n \in N_0$. Since $\{a_{mn}\}$ and $\{b_{mn}\}$ are not identically zero for $m, n \in N_0$, as noted earlier, $\{x_{mn}\}$ and $\{y_{mn}\}$ are eventually of one sign. So, without loss of generality, assume that $x_{mn} \neq 0$ for all $m, n \in N_0$. Furthermore, observe that the substitutions $z_{mn} = -x_{mn}$ and $w_{mn} = -y_{mn}$ transform system (12.1) into the following:

$$\begin{cases} \Delta_1(z_{mn}) = b_{mn}\bar{g}(w_{mn}), \\ T(\Delta_1, \Delta_2)(w_{mn}) = -a_{mn}\bar{f}(z_{mn}), \end{cases} \quad m, n \in N_0, \tag{12.27}$$

where $\bar{f}(u) = -f(u)$, $u \in R$, and $\bar{g}(v) = -g(v)$, $v \in R$. So, functions $\bar{f}$ and $\bar{g}$ are subjected to the condition imposed on f and g, respectively. For this reason, discussion is restricted only to the case where $\{x_{mn}\}$ is eventually positive for $m \geq m_0, n \geq n_0$ and either $y_{mn} \geq 0$ or $y_{mn} \leq 0$, for $m \geq m_0, n \geq n_0$.

First, consider the case of $y_{mn} \leq 0$ for $m \geq m_0$, $n \geq n_0$. The second equation in (12.1) implies that $\{ y_{mn}\}$ is decreasing. Since $y_{mn} \leq 0$, it approaches either $-\infty$ or a finite negative value as $m, n \to \infty$. Likewise, $g(y_{mn}) \to -\infty$ or to a finite negative value as $m, n \to \infty$. This, in view of condition (c_2), implies that

$$\sum_{m=m_0}^{\infty} \sum_{n=n_0}^{\infty} b_{mn} g(y_{mn}) = -\infty. \tag{12.28}$$

Now, taking m, n sufficiently large, summing the first equation on both sides of (12.1) from $m-1, n \to \infty$, we find that

$$\begin{aligned}\sum_{j=n_0}^{n-1} (x_{mj} - x_{m_0 j}) &= \sum_{i=m_0}^{m-1} \sum_{j=n_0}^{n-1} (x_{i+1,j} - x_{ij}) \\ &= \sum_{i=m_0}^{m-1} \sum_{j=n_0}^{n-1} \Delta_1(x_{mn}) = \sum_{i=m_0}^{m-1} \sum_{j=n_0}^{n-1} b_{ij} g(y_{i\,j}),\end{aligned}$$

so that

$$x_{m,n-1} \leq \sum_{j=n_0}^{n-1} x_{mj} = \sum_{j=n_0}^{n-1} x_{m_0 j} + \sum_{i=m_0}^{m-1} \sum_{j=n_0}^{n-1} b_{ij} g(y_{i\,j}). \tag{12.29}$$

Using c_7 and (12.29), we obtain

$$x_{m,n-1} \leq \sum_{j=n_0}^{n-1} x_{m_0 j} + \sum_{i=m_0}^{m-1} \sum_{j=n_0}^{n-1} b_{ij} g(y_{i\,j}) \to -\infty, \text{ as } m, n \to \infty,$$

which is a contradiction to the assumption that $x_{mn} > 0$ for all $m \geq m_0, n \geq n_0$.

Next, consider the case where $y_{mn} \geq 0$ for all $m \geq m_0, n \geq n_0$. From system (12.1), it is clear that $\{x_{mn}\}$ is increasing and $\{y_{mn}\}$ is decreasing in m, n. Thus, applying (c_6), (c_7) and summing the first equation in (12.1) yield

$$\begin{aligned}&(n-n_0)x_{m,n-1} - \sum_{j=n_0}^{n-1} x_{m_0 j} \\ &\geq \sum_{j=n_0}^{n-1} x_{mj} - \sum_{j=n_0}^{n-1} x_{m_0 j} = \sum_{j=n_0}^{n-1} (x_{mj} - x_{m_0 j}) \\ &= \sum_{i=m_0}^{m-1} \sum_{j=n_0}^{n-1} (x_{i+1,j} - x_{ij}) = \sum_{i=m_0}^{m-1} \sum_{j=n_0}^{n-1} b_{ij} g(y_{i\,j}).\end{aligned}$$

Hence,

$$\begin{aligned}
x_{m,n-1} &\geq \frac{1}{(n-n_0)}\left[\sum_{j=n_0}^{n-1} x_{m_0 j} + \sum_{i=m_0}^{m-1}\sum_{j=n_0}^{n-1} b_{ij} g(y_{i\,j})\right] \\
&\geq \frac{1}{(n-n_0)}\left[\sum_{i=m_0}^{m-1}\sum_{j=n_0}^{n-1} b_{ij} g(y_{i\,j})\right] \\
&\geq \frac{1}{(n-n_0)}\left[\sum_{i=m_0}^{m-1}\sum_{j=n_0}^{n-1} b_{ij} g(y_{m-1,n-1})\right] \\
&= \frac{1}{(n-n_0)} B_{mn} g(y_{m-1,n-1}) \\
&\geq \frac{1}{(n-n_0)} B_{mn} g(y_{mn}). \qquad (12.30)
\end{aligned}$$

Next, we can show that

$$\Delta_2\,(x_{mn}) \geq 0 \text{ for } m \geq m_0,\ n \geq n_0. \qquad (12.31)$$

In fact, if there would exist $m_1 \geq m_0$ and $n_1 \geq n_0$ such that $\Delta_2(x_{m_1 n_1}) = c < 0$, then $\Delta_2(x_{mn}) \leq c$ for $m \geq m_1, n \geq n_1$, so that

$$\sum_{i=m_1+1}^{m} (x_{in} - x_{in_1}) = \sum_{i=m_1+1}^{m}\sum_{j=n_1+1}^{n} \Delta_2(x_{ij}) \leq \sum_{i=m_1+1}^{m}\sum_{j=n_1+1}^{n} c = c(m-m_1)(n-n_1).$$

Note that c_7 yields

$$x_{mn} \leq \sum_{i=m_2+1}^{m} x_{in} \leq \sum_{i=m_2+1}^{m} x_{in_1} + c(m-m_1)(n-n_1) \to -\infty \text{ as } m, n \to \infty,$$

which contradicts the fact that $x_{mn} > 0$ for $m \geq m_1, n \geq n_1$. Thus, (12.31) holds, that is, $\{x_{mn}\}$ is also increasing in n. Therefore, it follows from (12.30) that

$$x_{mn} \geq x_{m,n-1} \geq \frac{1}{(n-n_0)} B_{mn} g(y_{mn}).$$

In view of (c_2), for all m, n large enough, $B_{mn} \geq 1$, so

$$f(x_{mn}) \geq f\left(\frac{1}{(n-n_0)} B_{mn}\right) f(g(y_{mn})) \geq \left(\frac{1}{(n-n_0)}\right) f(B_{mn}) f(g(y_{mn})).$$

The second equation in (12.1) leads to

$$T(\Delta_1, \Delta_2)(y_{mn}) \le -a_{mn} f\left(\frac{1}{(n-n_0)}\right) f(B_{mn}) f(g(y_{mn})),$$

that is,

$$\frac{T(\Delta_1, \Delta_2)(y_{mn})}{f(g(y_{mn}))} \le -a_{mn} f\left(\frac{1}{(n-n_0)}\right) f(B_{mn}). \tag{12.32}$$

Moreover, using (12.32), we have

$$\begin{aligned}
&-\frac{\Delta_1(y_{mn})}{f(g(y_{mn}))} - \frac{\Delta_2(y_{mn})}{f(g(y_{mn}))}\\
&= -\frac{\Delta_1(y_{mn}) + \Delta_2(y_{mn})}{f(g(y_{mn}))} = -\frac{(y_{m+1,n} - y_{mn}) + (y_{m,n+1} - y_{mn})}{f(g(y_{mn}))}\\
&= -\frac{y_{m+1,n} + y_{m,n+1} - 2y_{mn}}{f(g(y_{mn}))} \ge -\frac{y_{m+1,n} + y_{m,n+1} - y_{mn}}{f(g(y_{mn}))}\\
&= -\frac{T(\Delta_1, \Delta_2)(y_{mn})}{f(g(y_{mn}))} \ge a_{mn} f\left(\frac{1}{(n-n_0)}\right) f(B_{mn}),
\end{aligned}$$

that is

$$-\frac{\Delta_1(y_{mn})}{f(g(y_{mn}))} - \frac{\Delta_2(y_{mn})}{f(g(y_{mn}))} \ge a_{mn} f\left(\frac{1}{(n-n_0)}\right) f(B_{mn}). \tag{12.33}$$

Summing the inequality on both sides of (12.33) from $m_0, n_0 \to m-1, n-1$, one obtains

$$\begin{aligned}
&-\sum_{s=m_0}^{m-1}\sum_{t=n_0}^{n-1} \frac{\Delta_1(y_{st})}{f(g(y_{st}))} - \sum_{s=m_0}^{m-1}\sum_{t=n_0}^{n-1} \frac{\Delta_2(y_{st})}{f(g(y_{st}))}\\
&\ge \sum_{s=m_0}^{m-1}\sum_{t=n_0}^{n-1} a_{st} f\left(\frac{1}{(t-n_0)}) f(B_{st}\right).
\end{aligned} \tag{12.34}$$

Observe that for $y_{m+1,n} \le s \le y_{mn}$, we have $f(g(y_{mn})) \ge f(g(s))$, that is, $\frac{1}{f(g(s))} \ge \frac{1}{f(g(y_{mn}))}$. Therefore,

$$-\frac{\Delta_1(y_{mn})}{f(g(y_{mn}))} \le \int_{y_{m+1,n+1}}^{y_{mn}} \frac{ds}{f(g(s))}. \tag{12.35}$$

Similarly,

$$-\frac{\Delta_2(y_{mn})}{f(g(y_{mn}))} \le \int_{y_{m+1,n+1}}^{y_{mn}} \frac{dt}{f(g(t))}. \tag{12.36}$$

It then follows from (12.35) and (12.36) that

$$-\frac{\Delta_1(y_{mn})}{f(g(y_{mn}))}-\frac{\Delta_2(y_{mn})}{f(g(y_{mn}))}\leq\int_{y_{m+1,n+1}}^{y_{mn}}\frac{ds}{f(g(s))}+\int_{y_{m+1,n+1}}^{y_{mn}}\frac{dt}{f(g(t))}. \tag{12.37}$$

Summing the inequality on both sides of (12.37) from $m_0, n_0 \to m-1, n-1$, we have

$$\begin{aligned}&-\sum_{s=m_0}^{m-1}\sum_{t=n_0}^{n-1}\frac{\Delta_1(y_{st})}{f(g(y_{st}))}-\sum_{s=m_0}^{m-1}\sum_{t=n_0}^{n-1}\frac{\Delta_2(y_{st})}{f(g(y_{st}))}\\&\leq\sum_{t=n_0}^{n-1}\sum_{s=m_0}^{m-1}\int_{y_{s+1,t+1}}^{y_{st}}\frac{ds}{f(g(s))}+\sum_{s=m_0}^{m-1}\sum_{t=n_0}^{n-1}\int_{y_{s+1,t+1}}^{y_{st}}\frac{dt}{f(g(t))}\\&=2\int_{y_{mn}}^{y_{m_0n_0}}\frac{ds}{f(g(s))}.\end{aligned} \tag{12.38}$$

Now, combining (12.34) and (12.38) yields

$$\sum_{s=m_0}^{m-1}\sum_{t=n_0}^{n-1}a_{st}f\left(\frac{1}{(t-n_0)}\right)f(B_{st})\leq 2\int_{y_{mn}}^{y_{m_0n_0}}\frac{ds}{f(g(s))},$$

which, in view of conditions (c_4) and (c_5), leads to a contradiction. This completes the proof of the theorem.

Theorem 12.2 *Suppose that conditions* (c_1), (c_4) *and* (c_8)*–*(c_{10}) *are satisfied. Then, system (12.1) is oscillatory.*

Proof Suppose, on the contrary, that system (12.1) is not oscillatory, so that $\{x_{mn}\}$ and $\{y_{mn}\}$ are of one sign for $m, n \in N_0$. Without loss of generality, assume that $x_{mn} \geq 0$. The case $y_{mn} \leq 0$ can be handled exactly as it was in the proof of Theorem 12.2.

Now, consider the case where $y_{mn} \geq 0$ for all $m \geq m_0, n \geq n_0$. Define

$$w_{mn}=\frac{g(y_{mn})}{f(x_{mn})}.$$

Applying system (12.1), inequality (12.31) and condition (c_4), we have

$$f(x_{m+1,n}) \geq f(x_{mn}) \text{ and } g(y_{m+1,n}) \leq g(y_{mn}). \tag{12.39}$$

Moreover, note that

$$\Delta_1(w_{mn}) = w_{m+1,n} - w_{mn} \tag{12.40}$$
$$= \left[\frac{g(y_{m+1,n})}{f(x_{m+1,n})} - \frac{g(y_{mn})}{f(x_{mn})}\right]$$
$$= \frac{f(x_{mn})\left[g(y_{m+1,n}) - g(y_{mn})\right] - g(y_{mn})\left[f(x_{m+1,n}) - f(x_{mn})\right]}{f(x_{m+1,n})\, f(x_{mn})}. \tag{12.41}$$

Combining (c_8) and the second equation in system (12.1) gives

$$\begin{aligned} g(y_{m+1,n}) - g(y_{mn}) &= (y_{m+1,n} - y_{mn}) g_1(y_{m+1,n}, y_{mn}) \\ &\le (y_{m+1,n} + y_{m,n+1} - y_{mn}) g_1(y_{m+1,n}, y_{mn}) \\ &= -a_{mn} f(x_{mn}) g_1(y_{m+1,n}, y_{mn}). \end{aligned} \tag{12.42}$$

Similarly,

$$f(x_{m+1,n}) - f(x_{mn}) \le b_{mn} g(y_{mn}) f_1(x_{m+1,n}, x_{mn}). \tag{12.43}$$

Substituting (12.42) and (12.43) into (12.41) gives

$$\begin{aligned} \Delta_1(w_{mn}) &\le \frac{-a_{mn} f^2(x_{mn}) g_1(y_{m+1,n}, y_{mn}) - b_{mn} g^2(y_{mn}) f_1(x_{m+1,n}, x_{mn})}{f(x_{m+1,n})\, f(x_{mn})} \\ &= -a_{mn} \frac{f(x_{mn})}{f(x_{m+1,n})} g_1(y_{m+1,n}, y_{mn}) - \frac{b_{mn} g^2(y_{mn}) f_1(x_{m+1,n}, x_{mn})}{2 f(x_{m+1,n})\, f(x_{mn})} \\ &\le -a_{mn} \left[\frac{f(x_{mn})}{f(x_{m+1,n})}\right] g_1(y_{m+1,n}, y_{mn}). \end{aligned} \tag{12.44}$$

Using (12.39), and in view of (12.44), we obtain

$$\Delta_1(w_{mn}) \le -a_{mn} g_1(y_{m+1,n}, y_{mn}) \le -a_{mn}\xi.$$

Multiplying both sides of this inequality by B_{mn}, summing from m_0, n_0 to $m-1, n-1$, and then applying Lemma 12.2, we obtain

$$B_{m-1,n-1} w_{m,n-1} - \sum_{s=m_0}^{m-1} \sum_{t=n_0}^{n-1} b_{st} w_{s+1,t} \le -\sum_{s=m_0}^{m-1} \sum_{t=n_0}^{n-1} \xi B_{st} a_{st}. \tag{12.45}$$

It follows from (12.39) and defination of w_{mn} that w_{mn} is decreasing in m, n, so that, in view of (12.45), we have

$$B_{m-1,n-1} w_{mn} - \sum_{s=m_0}^{m-1} \sum_{t=n_0}^{n-1} b_{st} w_{s+1,t} \le -\sum_{s=m_0}^{m-1} \sum_{t=n_0}^{n-1} \xi B_{st} a_{st}. \tag{12.46}$$

In view of condition (c_{10}) and the positivity of $b_{mn}w_{mn}$, to complete the proof of the theorem it suffices to show that the second term on the left-hand side of (12.46) is bounded.

Using the first equation in (12.1) and also (12.39), it is clear that

$$\sum_{s=m_0}^{m-1}\sum_{t=n_0}^{n-1} b_{st}w_{s+1,t} = \sum_{s=m_0}^{m-1}\sum_{t=n_0}^{n-1} b_{st}\frac{g(y_{s+1,t})}{f(x_{s+1,t})} \\ \leq \sum_{s=m_0}^{m-1}\sum_{t=n_0}^{n-1} b_{st}\frac{g(y_{st})}{f(x_{s+1,t})} = \sum_{s=m_0}^{m-1}\sum_{t=n_0}^{n-1} \frac{\Delta_1(x_{st})}{f(x_{s+1,t})}. \tag{12.47}$$

Observe that for $x_{mn} \leq s \leq x_{m+1,n}$, one has $1/f(s) \geq 1/f(x_{m+1,n})$, and, therefore,

$$\int_{x_{mn}}^{x_{m+1,n+1}} \frac{ds}{f(s)} \geq \frac{\Delta_1(x_{st})}{f(x_{s+1,t})}. \tag{12.48}$$

Summing the both sides of (12.48) from $m_0, n_0 \to m-1, n-1$, we obtain

$$\int_{x_{m_0n_0}}^{x_{mn}} \frac{ds}{f(s)} = \sum_{s=m_0}^{m-1}\sum_{t=n_0}^{n-1} \int_{x_{st}}^{x_{s+1,t+1}} \frac{ds}{f(s)} \geq \sum_{s=m_0}^{m-1}\sum_{t=n_0}^{n-1} \frac{\Delta_1(x_{st})}{f(x_{s+1,t})}. \tag{12.49}$$

Combining (12.47) and (12.49), we finally obtain

$$\infty > \int_{x_{m_0n_0}}^{x_{mn}} \frac{ds}{f(s)} \geq \sum_{s=m_0}^{m-1}\sum_{t=n_0}^{n-1} b_{st}w_{s+1,t}.$$

This completes the proof of the theorem.

Theorem 12.3 *Suppose that conditions (c_1)–(c_7) and the following relationship:*

$$uf(u) > 0 \textit{ and } ug(u) > 0 \textit{ for all } u \neq 0 \tag{12.50}$$

are satisfied. Then system (12.4) is oscillatory.

Proof Suppose that the system (12.4) is not oscillatory so that $\{x_{mn}\}$ and $\{y_{mn}\}$ are of one sign for $m, n \in N_0$. Without loss of generality, one may assume that $x_{mn} \neq 0$ for all $m, n \in N_0$. Furthermore, observe that the substitution $z_{mn} = -x_{mn}$ and $w_{mn} = -y_{mn}$ transform (12.4) into the following:

$$\begin{cases} T(\nabla_1, \nabla_2)(z_{mn}) = -b_{mn}\bar{g}(w_{mn}), \\ T(\Delta_1, \Delta_2)(w_{mn}) = -a_{mn}\bar{f}(z_{mn}), \end{cases} \quad (m, n) \in N_0, \tag{12.51}$$

where, $\bar{f}(u) = -f(u)$, $u \in R$ and $\bar{g}(v) = -g(v)$, $v \in R$. So, functions $\bar{f}$ and $\bar{g}$ are subject to the condition imposed on f and g, respectively. For this reason, discussion is restricted only to the case where $\{x_{mn}\}$ is eventually positive for $m \ge m_0, n \ge n_0$ and either $y_{mn} \ge 0$ or $y_{mn} \le 0$ for $m \ge m_0, n \ge n_0$.

First, consider the case where $y_{mn} \ge 0$ for all $m \ge m_0, n \ge n_0$. From the system (12.4), it is clear that $\{x_{mn}\}$ is increasing and $\{y_{mn}\}$ is decreasing in m, n. Thus, applying (c_4) and summing the first equation in (12.4) yield

$$\begin{aligned}-x_{mn} &\le \sum_{i=m_0+1}^{m-1} \sum_{j=n_0+1}^{n} x_{i,j-1} + \sum_{j=n_0+1}^{n} x_{m_0,j} + x_{mn_0} - x_{mn} \\ &= \sum_{i=m_0+1}^{m} \sum_{j=n_0+1}^{n} T(\nabla_1, \nabla_2)(x_{mn}) \\ &= -\sum_{i=m_0+1}^{m} \sum_{j=n_0+1}^{n} b_{ij} g(y_{ij}) \le -\sum_{i=m_0+1}^{m} \sum_{j=n_0+1}^{n} b_{ij} g(y_{mn}).\end{aligned}$$

Hence

$$x_{mn} \ge \sum_{i=m_0+1}^{m} \sum_{j=n_0+1}^{n} b_{ij} g(y_{mn}) = B_{mn} g(y_{mn}).$$

In view of (c_5), for all m, n large enough, $B_{mn} \ge 1$, so

$$f(x_{mn}) \ge f(B_{mn} g(y_{mn})) \ge f(B_{mn}) f(g(y_{mn})). \tag{12.52}$$

The second equation in (12.4) leads to

$$T(\Delta_1, \Delta_2)(y_{mn}) \le -a_{mn} f(B_{mn}) f(g(y_{mn})). \tag{12.53}$$

Moreover, applying (12.53), we have

$$\begin{aligned}&-\frac{\Delta_1(y_{mn})}{f(g(y_{mn}))} - \frac{\Delta_2(y_{mn})}{f(g(y_{mn}))} \\ &= -\frac{\Delta_1(y_{mn}) + \Delta_2(y_{mn})}{f(g(y_{mn}))} = -\frac{(y_{m+1,n} - y_{mn}) + (y_{m,n+1} - y_{mn})}{f(g(y_{mn}))} \\ &= -\frac{y_{m+1,n} + y_{m,n+1} - 2y_{mn}}{f(g(y_{mn}))} \ge -\frac{y_{m+1,n} + y_{m,n+1} - y_{mn}}{f(g(y_{mn}))} \\ &= -\frac{T(\Delta_1, \Delta_2)(y_{mn})}{f(g(y_{mn}))} \ge a_{mn} f(B_{mn}),\end{aligned}$$

that is

$$-\frac{\Delta_1(y_{mn})}{f(g(y_{mn}))} - \frac{\Delta_2(y_{mn})}{f(g(y_{mn}))} \ge a_{mn} f(B_{mn}), \tag{12.54}$$

summing the inequality on both sides of (12.54) from $m_0, n_0 \to m-1, n-1$, we obtain

$$-\sum_{s=m_0}^{m-1}\sum_{t=n_0}^{n-1}\frac{\Delta_1(y_{st})}{f(g(y_{st}))}-\sum_{s=m_0}^{m-1}\sum_{t=n_0}^{n-1}\frac{\Delta_2(y_{st})}{f(g(y_{st}))}\geq\sum_{s=m_0}^{m-1}\sum_{t=n_0}^{n-1}a_{st}f(B_{st}). \tag{12.55}$$

Observe that for $y_{m+1,n}\leq s\leq y_{mn}$ and note that (c_4), $f(g(y_{mn}))\geq f(g(s))$. That is, $\frac{1}{f(g(y_{mn}))}\leq\frac{1}{f(g(s))}$. Therefore

$$-\frac{\Delta_1(y_{mn})}{f(g(y_{mn}))}\leq\int_{y_{m+1,n+1}}^{y_{mn}}\frac{ds}{f(g(s))}. \tag{12.56}$$

Similarly,

$$-\frac{\Delta_2(y_{mn})}{f(g(y_{mn}))}\leq\int_{y_{m+1,n+1}}^{y_{mn}}\frac{dt}{f(g(t))}. \tag{12.57}$$

It then follows from (12.56) and (12.57) that

$$-\frac{\Delta_1(y_{mn})}{f(g(y_{mn}))}-\frac{\Delta_2(y_{mn})}{f(g(y_{mn}))}\leq\int_{y_{m+1,n+1}}^{y_{mn}}\frac{ds}{f(g(s))}+\int_{y_{m+1,n+1}}^{y_{mn}}\frac{dt}{f(g(t))}, \tag{12.58}$$

summing the inequality on both sides of (12.58) from $m_0, n_0 \to m-1, n-1$, we obtain

$$\begin{aligned}&-\sum_{s=m_0}^{m-1}\sum_{t=n_0}^{n-1}\frac{\Delta_1(y_{st})}{f(g(y_{st}))}-\sum_{s=m_0}^{m-1}\sum_{t=n_0}^{n-1}\frac{\Delta_2(y_{st})}{f(g(y_{st}))}\\ &\leq\sum_{t=n_0}^{n-1}\sum_{s=m_0}^{m-1}\int_{y_{s+1,t+1}}^{y_{st}}\frac{ds}{f(g(s))}+\sum_{s=m_0}^{m-1}\sum_{t=n_0}^{n-1}\int_{y_{s+1,t+1}}^{y_{st}}\frac{dt}{f(g(t))}=2\int_{y_{mn}}^{y_{m_0n_0}}\frac{ds}{f(g(s))}.\end{aligned} \tag{12.59}$$

Now, combining (12.55) and (12.59) yields

$$\sum_{s=m_0}^{m-1}\sum_{t=n_0}^{n-1}a_{st}f(B_{st})\leq 2\int_{y_{mn}}^{y_{m_0n_0}}\frac{ds}{f(g(s))}$$

which, in view of conditions (c_3) and (c_6), leads to a contradiction. This completes the proof of the theorem.

Next, consider the case $y_{mn}<0$ for $m\geq m_0$, $n\geq n_0$. The second equation in (12.4) implies that $\{\ y_{mn}\}$ is decreasing. Since $y_{mn}<0$, it approaches either $-\infty$ or a finite negative value as $m, n\to\infty$. Since $\{x_{mn}\}$ is eventually positive for $m\geq m_0, n\geq n_0$, Likewise, $f(x_{mn})\to+\infty$ or to a finite positive value as $m, n\to\infty$. This, in view of condition (c_2), implies that

$$\sum_{m=m_0}^{\infty}\sum_{n=n_0}^{\infty} a_{mn} f(x_{mn}) = +\infty. \tag{12.60}$$

Now, taking m, n sufficiently large, summing the second equation on both sides of (12.4) from $m_0, n_0 \to m, n$, we find that

$$\begin{aligned}
&[(m-m_0)(n-n_0)+(m-m_0+2)]\, y_{m_0n_0} - y_{mn} \\
&\le \sum_{i=m_0+1}^{m+1}\sum_{j=n_0+1}^{n} y_{ij} + \sum_{i=m_0}^{m} y_{i,n+1} - y_{mn} + y_{m+1,n_0} \\
&\le \sum_{i=m_0+1}^{m+1}\sum_{j=n_0+1}^{n} y_{ij} + \sum_{i=m_0}^{m} y_{i,n+1} - y_{m_0,n_0} + y_{m+1,n_0} \\
&= \sum_{i=m_0}^{m-1}\sum_{j=n_0}^{n-1} (y_{i+1,j} + y_{i,j+1} - y_{ij}) \\
&= \sum_{i=m_0}^{m-1}\sum_{j=n_0}^{n-1} T(\Delta_1, \Delta_2)(y_{ij}) = -\sum_{i=m_0}^{m-1}\sum_{j=n_0}^{n-1} a_{ij} f(x_{ij}),
\end{aligned} \tag{12.61}$$

so that

$$y_{mn} \ge [(m-m_0)(n-n_0)+(m-m_0+2)]\, y_{m_0n_0} + \sum_{i=m_0}^{m-1}\sum_{j=n_0}^{n-1} a_{ij} f(x_{ij}).$$

Using (c_7), we obtain

$$\begin{aligned}
y_{mn} &\ge [(m-m_0)(n-n_0)+(m-m_0+2)]\, y_{m_0n_0} + \\
&\sum_{i=m_0}^{m-1}\sum_{j=n_0}^{n-1} a_{ij} f(x_{ij}) \to +\infty, \text{ as } m, n \to +\infty,
\end{aligned}$$

which is a contradiction to the assumption that $y_{mn} < 0$ for all $m \ge m_0, n \ge n_0$. This completes the proof of the theorem.

Theorem 12.4 . *Suppose that conditions* (c_1), (c_4), (c_8)–(c_{11}) *and* (12.62)

$$uf(u) > 0 \text{ and } ug(u) > 0 \text{ for all } u \neq 0 \tag{12.62}$$

are satisfied. Then, system (12.4) *is oscillatory.*

Proof Suppose, on the contrary, that system (12.4) is not oscillatory, so that $\{x_{mn}\}$ and $\{y_{mn}\}$ are of one sign for $m, n \in N_0$. Without loss of generality, assume that $x_{mn} \ge 0$. The case $y_{mn} < 0$ can be handled exactly as it was in the proof of Theorem 12.4.

Next, consider the case where $y_{mn} \geq 0$ for all $m \geq m_0$, $n \geq n_0$. Define

$$w_{mn} = \frac{g(y_{mn})}{f(x_{mn})}.$$

Applying (c_4), we have

$$f(x_{m+1,n}) \geq f(x_{mn}) \text{ and } g(y_{m+1,n}) \leq g(y_{mn}). \tag{12.63}$$

Moreover, note that

$$\begin{aligned}
&T(\Delta_1, \Delta_2)(w_{mn}) \\
&= w_{m+1,n} + w_{m,n+1} - w_{mn} = \left(w_{m+1,n} - \frac{1}{2}w_{mn}\right) + \left(w_{m,n+1} - \frac{1}{2}w_{mn}\right) \\
&= \left[\frac{g(y_{m+1,n})}{f(x_{m+1,n})} - \frac{1}{2}\frac{g(y_{mn})}{f(x_{mn})}\right] + \left[\frac{g(y_{m,n+1})}{f(x_{m,n+1})} - \frac{1}{2}\frac{g(y_{mn})}{f(x_{mn})}\right] \\
&= \frac{f(x_{mn})\left[2g(y_{m+1,n}) - g(y_{mn})\right] - g(y_{mn})\left[f(x_{m+1,n}) - f(x_{mn})\right]}{2f(x_{m+1,n})\, f(x_{mn})} + \\
&\quad \frac{f(x_{mn})\left[2g(y_{m,n+1}) - g(y_{mn})\right] - g(y_{mn})\left[f(x_{m,n+1}) - f(x_{mn})\right]}{2f(x_{m,n+1}) f(x_{mn})} \\
&\leq \frac{f(x_{mn})\left[g(y_{m+1,n}) - g(y_{mn})\right] - g(y_{mn})\left[f(x_{m+1,n}) - f(x_{mn})\right]}{2f(x_{m+1,n})\, f(x_{mn})} + \\
&\quad \frac{f(x_{mn})\left[g(y_{m,n+1}) - g(y_{mn})\right] - g(y_{mn})\left[f(x_{m,n+1}) - f(x_{mn})\right]}{2f(x_{m,n+1}) f(x_{mn})} \\
&\leq -\frac{g(y_{mn})\left[f(x_{m+1,n}) - f(x_{mn})\right]}{2f(x_{m+1,n})\, f(x_{mn})} - \frac{g(y_{mn})\left[f(x_{m,n+1}) - f(x_{mn})\right]}{2f(x_{m,n+1}) f(x_{mn})} \\
&\leq -\frac{g(y_{mn})\left[f(x_{m-1,n}) - f(x_{mn})\right]}{2f(x_{m+1,n})\, f(x_{mn})} - \frac{g(y_{mn})\left[f(x_{m,n-1}) - f(x_{mn})\right]}{2f(x_{m,n+1}) f(x_{mn})}.
\end{aligned} \tag{12.64}$$

Combining (c_8) and the first equation in system (12.3) gives

$$\begin{aligned}
f(x_{m-1,n}) - f(x_{mn}) &= (x_{m-1,n} - x_{mn}) f_1(x_{m,n-1}, x_{mn}), \\
&\leq (x_{m-1,n} + x_{m,n-1} - x_{mn}) f_1(x_{m,n-1}, x_{mn}) \\
&= -b_{mn} g(y_{mn}) f_1(x_{m,n-1}, x_{mn}).
\end{aligned} \tag{12.65}$$

Similarly

$$f(x_{m,n-1}) - f(x_{mn}) \leq -b_{mn} g(y_{mn}) f_1(x_{m-1,n}, x_{mn}). \tag{12.66}$$

Substituting (12.65) and (12.66) into (12.64) and using (c_4), (c_8) and (c_{11}), we obtain

$$
\begin{aligned}
&T(\Delta_1, \Delta_2)(w_{mn}) \\
&\leq \frac{b_{mn}}{2}\left(\frac{g^2(y_{mn})}{f(x_{m+1,n})f(x_{mn})}f_1(x_{m,n-1}, x_{mn}) + \frac{g^2(y_{mn})}{f(x_{m,n+1})f(x_{mn})}f_1(x_{m-1,n}, x_{mn})\right) \\
&< \frac{b_{mn}}{2}\left(\frac{g^2(y_{mn})}{f^2(x_{mn})}\eta + \frac{g^2(y_{mn})}{f^2(x_{mn})}\eta\right) = \eta b_{mn}\left(\frac{g(y_{mn})}{f(x_{mn})}\right)^2 \qquad (12.67) \\
&= \eta b_{mn}\left(-\frac{T(\Delta_1', \Delta_2')(x_{mn})}{b_{mn}f(x_{mn})}\right)^2 = \eta b_{mn}\left(\frac{x_{m,n-1} + x_{m-1,n} - x_{mn}}{b_{mn}f(x_{mn})}\right)^2 \\
&\leq \eta b_{mn}\left(\frac{x_{m,n-1} + x_{m-1,n}}{b_{mn}f(x_{mn})}\right)^2 \leq \eta b_{mn}\left(\frac{2x_{mn}}{b_{mn}f(x_{mn})}\right)^2 = \frac{4\eta}{b_{mn}}\left(\frac{x_{mn}}{f(x_{mn})}\right)^2,
\end{aligned}
$$

that is

$$
T(\Delta_1, \Delta_2)(w_{mn}) < \frac{4\eta}{b_{mn}}\delta^2. \qquad (12.68)
$$

Multiplying both sides of this inequality by B_{mn}, summing from m_0, n_0 to $m-1, n-1$, and then applying Lemma 12.4, one obtains

$$
\begin{aligned}
&B_{m-1,n-1}(w_{m,n-1} + w_{m-1,n}) - \sum_{s=m_0}^{m-1}\sum_{t=n_0}^{n-1} b_{st}(w_{s+1,t} + w_{s,t+1}) \\
&< \sum_{s=m_0}^{m-1}\sum_{t=n_0}^{n-1} 2\eta\frac{B_{st}}{b_{st}}\delta^2. \qquad (12.69)
\end{aligned}
$$

In view of condition (c_{10}) and the positivity of $b_{mn}w_{mn}$ to complete the proof of the theorem it suffices to show that the second term on the left-hand side of (12.69) is bounded.

Using the first equation in (12.4) and also (12.63), it is clear that

$$
\begin{aligned}
\sum_{s=m_0}^{m-1}\sum_{t=n_0}^{n-1} b_{st}w_{s+1,t} &= \sum_{s=m_0}^{m-1}\sum_{t=n_0}^{n-1} b_{st}\frac{g(y_{s+1,t})}{f(x_{s+1,t})} \leq \sum_{s=m_0}^{m-1}\sum_{t=n_0}^{n-1} b_{st}\frac{g(y_{st})}{f(x_{s+1,t})} \\
&= -\sum_{s=m_0}^{m-1}\sum_{t=n_0}^{n-1} \frac{T(\nabla_1, \nabla_2)(x_{st})}{f(x_{s+1,t})} = -\sum_{s=m_0}^{m-1}\sum_{t=n_0}^{n-1} \frac{x_{s-1,t} + x_{s,t-1} - x_{st}}{f(x_{s+1,t})} \\
&= \sum_{s=m_0}^{m-1}\sum_{t=n_0}^{n-1} \frac{x_{st} - x_{s-1,t} - x_{s,t-1}}{f(x_{s+1,t})} \leq \sum_{s=m_0}^{m-1}\sum_{t=n_0}^{n-1} \frac{x_{st} - x_{s-1,t}}{f(x_{s+1,t})} \\
&= \sum_{s=m_0}^{m-1}\sum_{t=n_0}^{n-1} \frac{\Delta_1(x_{s-1,t})}{f(x_{s+1,t})}. \qquad (12.70)
\end{aligned}
$$

Similarly,

$$\sum_{s=m_0}^{m-1}\sum_{t=n_0}^{n-1} b_{st} w_{s,t+1} \le \sum_{s=m_0}^{m-1}\sum_{t=n_0}^{n-1} \frac{\Delta_2(x_{s,t-1})}{f(x_{s,t+1})}. \tag{12.71}$$

Combining (12.70) and (12.71), we obtain

$$\sum_{s=m_0}^{m-1}\sum_{t=n_0}^{n-1} b_{st}(w_{s+1,t} + w_{s,t+1}) \le \sum_{s=m_0}^{m-1}\sum_{t=n_0}^{n-1} \left(\frac{\Delta_1(x_{s-1,t})}{f(x_{s+1,t})} + \frac{\Delta_2(x_{s,t-1})}{f(x_{s,t+1})}\right). \tag{12.72}$$

Observe that for $x_{m-1,n} \le s \le x_{mn}$, one has $1/f(s) \ge 1/f(x_{mn})$, and therefore,

$$\int_{x_{m-1,n}}^{x_{mn}} \frac{ds}{f(s)} \ge \frac{\Delta_1(x_{m-1,n})}{f(x_{mn})} \ge \frac{\Delta_1(x_{m-1,n})}{f(x_{m+1,n})}.$$

Likewise,

$$\int_{x_{m,n-1}}^{x_{mn}} \frac{dt}{f(t)} \ge \frac{\Delta_2(x_{m,n-1})}{f(x_{mn})} \ge \frac{\Delta_2(x_{m,n-1})}{f(x_{m,n+1})},$$

hence

$$\int_{x_{m-1,n}}^{x_{mn}} \frac{ds}{f(s)} + \int_{x_{m,n-1}}^{x_{mn}} \frac{dt}{f(t)} \ge \frac{\Delta_1(x_{m-1,n})}{f(x_{m+1,n})} + \frac{\Delta_2(x_{m,n-1})}{f(x_{m,n+1})}, \tag{12.73}$$

summing the both sides of (12.73) from $m_0, n_0 \to m-1, n-1$, we obtain

$$\begin{aligned}
2\int_{x_{m_0-1,n_0-1}}^{x_{m-1,n-1}} \frac{ds}{f(s)} &= 2\sum_{s=m_0}^{m-1}\sum_{t=n_0}^{n-1} \int_{x_{s-1,t-1}}^{x_{st}} \frac{ds}{f(s)} \\
&= \sum_{s=m_0}^{m-1}\sum_{t=n_0}^{n-1} \left(\int_{x_{s-1,t}}^{x_{st}} \frac{ds}{f(s)} + \int_{x_{s,t-1}}^{x_{st}} \frac{dt}{f(t)}\right) \\
&= \sum_{t=n_0}^{n-1} \left[\frac{\Delta_1(x_{s-1,t})}{f(x_{s+1,t})} + \frac{\Delta_2(x_{s,t-1})}{f(x_{s,t+1})}\right].
\end{aligned} \tag{12.74}$$

Combining (12.74) and (12.72), we finally have

$$\sum_{s=m_0}^{m-1}\sum_{t=n_0}^{n-1} b_{st}(w_{s+1,t} + w_{s,t+1}) \le 2\int_{x_{m_0-1,n_0-1}}^{x_{m-1,n-1}} \frac{ds}{f(s)} < \infty.$$

This completes the proof of the theorem.

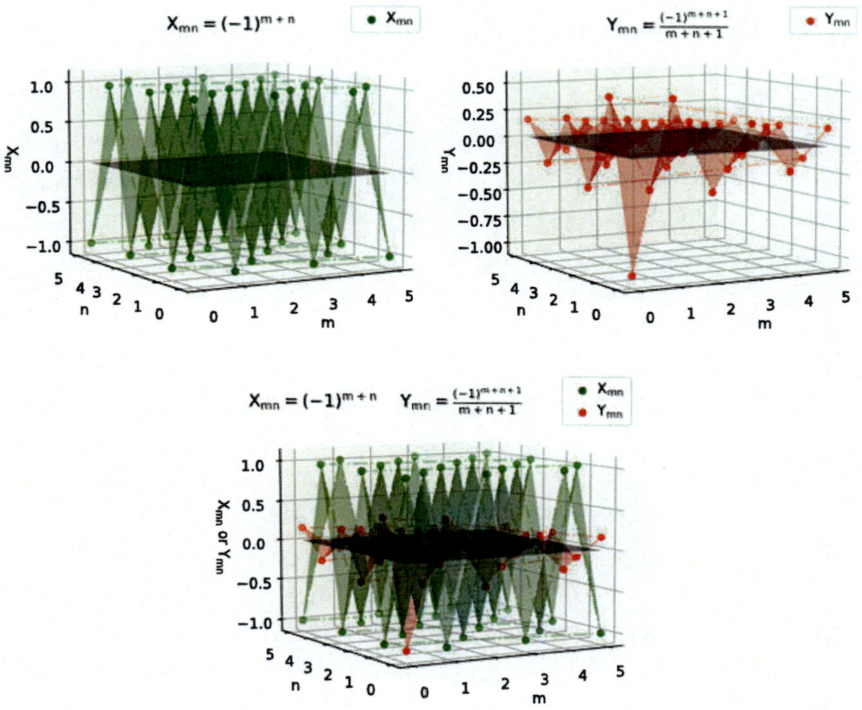

Fig. 12.1 Oscillatory behavior of system (12.76)

Remark 12.1 Similarly, we can consider the following system:

$$\begin{cases} \Delta_1(x_{mn}) - b_{mn}g(y_{m+1,n}) = 0, & (m,n) \in N_i = \{i, i+1, \dots, \}, \\ T(\Delta_1, \Delta_2)(y_{mn}) + a_{mn}f(x_{mn}) = 0, & i = 0, 1, 2, \dots . \end{cases} \tag{12.75}$$

Theorems 12.1 and 12.2 also hold for system (12.75).

Note also that the following systems can also be studied in a similar way:

$$\begin{cases} \Delta_2(x_{mn}) - b_{mn}g(y_{mn}) = 0, & (m,n) \in N_i = \{i, i+1, \dots, \}, \\ T(\Delta_1, \Delta_2)(y_{mn}) + a_{mn}f(x_{mn}) = 0, & i = 0, 1, 2, \dots , \end{cases}$$

and

$$\begin{cases} \Delta_2(x_{mn}) - b_{mn}g(y_{m+1,n}) = 0, & (m,n) \in N_i = \{i, i+1, \dots, \}, \\ T(\Delta_1, \Delta_2)(y_{mn}) + a_{mn}f(x_{mn}) = 0, & i = 0, 1, 2, \dots , \end{cases}$$

where $\Delta_2(x_{mn}) = x_{m,n+1} - x_{mn}$, for which similar results can be obtained.

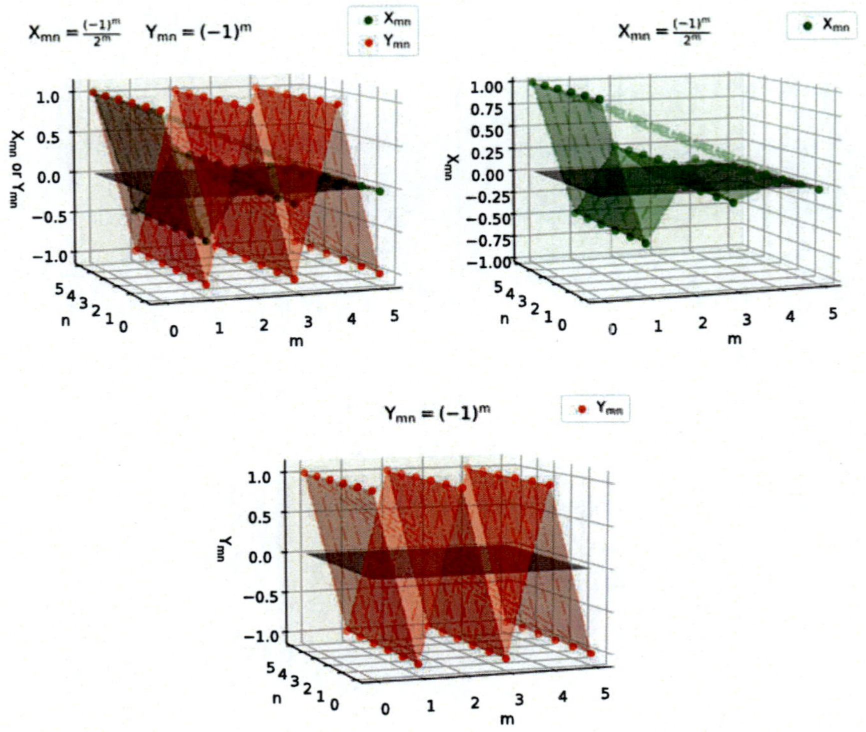

Fig. 12.2 Oscillatory behavior of system (12.77)

12.2.1 Examples

Example 12.1 Consider the partial difference system

$$\begin{cases} \Delta_1(x_{mn}) - 2(m+n+1)^{\frac{1}{3}} y_{mn}^{\frac{1}{3}} = 0, \\ T(\Delta_1, \Delta_2)(y_{mn}) + \frac{3(m+n)+4}{(m+n+2)(m+n+1)} x_{mn}^{\frac{5}{3}} = 0, \end{cases} \tag{12.76}$$

where, $a_{mn} = \frac{3(m+n)+4}{(m+n+2)(m+n+1)}$, $b_{mn} = 2(m+n+1)^{\frac{1}{3}}$, $g(u) = u^{\frac{1}{3}}$, $f(v) = v^{\frac{5}{3}}$. All conditions of Theorem 12.1 are satisfied and, hence, system (12.76) is oscillatory. In fact, $(\{x_{mn}\}, \{y_{mn}\}) = \left(\{(-1)^{m+n}\}, \{\frac{(-1)^{m+n+1}}{m+n+1}\}\right)$ is such an oscillatory solution of this system as shown in Fig. 12.1.

Example 12.2 Consider the partial difference system

$$\begin{cases} T(\nabla_1, \nabla_2)(x_{mn}) + \frac{1}{2^m}(y_{mn}^3 + y_{mn}) = 0, \\ T(\Delta_1, \Delta_2)(y_{mn}) + 2^{3m} x_{mn}^3 = 0, \end{cases} \tag{12.77}$$

where, $(m, n) \in N_0$, $a_{mn} = 2^{3m}$, $b_{mn} = \frac{1}{2^m}$, $g(u) = u^3 + u$, and $f(v) = v^3$. All conditions of Theorem 12.2 are satisfied and hence, system (12.77) is oscillatory. Indeed, $(\{x_{mn}\}, \{y_{mn}\}) = \left(\{\frac{(-1)^m}{2^m}\}, \{(-1)^m\}\right)$ is such an oscillatory solution of this system as shown in Fig. 12.2.

12.3 The First Wave Behavior of Uncoupled Nonlinear Dynamical System

12.3.1 Dynamical System

In this part, we focus on the wave behavior of the following system:

$$T(\Delta_1, \Delta_2)[c_{mn}\, \Delta_1\, (y_{mn})] + \sum_{i=1}^{s} a_i(m, n) f_i(y_{m+1,n}, \Delta_1(y_{mn})) = 0 \tag{12.78}$$

where $a_1(m, n), a_2(m, n), \ldots, a_s(m, n)$ are real double sequences, s is a positive integer, the double sequence $\{c_{mn}\}_{(m,n)\in N_0^2}$ is assumed to be positive. Actually, its continuous form is listed as following:

$$c(x, y)\frac{\partial^2 v(x,y)}{\partial y^2} + c(x, y+1) - c(x, y))\frac{\partial v(x,y)}{\partial y} + v(x, y)$$
$$+\sum_{i=1}^{s} a_i(x, y) f_i(\frac{\partial v(x,y)}{\partial y} + v(x, y), \frac{\partial v(x,y)}{\partial y}) = 0.$$

The following conditions will be assumed:

(c_1) $a_i(m, n) \geq 0$ for $m \geq m_0 \geq 0, n \geq n_0 \geq 0,\ i = 1, 2, \ldots, s$;
(c_2) $f_i : R^2 \to R, uf_i(u, v) > 0$ for all $u \neq 0,\ i = 1, 2, \ldots, s$.
(c_3) c_{mn} is a bounded sequence, that is, there exists a P and Q such that $1 \leq P < c_{mn} < Q$ for $m, n \in N_0$.

By a solution of (12.78) we mean a real double sequence $\{y_{mn}\}$ satisfying Eq. (12.78) for $m, n \in N_0$. We consider only such solutions which are nontrivial for all large m, n.

The purpose of this note is to give some criteria (sufficient conditions) for oscillation of all solutions of (12.78).

12.3.2 Main Results for the Uncoupled System

Theorem 12.5 *Suppose there exists an index j such that* (1^0) $f_j(\lambda u, \lambda v) = \lambda^{2\beta+1} f_j(u, v)$ *for all* $(u, v) \in R^2$,

λ is a real number different from 0, and β is a nonnegative integer.
(2^0) *$f_j(1, v)$ is continuous for $v = 0$.*
(3^0)

$$\sum_{i=m_0}^{\infty}\sum_{j=n_0}^{\infty} a_k(i, j) = \infty.$$

(4^0) *For initial value m_0, $n_0$define $y_{m_0,n_0} \equiv c > 0$. Then every bounded solution of (12.78) is oscillatory.*

Proof Suppose there exists a bounded nonoscillatory $\{y_{mn}\}$ of (12.78). Assume that $y_{mn} > 0$ for $m \geq m_1 \geq 0, n \geq n_1 \geq 0$. Then, by conditions ($c_1$) and ($c_2$), we see that $T(\Delta_1, \Delta_2)[c_{mn}\,\Delta_1\,(y_{mn})] < 0$, i.e., $\{c_{mn}\,\Delta_1\,(y_{mn})\}$ is nonincreasing sequence for $m \geq m_1, n \geq n_1$. Now, we can derive that

$$\Delta_1\,(y_{mn}) \geq 0 \text{ for } m \geq m_1,\ n \geq n_1. \tag{12.79}$$

In fact, if there would exist $m_2 \geq m_1$ and $n_2 \geq n_1$ such that $\Delta_1(y_{m_2n_2}) = c < 0$, then $\Delta_1(y_{m_2n_2}) \leq c$ for $m \geq m_2, n \geq n_2$ and hence

$$\sum_{j=n_2+1}^{n}(y_{mj} - y_{m_{2j}}) = \sum_{i=m_2}^{m-1}\sum_{j=n_2+1}^{n}\Delta_1(y_{ij}) \leq \sum_{i=m_2}^{m-1}\sum_{j=n_2+1}^{n} c = c(m - m_2)(n - n_2).$$

Note that based on the boundedness of $\{y_{mn}\}$ we have

$$y_{mn} \leq \sum_{j=n_2+1}^{n} y_{mj} \leq \sum_{j=n_2+1}^{n} y_{m_{2j}} + c(m - m_2)(n - n_2) \to -\infty \text{ as } m, n \to \infty,$$

which contradicts the fact that $y_{mn} \geq 0$ for $m \geq m_1, n \geq n_1$. Thus (12.79) holds, that is

$$y_{m+1,n} \geq y_{mn} \text{ for } m \geq m_1,\ n \geq n_1. \tag{12.80}$$

Using a similar argument as before, we also have

$$\Delta_2\,(y_{mn}) \geq 0 \text{ or } y_{m,n+1} \geq y_{mn} \text{ for } m \geq m_1,\ n \geq n_1. \tag{12.81}$$

From (12.80) and (12.81), that is, sequence $\{y_{mn}\}$ is monotone increasing in m and n for $m \geq m_1, n \geq n_1$. Note that $y_{m+1,n} = y_{mn} + \Delta_1(y_{mn})$, thus, $y_{m+1,n} \geq \Delta_1(y_{mn})$, that is $0 \leq \frac{\Delta_1(y_{mn})}{y_{m+1,n}} \leq 1$. Therefore

$$0 \leq \frac{\Delta_1(y_{mn})}{y_{m+1,n}} \leq \frac{\Delta_1(y_{mn})}{y_{mn}} < \frac{P\,\Delta_1\,(y_{mn})}{y_{mn}}, \tag{12.82}$$

let

$$Q_{mn} = \frac{P\,\Delta_1\,(y_{mn})}{y_{mn}}, \tag{12.83}$$

and $Q_{mn} \geq 0$, the following proof Q_{mn} is a decreasing sequence, note only that $\Delta_1(y_{mn}) = y_{m+1,n} - y_{mn} \geq 0$, thus we have also

$$\Delta_1\,(y_{m,n+1}) = y_{m+1,n+1} - y_{m,n+1} \geq 0. \tag{12.84}$$

From $Q_{mn} \geq 0$ and note that

$$\Delta_1 \left[\frac{c_{mn}\,\Delta_1\,(y_{mn})}{y_{mn}} \right] = \frac{y_{mn}\Delta_1(c_{mn}\,\Delta_1\,(y_{mn})) - c_{mn}[\Delta_1(y_{mn})]^2}{y_{mn}y_{m+1,n}}. \tag{12.85}$$

Using (12.84) and (12.85), an easy computation leads to

$$\Delta_1(Q_{mn}) \leq \Delta_1 \left[\frac{P\,\Delta_1\,(y_{mn})}{y_{mn}} \right] \leq \Delta_1 \left[\frac{c_{mn}\,\Delta_1\,(y_{mn})}{y_{mn}} \right] \tag{12.86}$$

$$= \frac{y_{mn}\Delta_1(c_{mn}\,\Delta_1\,(y_{mn})) - c_{mn}[\Delta_1(y_{mn})]^2}{y_{mn}y_{m+1,n}}$$

$$= \frac{y_{mn}\{c_{m+1,n}\,\Delta_1\,(y_{m+1,n}) - c_{mn}\,\Delta_1\,(y_{mn})\} - c_{mn}[\Delta_1(y_{mn})]^2}{y_{mn}y_{m+1,n}}$$

$$\leq \frac{y_{mn}\{c_{m+1,n}\,\Delta_1\,(y_{m+1,n}) + c_{m,n+1}\,\Delta_1\,(y_{m,n+1}) - \Delta_1(y_{mn})\}}{y_{mn}y_{m+1,n}} \tag{12.87}$$

$$- \frac{[\Delta_1(y_{mn})]^2}{y_{mn}y_{m+1,n}}$$

$$= \frac{1}{y_{m+1,n}}\{T(\Delta_1, \Delta_2)[c_{mn}\,\Delta_1\,(y_{mn})]\} - \frac{[\Delta_1(y_{mn})]^2}{y_{mn}y_{m+1,n}}$$

$$\leq -\frac{1}{y_{m+1,n}} \sum_{i=1}^{u} a_i(m,n) f_i(y_{m+1,n}, \Delta_1(y_{mn})) - \frac{[\Delta_1(y_{mn})]^2}{y_{mn}y_{m+1,n}} \tag{12.88}$$

$$\leq -\frac{a_k(m,n)}{y_{m+1,n}} f_k(y_{m+1,n}, \Delta_1(y_{mn})) \leq 0, \; m \geq m_0, n \geq n_0.$$

Using a similar argument as before, we also have $\Delta_2(Q_{mn}) \leq 0$, hence, the sequence $\{Q_{mn}\}$ is decreasing in m and n. Then, for sufficiently large m, n, we have

$$\lim_{\substack{n\to\infty \\ m\to\infty}} Q_{mn} = \lim_{\substack{n\to\infty \\ m\to\infty}} \frac{P\,\Delta_1\,(y_{mn})}{y_{mn}} = 0. \tag{12.89}$$

Combining (12.82), (12.83) and (12.89), we obtain

$$\lim_{\substack{n\to\infty \\ m\to\infty}} \frac{\Delta_1(y_{mn})}{y_{m+1,n}} = 0. \tag{12.90}$$

From (12.86) again, we have

$$\Delta_1\left(Q_{mn}\right) \le -\frac{a_k(m,n)}{y_{m+1,n}} f_k(y_{m+1,n}, \Delta_1(y_{mn})),\ m \ge m_0, n \ge n_0. \tag{12.91}$$

By (12.78) and the fact that $\{y_{mn}\}$ is nondecreasing, and

$$\begin{aligned} f_k\left(1, \frac{\Delta_1(y_{mn})}{y_{m+1,n}}\right) &= f_k\left(\frac{1}{y_{m+1,n}} y_{m+1,n}, \frac{1}{y_{m+1,n}} \Delta_1\left(y_{mn}\right)\right) \\ &= \left(\frac{1}{y_{m+1,n}}\right)^{2\beta+1} f_k(y_{m+1,n}, \Delta_1(y_{mn})), \end{aligned}$$

that is,

$$y_{m+1,n}^{2\beta+1} f_k\left(1, \frac{\Delta_1(y_{mn})}{y_{m+1,n}}\right) = f_k(y_{m+1,n}, \Delta_1(y_{mn})), \tag{12.92}$$

on using (12.80) and substituting (12.92) into (12.91), we have

$$\begin{aligned} \Delta_1(Q_{mn}) &\le -y_{m+1,n}^{2\beta} a_k(m,n) f_k\left(1, \frac{\Delta_1(y_{mn})}{y_{mn}}\right) \\ &\le -y_{m_0n_0}^{2\beta} a_k(m,n) f_k\left(1, \frac{\Delta_1(y_{mn})}{y_{mn}}\right) \\ &\le -c^{2\beta} a_k(m,n) f_k\left(1, \frac{\Delta_1(y_{mn})}{y_{m+1,n}}\right), \end{aligned}$$

for $m \ge m_0, n \ge n_0$. From (12.90), (2^0), and (c_2) we have

$$\lim_{m,n\to\infty} f_k\left(1, \frac{\Delta_1(y_{mn})}{y_{m+1,n}}\right) = f_k(1,0) > 0,$$

hence $f_k\left(1, \frac{\Delta_1(y_{mn})}{y_{m+1,n}}\right) \ge \frac{1}{2} f_k(1,0)$ for $m \ge m_3 \ge m_0, n \ge n_3 \ge n_0$. Therefore, we have

$$\Delta_1\left(Q_{mn}\right) \le -\frac{1}{2} c^{2\beta} a_k(m,n) f_k(1,0),\ m \ge m_3, n \ge n_3. \tag{12.93}$$

By (3^0), summing on both sides of (12.93) from m_3 to $m-1$ and n_3 to n, we obtain

$$\begin{aligned} Q_{mn} - \sum_{j=n_3}^{n} Q_{m_0 j} \le \sum_{j=n_3}^{n} Q_{mj} - \sum_{j=n_3}^{n} Q_{m_0 j} &= \sum_{i=m_3}^{m-1} \sum_{j=n_3}^{n} \Delta_1(Q_{ij}) \\ &\le -\frac{1}{2} \sum_{i=m_3}^{m-1} \sum_{j=n_3}^{n} f_k(1,0) c^{2\beta} a_k(i,j) \to -\infty \text{ as } m,n \to \infty. \end{aligned}$$

From (4^0) and (12.89), we see that $Q_{m_0 j}$ is bounded. Hence,

$$Q_{mn} \le \sum_{j=n_3}^{n} Q_{m_0 j} - \frac{1}{2}\sum_{i=m_3}^{m}\sum_{j=n_3}^{n} f_k(1,0)c^{2\beta}a_k(i,j)$$
$$\le \sum_{j=n_3}^{n} Q_{m_0 j} - \frac{1}{2} f_j(1,0)c^{2\beta}\sum_{i=m_3}^{m}\sum_{j=n_3}^{n} a_k(i,j) \to -\infty \text{ as } m,n \to \infty,$$

which contradicts the fact that $Q_{mn} \ge 0$ for $m \ge m_0, n \ge n_0$.

A similar argument applies to eventually negative solutions. The proof is complete.

Lemma 12.5 *Consider the partial difference equation*

$$T(\Delta_1,\Delta_2)\left[\left(3+\frac{e^{2mn}}{m+n}\right)\Delta_1(y_{mn})\right] + \left(\frac{m+n}{5}\right)^2 \frac{y_{m+1,n}[\Delta_1(y_{mn})]^2}{1+y_{m+1,n}^2} = 0 \tag{12.94}$$

where $f_1(u,v) = \frac{uv^2}{1+u^2}$, $c_{mn} = 3 + \frac{e^{2mn}}{m+n}$, $a_1(m,n) = (\frac{m+n}{5})^2$ *and tanking* $\lambda = 1, \beta = 0$. *It is easy to see that all assumptions of Theorem 12.5 hold. Therefor every solution of (12.94) oscillates. In fact,* $\{y_{mn}\} = \{\frac{(-1)^n}{e^n}\}$ *is such a solution.*

Taking $c_{mn} \equiv 1$, then we have the following result:

Theorem 12.6 *Assume that the following conditions hold:*

(5^0) *Let the boundary condition be* $y_{m_0 n} \le C_0$ *for all* $n \in N_0$,

(6^0) $f_k(-u,-v) = -f_k(u,v)$ *for every* $(u,v) \in R^2$, $k = 1,2,...,s$,

(7^0) $I \ne \Phi$, *where I denotes the set of all indices k for which the function* $f_k(u,v)$ *is nondecreasing with respect to u and with respect to v on R, as well as the function* $[f_k(u,0)]/u$ *is nonincreasing on* $(0,\infty)$,

(8^0) *there exists a positive sequence* $\{h_{mn}\}$ *for* $m \ge m_0, n \ge n_0$, *such that*

$$\sum_{i=m_0}^{\infty}\sum_{j=n_0}^{\infty}\left[h_{ij}\frac{1}{\alpha ij}\sum_{k\in I} a_k(i,j) f_k(ij,0) - \frac{[\Delta_1(h_{ij})]^2}{4h_{ij}}\right] = \infty$$

for every $\alpha \ge 1$, *where, for initial value* m_0 *and* $\{h_{mn}\}$, *we have the following boundary value:*

$$h_{m_0 n} \text{ is a constant } h_0 \text{ and } h_0 > 0, \text{ for } n \in N_0. \tag{12.95}$$

Then every bounded solution of (12.78) is oscillatory.

Proof If $\{y_{mn}\}$ is a bounded solution of (12.78), then by condition (6^0), $\{-y_{mn}\}$ is again a bounded solution of (12.78). Thus as in the proof of Theorem 12.5, the existence of a nonoscillatory solution $\{y_{mn}\}$ of (12.78) leads to $y_{mn} > 0$, $\Delta_1(y_{mn}) \ge 0$, and $T(\Delta_1,\Delta_2)[c_{mn}\Delta_1(y_{mn})] < 0$, that is, $\{\Delta_1(y_{mn})\}$ is nonincreasing sequence for $m \ge m_0, n \ge n_0$.

Let us denote

$$R_{mn} = \frac{h_{mn}\Delta_1(y_{mn})}{y_{mn}}, \ m \geq m_0, n \geq n_0. \tag{12.96}$$

Moreover, note that

$$\Delta_1[h_{mn}\Delta_1(y_{mn})] = \Delta_1(h_{mn})\,\Delta_1\,(y_{m+1,n}) + h_{mn}\,\Delta_1^2\,(y_{mn}), \tag{12.97}$$

then from Eq. (12.78) and using (12.85), (12.97) and (12.95), we get

$$\begin{aligned}
&\Delta_1(R_{mn}) = \Delta_1\left[\frac{h_{mn}\Delta_1(y_{mn})}{y_{mn}}\right] \\
&= \frac{\Delta_1(h_{mn})\,\Delta_1\,(y_{m+1,n})}{y_{m+1,n}} + \frac{h_{mn}\,\Delta_1^2\,(y_{mn})}{y_{m+1,n}} - \frac{h_{mn}[\Delta_1(y_{mn})]^2}{y_{mn}y_{m+1,n}} \\
&\leq \frac{\Delta_1(h_{mn})\,\Delta_1\,(y_{m+1,n})}{y_{m+1,n}} + \frac{h_{mn}(\Delta_1(y_{m+1,n}) + \Delta_1(y_{m,n+1}) - \Delta_1(y_{mn}))}{y_{m+1,n}} \\
&- \frac{h_{mn}[\Delta_1(y_{mn})]^2}{y_{mn}y_{m+1,n}} \\
&= \frac{h_{mn}}{y_{m+1,n}}\{T(\Delta_1, \Delta_2)[\Delta_1(y_{mn})]\} + \frac{\Delta_1(h_{mn})\,\Delta_1\,(y_{m+1,n})}{y_{m+1,n}} - \frac{h_{mn}[\Delta_1(y_{mn})]^2}{y_{mn}y_{m+1,n}} \\
&= -\frac{h_{mn}}{y_{m+1,n}}\sum_{i=1}^{s} a_i(m,n) f_i(y_{m+1,n}, \Delta_1(y_{mn})) + \frac{\Delta_1(h_{mn})\,\Delta_1\,(y_{m+1,n})}{y_{m+1,n}} \\
&- \frac{h_{mn}[\Delta_1(y_{mn})]^2}{y_{mn}y_{m+1,n}}.
\end{aligned}$$

Since, $\{\Delta_1(y_{mn})\}$ is nonnegative and nonincreasing sequence, therefore, $[\Delta_1(y_{mn})]^2 \geq [\Delta_1(y_{m+1,n})]^2$ and in view of monotonicity of y_{mn}, and (7^0) we see that

$$\begin{aligned}
&\Delta_1(R_{mn}) \\
&\leq -\frac{h_{mn}}{y_{m+1,n}}\sum_{i=1}^{s} a_i(m,n) f_i(y_{m+1,n}, \Delta_1(y_{mn})) \\
&+ \frac{\Delta_1(h_{mn})\,\Delta_1\,(y_{m+1,n})}{y_{m+1,n}} - h_{mn}\left[\frac{\Delta_1(y_{m+1,n})}{y_{m+1,n}}\right]^2 \\
&\leq -\frac{h_{mn}}{y_{m+1,n}}\sum_{i\in I} a_i(m,n) f_i(y_{m+1,n}, 0) + R_{m+1,n}\frac{\Delta_1 h_{mn}}{h_{m+1,n}} - R_{m+1,n}^2\frac{h_{mn}}{h_{m+1,n}^2} \\
&= -\frac{h_{mn}}{y_{m+1,n}}\sum_{i\in I} a_i(m,n) f_i(y_{m+1,n}, 0) - \frac{h_{mn}}{h_{m+1,n}^2}\left[R_{m+1,n} - \frac{\Delta_1(h_{mn}).h_{m+1,n}}{2h_{mn}}\right]^2 \\
&+ \frac{[\Delta_1(h_{mn})]^2}{4h_{mn}}.
\end{aligned}$$

That is,

$$\Delta_1(R_{mn}) \le -\frac{h_{mn}}{y_{m+1,n}} \sum_{i\in I} a_i(m,n) f_i(y_{m+1,n}, 0) + \frac{[\Delta_1(h_{mn})]^2}{4h_{mn}}, \quad (12.98)$$

$$m \ge m_0, n \ge n_0. \quad (12.99)$$

On the other hand, since

$$\sum_{j=n_0}^{n} \sum_{i=m_0}^{m} \Delta_1(y_{ij}) = y_{m+1,n} + \sum_{j=n_0}^{n-1} y_{m+1,j} - \sum_{j=n_0}^{n} y_{m_0 j} \ge y_{m+1,n} - \sum_{j=n_0}^{n} y_{m_0 j},$$

that is,

$$y_{m+1,n} \le \sum_{j=n_0}^{n} y_{m_0 j} + \sum_{j=n_0}^{n} \sum_{i=m_0}^{m} \Delta_1(y_{ij}), \quad (12.100)$$

from (12.100), using the nonnegative and nonincreasing character of $\{\Delta_1(y_{mn})\}$ and (5^0), we have

$$\begin{aligned} y_{m+1,n} &\le C_0(n-n_0+1) + \Delta_1(y_{m_0 n_0}) \sum_{j=n_0}^{n} \sum_{i=m_0}^{m} 1 \\ &= C_0(n-n_0+1) + \Delta_1(y_{m_0 n_0})(n-n_0+1)(m-m_0+1) \end{aligned}$$

and consequently there exists $m_1 \ge m_0, n_1 \ge n_0$ and $\alpha \ge 1$ such that

$$y_{m+1,n} \le \alpha mn \text{ for } m \ge m_1, n \ge n_1.$$

Thus returning to (12.99), by condition (7^0), we obtain

$$\Delta_1(R_{mn}) \le -h_{mn} \sum_{i\in I} a_i(m,n) \frac{f_i(mn,0)}{\alpha mn} + \frac{[\Delta_1(h_{mn})]^2}{4h_{mn}}, \quad m \ge m_0, n \ge n_0. \quad (12.101)$$

Summing on both sides of (12.101) from m_0 to $m-1$ and n_0 to n, we obtain

$$\begin{aligned} R_{mn} - \sum_{j=n_0}^{n} R_{m_0 j} \le \sum_{j=n_0}^{n} R_{mj} - \sum_{j=n_0}^{n} R_{m_0 j} &= \sum_{i=m_0}^{m-1} \sum_{j=n_0}^{n} \Delta_1(R_{ij}) \\ &\le \sum_{i=m_0}^{m-1} \sum_{j=n_0}^{n} \left[-h_{ij} \sum_{k\in I} a_k(i,j) \frac{f_i(ij,0)}{\alpha ij} + \frac{[\Delta_1(h_{ij})]^2}{4h_{ij}} \right], \end{aligned}$$

and then

$$R_{mn} \le \sum_{j=n_0}^{n} R_{m_0 j} + \sum_{i=m_0}^{m-1}\sum_{j=n_0}^{n}\left[-h_{ij}\frac{1}{\alpha ij}\sum_{k\in I} a_k(i,j) f_k(ij,0) + \frac{[\Delta_1(h_{ij})]^2}{4h_{ij}}\right].$$

Combining (5^0), (12.95) and (12.96), $R_{m_0 j}$ is bounded, hence, now, by (8^0), it is easy to see that $\{R_{mn}\}$ is eventually negative, which is a contradiction, since $R_{mn} \ge 0$, for $m \ge m_0, n \ge n_0$. This completes the proof.

A close look at the proof of Theorem 12.6 ensures the validity of the following:

Theorem 12.7 *If, in addition to condition* (4^0), (16) *and* $c_{mn} \equiv 1$, *we assume that*

(9^0) $I \ne \Phi$, *where I denotes the set of all indices k for which the function* $f_k(u, v)$ *is nondecreasing with respect to* v *on R, as well as the function* $[f_k(u, 0)]/u$ *is nonincreasing on* $(0.\infty)$,

(10^0) *there exists a positive sequence* $\{h_{mn}\}$ $m \ge m_0, n \ge n_0$, *such that*

$$\sum_{i=m_0}^{\infty}\sum_{j=n_0}^{\infty}\left[h_{ij}\frac{1}{\alpha ij}\sum_{k\in I} a_k(i,j) f_k(\alpha ij,0) - \frac{[\Delta_1(h_{ij})]^2}{4h_{ij}}\right] = +\infty$$

for every $\alpha \ge 1$.

Then every bounded solution of (12.78) *is oscillatory.*

In the above theorems, conditions (8^0) and (10^0) depend on a parameter $\alpha \ge 1$. Because of this, several difficulties may appear in the verification of (8^0) or (10^0). For this reason, it is important to relax these conditions by requiring (8^0) only for $\alpha = 1$. More precisely, the following theorem is valid.

Theorem 12.8 *Assume that* (12.95) *holds and*

(11^0) $I \ne \Phi$, *where I denotes the set of all indices k for which the function* $f_k(u, v)$ *is nondecreasing with respect to* v *on R and* $f_k(\lambda u, 0) = \lambda f_k(u, 0)$ *for every* $u \in R$ *and real* $\lambda \ne 0$.

(12^0) *there exists a positive sequence* $\{h_{mn}\}$ $m \ge m_0, n \ge n_0$, *such that*

$$\sum_{i=m_0}^{\infty}\sum_{j=n_0}^{\infty}\left[h_{ij}\sum_{k\in I} a_k(i,j) f_i(1,0) - \frac{[\Delta_1(h_{ij})]^2}{4h_{ij}}\right] = +\infty.$$

Then every bounded solution of (12.78) *is oscillatory.*

Proof Assuming the existence of nonoscillatory solution $y_{mn} > 0$ for $m \ge m_0, n \ge n_0$ and arguing as in the proof of theorem 2 we obtain the inequality (12.101). This in turn, by (11^0), implies

$$\Delta_1(R_{mn}) \le -h_{mn}\sum_{i\in I} a_i(m,n) f_i(1,0) + \frac{[\Delta_1(h_{mn})]^2}{4h_{mn}}, \quad m \ge m_0, \; n \ge n_0.$$

Thus our assertion follows now exactly the same way as the previous one. A similar argument holds in the case of an eventually negative solution.

Example 12.3 Consider the partial difference equation

$$T(\Delta_1, \Delta_2)[\Delta_1(y_{mn})] + (m+n)^2 y_{m+1,n}[\Delta_1(y_{mn})]^2 = 0 \tag{12.102}$$

where $f_1(u, v) = uv^2$, $c_{mn} = 1$ and $a_1(m, n) = (m+n)^2$. It is easy to see that all assumptions of Theorem 12.6 hold. Therefore every solution of (12.102) oscillates. In fact, $\{y_{mn}\} = \{\frac{(-1)^n}{3^n}\}$ is such a solution.

Example 12.4 Consider the partial difference equation

$$T(\Delta_1, \Delta_2)\left[\left(3 + \frac{e^{2mn}}{m+n}\right)\Delta_1(y_{mn})\right] + \left(\frac{m+n}{5}\right)^2 \frac{y_{m+1,n}[\Delta_1(y_{mn})]^2}{1 + y_{m+1,n}^2} = 0 \tag{12.103}$$

where $f_1(u, v) = \frac{uv^2}{1+u^2}, c_{mn} = 3 + \frac{e^{2mn}}{m+n}, a_1(m, n) = (\frac{m+n}{5})^2$ and tanking $\lambda = 1, \beta = 0$. It is easy to see that all assumptions of Theorem 12.6 hold. Therefor every solution of (E_1) oscillates. In fact, $\{y_{mn}\} = \{\frac{(-1)^n}{e^n}\}$ is such a solution.

12.4 The Second Wave Behavior of Uncoupled Nonlinear Dynamical System

12.4.1 The Second-Order Nonlinear Dynamical System

In this part, we consider the wave behaviors of the second-order nonlinear dynamical system

$$T(\Delta_1, \Delta_2)\,[c_{mn}T(\Delta_1, \Delta_2)(y_{mn})] + p(m, n)(y_{m+1,n} + y_{m,n+1})^\nu = 0, \tag{12.104}$$

where ν is a quotient of odd positive integers, $m, n \in N_i$. Some sufficient conditions for of all solutions of the above equation with $\nu > 1$ and $\nu < 1$ are obtained.

By a solution of Eq. (12.104), we mean a real double sequence $\{y_{mn}\}$ satisfying (12.104) for $m, n \in N_0$. We consider only such solutions that are nontrivial for all large m, n. A solution $\{y_{mn}\}$ of (12.104) is called *nonoscillatory* if it is eventually positive or eventually negative; otherwise, it is called *oscillatory* (Figs. 12.3 and 12.4).

12.4.2 Main Results for the Second-Order Nonlinear Dynamical System

The following elementary identity for double sequences will be used later.

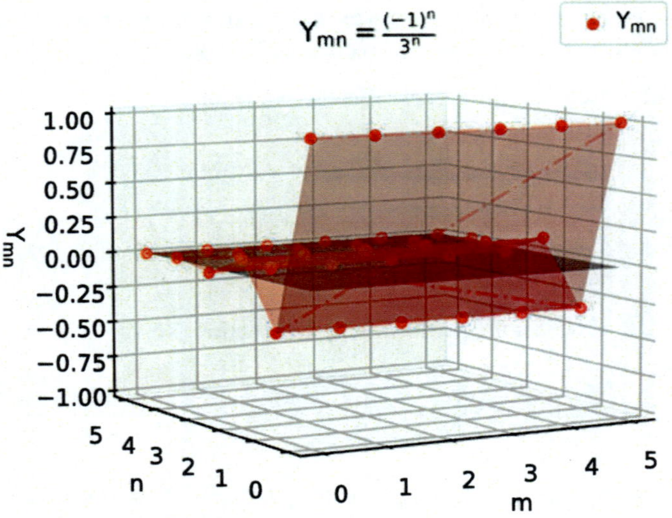

Fig. 12.3 Oscillatory behavior of system (12.102)

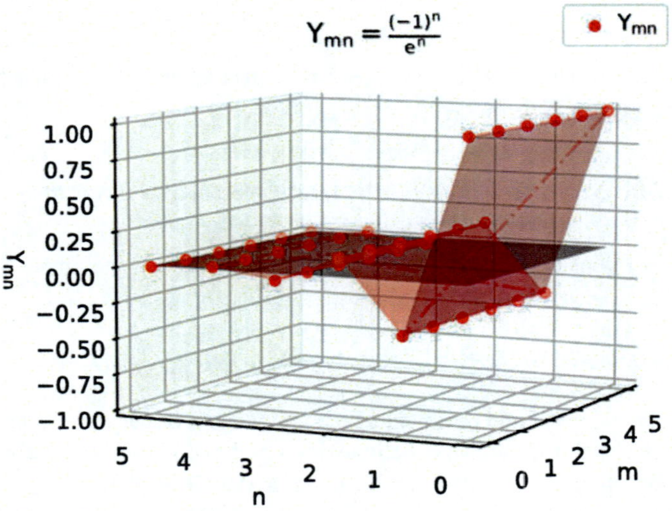

Fig. 12.4 Oscillatory behavior of system (12.103)

We consider the case where $c_{ij} > 0$ for all $i \geq 0,\ j \geq 0$ and

$$\sum_{i=0}^{\infty}\sum_{j=0}^{\infty} 1/c_{ij} < \infty. \tag{12.105}$$

Theorem 12.9 *Assume that $c_{ij} > 0$ for all $i \geq 0,\ j \geq 0$ and (12.105) holds. Further, assume that $\nu > 1$ and $p_{ij} > 0$ for all $i \geq 0,\ j \geq 0$, and*

$$\sum_{i=M}^{\infty}\sum_{j=N}^{\infty} \left(\frac{1}{2}\right)^{i+j} p_{ij}\rho_{i+1,j}^{\nu} = \infty, \tag{12.106}$$

where

$$\rho_{mn} = \sum_{i=m}^{\infty}\sum_{j=n}^{\infty} 1/c_{ij}. \tag{12.107}$$

Then all solutions of Eq. (12.104) oscillate.

Proof Assume the contrary, namely, there exists a nonosillatory solution $\{y_{mn}\}$. Without loss of generality, assume that $y_{mn} > 0$ for $m \geq M, n \geq N$. Then

$$T(\Delta_1, \Delta_2)[c_{mn}T(\Delta_1, \Delta_2)(y_{mn})] \leq 0 \text{ for } m \geq M, n \geq N. \tag{12.108}$$

In view of (12.108), we have

$$\begin{aligned} c_{m+1,n}T(\Delta_1, \Delta_2)(y_{m+1,n}) &\leq c_{mn}T(\Delta_1, \Delta_2)(y_{mn}), \\ c_{m,n+1}T(\Delta_1, \Delta_2)(y_{m,n+1}) &\leq c_{mn}T(\Delta_1, \Delta_2)(y_{mn}), \end{aligned}$$

that is, $\{c_{mn}T(\Delta_1, \Delta_2)(y_{mn})\}$ is a noincreasing, thus

$$c_{mn}T(\Delta_1, \Delta_2)(y_{mn}) \leq c_{MN}T(\Delta_1, \Delta_2)(y_{MN}) \text{ for } m \geq M,\ n \geq N, \tag{12.109}$$

or

$$T(\Delta_1, \Delta_2)(y_{mn}) \leq c_{MN}T(\Delta_1, \Delta_2)(y_{MN})/c_{mn} \quad \text{for } m \geq M, n \geq N.$$

Applying Lemma 12.5 and summing the above inequality from M, N to m, n, we obtain

$$
\begin{aligned}
y_{mn} - y_{MN} &\le y_{mn} + \left(\sum_{i=M+1}^{m-1} \sum_{j=N+1}^{n} y_{ij} + \sum_{j=N+1}^{n-1} y_{mj} + \dots \right) - y_{MN} \\
&= \sum_{i=M+1}^{m+1} \sum_{j=N+1}^{n} y_{ij} + \sum_{i=M}^{m} y_{i,n+1} + y_{m+1,N} - y_{MN} \\
&\le [c_{MN} T(\Delta_1, \Delta_2)(y_{MN})] \sum_{i=M}^{m} \sum_{j=N}^{n} 1/c_{ij},
\end{aligned}
$$

that is,

$$
y_{mn} - y_{MN} \le [c_{MN} T(\Delta_1, \Delta_2)(y_{MN})] \sum_{i=M}^{m} \sum_{j=N}^{n} 1/c_{ij} \text{ for } m \ge M, n \ge N. \tag{12.110}
$$

Hence, y_{mn} is bounded above. From (12.110) we have

$$
y_{MN} \ge -[c_{MN} T(\Delta_1, \Delta_2)(y_{MN})] \sum_{i=M}^{m} \sum_{j=N}^{n} 1/c_{ij} \text{ for } m \ge M, n \ge N. \tag{12.111}
$$

Letting $m, n \to \infty$ gives

$$
y_{MN} \ge -[c_{MN} T(\Delta_1, \Delta_2)(y_{MN})] \rho_{MN}, \tag{12.112}
$$

where ρ_{MN} is defined by (12.107) and M, N are two sufficiently large numbers.

It follows from (12.108) that there are two possible cases of $T(\Delta_1, \Delta_2)(y_{mn})$. First, we consider the case where $T(\Delta_1, \Delta_2)(y_{mn}) \ge 0$ for $m \ge M, n \ge N$. Summing (12.104) from M, N to m, n, we obtain

$$
\sum_{i=M}^{m} \sum_{j=N}^{n} T(\Delta_1, \Delta_2) \left[c_{ij} T(\Delta_1, \Delta_2)(y_{ij}) \right] + \sum_{i=M}^{m} \sum_{j=N}^{n} p_{ij} (y_{m+1,n} + y_{m,n+1})^{\nu} = 0. \tag{12.113}
$$

Applying Lemma 12.104 again, we have

$$
\begin{aligned}
&c_{mn} T(\Delta_1, \Delta_2)(y_{mn}) - c_{MN} T(\Delta_1, \Delta_2)(y_{MN}) \\
&+ \sum_{i=M}^{m} \sum_{j=N}^{n} p_{ij} (y_{m+1,n} + y_{m,n+1})^{\nu} \le 0,
\end{aligned}
$$

or

$$
\sum_{i=M}^{m} \sum_{j=N}^{n} p_{ij} (y_{m+1,n} + y_{m,n+1})^{\nu} \le c_{MN} T(\Delta_1, \Delta_2)(y_{MN}).
$$

Letting $m, n \to \infty$, we have

$$\sum_{i=M}^{\infty}\sum_{j=N}^{\infty} p_{ij}(y_{m+1,n} + y_{m,n+1})^{\nu} < \infty. \tag{12.114}$$

Since $c_{mn}T(\Delta_1, \Delta_2)(y_{mn}) \geq 0$ for $m \geq M, n \geq N$, that is,

$$y_{m+1,n} + y_{m,n+1} \geq y_{mn},$$

there exists a positive number, c, such that $y_{mn} > c > 0$ for $m > M, n \geq N$. Thus, there exist $M_1 \geq M$, $N_1 \geq N$, such that

$$y_{m+1,n} + y_{m,n+1} \geq y_{m+1,n} \geq \rho_{m+1,n} \text{ for } m \geq M_1,\ n \geq N_1, \tag{12.115}$$

since $\rho_{mn} \to 0$ as $m, n \to \infty$. Combining (12.114) and (12.115), we have

$$\sum_{i=M}^{\infty}\sum_{j=N}^{\infty} p_{ij}\rho_{i+1,j}^{\nu} < \infty, \tag{12.116}$$

that is,

$$\sum_{i=M}^{\infty}\sum_{j=N}^{\infty}\left(\frac{1}{2}\right)^{i+j} p_{ij}\rho_{i+1,j}^{\nu} < \infty, \tag{12.117}$$

which contradicts (12.106).

Now, we consider the other case where

$$c_{mn}T(\Delta_1, \Delta_2)(y_{mn}) < 0, \text{ for } m \geq M, n \geq N.$$

We have

$$\begin{aligned}
&T(\Delta_1, \Delta_2)\left[\left(\frac{1}{2}\right)^{m+n-1}(c_{mn}T(\Delta_1, \Delta_2)(y_{mn}))^{-\nu+1}\right] \\
&= \left(\frac{1}{2}\right)^{m+n}\left(c_{m+1,n}T(\Delta_1, \Delta_2)(y_{m+1,n})\right)^{-\nu+1} \\
&\quad + \left(\frac{1}{2}\right)^{m+n}\left(c_{m,n+1}T(\Delta_1, \Delta_2)(y_{m,n+1})\right)^{-\nu+1} \\
&\quad - \left(\frac{1}{2}\right)^{m+n-1}(c_{mn}T(\Delta_1, \Delta_2)(y_{mn}))^{-\nu+1} \\
&= \left(\frac{1}{2}\right)^{m+n}\left\{\begin{array}{c}\left(c_{m+1,n}T(\Delta_1, \Delta_2)(y_{m+1,n})\right)^{-\nu+1} + \left(c_{m,n+1}T(\Delta_1, \Delta_2)(y_{m,n+1})\right)^{-\nu+1} \\ -2\left(c_{mn}T(\Delta_1, \Delta_2)(y_{mn})\right)^{-\nu+1}\end{array}\right\} \\
&= \left(\frac{1}{2}\right)^{m+n}\left\{\begin{array}{l}\left(c_{m+1,n}T(\Delta_1, \Delta_2)(y_{m+1,n})\right)^{-\nu+1} - (c_{mn}T(\Delta_1, \Delta_2)(y_{mn}))^{-\nu+1} \\ +\left(c_{m,n+1}T(\Delta_1, \Delta_2)(y_{m,n+1})\right)^{-\nu+1} - (c_{mn}T(\Delta_1, \Delta_2)(y_{mn}))^{-\nu+1}\end{array}\right\}
\end{aligned}$$

$$
\begin{aligned}
&= (-\nu+1)\left(\frac{1}{2}\right)^{m+n}\left\{\begin{array}{l}\xi^{-\nu}\left(c_{m+1,n}T(\Delta_1,\Delta_2)(y_{m+1,n}) - c_{mn}T(\Delta_1,\Delta_2)(y_{mn})\right)\\ +\eta^{-\nu}\left(c_{m,n+1}T(\Delta_1,\Delta_2)(y_{m,n+1}) - c_{mn}T(\Delta_1,\Delta_2)(y_{mn})\right)\end{array}\right\}\\
&\leq (-\nu+1)\xi^{-\nu}\left(\frac{1}{2}\right)^{m+n}\left(\begin{array}{c}c_{m+1,n}T(\Delta_1,\Delta_2)(y_{m+1,n}) + c_{m,n+1}T(\Delta_1,\Delta_2)(y_{m,n+1})\\ -2c_{mn}T(\Delta_1,\Delta_2)(y_{mn})\end{array}\right)\\
&\leq (-\nu+1)\xi^{-\nu}\left(\frac{1}{2}\right)^{m+n}\left(\begin{array}{c}c_{m+1,n}T(\Delta_1,\Delta_2)(y_{m+1,n}) + c_{m,n+1}T(\Delta_1,\Delta_2)(y_{m,n+1})\\ -c_{mn}T(\Delta_1,\Delta_2)(y_{mn})\end{array}\right)\\
&= (-\nu+1)\xi^{-\nu}\left(\frac{1}{2}\right)^{m+n}\left\{T(\Delta_1,\Delta_2)\Big[c_{mn}T(\Delta_1,\Delta_2)(y_{mn})\Big]\right\}\\
&= (-\nu+1)\xi^{-\nu}\left(\frac{1}{2}\right)^{m+n}\left[-\,p_{mn}(y_{m+1,n}+y_{m,n+1})^{\nu}\right],
\end{aligned}
$$

where

$$c_{m+1,n}T(\Delta_1,\Delta_2)(y_{m+1,n}) < \xi < c_{mn}T(\Delta_1,\Delta_2)(y_{mn}),$$

$$c_{m,n+1}T(\Delta_1,\Delta_2)(y_{m,n+1}) \leq \eta \leq c_{mn}T(\Delta_1,\Delta_2)(y_{mn}),$$

and, without loss of generality, $\xi^{-\nu} = \min(\xi^{-\nu}, \eta^{-\nu})$.

We note that $y_{mn} \geq -[c_{mn}T(\Delta_1,\Delta_2)(y_{mn})]\rho_{mn}$, for $m \geq M, n \geq N$ by (12.112). Hence, we also have

$$y_{m+1,n} \geq -[c_{m+1,n}T(\Delta_1,\Delta_2)(y_{m+1,n})]\rho_{m+1,n},$$

so that

$$y_{m,n+1} + y_{m+1,n} \geq y_{m+1,n} \geq -[c_{m+1,n}T(\Delta_1,\Delta_2)(y_{m+1,n})]\rho_{m+1,n}. \tag{12.118}$$

Equation (12.118) implies that

$$
\begin{aligned}
&T(\Delta_1,\Delta_2)\left[\left(\frac{1}{2}\right)^{m+n-1}(c_{mn}T(\Delta_1,\Delta_2)(y_{mn}))^{-\nu+1}\right]\\
&\leq \left(\frac{1}{2}\right)^{m+n}(-\nu+1)\xi^{-\nu}\left[-\,p_{mn}(y_{m+1,n}+y_{m,n+1})^{\nu}\right]\\
&\leq \left(\frac{1}{2}\right)^{m+n}(-\nu+1)\xi^{-\nu}\left\{p_{mn}\left[\left(c_{m+1,n}T(\Delta_1,\Delta_2)(y_{m+1,n})\right)\rho_{m+1,n}\right]^{\nu}\right\}\\
&\leq \left(\frac{1}{2}\right)^{m+n}(-\nu+1)\left\{p_{mn}\left[\left(c_{m+1,n}T(\Delta_1,\Delta_2)(y_{m+1,n})\right)\rho_{m+1,n}\right]^{\nu}\right\}\\
&\quad\cdot\left[c_{m+1,n}T(\Delta_1,\Delta_2)(y_{m+1,n})\right]^{-\nu}.\\
&= -(\nu-1)\left(\frac{1}{2}\right)^{m+n} p_{mn}\rho^{\nu}_{m+1,n}.
\end{aligned}
$$

Hence,

$$T(\Delta_1, \Delta_2)\left[\left(\frac{1}{2}\right)^{m+n-1}(c_{mn}T(\Delta_1, \Delta_2)(y_{mn}))^{-\nu+1}\right]$$
$$\leq -(\nu-1)\left(\frac{1}{2}\right)^{m+n} p_{mn}\rho^{\nu}_{m+1,n}. \tag{12.119}$$

Using Lemma 12.1 and summing (12.119) from M, N to m, n, we obtain

$$\begin{aligned}
&\left(\frac{1}{2}\right)^{M+n}\left(c_{M,n+1}T(\Delta_1, \Delta_2)(y_{M,n+1})\right)^{-\nu+1} \\
&-\left(\frac{1}{2}\right)^{M+N-1}(c_{MN}T(\Delta_1, \Delta_2)(y_{MN}))^{-\nu+1} \\
&\leq \sum_{i=M+1}^{m}\sum_{j=N}^{n}\left[\left(\frac{1}{2}\right)^{i+j}\left(c_{i,j+1}T(\Delta_1, \Delta_2)(y_{i,j+1})\right)^{-\nu+1}\right] \\
&\quad+\sum_{j=N}^{n}\left[\left(\frac{1}{2}\right)^{m+j}\left(c_{m+1,j}T(\Delta_1, \Delta_2)(y_{m+1,j})\right)^{-\nu+1}\right] \\
&\quad+\left(\frac{1}{2}\right)^{M+n}\left(c_{M,n+1}T(\Delta_1, \Delta_2)(y_{M,n+1})\right)^{-\nu+1} \\
&\quad-\left(\frac{1}{2}\right)^{M+N-1}(c_{MN}T(\Delta_1, \Delta_2)(y_{MN}))^{-\nu+1} \\
&=\sum_{i=M}^{m}\sum_{j=N}^{n}T(\Delta_1, \Delta_2)\left[\left(\frac{1}{2}\right)^{i+j-1}\left(c_{ij}T(\Delta_1, \Delta_2)(y_{ij})\right)^{-\nu+1}\right] \\
&\leq -(\nu-1)\sum_{i=M}^{m}\sum_{j=N}^{n}\left(\frac{1}{2}\right)^{i+j} p_{ij}\rho^{\nu}_{i+1,j},
\end{aligned}$$

that is,

$$\begin{aligned}
&-\left(\frac{1}{2}\right)^{M+N-1}(c_{MN}T(\Delta_1, \Delta_2)(y_{MN}))^{-\nu+1} \\
&\geq (\nu-1)\sum_{i=M}^{m}\sum_{j=N}^{n}\left(\frac{1}{2}\right)^{i+j} p_{ij}\rho^{\nu}_{i+1,j} - \left(\frac{1}{2}\right)^{M+N-1}(c_{MN}T(\Delta_1, \Delta_2)(y_{MN}))^{-\nu+1}.
\end{aligned}$$

So, letting $m, n \to \infty$, we have

$$\sum_{i=M}^{m}\sum_{j=N}^{n}\left(\frac{1}{2}\right)^{i+j} p_{ij}\rho^{\nu}_{i+1,j} < \infty,$$

which contradicts (12.106). This completes the proof of the theorem.

Now, we consider the sublinear case, i.e., $0 < \nu < 1$.

Theorem 12.10 *Assume that* $c_{ij} > 0$ *for all* $i \geq 0$, $j \geq 0$, *and* (12.105) *holds. Further, assume that* $0 < \nu < 1$ *and* $p_{ij} > 0$ *for all* $i \geq 0$, $j \geq 0$, *and*

$$\sum_{i=M}^{\infty}\sum_{j=N}^{\infty}\left(\frac{1}{2}\right)^{i+j} p_{ij}\rho_{i+1,j} = \infty. \tag{12.120}$$

Then, all solutions of Eq. (12.104) *oscillate.*

Proof Suppose the contrary, namely, there exists a nonoscillatory solution, $\{y_{mn}\}$. Without loss of generality, assume the $y_{mn} > 0$ for $m \geq M, n \geq N$. Then

$$T(\Delta_1, \Delta_2)[c_{mn}T(\Delta_1, \Delta_2)(y_{mn})] \leq 0 \text{ for } m \geq M, n \geq N.$$

If $c_{mn}T(\Delta_1, \Delta_2)(y_{mn}) \geq 0$ for $m \geq M, n \geq N$, we have (12.114) and (12.116). For large m, n, we have $\rho_{mn} \leq 1$ and $\rho_{mn}^{\nu} \geq \rho_{mn}$. Therefore, from (12.116), we have

$$\sum_{i=M}^{\infty}\sum_{j=N}^{\infty}\left(\frac{1}{2}\right)^{i+j} p_{ij}\rho_{i+1,j} < \infty, \tag{12.121}$$

which contradicts (12.120).

For the case where $c_{mn}T(\Delta_1, \Delta_2)(y_{mn}) < 0$ for $m \geq M, n \geq N$, using Lemma 12.5 and summing (12.104) from M, N to m, n, we obtain

$$\sum_{i=M}^{m}\sum_{j=N}^{n} T(\Delta_1, \Delta_2)\left[c_{ij}T(\Delta_1, \Delta_2)(y_{ij})\right] + \sum_{i=M}^{m}\sum_{j=N}^{n} p_{ij}(y_{i+1,j} + y_{i,j+1})^{\nu} = 0,$$

that is,

$$\begin{aligned}
&\sum_{i=M+1}^{m+1}\sum_{j=N+1}^{n}\left[c_{ij}T(\Delta_1, \Delta_2)(y_{ij})\right] + \sum_{i=M}^{m}\left[c_{i,n+1}T(\Delta_1, \Delta_2)(y_{i,n+1})\right] \\
&+\left[c_{m+1,N}T(\Delta_1, \Delta_2)(y_{m+1,N})\right] - \left[c_{MN}T(\Delta_1, \Delta_2)(y_{MN})\right] \\
&+\sum_{i=M}^{m}\sum_{j=N}^{n} p_{ij}(y_{i+1,j} + y_{i,j+1})^{\nu} = 0.
\end{aligned}$$

Hence,

$$
\begin{aligned}
&-\left[c_{m+1,n}T(\Delta_1,\Delta_2)(y_{m+1,n})\right]\\
&\geq -\left\{\begin{array}{l}\sum_{i=M+1}^{m+1}\sum_{j=N+1}^{n}\left[c_{ij}T(\Delta_1,\Delta_2)(y_{ij})\right]+\sum_{i=M}^{m}\left[c_{i,n+1}T(\Delta_1,\Delta_2)(y_{i,n+1})\right]\\ +\left[c_{m+1,N}T(\Delta_1,\Delta_2)(y_{m+1,N})\right]\end{array}\right\}\\
&\geq \sum_{i=M}^{m}\sum_{j=N}^{n}p_{ij}(y_{i+1,j}+y_{i,j+1})^{\nu}\ \text{for}\ m\geq M,\ n\geq N,
\end{aligned}
$$

that is,

$$
-\left[T(\Delta_1,\Delta_2)(y_{m+1,n})\right]\geq \frac{1}{c_{m+1,n}}\sum_{i=M}^{m}\sum_{j=N}^{n}p_{ij}(y_{i+1,j}+y_{i,j+1})^{\nu}\ \text{for}\ m\geq M,\ n\geq N. \tag{12.122}
$$

We consider the partial difference $T(\Delta_1,\Delta_2)\left[(\frac{1}{2})^{m+n-1}y_{m+1,n}^{2\varepsilon}\right]$, where $\varepsilon>0$, such that $2\varepsilon<1-\nu$. Note that y_{mn} is nonincreasing. Thus,

$$
\begin{aligned}
&-T(\Delta_1,\Delta_2)\left[\left(\frac{1}{2}\right)^{m+n-1}y_{m+1,n}^{2\varepsilon}\right]\\
&=-\left(\frac{1}{2}\right)^{m+n}\left[y_{m+2,n}^{2\varepsilon}+y_{m+1,n+1}^{2\varepsilon}-2y_{m+1,n}^{2\varepsilon}\right]\\
&=-\left(\frac{1}{2}\right)^{m+n}\left[\left(y_{m+2,n}^{2\varepsilon}-y_{m+1,n}^{2\varepsilon}\right)+\left(y_{m+1,n+1}^{2\varepsilon}-y_{m+1,n}^{2\varepsilon}\right)\right]\\
&=-2\varepsilon\left(\frac{1}{2}\right)^{m+n}\left[\lambda^{2\varepsilon-1}\left(y_{m+2,n}-y_{m+1,n}\right)+\mu^{2\varepsilon-1}\left(y_{m+1,n+1}-y_{m+1,n}\right)\right]\\
&\geq -2\varepsilon\left(\frac{1}{2}\right)^{m+n}\lambda^{2\varepsilon-1}\left(y_{m+2,n}+y_{m+1,n+1}-2y_{m+1,n}\right)\\
&\geq -2\varepsilon\left(\frac{1}{2}\right)^{m+n}\lambda^{2\varepsilon-1}\left(y_{m+2,n}+y_{m+1,n+1}-y_{m+1,n}\right)\\
&=2\varepsilon\left(\frac{1}{2}\right)^{m+n}\lambda^{2\varepsilon-1}\left[-T(\Delta_1,\Delta_2)(y_{m+1,n})\right],
\end{aligned}
\tag{12.123}
$$

where $y_{m+2,n}\leq\lambda\leq y_{m+1,n}$, $y_{m+1,n+1}\leq\eta\leq y_{m+1,n}$, and without loss of generality, $\lambda^{2\varepsilon-1}=\min(\lambda^{2\varepsilon-1},\mu^{2\varepsilon-1})$. Substituting (12.122) into (12.123) gives

$$
\begin{aligned}
&-T(\Delta_1,\Delta_2)\left[\left(\frac{1}{2}\right)^{m+n-1}y_{m+1,n}^{2\varepsilon}\right]\\
&\geq 2\varepsilon\left(\frac{1}{2}\right)^{m+n}\lambda^{2\varepsilon-1}\frac{1}{c_{m+1,n}}\sum_{i=M}^{m}\sum_{j=N}^{n}p_{ij}(y_{i+1,j}+y_{i,j+1})^{\nu}\\
&\geq 2\varepsilon\left(\frac{1}{2}\right)^{m+n}y_{m+1,n}^{2\varepsilon-1}\frac{1}{c_{m+1,n}}\sum_{i=M}^{m}\sum_{j=N}^{n}p_{ij}(y_{i+1,j}+y_{i,j+1})^{\nu}
\end{aligned}
$$

$$\geq 2\varepsilon\left(\frac{1}{2}\right)^{m+n}\frac{(y_{m+1,n}+y_{m,n+1})^{2\varepsilon-1}}{c_{m+1,n}}\sum_{i=M}^{m}\sum_{j=N}^{n}p_{ij}(y_{i+1,j}+y_{i,j+1})^{\nu}$$

$$\geq\left(\frac{1}{2}\right)^{m+n}\frac{2\varepsilon}{c_{m+1,n}}\sum_{i=M}^{m}\sum_{j=N}^{n}p_{ij}(y_{i+1,j}+y_{i,j+1})^{\nu+2\varepsilon-1}.$$

Since $H \geq y_{mn} > 0$ for $m \geq M, n \geq N$, where $H > 0$ is a constant, there exists a positive number K such that

$$-T(\Delta_1,\Delta_2)\left[\left(\frac{1}{2}\right)^{m+n-1}y_{m+1,n}^{2\varepsilon}\right]\geq\frac{K}{c_{m+1,n}}\left(\frac{1}{2}\right)^{m+n}\sum_{i=M}^{m}\sum_{j=N}^{n}p_{ij}.$$

Summing both sides of this inequality it gives

$$\begin{aligned}&-\left(\sum_{i=M+1}^{m}\sum_{j=N}^{n}\left(\frac{1}{2}\right)^{i+j}y_{i,j+1}^{2\varepsilon}+\sum_{j=N}^{n}\left(\frac{1}{2}\right)^{m+j}y_{m+1,j}^{2\varepsilon}\right.\\&\left.+\left(\frac{1}{2}\right)^{M+n}y_{M,n+1}^{2\varepsilon}-\left(\frac{1}{2}\right)^{M+N-1}y_{MN}^{2\varepsilon}\right)\\&=-\sum_{i=M}^{m}\sum_{j=N}^{n}T(\Delta_1,\Delta_2)\left[\left(\frac{1}{2}\right)^{i+j-1}y_{i+1,j}^{2\varepsilon}\right]\\&\geq K\sum_{i=M}^{m}\sum_{j=N}^{n}\frac{1}{c_{i+1,j}}\left(\frac{1}{2}\right)^{i+j}\sum_{u=M}^{i}\sum_{v=N}^{j}p_{uv}.\end{aligned}$$

Hence,

$$\left(\frac{1}{2}\right)^{M+N-1}y_{MN}^{2\varepsilon}-\left(\frac{1}{2}\right)^{m+n}y_{m+1,n}^{2\varepsilon}\geq K\sum_{i=M}^{m}\sum_{j=N}^{n}\frac{1}{c_{i+1,j}}\left(\frac{1}{2}\right)^{i+j}\sum_{u=M}^{i}\sum_{v=N}^{j}p_{uv}.$$

By rearranging the double sum, we have

$$\left(\frac{1}{2}\right)^{M+N-1}y_{MN}^{2\varepsilon}-\left(\frac{1}{2}\right)^{m+n}y_{m+1,n}^{2\varepsilon}\geq K\sum_{u=M}^{m}\sum_{v=N}^{n}\left(\frac{1}{2}\right)^{u+v}p_{uv}\sum_{i=u}^{m}\sum_{j=v}^{n}\frac{1}{c_{i+1,j}},$$

and so letting $m, n \to \infty$ yields

$$\sum_{u=M}^{\infty}\sum_{v=N}^{\infty}\left(\frac{1}{2}\right)^{u+v}p_{uv}\rho_{u+1,v}<\infty,$$

which is a contradiction.

12.4.3 Examples

Example 12.5 Consider the partial difference equation

$$T(\Delta_1, \Delta_2)\left(2^{m+n}T(\Delta_1, \Delta_2)(y_{mn})\right) + 3 \times 2^{3m+n+1}(y_{m+1,n} + y_{m,n+1})^3 = 0, \tag{12.124}$$

where $c_{mn} = 2^{m+n}$, $p_{mn} = 3 \times 2^{3m+n+1}$, and $\nu = 3$. It is easy to see that all assumptions of Theorem 12.9 hold. So, Eq. (12.124) has a oscillatory solution $\{y_{mn}\}$. In fact, $\{y_{mn}\} = \{\frac{(-1)^n}{2^m}\}$ is also such a solution.

Example 12.6 Consider the partial difference equation

$$T(\Delta_1, \Delta_2)\left(e^{m+n}T(\Delta_1, \Delta_2)(y_{mn})\right) + \left(\frac{3e}{4} - 1\right)4^{-\frac{m+1}{3}}e^{m+n}(y_{m+1,n} + y_{m,n+1})^{\frac{1}{3}} = 0, \tag{12.125}$$

where $c_{mn} = e^{m+n}$, $p_{mn} = \left(\frac{3e}{4} - 1\right)4^{-\frac{m+1}{3}}e^{m+n}$ and $\nu = \frac{1}{3}$. It is easy to see that all assumptions of Theorem 12.10 hold. So, Eq. (12.125) has a oscillatory solution $\{y_{mn}\}$. In fact, $\{y_{mn}\} = \{\frac{(-1)^m}{2^m}\}$ is such a solution.

12.5 The Third Wave Behavior of Uncoupled Nonlinear Dynamical System

12.5.1 Target System

The following system is our target system in this part:

$$T^2(\Delta_1, \Delta_2)(y_{mn}) + \sum_{i=1}^{s} a_i(m,n) f_i\left(y_{m+1,n} + y_{m,n+1}, T(\Delta_1, \Delta_2)(y_{mn})\right) = 0, \tag{12.126}$$

where, $T^2(\Delta_1, \Delta_2) = T\left(T(\Delta_1, \Delta_2)\right)$.

The following conditions will be assumed:

(c_1) $a_i(m,n) \geq 0$ for $m \geq m_0 \geq 0, n \geq n_0 \geq 0,\ i = 1, 2, \ldots, s$;
(c_2) $f_i : R^2 \to R, uf_i(u,v) > 0$ for all $u \neq 0,\ i = 1, 2, \ldots, s$.

12.5.2 Main Results

The following elementary identity for double sequences will be used later (Figs. 12.5 and 12.6).

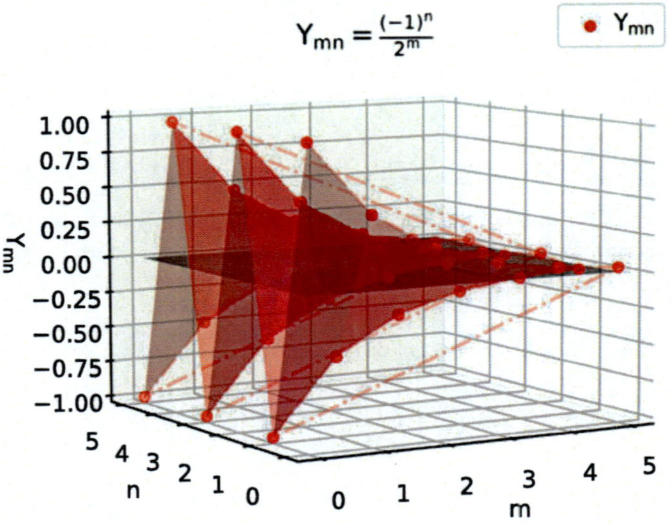

Fig. 12.5 Oscillatory behavior of system (12.124)

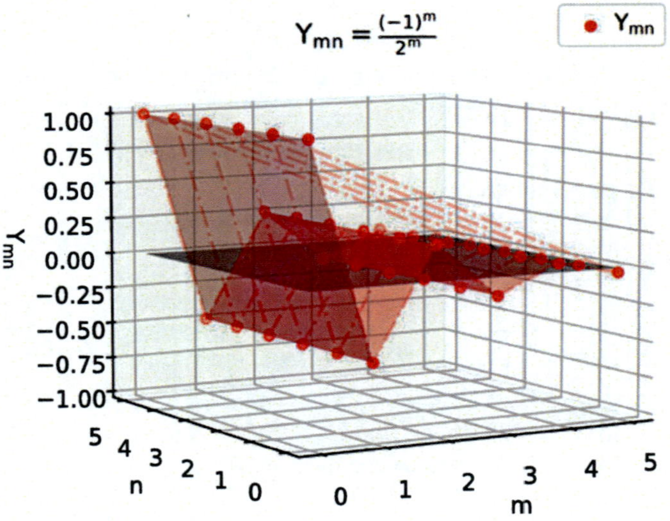

Fig. 12.6 Oscillatory behavior of system (12.125)

Theorem 12.11 *Suppose there exists an index j such that* (1^0) $f_j(\lambda u, v) = \lambda^{\beta+1} f_j(u, v)$ *for all* $(u, v) \in R^2$, λ *is a real number different from* 0, *and* β *is a nonnegative integer.*
(2^0) $f_j(1, v)$ *is continuous for* $v = 0$.
(3^0)

$$\sum_{i=m_0}^{\infty}\sum_{j=n_0}^{\infty}\frac{1}{2^{(\beta+1)i}}a_k(i, j) = \infty.$$

(4^0) *For initial value* y_{00} *define* $y_{00} \equiv c > 0$. *Then every solution of (12.126) is oscillatory.*

Proof Suppose there exists a nonoscillatory $\{y_{mn}\}$ of (12.126). Assume that $y_{mn} > 0$ for $m \geq m_1 \geq 0, n \geq n_1 \geq 0$. Then, by conditions (c_1) and (c_2), we see that $T^2(\Delta_1, \Delta_2)(y_{mn}) = T\,(T(\Delta_1, \Delta_2)(y_{mn})) < 0$, i.e., $\{T(\Delta_1, \Delta_2)(y_{mn})\}$ is nonincreasing sequence for $m \geq m_1, n \geq n_1$. Now, we can derive that

$$T(\Delta_1, \Delta_2)(y_{mn}) \geq 0 \text{ for } m \geq m_1,\ n \geq n_1. \tag{12.127}$$

In fact, if there would exist $m_2 \geq m_1$ and $n_2 \geq n_1$ such that $T(\Delta_1, \Delta_2)(y_{m_2n_2}) = \sigma < 0$, then $T(\Delta_1, \Delta_2)(y_{mn}) \leq \sigma$ for $m \geq m_2, n \geq n_2$ and by Lemma, we have

$$\begin{aligned} y_{mn} - y_{m_2n_2} &\leq y_{mn} + \left(\sum_{i=m_2+1}^{m-1}\sum_{j=n_2+1}^{n-1} y_{ij} + \cdots + \sum_{i=m_2}^{m-1} y_{i,n+1} + y_{m+1,n_0}\right) - y_{m_2n_2} \\ &= \sum_{i=m_2+1}^{m}\sum_{j=n_2+1}^{n} y_{ij} + \sum_{i=m_2}^{m-1} y_{i,n+1} - y_{m_2n_2} + y_{m+1,n_0} \\ &= \sum_{i=m_2}^{m-1}\sum_{j=n_2}^{n} T(\Delta_1, \Delta_2)(y_{ij}) \leq \sum_{i=m_2}^{m-1}\sum_{j=n_2}^{n} \sigma = \sigma(m - m_2)(n - n_2 + 1), \end{aligned}$$

that is

$$y_{mn} \leq y_{m_2n_2} + \sigma(m - m_2)(n - n_2) \to -\infty \text{ as } m, n \to \infty,$$

which contradicts the fact that $y_{mn} \geq 0$ for $m \geq m_1, n \geq n_1$. Thus (12.127) holds, that is

$$y_{m+1,n} + y_{m,n+1} \geq y_{mn} \text{ for } m \geq m_1,\ n \geq n_1. \tag{12.128}$$

Hence, we have

$$\begin{aligned} y_{m+1,n} + y_{m,n+1} &\geq y_{mn} = \frac{1}{2}(2y_{mn}) = \frac{1}{2}\,(y_{mn} + y_{mn}) \\ &\geq \frac{1}{2}y_{m-1,n-1} = \left(\frac{1}{2}\right)^2(2y_{m-1,n-1}) = \left(\frac{1}{2}\right)^2(y_{m-1,n-1} + y_{m-1,n-1}) \end{aligned}$$

$$\geq \left(\frac{1}{2}\right)^2 y_{m-2,n-2} \geq \cdots \geq \left(\frac{1}{2}\right)^m y_{00} = \left(\frac{1}{2}\right)^m c,$$

that is

$$y_{m+1,n} + y_{m,n+1} \geq \left(\frac{1}{2}\right)^m c. \tag{12.129}$$

On the other hand, let

$$Q_{mn} = T(\Delta_1, \Delta_2)(y_{mn}), \tag{12.130}$$

then, by (12.127), we have $Q_{mn} \geq 0$, the following proof Q_{mn} is a decreasing sequence. Due to, then an easy computation leads to

$$\begin{aligned} T(\Delta_1, \Delta_2)(Q_{mn}) &= T(\Delta_1, \Delta_2)\left(T(\Delta_1, \Delta_2)(y_{mn})\right) = T^2(\Delta_1, \Delta_2)(y_{mn}) \\ &= -\sum_{i=1}^{s} a_i(m, n) f_i\left(y_{m+1,n} + y_{m,n+1}, T(\Delta_1, \Delta_2)(y_{mn})\right) \\ &\leq -a_k(m, n) f_k\left(y_{m+1,n} + y_{m,n+1}, T(\Delta_1, \Delta_2)(y_{mn})\right) \leq 0. \end{aligned} \tag{12.131}$$

Hence, the sequence $\{Q_{mn}\}$ is decreasing in m and n. Then, for sufficiently large m, n, therefore

$$\lim_{n\to\infty, m\to\infty} Q_{mn} = \alpha \geq 0. \tag{12.132}$$

By (12.126), (12.128) and note that

$$\begin{aligned} & f_k(1, T(\Delta_1, \Delta_2)(y_{mn})) \\ &= f_k\left(\frac{1}{y_{m+1,n} + y_{m,n+1}}(y_{m+1,n} + y_{m,n+1}), T(\Delta_1, \Delta_2)(y_{mn})\right) \\ &= \left(\frac{1}{y_{m+1,n} + y_{m,n+1}}\right)^{\beta+1} f_k\left(y_{m+1,n} + y_{m,n+1}, T(\Delta_1, \Delta_2)(y_{mn})\right), \end{aligned}$$

that is,

$$(y_{m+1,n} + y_{m,n+1})^{\beta+1} f_k\left(1, T(\Delta_1, \Delta_2)(y_{mn})\right) = f_k\left(y_{m+1,n} + y_{m,n+1}, T(\Delta_1, \Delta_2)(y_{mn})\right), \tag{12.133}$$

using (12.129) and substituting (12.133) into (12.131), we get

$$\begin{aligned} T(\Delta_1, \Delta_2)(Q_{mn}) &= -a_k(m, n)(y_{m+1,n} + y_{m,n+1})^{\beta+1} f_k\left(1, T(\Delta_1, \Delta_2)(y_{mn})\right) \\ &\leq -a_k(m, n)\left(\frac{c}{2^m}\right)^{\beta+1} f_k\left(1, T(\Delta_1, \Delta_2)(y_{mn})\right). \end{aligned}$$

For $m \geq m_0, n \geq n_0$, from (12.132), (2^0) and (c_2) we have $\lim_{m,n\to\infty} f_k(1, T(\Delta_1, \Delta_2)(y_{mn}) = f_k(1, \alpha) > 0$, hence $f_k(1, T(\Delta_1, \Delta_2)(y_{mn})) \geq \frac{1}{2} f_k(1, \alpha)$ for $m \geq m_3 \geq m_0, n \geq n_3 \geq n_0$, then we have

$$T(\Delta_1,\Delta_2)(Q_{mn}) \le -\frac{1}{2}\left(\frac{c}{2^m}\right)^{\beta+1} a_k(m,n) f_k(1,\alpha),\ m \ge m_3, n \ge n_3. \quad (12.134)$$

By (4^0), summing on both sides of (12.134) from m_3 to $m-1$ and n_3 to n, we obtain

$$\begin{aligned} Q_{mn} - Q_{m_3n_3} &\le Q_{mn} + \left(\sum_{i=m_3+1}^{m-1}\sum_{j=n_3+1}^{n-1} Q_{ij} + \cdots + \sum_{i=m_3}^{m-1} Q_{i,n+1} - Q_{m_3n_3} + Q_{m+1,n_3}\right) \\ &= \sum_{i=m_3}^{m-1}\sum_{j=n_3}^{n} T(\Delta_1,\Delta_2)(Q_{ij}) \\ &\le -\frac{c^{\beta+1} f_k(1,\alpha)}{2}\sum_{i=m_3}^{m-1}\sum_{j=n_3}^{n}\frac{1}{2^{(\beta+1)i}} a_k(i,j). \end{aligned}$$

By (3^0), we obtain,

$$Q_{mn} \le Q_{m_3n_3} - \frac{c^{2\beta+1} f_k(1,\alpha)}{2}\sum_{i=m_3}^{m-1}\sum_{j=n_3}^{n}\frac{1}{2^{(\beta+1)i}} a_k(i,j) \to -\infty \text{ as } m,n\to\infty,$$

which contradicts the fact that $Q_{mn} \ge 0$ for $m \ge m_0, n \ge n_0$.

A similar argument applies to eventually negative solutions. The proof is complete.

Lemma 12.6 *Consider the partial difference equation*

$$T^2(\Delta_1,\Delta_2)(y_{mn})] + \frac{2e+1}{e(e+1)}\frac{e^{2(n+1)}(y_{m+1,n}+y_{m+1,n})}{e^{2(n+1)}+(2e+1)^2}\left[1+(T(\Delta_1,\Delta_2)(y_{mn}))^2\right] = 0, \quad (12.135)$$

where $f_1(u,v) = u(1+v^2)$, $a_1(m,n) = \frac{2e+1}{e(e+1)}\frac{e^{2(n+1)}}{e^{2(n+1)}+(2e+1)^2}$ *and tanking* $\lambda = 1$, $\beta = 0$. *It is easy to see that all assumptions of Theorem 12.11 hold. Therefor every solution of* (E_1) *oscillates. In fact,* $\{y_{mn}\} = \{\frac{(-1)^{m+n}}{e^n}\}$ *is such a solution.*

Theorem 12.12 *Assume that the following conditions hold:*

(5^0) *Let the initial value condition be* $y_{m_0n_0} \le C_0$ *for all* $n \in N_0$,

(6^0) $f_k(-u,-v) = -f_k(u,v)$ *for every* $(u,v) \in R^2$, $k = 1, 2, \ldots, s$,

(7^0) $I \ne \Phi$, *where* I *denotes the set of all indices* k *for which the function* $f_k(u,v)$ *is nondecreasing with respect to* u *and with respect to* v *on* R, *as well as the function* $f_k(u,0)/u$ *is nonincreasing on* $(0,\infty)$,

(8^0) *there exists a positive decreasing sequence* $\{h_{mn}\}$ *for* $m \ge m_0, n \ge n_0$, *such that*

$$\sum_{i=m_0}^{m-1}\sum_{j=n_0}^{n}\frac{1}{\rho}\frac{h_{ij}}{ij}\sum_{k\in I} a_k(i,j) f_k(\rho ij, 0) = \infty \quad (12.136)$$

for every $\rho \ge 1$. *Then every solution of (12.126) is oscillatory.*

Proof If $\{y_{mn}\}$ is a solution of (12.126), then by condition (6^0), $\{-y_{mn}\}$ is again a solution of (12.126). Thus as in the proof of Theorem 12.11, the existence of a nonoscillatory solution $\{y_{mn}\}$ of (12.126) leads to $y_{mn} > 0$, $T(\Delta_1, \Delta_2)(y_{mn}) \geq 0$, and $T(\Delta_1, \Delta_2)[T(\Delta_1, \Delta_2)(y_{mn})] < 0$, that is, $\{T(\Delta_1, \Delta_2)(y_{mn})\}$ is nonincreasing sequence for $m \geq m_0, n \geq n_0$.

Let us denote

$$R_{mn} = h_{mn}T(\Delta_1, \Delta_2)(y_{mn}), \; m \geq m_0, n \geq n_0. \tag{12.137}$$

Then from Eq. (12.126), we get

$$\begin{aligned}
& T(\Delta_1, \Delta_2)(R_{mn}) = T(\Delta_1, \Delta_2)\left(h_{mn}T(\Delta_1, \Delta_2)(y_{mn})\right) \\
&= h_{m,n+1}T(\Delta_1, \Delta_2)(y_{m,n+1}) + h_{m+1,n}T(\Delta_1, \Delta_2)(y_{m+1,n}) - h_{mn}T(\Delta_1, \Delta_2)(y_{mn}) \\
&\leq h_{mn}\left(T(\Delta_1, \Delta_2)(y_{m,n+1}) + T(\Delta_1, \Delta_2)(y_{m+1,n}) - T(\Delta_1, \Delta_2)(y_{mn})\right) \\
&= h_{mn}T^2(\Delta_1, \Delta_2)(y_{mn}) \\
&= -h_{mn}\sum_{i=1}^{s} a_i(m, n) f_i\left(y_{m+1,n} + y_{m,n+1}, T(\Delta_1, \Delta_2)(y_{mn})\right) \\
&\leq -h_{mn}\sum_{i\in I} a_i(m, n) f_i(y_{m+1,n} + y_{m,n+1}, 0) \\
&\leq -h_{mn}\sum_{i\in I} a_i(m, n) \frac{f_i(y_{m+1,n} + y_{m,n+1}, 0)}{y_{m+1,n} + y_{m,n+1}},
\end{aligned}$$

that is,

$$T(\Delta_1, \Delta_2)(R_{mn}) \leq -h_{mn}\sum_{i\in I} a_i(m, n) \frac{f_i(y_{m+1,n} + y_{m,n+1}, 0)}{y_{m+1,n} + y_{m,n+1}}, \tag{12.138}$$

$$m \geq m_0, \; n \geq n_0. \tag{12.139}$$

On the other hand, since

$$\begin{aligned}
\sum_{j=n_0}^{n}\sum_{i=m_0}^{m} T(\Delta_1, \Delta_2)(y_{ij}) &= \sum_{i=m_0+1}^{m+1}\sum_{j=n_0+1}^{n} y_{ij} + \sum_{i=m_0}^{m} y_{i,n+1} - y_{m_0n_0} + y_{m+1,n_0} \\
&\geq y_{m+1,n} + y_{m,n+1} - y_{m_0n_0},
\end{aligned}$$

hence

$$y_{m+1,n} + y_{m,n+1} \leq y_{m_0n_0} + \sum_{j=n_0}^{n}\sum_{i=m_0}^{m} T(\Delta_1, \Delta_2)(y_{ij}). \tag{12.140}$$

From (12.140), using the nonnegative and nonincreasing character of $\{T(\Delta_1, \Delta_2)(y_{mn})\}$ and (5^0), we have

$$\begin{aligned} y_{m+1,n} + y_{m,n+1} &\le C_0 + T(\Delta_1, \Delta_2)(y_{m_0n_0}) \sum_{j=n_0}^{n} \sum_{i=m_0}^{m} 1 \\ &= C_0 + T(\Delta_1, \Delta_2)(y_{m_0n_0})(n - n_0 + 1)(m - m_0 + 1) \end{aligned}$$

and consequently there exists $m_1 \ge m_0, n_1 \ge n_0$ and $\rho \ge 1$ such that

$$y_{m+1,n} + y_{m,n+1} \le \rho mn \text{ for } m \ge m_1, n \ge n_1.$$

Thus returning to (12.139), by condition (7^0), we obtain

$$T(\Delta_1, \Delta_2)(R_{mn}) \le -h_{mn} \sum_{i\in I} a_i(m, n) \frac{f_i(\rho mn, 0)}{\rho mn}, \quad m \ge m_0, n \ge n_0. \quad (12.141)$$

Summing on both sides of (12.141) form m_0 to $m - 1$ and n_0 to n, we obtain

$$\begin{aligned} R_{mn} - R_{m_0n_0} &\le \sum_{i=m_0+1}^{m} \sum_{j=n_0+1}^{n} R_{ij} + \sum_{i=m_0}^{m} R_{i,n+1} - R_{m_0n_0} + R_{m+1,n_0} \\ &= \sum_{i=m_0}^{m-1} \sum_{j=n_0}^{n} T(\Delta_1, \Delta_2)(R_{ij}) \\ &\le - \sum_{i=m_0}^{m-1} \sum_{j=n_0}^{n} \left(h_{ij} \sum_{k\in I} a_k(i, j) \frac{f_i(\rho ij, 0)}{\rho ij} \right), \end{aligned}$$

and then

$$R_{mn} \le R_{m_0n_0} - \sum_{i=m_0}^{m-1} \sum_{j=n_0}^{n} \frac{1}{\rho} \frac{h_{ij}}{ij} \sum_{k\in I} a_k(i, j) f_k(\rho ij, 0).$$

Now, by (8^0), it is easy to see that $\{R_{mn}\}$ is eventually negative, which is a contradiction, since $R_{mn} \ge 0$, for $m \ge m_0, n \ge n_0$. This completes the proof.

Example 12.7 Consider the partial difference equation

$$\begin{aligned} &T^2(\Delta_1, \Delta_2)(y_{mn}) \\ &\quad + \frac{9}{2[1 + 3(-1)^{m+n+1}]^2} (y_{m+1,n} + y_{m,n+1})[1 + T(\Delta_1, \Delta_2)(y_{mn})]^2 = 0 \end{aligned} \quad (12.142)$$

where $m > 0, n > 0$, $f_1(u, v) = u(1 + v)^2$, $a_1(m, n) = \frac{3}{2[1+3(-1)^{m+n+1}]^2}$ and taking $h_{mn} = \frac{1}{mn}$. It is easy to see that all assumptions of Theorem 12.12 hold. Therefor every solution of (E_2) oscillates. In fact, $\{y_{mn}\} = \{(-1)^{m+n}\}$ is such a solution.

A close look at the proof of Theorem 12.12 ensures the validity of the following:

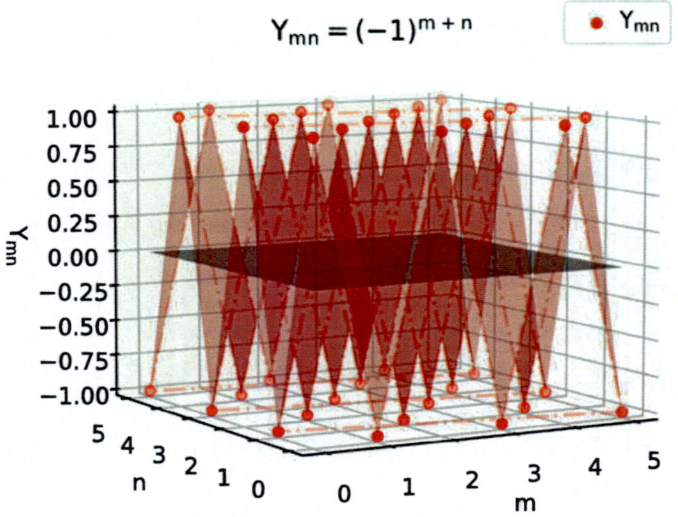

Fig. 12.7 Oscillatory behavior of system (12.142)

Theorem 12.13 *If, in addition to condition* (6^0), *we assume that*

(9^0) $I \neq \Phi$, *where I denotes the set of all indices k for which the function* $f_k(u, v)$ *is nondecreasing with respect to* v *on* R, *as well as the function* $f_k(u, 0)/u$ *is nonincreasing on* $(0.\infty)$.

(10^0) *there exists a positive decreasing sequence* $\{h_{mn}\}$ *for* $m \geq m_0, n \geq n_0$, *such that*

$$\sum_{i=m_0}^{\infty}\sum_{j=n_0}^{\infty}\frac{1}{\rho}\frac{h_{ij}}{ij}\sum_{k\in I}a_k(i,j)f_k(\rho ij,0)=+\infty$$

for every $\rho \geq 1$.
Then every solution of (12.126) is oscillatory.

In the above theorems, conditions (8^0) and (10^0) depend on a parameter $\rho \geq 1$. Because of this, several difficulties may appear in the verification of (8^0) or (10^0). For this reason, it is important to relax these conditions by requiring (8^0) only for $\rho = 1$. More precisely, the following theorem is valid (Figs. 12.7 and 12.8).

Theorem 12.14 *Assume that*

(11^0) $I \neq \Phi$, *where I denotes the set of all indices k for which the function* $f_k(u, v)$ *is nondecreasing with respect to* v *on* R *and* $f_k(\lambda u, 0) = \lambda f_k(u, 0)$ *for every* $u \in R$ *and real* $\lambda \neq 0$.

(12^0) *there exists a positive decreasing sequence* $\{h_{mn}\}$ $m \geq m_0, n \geq n_0$, *such that*

$$\sum_{i=m_0}^{\infty}\sum_{j=n_0}^{\infty}h_{ij}\sum_{k\in I}a_k(i,j)f_i(1,0)=+\infty.$$

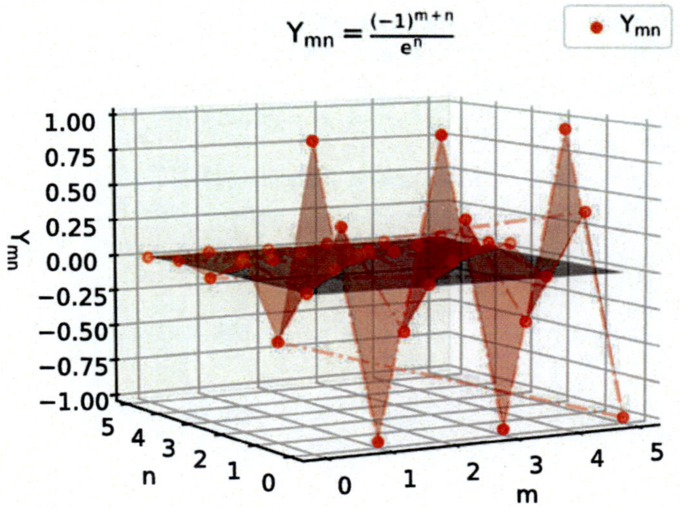

Fig. 12.8 Oscillatory behavior of system (12.143)

Then every solution of (12.126) is oscillatory.

Proof Assuming the existence of nonoscillatory solution $y_{mn} > 0$ for $m \geq m_0, n \geq n_0$ and arguing as in the proof of Theorem 12.12 we obtain the inequality (12.141). This in turn, by (11^0), implies

$$T(\Delta_1, \Delta_2)(R_{mn}) \leq -h_{mn} \sum_{i \in I} a_i(m, n) f_i(1, 0), \ m \geq m_0, \ n \geq n_0.$$

Thus our assertion follows now exactly the same way as the previous one. A similar argument holds in the case of an eventually negative solution.

12.6 Examples

Example 12.8 Consider the partial difference equation

$$T^2(\Delta_1, \Delta_2)(y_{mn})] + \frac{2e+1}{e(e+1)} \frac{e^{2(n+1)}(y_{m+1,n} + y_{m+1,n})}{e^{2(n+1)} + (2e+1)^2} \left[1 + (T(\Delta_1, \Delta_2)(y_{mn}))^2\right] = 0, \quad (12.143)$$

where $f_1(u,v)=u(1+v^2), a_1(m,n)=\frac{2e+1}{e(e+1)}\frac{e^{2(n+1)}}{e^{2(n+1)}+(2e+1)^2}$ and tanking $\lambda=1, \beta=0$. It is easy to see that all assumptions of Theorem 1 hold. Therefor every solution of (E_1) oscillates. In fact, $\{y_{mn}\}=\{\frac{(-1)^{m+n}}{e^n}\}$ is such a solution.

Part IV
Applications of Surface Chaos

Chapter 13
Nuclear Fission and Surface Chaos

Check for updates

13.1 Preliminaries

The nuclear fission in the nuclear reactor usually makes the nucleus of the fissionable material fission into two medium-mass nuclei (which is called fission fragments). Simultaneously, an average of two or more new fission neutrons can be generated, releasing the nuclear energy held in interior atomic nucleus. In some proper conditions, these new fission neutrons will cause the fission of other surrounding fissile isotopes, and repeat this activity in the same way without any halt, which is the self-sustaining chain fission [375–377].

It will be worth noticing that, the physical process of the self-sustaining chain fission bears some relation to with the motility of neutron energy groups inside the reactor system and the space-energy distribution of neutron density inside of the core, but the physical analysis system shows neutron distribution and behavior law in the reactor, i.e., neutron transport equation or it is called a Boltzmann Equation [378], that is,

$$\frac{\partial n}{\partial t} = \text{Supply rate of neutron } (S) - \text{Leakage rate } (L) - \text{Absorption rate } (A), \tag{13.1}$$

where $\frac{\partial n}{\partial t}$ is the variation rate of the neutron density as the time passes, which could be further written as:

$$\begin{aligned} &\frac{1}{v}\frac{\partial \phi}{\partial t} + \Omega \nabla \phi + \sum_k \left(\vec{r}, E\right)\phi \\ &= \int_0^\infty \int_{\Omega'} \sum_s \left(\vec{r}, E'\right) f\left(\vec{r}, E' \to E, \Omega' \longrightarrow \Omega\right) \\ &\quad \times \phi\left(\vec{r}, E', \Omega', t\right) dE' d\Omega' + S\left(\vec{r}, E, \Omega, t\right). \end{aligned} \tag{13.2}$$

S. T. Liu and L. Zhang, *Surface Chaos and Its Applications*,
https://doi.org/10.1007/978-981-16-8229-2_13

So it is none other than system (13.2) forming up the foundation of reactor physical analysis and neutron transport theory, of which the physical significance and representation of each symbol are listed in Ref. [378].

As we can see, the system (13.2) is actually a nonlinear distribution parameter system, whose physical form is the dynamical behavior in space. In this chapter, through analyzing the spatial dynamical behavior of system (13.2), we elaborate the relation between nuclear fission and spatial chaos; the relation between neutron effective multiplication coefficient k and the real parameter μ of spatial chaos bifurcation in self-sustaining chain fission control; concentration rate of uranium; the relation between neutron multiplication theory and spatial bifurcation, etc., all of which indicate that the nuclear self-sustaining chain fission and nuclear fusion are actually spatial nonlinear dynamical behavior [71]. Furthermore, through analyzing the spatial nonlinear system, we also obtain that the absolutely stable area of nuclear reactor is $0 < k < 3/4$, but not the formerly mentioned $k < 1$.

13.2 Nuclear Fission and Neutron Transport System

As we know, the self-sustaining chain fission of uranium includes two aspects of content: one is that the atomic nucleus of fissionable material fissions into two medium-mass nuclei and discharges energy; the other is that two or three new fission neutrons will be generated when the nucleus disintegrates. Actually, the two aspects are contained in the same thing, occurring simultaneously, existing mutually, interdependent with each other, and continually repeating the same. Therefore, the physical form of nuclear fission and releasing nuclear energy represents the distribution of the neutron groups inside the system during all process of fission; on the contrary, what about the diffusion and distribution of the neutrons in reactor? It also explains the physical form of fissionable material and energy, which are just alternative descriptions of the same question. Therefore, in the nuclear fission chain reaction, we could utilize the neutron transport theory about the scattering and absorbing of reacting neutrons in the medium to explain the fission of fissionable materials and the physical form of energy in the nuclear reactor. Evidently, these physical forms of nuclear fission also reflect naturally in the spatial dynamical behavior of nonlinear neutron transport. We just utilize these nonlinear dynamical characteristics of neutron transport system to explain the quantitative characteristics of each physical form in the nuclear fission reaction.

13.3 Nuclear Fission and Surface Chaos

On the basis of neutron transport theory, i.e., system (13.2), we can see that, it is very difficult to obtain the consecutive dependency of neutron flux on energy through analyzing this system. In practice, it is almost impossible to carry it out, and the traditional approach is to apply "group method" to get a solution approximately, i.e., using multigroup diffusion equation [380]:

$$\begin{cases} -\nabla D_g \nabla \phi_g(\vec{r}) + \sum_{t,g} \phi_g(\vec{r}) - \sum_{g'=1}^{G} \sum_{g-g'} \phi_{g'}(\vec{r}), \\ = \frac{X_g}{k} Q(\vec{r}), g = 1, 2, \cdots, G, \\ Q(\vec{r}) = \sum_{g'=1}^{G} \left(v \sum_f\right)_{g'} \phi_{g'}(\vec{r}), \end{cases} \tag{13.3}$$

where g is the neutron flux of the g-th g energy groups, sign as

$$\phi_g(\vec{r}) = \int_{E_g}^{E_g - 1} \phi\left(\vec{r}, E\right) dE, \tag{13.4}$$

in (13.3) and (13.4), other signs and their representative physical meanings refer to [380]. If we consider the diffusion equation under two-dimensional case, then (13.3) reduces to the following form [380]:

$$-\frac{\partial}{\partial x}\left(D\frac{\partial \phi}{\partial x}\right) - \frac{\partial}{\partial y}\left(D\frac{\partial \phi}{\partial y}\right) + \sum\Big|_R (x, y)\phi(x, y) = S(x, y), \tag{13.5}$$

and (13.5) could be changed into the nonhomogeneous two-dimensional discrete dynamical system of variable coefficient

$$\begin{aligned} &a_{1mn}\phi_{m+1,n} + a_{2mn}\phi_{m,n+1} + a_{3mn}\phi_{m-1,n} + a_{4mn}\phi_{m,n-1} + a_{5mn}\phi_{mn} \\ &= S_{mn}, \end{aligned} \tag{13.6}$$

where $(m, n) \in N_0^2 = \{0, 1, 2, ..., \}^2$. For (13.6), taking

$$\begin{aligned} &C_{mn}(r) = a_{1mn}\phi_{m+r,n} + a_{2mn}\phi_{m,n-r} + a_{3mn}\phi_{m-r,n} + a_{4mn}\phi_{m,n-r}, \\ &r = 0, 1, 2, \cdots \end{aligned} \tag{13.7}$$

Specially, let $r = 0$, then

$$C_{mn}(0) = \left(\sum_{i=1}^{4} a_{imn}\right)\phi_{mn}. \tag{13.8}$$

And from [257], any requirement regarding system stability usually assumes $-1 < \sum_{i=1}^{4} a_{imn} < 1$, and $\sum_{i=1}^{4} a_{imn} \neq 0$. Using (13.7), then (13.6) reduces to

$$C_{mn}(1) = S_{mn} - e_{mn}\phi_{mn}. \tag{13.9}$$

In addition, from (13.6), it is a linear nonhomogeneous two-dimensional partial difference dynamical system. And it is supposed that under some conditions such as all movements of neutrons are unrelated mutually in the medium [380], that is to say, the scattering of neutrons is homogeneous inside the nuclear reactor, media are infinite and even etc. This system is deduced to outcome. Consequently, it makes the neutron transport equation become a nonhomogeneous linear equation. But it will be worth noticing that, in fact, the angular distribution of scattered neutron is usually different in nature with each other. If the nucleon mass number becomes smaller, the energy of neutron will be higher, and scattered isomerism will be more powerful [381], which means that the movement of neutrons is disturbed by all kinds of factors. However, from the mathematical and physical analysis, we can easily conclude that, all the disturbance factors should be considered together, i.e., in the usual inhomogeneous system there is to be a disturbance item, or, it is called a forced term. Without loss of generality, we could assume

$$g\left(a_0, b_0, x\right), \tag{13.10}$$

because (13.10) is a noise disturbance term. So it is usually a nonlinear function. And without loss of generality, which is assumed as a continuous function, using *Weierstrass* polynomial uniform approximation theorem [393], then $g\left(a_0, b_0, x\right)$ could utilize a polynomial function to uniformly approach, hence, for the purpose of convenient handling, we can take

$$g\left(a_0, b_0, x\right) = a_0 + b_0 x - \beta\left[\left(\sum_{i=1}^{4} a_{imn}\right) x\right]^2,$$

where $a_0\beta \neq 0$. Then,

$$g\left(a_0, b_0, \phi_{mn}\right) = a_0 + b_0\phi_{mn} - \beta\left[\left(\sum_{i=1}^{4} a_{imn}\right)\phi_{mn}\right]^2. \tag{13.11}$$

Using the above analyses, from (13.6) and (13.9), we obtain

$$C_{mn}(1) = a_0 + S_{mn} + b_0\phi_{mn} - \beta\left[\left(\sum_{i=1}^{4} a_{imn}\right)\phi_{mn}\right]^2. \tag{13.12}$$

And using (13.8) and (13.12), we have

$$\begin{aligned} &C_{mn}(1) - \left(a_0 + S_{mn}\right) \\ =&\beta C_{mn}(0)\left[\frac{1}{\beta}\left(b_0 - e_{mn}\right)\Big/\left(\sum_{i=1}^{4} a_{imn}\right) - C_{mn}(0)\right]. \end{aligned} \tag{13.13}$$

Particularly, in (13.13), let

$$\begin{cases} \alpha = (b_0 - e_{mn}) \Big/ \left(\sum\limits_{i=1}^{4} a_{imn} \right), \\ y = \left(\dfrac{\alpha^2}{4\beta} - \dfrac{\alpha}{2\beta} + a_0 + S_{mn} \right) x + \dfrac{\alpha}{2\beta}, \\ \mu = \dfrac{\alpha}{2}\left(\dfrac{\alpha}{2} - 1 \right) + \beta(a_0 + S_{mn}), \end{cases} \tag{13.14}$$

from (13.13) and (13.14), we obtain

$$C_{mn}(1) = 1 - \mu C_{mn}^2(0) \,. \tag{13.15}$$

We can see (13.15) is similar to the familiar one-dimensional Logistic system. Additionally, it needs to be explained detailedly that, the real parameter μ in (13.15) have already gotten comprehensively universal meanings, the reason of which is shown in the following: with respect to any disturbance term (13.11) of the transport equation, it could be approached with a polynomial by using *Weierstrass* theorem [393]. Thus, the form of (13.15) holds always. Consequently, parameter μ has a universal sense. Moreover, because S_{mn} refers to fission source term of nuclear reactor, generally we start the iteration from a certain constant. Hence, without loss of generality, let us take $S_{mn} = C - a_0$, and $S_{mn} > 0$. Easily, when

$$\mu = \frac{\alpha}{2}\left(\frac{\alpha}{2} - 1 \right) + \beta(a_0 + S_{mn}) \geq 1.55,$$

then system (13.15) will be in a state of spatial chaos. For example, taking $(a_{1mn}, a_{2mn}, a_{3mn}, a_{4mn}) = (-0.202, 1, -0.07, -0.5)$ and $(a_0 + S_{mn}, b_0 - e_{mn}, \beta) = (2, 0.456, 0.97)$, then $\mu = 1.94 \geq 1.55$, and the bifurcating and chaotic behavior is confirmed by the following simulation Figs. 13.1, 13.2, 13.3 and 13.4.

Because (13.15) is the variance situation of neutron flux in the nuclear fission reaction, so the distribution situation of neutron flux in the reactor reflects the physical case of chain fission and energy situation in the nuclear reactor. Hence, as the real parameter μ varies and augments, the neutron flux changes from stability to spatial chaos gradually, which is also to say when $\mu > 1.55$, the fission and energy in the nuclear reactor is in uncontrollable chaos. On the other hand, from [394], consider the case of the surface with 2^r – period–doubling bifurcation in space in the 2-D system (13.15), $r = 0, 1, 2, \ldots$, then we obtain the critical and bifurcating fixed curve on system (13.15): $\mu_0 = \frac{3}{4}$, $\mu_1 = \frac{5}{4}$, $\mu_2 = 1.3681$, $\mu_3 = 1.394, \cdots$. Therefore, the chaotic behavior of nuclear fission can be drawn from the following Table 13.1 specifically:

And note that we can observe the bifurcation case of nuclear fission and neutron flux from the following schematic (Figs. 13.5 and 13.7.)

In system (13.15), when taking $(a_{1mn}, a_{2mn}, a_{3mn}, a_{4mn}) = (1, 0, 0, 0)$, $(a_0 + S_{mn}, b_0 - e_{mn}, \beta) = (2, 0, 0.93)$, $n = n_0$ and use notation $u_{mn_0} = u_m$, then system (3.13) reduces to $\phi_{m+1} = 1 - 1.86\phi_m^2$, which is a familiar one-dimensional Logistic discrete

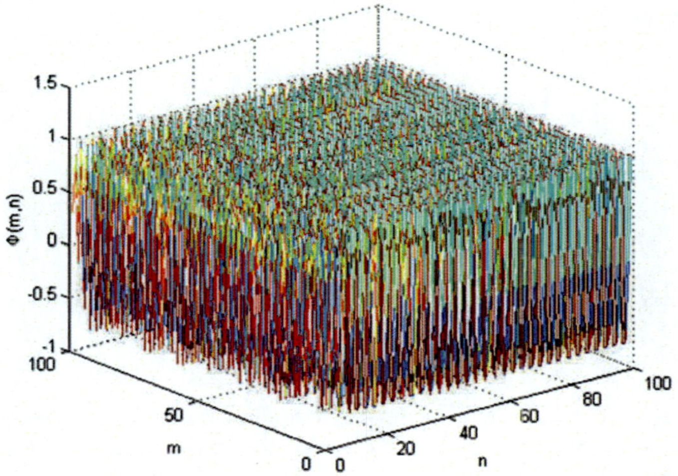

Fig. 13.1 A spatial chaos section of neutron transport system

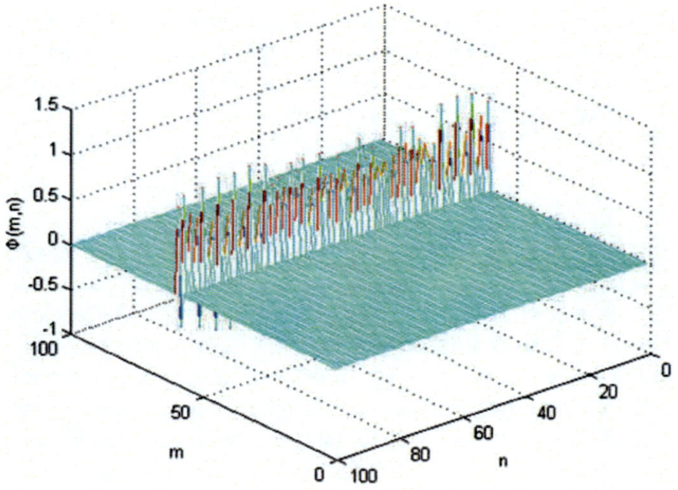

Fig. 13.2 A spatial bifurcation section of neutron transport system

dynamical system, and because $\lambda = 1.86 > 1.55$, then the system is in chaos [245]. The bifurcation behavior of this system is demonstrated in Fig. 13.6. The chaotic behavior of this system is demonstrated in Fig. 13.7.

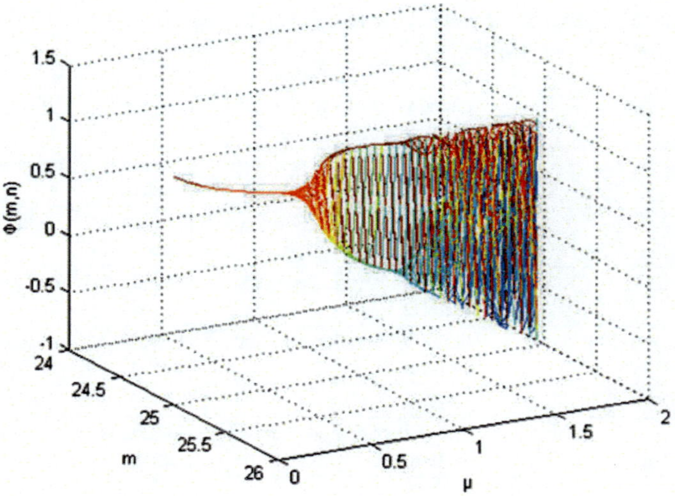

Fig. 13.3 Nuclear chain fission from stability to chaos in space

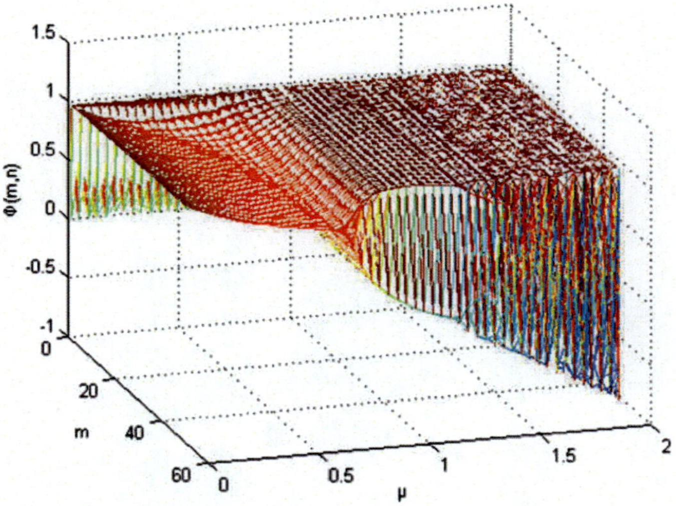

Fig. 13.4 The spatial bifurcation case of neutron transport system (13.15) from stability to chaos

Table 13.1 The spatial nonlinear dynamical behavior of neutron flux and nuclear fission as transport system parameter μ enlarges gradually:

μ	The spatial bifurcation phenomena of neutron flux ϕ_{mn}	The fission case
$0 < \mu < \frac{3}{4}$	Stable 2^0 iteration periodic surface	Fission to 2^0 stable uranium nuclear
$\frac{3}{4} \leq \mu < \frac{5}{4}$	Stable 2^1 iteration periodic surfaces	Fission to 2^1 medium-mass nuclei
$\frac{5}{4} \leq \mu < 1.368$	Stable 2^2 iteration periodic surfaces	Fission to 2^2 medium-mass nuclei
$1.368 \leq \mu < 1.394$	Stable 2^3 iteration periodic surfaces	Fission to 2^3 medium-mass nuclei
...	...	...
$\mu \geq 1.55$	The case of spatial chaos of system	Fission is in uncontrollable chaos

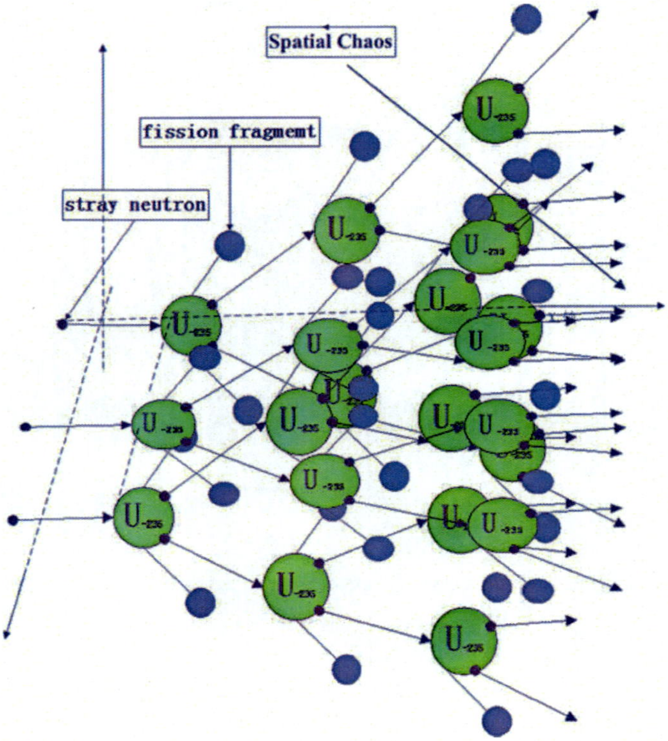

Fig. 13.5 The bifurcation section of neutron flux changing from stability into chaos in one-dimension

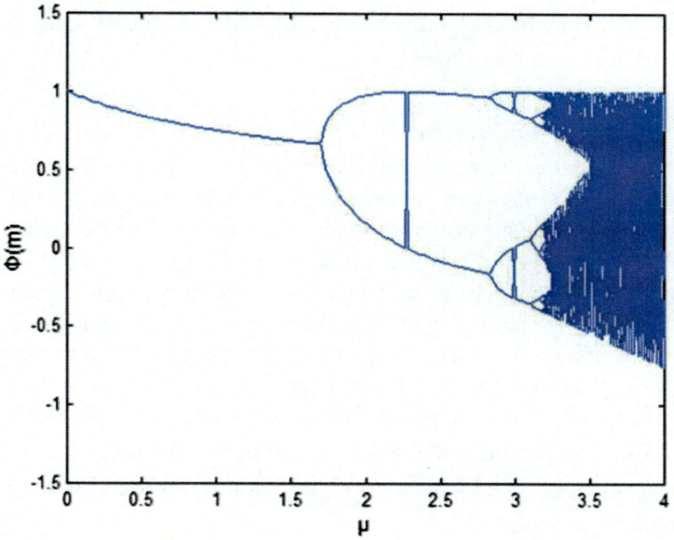

Fig. 13.6 Chain fission schematic figure of uranium in one dimension

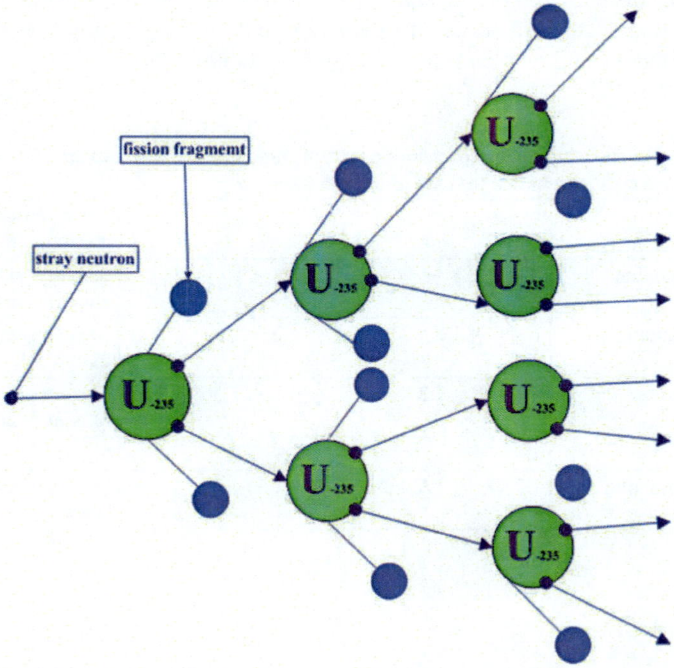

Fig. 13.7 The spatial chaos case of neutron transport equation when $\mu = 1.94$

13.4 Absolutely Safe Area, Concentration Rate of Uranium and System Parameter

We can see from the third part that, when $0 < \mu < \frac{3}{4}$, the neutron flux density has only 2^0 stable iteration periodic orbit. And as long as $\frac{3}{4} \le \mu < \frac{5}{4}$, the neutron flux density has 2^1-period iteration periodic orbits. Without any outside restriction, it shows the starting of chain fission of fissionable material. Therefore, the parameter μ of dynamical system (13.15) actually represents the effective multiplication factor k of the reactor. Thus it can be seen, in order to ensure the safe operation and control the chain reaction, it is necessary to make the neutron effective multiplication factor k satisfy the condition $0 < k = \mu < \frac{3}{4}$, which has strictly different meanings from the usually rough saying $k = \mu < 1$. Particularly, we assume:

(1) Let $F_1 = \left(0, \frac{3}{4}\right)$, it is called absolutely stable area of nuclear reactor.

(2) For $F_2 = \left[\frac{3}{4}, \frac{5}{4}\right)$, it is called prime area of nuclear fission.

(3) Let $\Omega = \left\{ \xi \mid \xi \ge \frac{5}{4} \right\}$ and $F_3 = \bigcup_{\sigma \in \Omega} \left[\frac{5}{4}, \sigma\right)$, it is called area of nuclear fission reaction trending to chaos.

Additionally, seen from another point of view, the parameter μ of system (13.15) is defined as the concentration rate of uranium correspondingly. It is because the higher the concentration rate of uranium is, the larger the resulting neutron effective multiplication factor will be, and the more drastic the corresponding fission reaction is. All of them could be seen from the following Table 13.2:

Table 13.2 Relationship among concentration rate of uranium (CRU), the variation case of effective multiplication coefficient k and the case of nuclear reactor μ

CRU	k	μ	The case of nuclear reactor
Natural uranium	$0 < k < 1$	$0 < \mu < \frac{3}{4}$	Stability without nuclear fission
1% of uranium	$1 \le k \le 1.1$	$\frac{3}{4} \le \mu < \frac{5}{4}$	Beginning of nuclear fission
2% of uranium	$1.1 < k \le 1.3$	$\frac{5}{4} \le \mu < 1.368$	The intensification of nuclear fission
...	...	...	...
100% of uranium	$1.3 < k \le 1.6$	$\mu > 1.55$	Fission in the spatial chaos

Table 13.3 System parameter and neutron multiplication case

μ	Bifurcation case	The case of neutron multiplication
$0 < \mu < \frac{3}{4}$	2^0 stable periodic surface	No new neutron generated
$\frac{3}{4} \leq \mu < \frac{5}{4}$	2^1 stable periodic surfaces	At least 2^1 new neutrons generated
$\frac{5}{4} \leq \mu < 1.368$	2^2 stable periodic surfaces	At least 2^2 new neutrons generated
$\cdots$	$\cdots$	$\cdots$
$\mu > 1.55$	2^∞ stable periodic surfaces $\Longrightarrow \cdots \Longrightarrow$ spatial chaos	At least 2^∞ new neutrons generated $\Longrightarrow \cdots \Longrightarrow$ spatial chaos

13.5 Relation Between Neutron Multiplication in the Nuclear Reactor and System Parameter μ

From the spatial period-doubling bifurcation of nonlinear system (13.15), we can easily see that, the periodic orbit of system kinematics is carried out by 2^r- period–doubling bifurcation, $r = 0, 1, 2, \ldots$ And as parameter μ augments gradually, the quantity of periodic orbit trends to 2^∞. Thereby, the system presents chaos in space, on the other hand, from [395], from the neutron multiplication factor k and neutron lifespan l, we can easily calculate out the neutron number $N(t)$ in the time t of nuclear reactor

$$N(t) = N_0 \exp\left[\left(\frac{k-1}{l}\right)t\right], \tag{13.16}$$

where, $N(0)$ is the initial neutron number at t time of reactor, from (13.16), as the time varies, the neutron density increases and decreases all conforming to the exponent law in the reactor. In particular, $N(0) > 1$, (it is obtained easily in nuclear chain reaction), with t increasing, it explains that the neutron number multiplies according to the exponent in the nuclear reactor, and its exponent multiplication law is in line with the variance of the system (13.15) parameter μ, which can be seen from the following Table 13.3.

Chapter 14
Uniformity and Surface Chaos of Spatial Physics Kinematic System

14.1 Uniformity of Spatial Physics Kinematic System

Spatial physics kinematic system, a physics kinematic system containing at least two independent variables, actually is surface or volume in space, or hypersurface and hypervolume in the hypersapce. These kinematic forms are actually typical kinematic forms of spatial physical movement. In this chapter, we summarize some classical uniformity laws of physics kinematic system from different aspects; and from the view of spatial dynamical system, and fully explain the surface chaos behavior of uniformity system and the relation between the coupled map lattice model which has attracted considerable interest currently (hereinafter referred to as "CML")

$$x_{m+1,n} = (1-\varepsilon) f(x_{m,n}) + \frac{\varepsilon}{2}[f(x_{m,n-1}) + f(x_{m,n+1})], \tag{14.1}$$

and 1-D nonlinear dynamical system:

$$x_{m+1} = f(\mu, x_m). \tag{14.2}$$

Consider the following spatial physics kinematic systems: where $F_i\ (\mu_i, u\,(x, y))$, $i = 1, 2, 3, 4$ are nonlinear functions, ω is a constant, μ_i, $i = 1, 2, 3, 4$ are real parameters (Table 14.1).

And reconsider their difference forms: because $f_i\ (\mu_i, u_{mn})$, $i = 1, 2, 3, 4$ are forced terms, therefore it is generally a nonlinear function. Let $N_0 = \{0, 1, 2, \ldots, \}$, then (14.7)–(14.10) holds on the N_0^2 for all $(m, n) \in N_0^2$. Furthermore, we can obtain its boundary value u_{m0} and u_{0n} from specific problems. In addition, the existence and uniqueness of the solution of (14.7)–(14.10) are also solved (Table 14.2).

If we also observe (14.7)–(14.10), then they can reduce to a uniform system

$$au_{m+1,n} + bu_{m,n+1} + cu_{m-1,n} + du_{m,n-1} = f\,(\mu, (a+b+c+d)\,u_{mn}), \tag{14.3}$$

S. T. Liu and L. Zhang, *Surface Chaos and Its Applications*,
https://doi.org/10.1007/978-981-16-8229-2_14

Table 14.1 Spatial physics kinematic systems

Name	Equation
1-D convection system with forced term	$\frac{\partial u}{\partial x} + \omega \frac{\partial u}{\partial y} = F_1(\mu_1, u(x, y))$ (14.3)
1-D wave equation with forced term	$\frac{\partial^2 u}{\partial x^2} - \omega^2 \frac{\partial^2 u}{\partial y^2} = F_2(\mu_2, u(x, y))$ (14.4)
1-D heat conduction equation with forced term	$\frac{\partial u}{\partial x} - \omega^2 \frac{\partial^2 u}{\partial y^2} = F_3(\mu_3, u(x, y))$ (14.5)
1-D harmonic equation with forced term	$\frac{\partial^2 u}{\partial x^2} + \frac{\partial^2 u}{\partial y^2} = F_4(\mu_4, ux, y))$ (14.6)

Table 14.2 Difference form of spatial physics kinematic systems

Name	Equation
Difference of (14.3)	$u_{m+1,n} + \omega u_{m,n+1} = f_1(\mu_1, u_{mn})$ (14.7)
Difference of (14.4)	$u_{m+1,n} - \omega^2 u_{m,n+1} + u_{m-1,n}$ $-\omega^2 u_{m,n-1} = f_2(\mu_2, u_{mn})$ (14.8)
Difference of (14.5)	$u_{m+1,n} - \omega^2 u_{m,n+1} - \omega^2 u_{m,n-1}$ $= f_3(\mu_3, u_{mn})$ (14.9)
Difference of (14.6)	$u_{m+1,n} + u_{m,n+1} + u_{m-1,n} + u_{m,n-1}$ $= f_4(\mu_4, u_{mn})$ (14.10)

where $a + b + c + d \neq 0$. And in (14.3), if we choose different values of real parameter a, b, c, d and $f(\mu, (a + b + c + d)\, u_{mn})$, then (14.7)–(14.10) is obtained respectively. For example, assume that $(a, b, c, d) = (1, \omega, 0, 0)$ and $f_1(\mu_1, u_{mn}) = f(\mu, (a + b + c + d))\, u_{mn} = \omega u_{mn} + F_1(\mu_1, u_{mn})$, then (14.3) reduces to (14.7). Similarly we can also obtain (14.8)–(14.10). Therefore, (14.3) is a uniform physical system of (14.7)–(14.10), which is said to be a *uniform physics system*. Hence, further investigation of the two-dimensional discrete dynamical system (14.3), not only provides some useful information for system (14.3)–(14.6), but also brings some important theoretical significance to nonlinear two-dimensional discrete dynamical system.

Furthermore, for (14.3), we usually write it in the following form

$$au_{m+1,n} + bu_{m,n+1} = f(\mu, (a + b + c + d)\, u_{mn}) - cu_{m-1,n} - du_{m,n-1}, \quad (14.4)$$

or more generally

$$au_{m+1,n} + bu_{m,n+1} = \sum_{i=1}^{k} p_{imn} g_i \left(\mu_i, u_{m-\sigma_i, n-\iota_i}\right), \quad (14.5)$$

where k is a positive integer, σ_i and τ_i, $i = 1, 2, \ldots, k$ are nonegative integers. When $k = p_{1mn} = 1$ and $g_1\left(\mu_1, u_{m-\sigma_1, n-\iota_1}\right) = f(\mu, (a + b + c + d)\, u_{mn}) - bu_{m-1,n} - cu_{m,n-1}$ then (14.5) reduces to (14.4). For (14.5), seen from mathematical analyses, there have been pretty rich results at the present time.

14.2 The Relation Between the Uniform Physics System and Coupled Map Lattice, One-Dimensional Discrete Dynamical System

As we know, as to chaos, bifurcation, control as well as application of CML (14.3), we have already obtained plentiful achievements. But it will be worth noticing that, in these researches many restrictions are attached to this system. Additionally, it is a duality function system. However, no one practically utilizes the characteristics of duality function to carry out the study. And still reduces it to one-dimensional difference equation systems for investigation, which restricts the natural revealment of its physical characteristics in space.

In fact, for CML mapping (3.1), it is a special case of uniform system (14.4), because CML model can be rewritten into the following form as System (3.1)

$$x_{m+1,m} - \frac{\varepsilon}{2} f\left(x_{m,n+1}\right) - \frac{\varepsilon}{2} f\left(x_{m,n-1}\right) = (1-\varepsilon)\, f\left(x_{mn}\right),$$

where, $f(x) = \mu x(1-x)$. For (3.1), we only make the following transformation in (14.3)

$$\begin{cases} \text{taking } a = 1, b = d = -\frac{\varepsilon}{2},\ c = 0 \text{ and } u_{mn} = x_{mn}, \\ \text{replace } x_{m,n+1} \text{ by } f(x_{m,n+1}), \\ \text{replace } x_{m,n-1} \text{ by } f(x_{m,n-1}), \\ \text{replace } f\left(\mu, (a+b+c+d)\, u_{mn}\right) \text{ by } (1-\varepsilon)\, f\left(x_{mn}\right). \end{cases}$$

Then the uniform system (14.3) reduces to (3.1), therefore, system (3.1) is only a particular case of (14.3). In addition, in (14.3), let $(a, b, c, d) = (1, 0, 0, 0)$, $n = n_0$ and use notation $u_{mn_0} = u_m$, then system (14.4) also reduces to the familiar one-dimensional Logistic model

$$u_{m+1} = f\left(\mu, u_{mn}\right). \tag{14.6}$$

And there have been rich results for (14.6) nonlinear dynamical behavior investigation. Therefore, typical Logistic system is even a particular case of (14.3). Thus we can determine again, system (14.3) is a fairly comprehensive system.

14.3 Physical Uniformity System and Surface Chaos

Seen from the right side of physics uniform system (14.3), it is easy to see $f\left(\mu, (a+b+c+d)\, u_{mn}\right)$ is a forced term disturbed by all kinds of noises. Therefore, it is usually a nonlinear function. And without loss of generality, assuming that the nonlinear function is continuous, by using Weierstrass polynomial uniform approximation theorem, we can succeed in using a polynomial function to approach this nonlinear forced function. In order to deal with the problem conveniently, without loss of generality, we assume that

$$f(\mu, (a+b+c+d)x) = a_0 + b_0 x - c_0 ((a+b+c+d)x)^2,$$

where $b_0 c_0 \neq 0$. Specially, we obtain $f(\mu, (a+b+c+d)u_{mn}) = a_0 + b_0 u_{mn} - c_0 ((a+b+c+d)u_{mn})^2$. And also

$$\begin{aligned} & u_{m+1,n} + a u_{m,n+1} + b u_{m-1,n} + c u_{m,n-1} \\ &= a_0 + b_0 u_{mn} - c_0 ((a+b+c+d)u_{mn})^2 . \end{aligned} \tag{14.7}$$

From (14.3), moreover let

$$C_{mn}(r) = a u_{m+r,n} + b u_{m,n+r} + c u_{m-r,n} + d u_{m,n-r}.$$

Thus

$$\begin{aligned} C_{mn}(1) &= a u_{m+1,n} + b u_{m,n+1} + c u_{m-1,n} + d u_{m,n-1}, \\ C_{mn}(0) &= (a+b+c+d) u_{mn}, \end{aligned} \tag{14.8}$$

using $a+b+c+d \neq 0$. Substituting (14.8) into (14.3), we obtain

$$C_{mn}(1) = a_0 + \frac{b_0}{a+b+c+d} C_{mn}(0) - c_0 C_{mn}(0)^2 . \tag{14.9}$$

Let

$$\begin{cases} \alpha = \frac{b_0}{a+b+c+d}, \\ y = \left(\frac{\alpha^2}{4c_0} - \frac{\alpha}{2c_0} + a_0\right) x + \frac{b_0}{2c_0}, \\ \lambda = \frac{\alpha^2}{4} - \frac{\alpha}{2} + a_0 c_0 . \end{cases} \tag{14.10}$$

Then (14.9) reduces to

$$C_{mn}(1) = 1 - \lambda C_{mn}(0)^2 . \tag{14.11}$$

In general, we consider the following system

$$C_{mn}(r+1) = 1 - \lambda C_{mn}(r)^2, r = 0, 1, 2, \cdots \tag{14.12}$$

Hence, when $r = 0$, then (14.12) reduces to (14.11). It is easy to see from (14.11) that it is fairly similar to the familiar one-dimensional Logistic system. And for (14.11), we obtain the physical uniform system (14.3) is chaotic in space when $\lambda = \frac{\alpha^2}{4} - \frac{\alpha}{2} + a_0 c_0 = 1.86 > 1.55$ for any $a, b, c, a_0, b_0, c_0 \in R$. For example, taking $(a, b, c, d) = (-0.202, 1, -0.07, -0.5)$ and $(a_0, b_0, c_0) = (2, 0.456, 0.93)$, then $\lambda = 1.86$. Hence system (14.11) reduces to

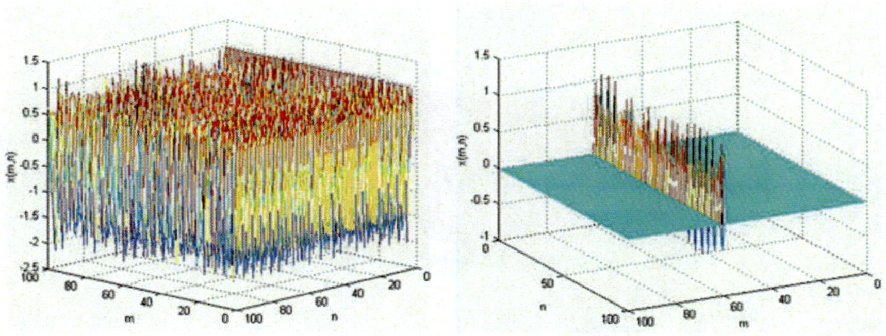

Fig. 14.1 **a** System (14.13) obtains chaos performance in space. **b** System (14.13) obtains a section of spatial chaos phenomena

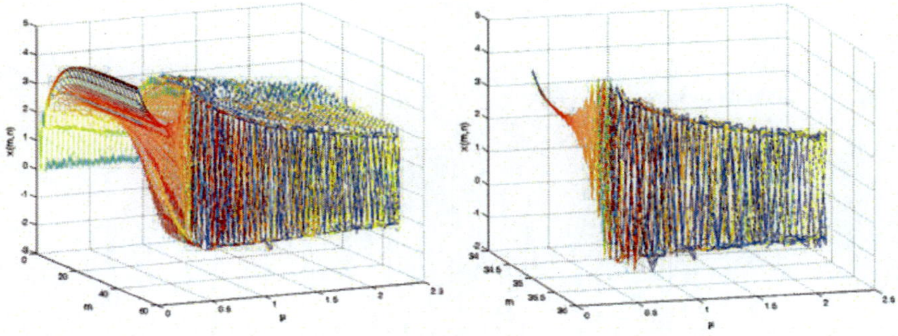

Fig. 14.2 **a** System (14.13) obtains surface bifurcation phenomena. **b** System (14.13) obtains a spatial section of surface bifurcation phenomena

$$C_{mn}(1) = 1 - 1.86C_{mn}(0)^2, \tag{14.13}$$

then system (14.13) is chaotic in space, the following simulations are the chaotic behavior and bifurcation phenomena obtained by (14.13) in space (Figs. 14.1 and 14.2).

Similarly, if we take $(a, b, c, d) = (-0.05, 1, -0.05, -0.05,)$ and $(a_0, b_0, c_0) = (a - 19.5, 4, 1)$, then $\lambda = 0.48$, and (14.13) reduces to the following form

$$C_{mn}(1) = 1 - 0.48C_{mn}(0)^2. \tag{14.14}$$

And when $\lambda = 0.48 < 1.55$, system (14.14) is stable in space, and the stability behavior of this system is demonstrated by Fig. 14.3.

Especially, in system (14.11), when $(a, b, c, d) = (1, 0, 0, 0)$, $(a_0, b_0, c_0) = (1.76, 2, 1)$ and $n = n_0$, and use notation $u_{mn_0} = u_m$, system (14.11) reduces to

$$u_{m+1} = 1 - 1.76u_m^2, \tag{14.15}$$

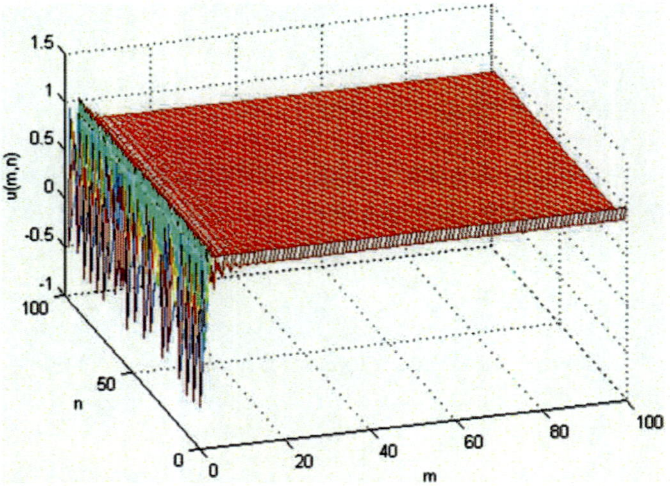

Fig. 14.3 Stability of system (14.14)

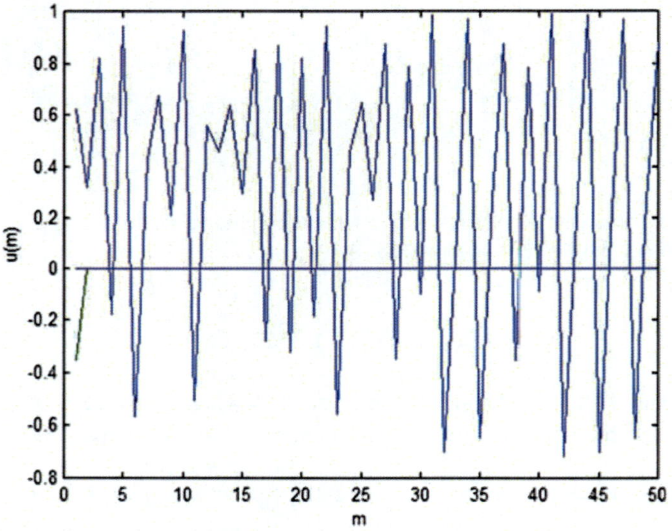

Fig. 14.4 Surface chaotic behavior of (14.15)

which is a familiar one-dimensional Logistic discrete dynamical system. And because $\lambda = 1.76 > 1.55$, then the system is in surface chaos. The surface chaotic behavior of this system is demonstrated in Figs. 14.4 and 14.5.

In addition, it needs to be pointed out that, in the uniform physics system (14.11), if we take proper values for parameters a, b, c, a_0, b_0, c_0, we can obtain the surface chaos and surface bifurcation behavior of one-dimensional wave equation, one-

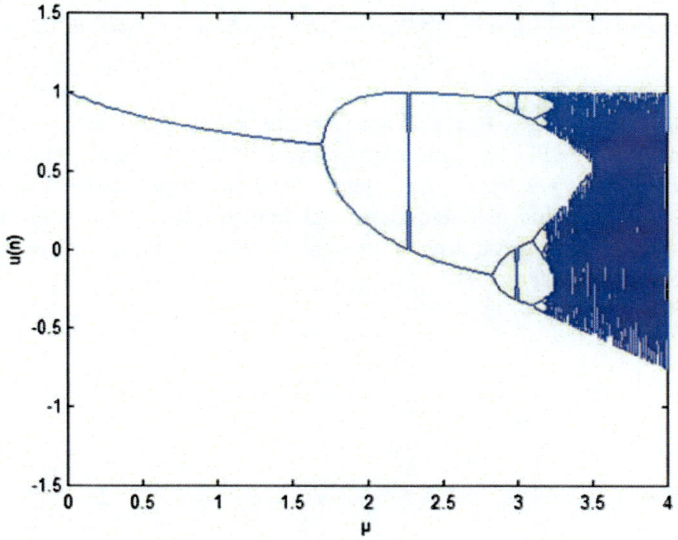

Fig. 14.5 Surface bifurcation case of (14.15)

dimensional heat conduction equation and harmonic equation. For example, taking $(a, b, c, d) = (1, 1, 1, 1)$ and $(a_0, b_0, c_0) = (0.83, 8, 1)$, then $\lambda = 1.83 > 1.55$, system (14.11) is also surface chaotic. And the corresponding is just the spatial one-dimensional discrete Laplace system, i.e.,

$$u_{m+1,n} + u_{m,n+1} + u_{m-1,n} + u_{m,n-1} = 1 - 1.83(4u_{mn})^2. \tag{14.16}$$

Furthermore the corresponding continuous system of (14.16) is exactly the one-dimensional Laplace equation

$$\frac{\partial^2 u}{\partial x^2} + \frac{\partial^2 u}{\partial y^2} = 1 - 1.83(4u\,(x, y))^2. \tag{14.17}$$

Therefore, the investigation of the spatial nonlinear dynamical behavior of (14.16) provides some useful information for the investigation of the continuous system of (14.17).

It also needs to be mentioned particularly that, for the spatial physics kinematic systems with two-dimensions or higher dimensions, we can also reduce them to a uniform form, to study its surface chaos phenomena and other spatial nonlinear dynamical property in a similar way. On the other hand, we also should point out that, even more generally, we can consider the surface chaos behavior and surface nonlinear dynamical property of

$$a_{mn}x_{m+1,n} + b_{mn}x_{m,n+1} + c_{mn}x_{m-1,n} + d_{mn}x_{m,n-1} + e_{mn}x_{mn} \\ = g\left(\mu, (a_{mn} + b_{mn} + c_{mn} + d_{mn})\, x_{mn}\right), \qquad (14.18)$$

where $a_{mn}, b_{mn}, c_{mn}, d_{mn}, e_{mn}$ are known functions and g is a nonlinear function. The system (14.18) will have significant applications in $neutron\ diffusion\ theory$ in $nuclear\ fission\ reaction$model. Hence, the uniform physics system is actually a natural popularization of one-dimensional Logistic discrete dynamical system, or one-dimensional Logistic discrete dynamical system is a special case of spatial physics kinematic system.

Chapter 15
Uniformity of Special Physical Motion Systems and Surface Chaos Behavior in the Sense of Li-Yorke

15.1 Preliminaries

In the case of system (4.1), assume that the function $f(z) : I \to I$ has the following properties:
(i) $f(z)$ is a piecewise continuously differentiable function defined on the interval I with a single jump discontinuity at $z = \xi \in I$, where, for $a > -1$, $I = [0, 1 + a]$; For $a < -1$, $I = [1 + a, 0]$;

(ii) For any $z \neq \xi$, $0 < \frac{df(z)}{dz} < +\infty$;

(iii) When $a > -1$, $\lim\limits_{z \to \xi^+} f(z) = 0$, $\lim\limits_{z \to \xi^-} f(z) = 1 + a$, $f(\xi) = 0$;

When $a > -1$, $\lim\limits_{z \to \xi^+} f(z) = 1 + a$, $\lim\limits_{z \to \xi^-} f(z) = 0$, $f(\xi) = 1 + a$.

Here we call convection system (4.1) satisfying (i)–(iii) convection system (4.1) with a piecewise continuous forced term. A typical plot of the function $f(z) : I \to I$ is shown in Figs. 15.1 and 15.2.

In the case of system (4.1), define the following unitary function, that is, for any variable $y \in R$, let

$$G(\mu, a, x, y) = f(\mu, (1 + a)x) : [0, 1] \to [0, 1 + a].$$

We see that $G(\mu, a, x, y)$ is a special dualistic function, and its spatial jump plot is shown in Figs. 15.3 and 15.4.

In fact, it is convenient to view the mapping $f(z) : I \to I$ as a special mapping of a circle into itself, and then use the rotation number, an important concept and property in the circle mapping, to study the behavior of surface chaos for convection system (4.1) with a piecewise continuous forced term.

The concept of the rotation number is given in the following:

S. T. Liu and L. Zhang, *Surface Chaos and Its Applications*,
https://doi.org/10.1007/978-981-16-8229-2_15

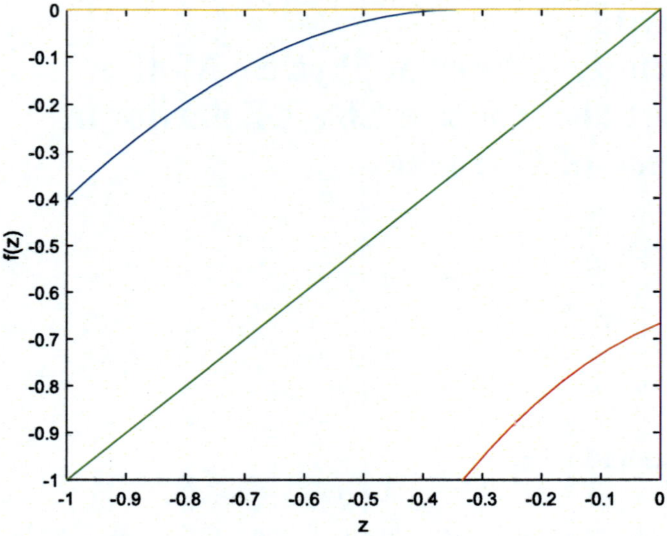

Fig. 15.1 The function f for $a < -1$

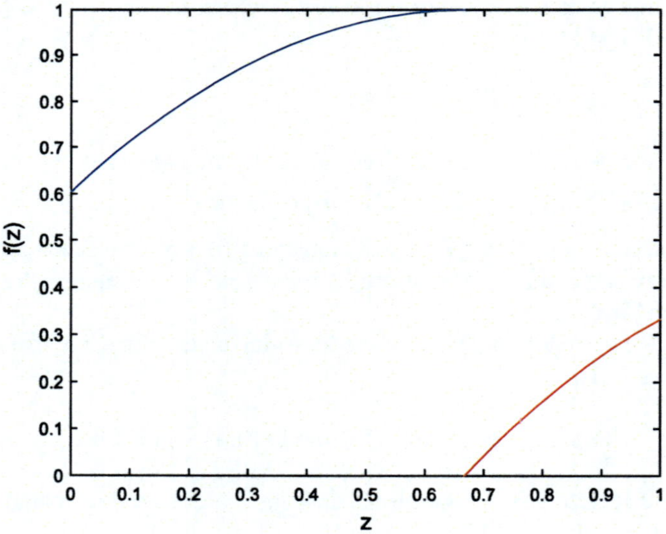

Fig. 15.2 The function f for $a > -1$

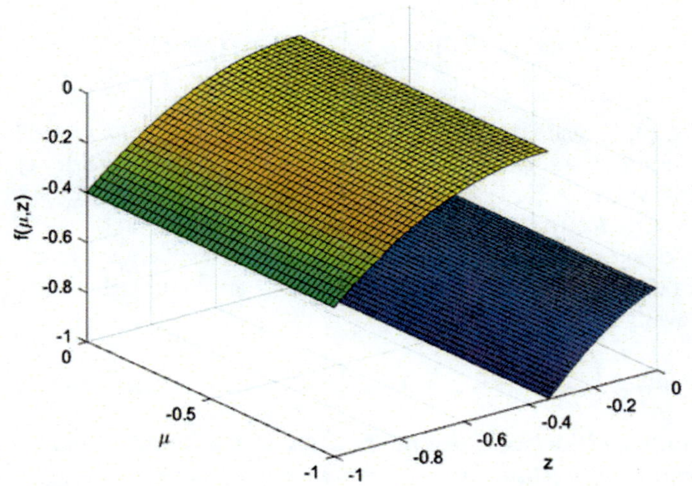

Fig. 15.3 The function G for $a < -1$

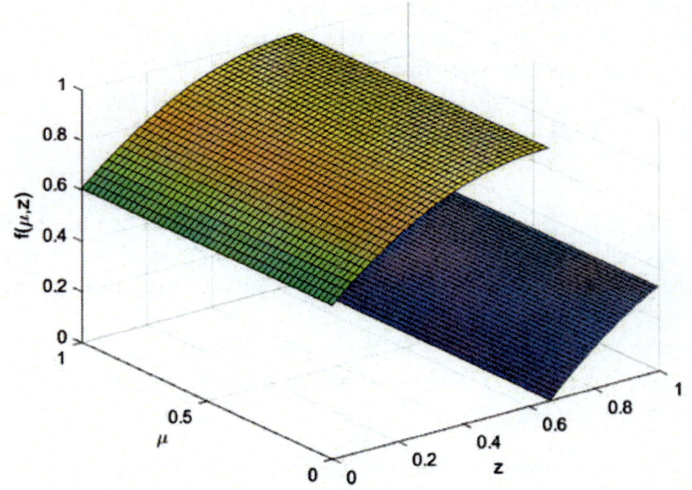

Fig. 15.4 The function G for $a > -1$

The Rotation Number

When the function f satisfies (i)–(iii), the kth iterate of system (4.1) is given by

$$x_{m+k} + ax_{m,n+k} = \underbrace{f(\mu, f(\mu, \ldots, f(\mu, x_{mn} + ax_{mn})\cdots))} = f^{(k)}(\mu, x_{mn} + ax_{mn}) \tag{2.1}$$

where $x_{mn} + ax_{mn} \in I$ is an initial value.

For any y, assume that the dualistic function is given by

$$\rho(x, y) = \lim_{k \to +\infty} \frac{1}{k} \sum_{i=1}^{k} X_\theta(f^{(i)}(\mu, (1+a)x)) \tag{15.1}$$

where $x \in [0, 1]$, that is, $z = (1 + a)x \in I$ is an initial value, $x = \theta \in [0, 1]$, $z = \xi = (1 + a)\theta \in I$, X_θ is the characteristic function for the interval $[\theta, 1]$. For each given initial value $x \in [0, 1]$ and any y, we define the function (15.1) to be the rotation number $\rho(x, y)$ for system (4.1).

The rotation number $\rho(x, y)$ as defined here counts the number of times, the iterates of each given initial value $x_{mn} + ax_{mn} \in I$ are on the right of $z = \xi = (1 + a)\theta$, following the iterate sequence

$$x_{mn} + ax_{mn},\ x_{m+1,n} + ax_{m,n+1},\ x_{m+2,n} + ax_{m,n+2},\ \ldots,\ x_{m+k,n} + ax_{m,n+k},\ \ldots$$

The Definition of Surface Chaos in The Sense of The Rotation Number Interval For system (4.1) that satisfies (i)–(iii) if the range of its rotation number $\rho(x, y)$ covers some nonzero interval $[\alpha, \beta]$, $\alpha \neq \beta$, for each point $x \in [0, 1]$, then the motion of convection system (4.1) with a piecewise continuous forced term is said to be surface chaotic in the sense of the rotation number interval.

In the case of system (4.1), assume that the function $f(z) : I \to I$ satisfies the following conditions at the same time:
(iv) When $a > -1$, $f(0) = 0$; When $a < -1$, $f(1 + a) = 1 + a$.
(v) When $a > -1$, $f(1 + a) < \xi$; When $a < -1$, $f(0) < \xi$.
(vi) For all $z \neq \xi$, $f(z) \neq z$.

Analyzing (iv)–(vi), we can see that system (4.1) has only one fixed point of the mapping x_1 : when $a > -1$, $f(0) = 0$; when $a < -1$, $f(1 + a) = 1 + a$. It follows from the definition of the rotation number $\rho(x, y)$ that this point has the rotation number $\rho(x_1, y) = 0$ for any y.

We can proof that system (4.1) has a periodic point x_{k+1} of period $k + 1$: $f^{(1+k)}(\mu, (1 + a)x_{k+1}) = (1 + a)x_{k+1}$, and this point x_{k+1} has the rotation number $\rho(x_{k+1}, y) = \frac{1}{1+k}$ for any y.

Lemma 15.1 *For system (4.1) satisfying (i)–(vi), when $a > -1$, let k be the smallest positive integer such that $\xi < f^{(k)}(1 + a) < 1 + a$; When $a < -1$, let k be the smallest positive integer such that $\xi < f^{(k)}(0) < 0$. Then there is a point $x_{k+1} \in [0, 1]$, that is, $(1 + a)x_{k+1} \in I$ with $\rho(x_{k+1}, y) = \frac{1}{1+k}$ for any y.*

Proof Only the proof of the case of $a < -1$ is given in the following, and the similar proof can be adapted to the case of $a > -1$.

Since when $a < -1$, k is the smallest positive integer such that $\xi < f^{(k)}(0) < 0$, $f^{(j)}(0) \in [1 + a, \xi]$, $j = 1, 2, \ldots, k - 1$. Let $\lambda \in (1 + a, 0)$ be any point for which $\xi < f^{(k)}(\lambda) < 0$. Then there is a point ε_0 such that for each $\varepsilon \in (0, \varepsilon_0)$, $f(\xi - \varepsilon) \geq \lambda$. We can see that f is continuous on each of the $k + 1$ intervals: $P_0 = [\xi, \lambda]$, $P_j = [1 + a, f^{(j)}(\lambda)]$, $j = 1, 2, \ldots, k - 1$ and $P_k = [1 + a, \xi - \varepsilon]$, and that $f(P_i) \supset P_{i+1}, i = 0, 1, 2, \ldots, k-1$ and $f(P_k) \supset P_k \cup P_0$. Thus, we obtain $f^{(k+1)}(P_0) \supset P_0$, which suggests that there is a spatially periodic point $(1+a)x_{k+1} \in P_0$ of period $k+1$ such that $f^{(1+k)}(\mu, (1+a)x_{k+1}) = (1+a)x_{k+1}$. It follows from the definition of the rotation number $\rho(x, y)$ that this point has the rotation number $\rho(x_{k+1}, y) = \frac{1}{1+k}$ for any y.

Theorem 15.1 *If system (4.1) satisfies (i)–(vi), then its rotation number $\rho(x, y)$ takes on at least all values in the interval $[0, \frac{1}{1+k}]$ for each point $x \in I$, that is, $(1+a)x \in I$, and any y, which shows that system (4.1) is in the state of spatial chaos, that is, the function f is said to be surface chaotic in the sense of the rotation number interval.*

Proof Using Lemma, we can see that the mapping f is equivalent to a shift on the sequence space $P_0, P_1, \ldots, P_{k-1}, P_k$. According to a shift on a sequence space (one definition of chaos), we obtain that there exists at least two points $x_j \in I,\ j = 1, 2$ for the sequence space $P_0, P_1, \ldots, P_{k-1}, \underbrace{P_k}_{1}$ such that $\rho_1(x_1, y) = \frac{0}{1+k*0} = 0$, $\rho_1(x_2, y) = \frac{1}{1+k*1} = \frac{1}{1+k}$. For the sequence space $P_0, P_1, \ldots, P_{k-1}, \underbrace{P_k, P_k}_{2}$ there exists at least three points $x_j \in I,\ j = 1, 2, 3$ such that $\rho_2(x_1, y) = \frac{0+0}{2+k*(0+0)} = 0$, $\rho_2(x_2, y) = \frac{0+1}{2+k*(0+1)} = \frac{1}{2+k}$, $\rho_2(x_3, y) = \frac{1+1}{2+k*(1+1)} = \frac{2}{2+2k}$. The rest may be deduced by analogy: After such additional : $P_0, P_1, \ldots, P_{k-1}, \underbrace{P_k, \ldots, P_k}_{n}$ there exists at least $n+1$ points $x_j \in I,\ j = 1, 2, \ldots, n+1$ such that $\rho_n(x_j, y) = \frac{\sum^{n} s_j}{n+k\sum^{n} s_j}$.

Take $n = 5$ for example. $\sum^{n} s_j$ refers to in turn $\sum^{5} s_1 = (0+0+0+0+0) = 0, \sum^{5} s_2 = (0+0+0+0+1) = 1, \sum^{5} s_3 = (0+0+0+1+1) = 2, \sum^{5} s_4 = (0+0+1+1+1) = 3, \sum^{5} s_5 = (0+1+1+1+1) = 4, \sum^{5} s_6 = (1+1+1+1+1) = 5$. In view of the difference between two neighboring elements on $\rho_n(x, y)$: $\rho_n(x_{j+1}, y) - \rho_n(x_j, y) = \frac{1+\sum^{n} s_j}{n+k+k\sum^{n} s_j} - \frac{\sum^{n} s_j}{n+k\sum^{n} s_j} < \frac{1}{n}$, we obtain that $\rho_n(x_{j+1}, y) \to \rho_n(x_j, y)$ as $n \to +\infty$. This suggests that every irrational or rational number contained in the interval $[0, \frac{1}{1+k}]$ is reached in the limit $n \to +\infty$. Moreover, we notice that none of the points to the right of the largest point of period $k+1$ are needed in the proof by choosing the right endpoints of the intervals to be points of period $k+1$. Therefore, the range of the rotation number $\rho(x, y)$ for system (4.1) at least covers the interval $[0, \frac{1}{1+k}]$ for each point $x \in I$, that is, $(1+a)x \in I$, which suggests that system (4.1) is in the state of spatial chaos, that is, the function f is said to be surface chaotic in the sense of the rotation number interval.

15.2 An Illustrative Example

Consider the following convection system with a piecewise continuous forced term

Example 15.1

$$
\begin{aligned}
& x_{m+1,n} + ax_{m,n+1} \\
&= f(\mu, (1+a)x_{mn}) \\
&= \begin{cases} -((1+a)x_{mn})^2 + \mu(1+a)x_{mn} \\ -((1+a)x_{mn})^2 + (\mu+0.875)(1+a)x_{mn} - 1.3854x_{mn} \\ < \theta x_{mn} \\ > \theta \end{cases},
\end{aligned} \tag{15.2}
$$

where $a > -1$ and the piecewise continuous function is given by

$$
\begin{aligned}
& G(\mu, a, x, y) \\
&= f(\mu, (1+a)x) \\
&= \begin{cases} -((1+a)x)^2 + \mu(1+a)x \\ -((1+a)x)^2 + (\mu+0.875)(1+a)x - 1.3854x \\ < \theta x > \theta. \end{cases}
\end{aligned} \tag{15.3}
$$

Take $a = -0.125$, $\mu = 2.0833$, $\theta = 0.667$. Then $f' = -2(1+a)^2x + \mu(1+a) = -1.5313x + 1.8229 > 0$ is incremental in the interval $[0, \theta]$, and it follows from Extremum Theorem that $maxf_{x=\theta} = 1 + a$, min $f_{x=0} = 0$; Similarly, $f' = -2(1+a)^2x + (\mu + 0.875)(1+a) = -1.5313x + 2.5889 > 0$ is also incremental in the interval $[\theta, 1]$, and $maxf_{x=1} = 0.4375$, min $f_{x=\theta} = 0$. Thus it can be seen that $G(\mu, a, x, y) = f(\mu, (1+a)x) : [0, 1] \to [0, 1+a]$ satisfies all conditions of surface chaos in the sense of the rotation number interval. Thereby system (15.2) is in the state of surface chaos and its behavior of surface chaos is demonstrated by Fig. 15.5.

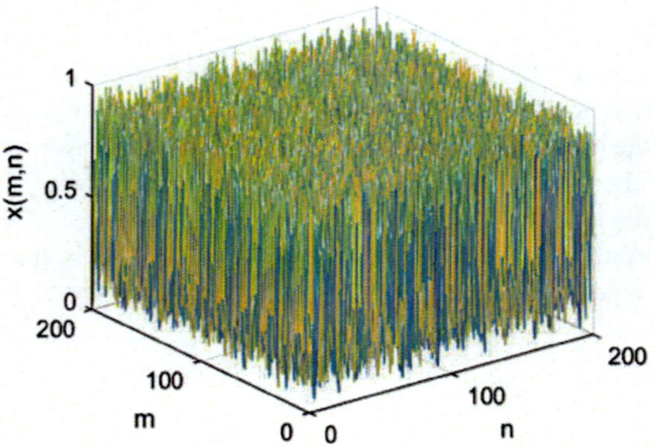

Fig. 15.5 The behavior of spatial chaos for system (15.2)

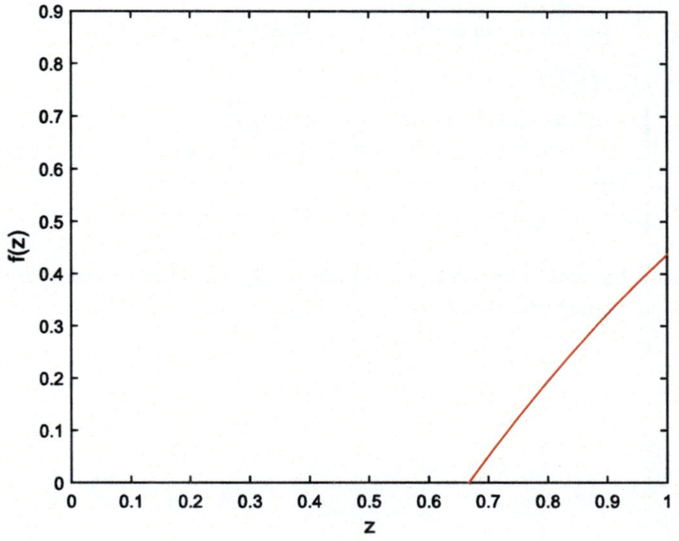

Fig. 15.6 The 1-D plot of $f(\mu, (1+a)x)$, where $\mu = 2.0833$, and $a = -0.125$

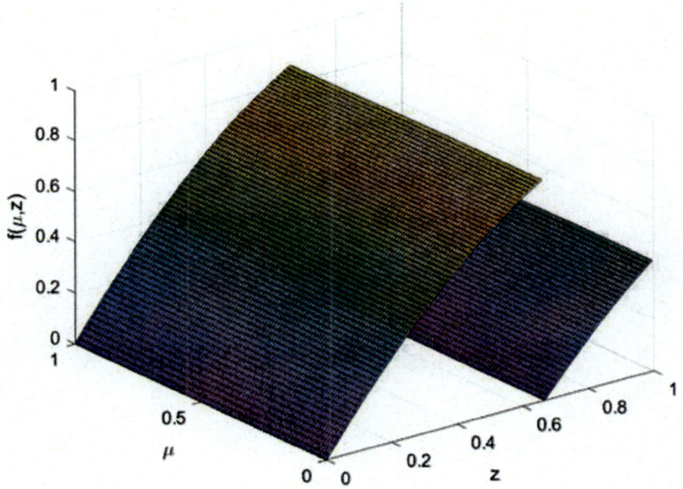

Fig. 15.7 The spatial plot of $G(\mu, a, x, y)$, where $\mu = 2.0833$, and $a = -0.125$

The plot of the piecewise continuous function f is shown in Figs. 15.6 and 15.7. In addition, the continuous analog of system (15.2) is given by

$$
\begin{aligned}
&\frac{\partial v}{\partial x}+a\frac{\partial v}{\partial y}+av(x,y)\\
&=f(\mu,(1+a)v(x,y))\\
&=\begin{cases}-((1+a)v(x,y))^2+\mu(1+a)v(x,y)\\ -((1+a)v(x,y))^2+(\mu+0.875)(1+a)v(x,y)-1.3854v\\ <\theta v\\ >\theta\end{cases}.
\end{aligned}
\tag{15.4}
$$

Therefore, the research on discrete system (15.2) could provide some useful information for analysing continuous system (15.2).

Chapter 16
Surface Chaos Behavior of Molecular Orbit

16.1 Introduction

From Hückel's molecular orbit theory [407, 408], each electron has its independent electronic orbit or its kinematical form, which is determined by a wave function. And the orbit of molecule i is the weighted sum of the orbits of all atoms in that molecule, i.e., if ψ_i is used to express the molecular orbit, then

$$\psi_i = a_{1i}\phi_1 + a_{2i}\phi_2 + \cdots + a_{ni}\phi_n, i = 1, 2, \ldots, n,$$

where a_{ji} is the orbit coefficient, ϕ_j is the atomic orbit (or atomic function). The orbit coefficient a_{ji} can be expressed as a function of the coordinates (q_1, q_2, q_3) of an atom. The function is called a carrier wave of molecular orbit and expressed by

$$a_{ji} = f_j(q_1, q_2, q_3), \ j = 1, 2, \ldots, n.$$

Then the molecular orbit ψ_i can be expressed by [409]

$$\psi_i = \sum_{j=1}^{n} f_j(q_1, q_2, q_3)\phi_j(q_1, q_2, q_3, x, y, z).$$

If the three coordinates of each atom are properly selected such that they are integers, $f_j(q_1, q_2, q_3)$ reduces to a function with three integer input variables.

The carrier wave function reflects the relation between orbit coefficients and atomic positions. Molecular orbit is the whole effect of all atomic orbits. The energy level of molecular orbit is $\varepsilon = \int \psi_i^* \bar{H} \psi_i dxdydz$ [409] obtained from electronic wave function, where $\psi_i^*$is the conjugated wave function of $\psi_i^*, \bar{H}$ is the Hamilton operator of a single electron. In addition, from Hückel 's molecular orbit theory, the orbit of a single electron with power $V(x, y, z)$ describes the status of the molecular

S. T. Liu and L. Zhang, *Surface Chaos and Its Applications*,
https://doi.org/10.1007/978-981-16-8229-2_16

system, regardless of electron spinning [401]. The orbits of electrons in a multi-electron system can be very complicated if electrons spin with inverse parallelism, which usually is satisfied in a nonlinear system. In this chapter, the spatial nonlinear dynamic behavior of molecular orbit is mainly studied, i.e., the surface chaos and surface bifurcation of molecular orbit. In addition, the relationship between molecular orbit and its corresponding energy level in a nonlinear dynamic system is also studied.

16.2 Spatial Nonlinear System of Molecular Orbit

Square-lattice conjugated molecules with rectangular boundary is taken into consideration, i.e., $m \times n$ atoms are arranged in a square lattice with rectangle boundary. The atom bondings are shown in Figs. 16.1 and 16.2.

If the rectangular axis $0 - mn$ is selected with unit bond length-d, then the two coordinates (m, n) of each atom are all integers, and carrier wave $A(m, n)$ [409] of molecular orbit is the function of two integer input variables. At that time, if we assume that $A(m, n)$ is not on the boundary, there are four atom bondings with it, i.e., $(m + 1, n)$, $(m, n + 1)$, $(m - 1, n)$ and $(m, n - 1)$, and the carrier waves corresponding to these four atoms are $A(m + 1, n)$, $A(m, n + 1)$, $A(m - 1, n)$ and $A(m, n - 1)$ respectively. A relationship among them can be obtained by using variational method:

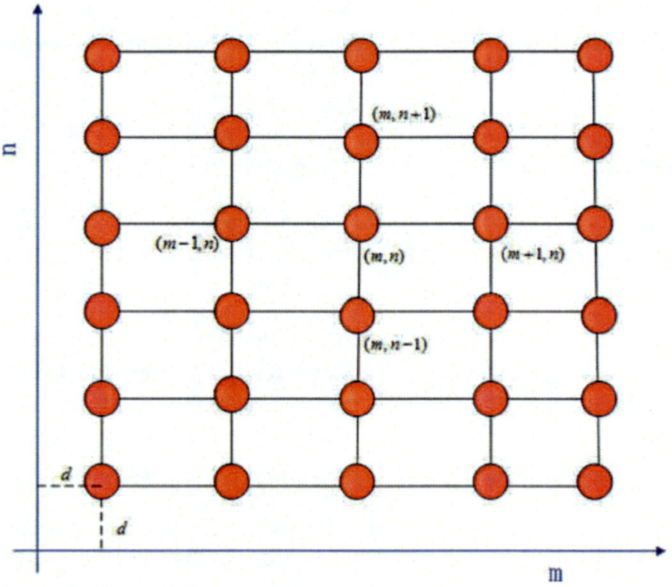

Fig. 16.1 Square-lattice conjugated molecules with rectangular boundary

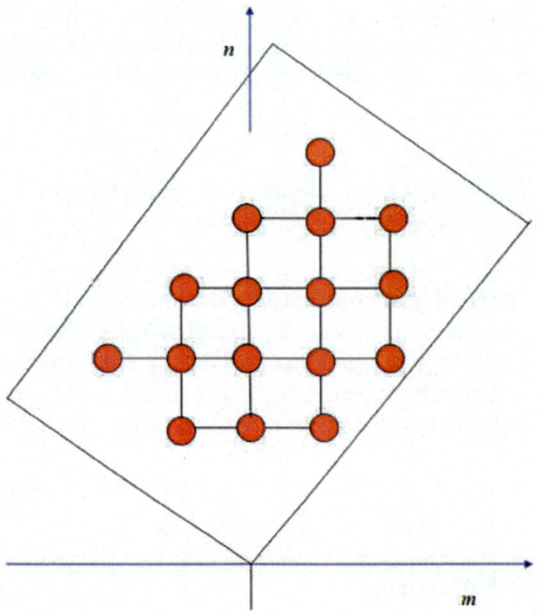

Fig. 16.2 Rectangular boundary of square-lattice conjugated molecules make an 45°

$$\beta\left(A(m+1,n)+A(m,n+1)+A(m-1,n)+A(m,n-1)\right)$$
$$=(\varepsilon-\alpha)A(m,n), \tag{16.1}$$

where ε is the energy level of a molecule, α is the *Coulomb* integral and β is the exchange integral [409]. Equation (16.1) is a non-homogeneous two-dimensional discrete dynamic system under the approximate **Hückel's** hypothesis. However, the behavior of electrons in a real multi-electron system is very complicated. When Hückel's hypothesis is incompletely considered, and electron spinning and various noise disturbance in the real physical world are also considered, the spatial dynamic system (16.1) usually becomes a nonlinear system

$$\beta\left(A(m+1,n)+A(m,n+1)+A(m-1,n)+A(m,n-1)\right)$$
$$=f\left(\mu(\varepsilon),A(m,n)\right), \tag{16.2}$$

where $\mu(\varepsilon)$ is a real parameter that is a function of the energy level ε, f is a nonlinear function called forced term. Equation (16.2) is usually called a molecular orbit system. Besides, the corresponding continuous system of (16.2) is

$$\beta\left(\frac{\partial^2 A}{\partial x^2}+\frac{\partial^2 A}{\partial y^2}+4A(x,y)\right)=f\left(\mu(\varepsilon),A(x,y)\right), \tag{16.3}$$

where $f(\mu(\varepsilon), A(x, y))$ is a nonlinear function. Since (16.3) is a one-dimensional harmonic equation, the study on the nonlinear dynamic behavior of the system (16.2) provides some useful information about the continuous system (16.3).

16.3 Surface Chaos and Bifurcation Behavior of Molecular Orbit

For nonlinear system (16.2), the forced term f can be assumed to be a continuous function without loss of generality. Then from the theorem of *Weretriss* polynomial uniform approximation [401], f can be always approximated by a polynomial without loss of generality. If $f(\mu(\varepsilon), x) = a_0 + a_1 x - a_2(2\beta x)^2$ and $a_1 = \alpha$ for the convenience of dealing with the problem, the following is obtained from (16.2):

$$\begin{aligned} &\beta\left(A(m+1, n) + A(m, n+1) + A(m-1, n) + A(m, n-1)\right) \\ &= a_0 + \varepsilon A_{mn} - a_2(4\beta A_{mn})^2. \end{aligned} \tag{16.4}$$

From (16.4), if

$$\begin{aligned} C_{m+r,n+r} &= \beta\left(A(m+r, n) + A(m, n+r) + A(m-r, n) + A(m, n-r)\right) \\ r &= 0, 1, 2, \ldots, \end{aligned} \tag{16.5}$$

and $r = 0, 1$, then $C_{m,n} = 4\beta A_{mn}$ and

$$C_{m+1,n+1} = \beta\left(A(m+1, n) + A(m, n+1) + A(m-1, n) + A(m, n-1)\right).$$

Thus (16.4) reduces to

$$C_{m+1,n+1} - a_0 = a_2 C_{m,n}\left(\frac{\varepsilon}{4a_2\beta} - C_{m,n}\right). \tag{16.6}$$

If

$$\begin{cases} \xi = \frac{\varepsilon}{4\beta}, \\ y = (\frac{\xi^2}{4a_2} - a_0)x + \frac{\xi}{2a_2}, \\ \mu = \frac{\xi^2}{4} + a_0 a_2, \end{cases} \tag{16.7}$$

then from (16.6) and (16.7), the following is obtained

$$C_{m+1,n+1} = 1 - \mu C_{m,n}^2. \tag{16.8}$$

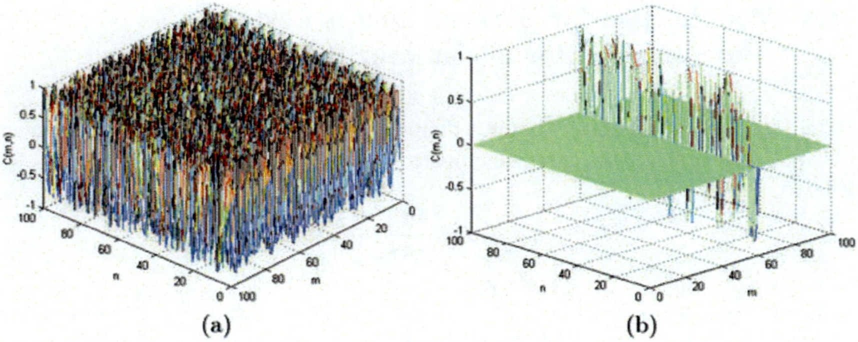

Fig. 16.3 Surface oscillatory behavior of system with m = 15

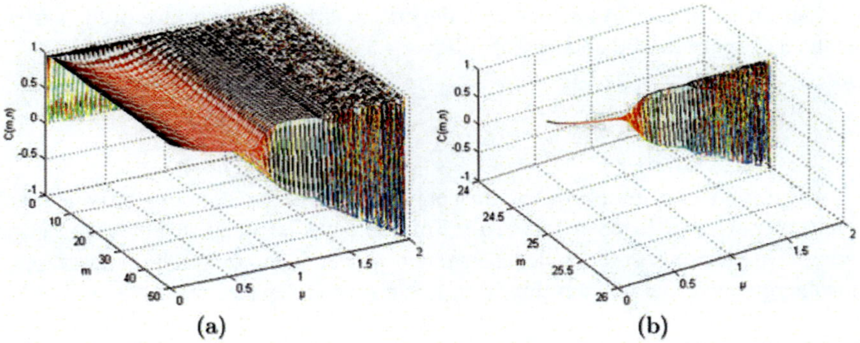

Fig. 16.4 **a** Bifurcation behavior of molecular orbit in 3-D space. **b** A section of spatial bifurcation behavior

Equation (16.8) is similar to our familiar one-dimensional Logistic system $C_{s+1} = 1 - \mu C_s^2$. Besides, from (16.5) and (16.8), we can consider the molecular system as

$$C_{m+r+1,n+r+1} = 1 - \mu C^2_{m+r,n+r}, \; r = 0, 1, 2, \ldots \; . \tag{16.9}$$

Then when $r = 0$, (16.9) reduces to (16.8). From (16.7), for proper values of $(\varepsilon, \beta, a_0, a_2)$ and $\mu = \frac{\xi^2}{4} + a_0a_2 > 1.55$, the system (16.8) is chaotic. For example, if $(\varepsilon, \beta, a_0, a_2) = (8\beta, \beta, 2, 0.43)$, then $\mu = \frac{\xi^2}{4} + a_0a_2 = 1.86 > 1.55$ and the system (16.8) reduces to $C_{m+1,n+1} = 1 - 1.86C^2_{m,n}$. The corresponding molecular orbit is chaotic as shown in Fig. 16.3a and b.

In addition, since $\mu(\varepsilon)$ is a variable real parameter, when ε is changed, bifurcation of molecular orbit can occur in the 3-D space as shown in Fig. 16.4a and b.

16.4 The Relationship Between Energy Level and Molecular Orbit in a Nonlinear Dynamical System

From (16.7), it is easy to see that in a definite system, since ξ, β and a_i are known, $i = 0, 2$ and ε is a variable parameter, the real parameter μ is a function of the energy level ε:

$$\mu(\varepsilon) = \mu = \frac{\varepsilon^2}{64\beta^2} + a_0 a_2.$$

Moreover, since $\mu'(\varepsilon) \geq 0$, $\mu(\varepsilon)$ is a monotonically increasing function of the energy level ε. Therefore, with the increase of the energy level ε, the coefficient of molecular orbit in the 3-D space has period doubling bifurcation, i.e., from the orbit surface with the periods of $2^0, 2^1, \ldots, 2^\infty$. In fact, this can be calculated from (16.8), i.e., for any y, if $G(\mu, y) = 1 - \mu y^2$, the critical bifurcation surface of the system in the 3-D space can be obtained as follows:

$$(\mu_0, \mu_1, \mu_2, \mu_3, \ldots) = \left(\frac{3}{4}, \frac{4}{5}, 1.368, 1.394, \cdots\right).$$

The change of parameter μ and the corresponding period doubling bifurcation of molecular orbit are listed in Table 16.1: This table well explains that the distribution density of electrons in the molecular system will be increased with the increase of the energy level, and will develop from stable status to chaos.

Table 16.1 The period doubling bifurcation of the molecular orbit

Range of parameter μ	Bifurcation of molecular orbit
$0 < \mu < 0.75$	2^0 stable periodic surface
$0.75 \leq \mu < 1.25$	2^1 stable periodic surfaces
$1.25 \leq \mu < 1.368$	2^2 stable periodic surfaces
$1.368 \leq \mu < 1.394$	2^3 stable periodic surfaces
...	...
$\mu \geq 1.55$	2^∞ stable periodic surfaces $\Longrightarrow \cdots \Longrightarrow$ spatial chaos

Chapter 17
Surface Chaotic Theory and the Growth of Harmful Algal Bloom

17.1 Introduction

Harmful algal blooms (HABs) are mainly produced by rapid growth of algal species named as the causative species with certain conditions and have severe impacts on public health and coastal ecosystems [410, 411]. One of the most famous reproduction mode of the causative species for the unicellular algae [412, 413] is the well-known fact that one cell divided into two cells by asexual reproduction. In addition, the rate of the growth is complicatedly and nonlinearly affected by different environmental factors such as the ratios of nitrogen to phosphorus [414], fluorescent and red light environments [415]. So the inherent regularity between the growth process of the causative species and the complex environmental factors still remains a mystery.

Nowadays, many works focus on the control of growth process for these causative species in HABs experimentally. For instance, the higher concentration of kanamycin was found to inhibit the growth velocity of unicellular green algae *Chlorella vulgaris* [416]. The effect of assemblage age on the uptake rate of algal species was studied for the incorporation of cells and eliminating differences in substrate roughness [417]. When three proteins named GsSPT1 (G. sulphuraria sugar and polyol transproter 1), GsSPT2 and GsSPT4 were studied, results were shown that the deletion of N-terminus of the protein in SPT1 affected the maximal transport velocity and released the dependency of the pH on sugar uptake, while the uptake by other two proteins was not active for the pH-dependence [418].

Moreover, according to the fact that the growth process of the causative species in reality becomes much more complex than that of growth under laboratory, the growth process can be described mathematically by dynamical systems and many interesting and important works have been obtained. For example, the dynamical behavior of algal growth with impulsive disturbance [419], such as chaos [420] and stability of attractors [421, 422], was studied in colony formation, in the interaction growth and in the allelopathic of phytoplankton. In the light of the experimental results of the growth interaction such as toxin or non-toxic algae, organic pollutants, nutrients

S. T. Liu and L. Zhang, *Surface Chaos and Its Applications*,
https://doi.org/10.1007/978-981-16-8229-2_17

and bacteria, the coupled dynamical models were used to simulate the interactions [423–427]. The effect of allelopathic competition between two algal species on the growth rate of algae was also reported by coupled nonlinear models [428–430].

However, there exist few results about the effect of the environmental factor on the growth process of the causative species from the point view of control theory in term of the experimental results, data collection in field and the current mathematical models of causative species. In this chapter, we concentrate on the control effect of the periodic environmental factor on the cell density and the growth rate of the causative species, in which the periodic environmental factor include variation of seasons, light intensity, temperature and some inner cycle clock, etc. [431, 432]. Our work can play vital roles for revealing the secret of the HABs.

The chapter is organized as follows. Firstly, the nonlinear controlled system for the causative species growth is established and the stable condition of the system is presented. Secondly, three theorems about the relationships between the periodic factor and the cell density and the growth rate of the causative species are studied, respectively. Thirdly, we give simulations and experimental data to illustrate the control effect of the periodic factor on the cell density and the growth rate of causative species theoretically and experimentally. Conclusions and discussions are given finally.

17.2 Problem Assumption

This section revolves around the conditions like exist and unique solution and stable theory for the nonlinear reaction-diffusion system with initial and boundary conditions based on the experimental results of growth of the causative species.

In general, the growth process of the causative species dynamically satisfies the following system:

$$\frac{\partial P(x,y,z,t)}{\partial t} = f(P(x,y,z,t), \alpha(x,y,z,t), \beta(x,y,z,t), u(x,y,z,t)), \quad (17.1)$$

where $P(x, y, z, t)$ is the cell density of the causative species,

$$P \in \textit{Hilbert space } H([\xi, \eta], \mathbb{R}^n),$$

$$(x, y, z, t) \in \mathbb{R}^3 \times (0, +\infty),$$

$$f(P(x,y,z,t), \alpha(x,y,z,t), \beta(x,y,z,t), u(x,y,z,t))$$

is the factor including the birth rate $\alpha(x, y, z, t)$, the death rate $\beta(x, y, z, t)$ and the controller as $u(x, y, z, t)$, like gene modulating factors and the strength of sunlight, etc.

Based on the models in Refs. [433, 434] and the fact that the main reproductive mode of single cell species in the causative species is binary fission and the growth velocity varies with different spatial location and the different environmental factors, the dynamical system (17.1) for the growth process of the causative species can be represented as follows:

$$\frac{\partial P(x,y,z,t)}{\partial t} = f(P(x, y, z, t), \alpha(x, y, z, t), \beta(x, y, z, t), u(x, y, z, t))$$
$$= (\alpha + u(t) - \beta(t))P(x, y, z, t) - (\alpha + u(t))\frac{P^2(x,y,z,t)}{P_A} + D\Delta P(x, y, z, t), \tag{17.2}$$

where

$$(\alpha + u(t) - \beta(t))P(x, y, z, t) - (\alpha + u(t))\frac{P^2(x, y, z, t)}{P_A}$$

means the binary fission term,

$$D\Delta P(x, y, z, t) = D\left(\frac{\partial^2 P(x, y, z, t)}{\partial x^2} + \frac{\partial^2 P(x, y, z, t)}{\partial y^2} + \frac{\partial^2 P(x, y, z, t)}{\partial z^2}\right)$$

stands for the spatial diffusion term. Since the velocity of the cell division of the causative species is affected by different periodic factors like strength of sunlight, temperature and inner cycle clock, etc., we introduce the controller as $u = Asin(\omega t + \theta)$, where A is the strength of the periodic factors. And ω and θ represent the amplitude and phase in periodic environmental factors like strength of sunlight, temperature and inner cycle clock, etc., respectively. D indicates the diffusion coefficient. In order to find the control effect of the controller u on the growth process of the causative species easily, we assume the following initial and boundary conditions for system (17.2):

$$\begin{cases} P(x, y, z, 0) = (\phi(x, y, z), 0, 0, \dots, 0), \\ P(0, y, z, t) = P(x, 0, z, t) = P(x, y, 0, t) = (0, 0, \dots, 0, a_0), \\ P(l, y, z, t) = P(x, l, z, t) = P(x, y, l, t) = (1, 1, \dots, 1, a_1), \end{cases}$$

where $\phi(x, y, z)$ is any continuous function, a_0 and a_1 are constant parameters and $a_1 > a_0 \geq 0$.

The stable condition will be discussed for the system (17.2) due to the exponential stability in this paragraph. On the basis of the condition of (17.2) and the reason is clear that the function

$$f_1(P, \alpha, \beta, u) = (\alpha + u - \beta)P - (\alpha + u)\frac{P^2}{P_A}$$

satisfies

$$f_1(P_0, \alpha, \beta, u) = (0, 0, \dots, 0)^T,$$

where $P_0 = (0, 0, \dots, 0)^T$ is the initial growth density. Then we assume $f_1(P, \alpha, \beta, u)$ is locally Lipschitz continuous, i.e. there exist positive real numbers K, K_0 for any $P_1, P_2 \in H$ satisfying the following condition:

$$max\{\| P_1 \|, \| P_2 \|\} \leq K.$$

It means that

$$\| f_1(P_1, \alpha, \beta, u) - f_1(P_2, \alpha, \beta, u) \| \leq K_0 \| P_1 - P_2 \|,$$

where $P_1,\ P_2 \in Hilbert\ space\ H([\xi, \eta], \mathbb{R}^n)$. In addition, the parameters of (17.2) satisfy the following condition:

$$K_0 \geq \left[(\alpha + u - \beta) - \frac{(\alpha + u) + P)}{\phi}\right]. \tag{17.3}$$

Because the norm theory will be used in the proof of the following theorems in next section, the following lemma is presented:

Lemma 17.1 *For different norm definitions:*

$$\|P\|_1 = \sum_{i=1}^{n}\sum_{j=1}^{n} |P_{i,j}|,$$

and

$$\|P\|_2 = \sum_{i=1}^{n}\sum_{j=1}^{n} |P_{i,j}|^2,$$

the two norms have the following inequality relationship:

$$\|P\|_1 \leq n\|P\|_2.$$

17.3 Main Theorem

Without loss of generality and in order to calculate easily, we omit the panel effect and consider only the water depth so that the system (17.2) becomes the following form

$$\begin{aligned}\frac{\partial P(z,t)}{\partial t} &= f(P(z,t), \alpha(z,t), \beta(z,t), u(z,t)) \qquad (17.4)\\ &= (\alpha(z,t) + u(z,t)) - \beta(z,t))P(z,t) - (\alpha(z,t) + u(z,t))\frac{P^2(z,t)}{P_A}\\ &+ D\frac{\partial^2 P(z,t)}{\partial z^2},\end{aligned}$$

with the initial and boundary conditions shown below as:

$$\begin{cases} P(z,0) = (0,0,\ldots,0,\phi(z)), \\ P(0,t) = (0,0,\ldots,0,a_0), \\ P(1,t) = (1,1,\ldots,0,a_1). \end{cases}$$

As the complexity of the form of f in system (17.4), it is difficult to solve the solution analytically. So we next use the finite difference method for the system of (17.4), which is the famous difference method for PDEs. Then system (17.4) is discretized as follows:

$$P_{k,j} - P_{k-1,j} = rP_{k-1,j+1} - 2rP_{k-1,j} + rP_{k-1,j-1} + 2(\alpha(k) + u - \beta)\tau P_{k,j} \\ -\frac{2(\alpha(k)+u(k))\tau}{P_A} P_{k,j}^2, \tag{17.5}$$

where $r = 2D\tau/h^2$ is the mesh ratio of the discretization process, h stands for the space step and τ means the time step. In order to compute easily, we assume $\tau = 1/2$, $D = h^2$. Then as the numerical process for system (17.5), we have

$$\frac{\alpha+u}{P_A}P_{22}^2 + (1-\alpha-u+\beta)P_{2,2} + P_{1,2} - P_{1,3} - P_{1,1} = 0,$$

$$\frac{\alpha+u}{P_A}P_{32}^2 + (1-\alpha-u+\beta)P_{3,2} + P_{2,2} - P_{2,3} - P_{2,1} = 0,$$

$\ldots,$

$$\frac{\alpha+u}{P_A}P_{n,n-1}^2 + (1-\alpha-u+\beta)P_{n,n-1} + P_{n-1,n-1} - P_{n-1,n} - P_{n-1,n-2} = 0.$$

When we define the following symbols as

$$M = \begin{pmatrix} -1 & 0 & \cdots, & 0 & 1\,(1-\alpha-u+\beta) & \cdots & 0 & -1 & 0 & \cdots & 0 \\ 0 & -1 & \cdots, & 0 & 0 \quad 1 & \cdots, & 0 & 0 & -1 & \cdots, & 0 \\ \vdots & \vdots & \vdots & \vdots & \vdots \quad \vdots & \vdots & \vdots & \vdots & \vdots & \vdots & \vdots \\ 0 & \cdots, & 0 & -1 & 0 \quad 0 & 1(1-\alpha-u+\beta) & 0 & \cdots, & 0 & -1 & 0 \end{pmatrix},$$

$$P = [P_{1,1}, P_{2,1}, \ldots, P_{n,1}, P_{1,2}, P_{2,2}, \ldots, P_{n,2}, \ldots, P_{1,n}, P_{2,n}, \ldots, P_{n,n}]^T,$$

$$F = [P_{2,2}^2, P_{3,2}^2, \ldots, P_{n,2}^2, P_{2,3}^2, P_{3,3}^2, \ldots, P_{n,3}^2, \ldots, P_{2,n-1}^2, P_{3,n-1}^2, P_{n,n-1}^2]^T,$$

the numerical process of the system (17.5) can be substituted as follows

$$MP = \frac{\alpha+u}{P_A}F. \tag{17.6}$$

Next we will discuss the different effects of the controller u on the growth process of the causative species.

Theorem 17.1 *The norm of the growth density* $||P||_2$ *will be controlled by the controller u according to the following relationship:*

$$0 \leq ||P||_2 \leq \frac{P_A n[4-\alpha-u+\beta]}{2(\alpha+u)} + \frac{P_A\sqrt{n^2[4-\alpha-u+\beta]^2+4c(\alpha+u)^2/P_A^2}}{2(\alpha+u)}, \tag{17.7}$$

where

$$c = \sum_{i=1}^{n}|P_{i,1}|^2 + \sum_{i=1}^{n}|P_{i,n}|^2 + \sum_{j=1}^{n}|P_{1,j}|^2.$$

Proof For (17.6), we choose the norms of

$$\|P\|_1 = \sum_{i=1}^{n}\sum_{j=1}^{n}|P_{ij}|,$$

$$\|P\|_2 = \sqrt{\sum_{i=1}^{n}\sum_{j=1}^{n}|P_{ij}|^2},$$

and

$$\|M\|_1 = max_{1\leq i\leq n}\sum_{j=1}^{n}|M_{ij}|.$$

In accordance with Lemma 17.1 and the system (17.6), we obtain

$$\|MP\|_1 \leq \|M\|_1\|P\|_1,$$

$$\|MP\|_1 = \frac{\alpha+u}{P_A}\parallel F\parallel_1,$$

and

$$\|P\|_1 \leq n\|P\|_2.$$

Later we have

$$\|M\|_1 \parallel P\parallel_1 \geq \|MP\parallel_1 = \frac{\alpha+u}{P_A}\parallel F\parallel_1 = \frac{\alpha+u}{P_A}\sum_{i=2}^{n}\sum_{j=2}^{n-1}|P_{i,j}|^2,$$

which means that

$$\|M\|_1 \parallel P\parallel_1 + \tfrac{\alpha+u}{P_A}[\textstyle\sum_{i=1}^{n}|P_{i,1}|^2 + \sum_{i=1}^{n}|P_{i,n}|^2 + \sum_{j=1}^{n}|P_{1,j}|^2]$$
$$\geq \tfrac{\alpha+u}{P_A}[\textstyle\sum_{i=2}^{n}\sum_{j=2}^{n-1}|P_{i,j}|^2 + \sum_{i=1}^{n}|P_{i,1}|^2 + \sum_{i=1}^{n}|P_{i,n}|^2 + \sum_{j=1}^{n}|P_{1,j}|^2].$$

Subsequently we get

$$\|M\|_1 \parallel P \parallel_1 + \frac{\alpha+u}{P_A}c \geq \frac{\alpha+u}{P_A}\sum_{i=1}^{n}\sum_{j=1}^{n}|P_{i,j}|^2 = \frac{\alpha+u}{P_A}\|P\|_2^2. \tag{17.8}$$

According to Lemma 17.1 and the inequality of (17.8), we gain

$$n||M||_1||P||_2 + \frac{\alpha+u}{P_A}c \geq \|M\|_1 \parallel P \parallel_1 + \frac{\alpha+u}{P_A}c \geq \frac{\alpha+u}{P_A}\|P\|_2^2,$$

which can be written as the following form:

$$\frac{\alpha+u}{P_A}||P||_2^2 - n||M||_1||P||_2 - \frac{\alpha+u}{P_A}c \leq 0. \tag{17.9}$$

Soon afterwards we have the inequality as the relationship of (17.9):

$$0 \leq ||P||_2 \leq \frac{P_A n[4-\alpha-u+\beta]}{2(\alpha+u)} + \frac{P_A\sqrt{n^2[4-\alpha-u+\beta]^2 + 4c(\alpha+u)^2/P_A^2}}{2(\alpha+u)}, \tag{17.10}$$

where

$$c = \sum_{i=1}^{n}|P_{i,1}|^2 + \sum_{i=1}^{n}|P_{i,n}|^2 + \sum_{j=1}^{n}|P_{1,j}|^2.$$

□

Remark 17.1 In the light of Theorem 17.1, it is clear that α, u and β play important role in the boundedness states of $||P||_2$ because of (17.10).

Remark 17.2 Moreover, based on the boundedness effect of Theorem 17.1, we next further consider other control effect as the following two theorems and the following system.

When we use the method for system (17.6) as follows:

$$C_i(k, j) = 2P_{k-i,j} - P_{k-i,j-i} - P_{k-i,j+i} - P_{k-i,j}, \tag{17.11}$$

and

$$\begin{aligned} &D_i(k, j) \\ &= \frac{4\alpha(u)}{P_A[(1-\alpha-u+\beta)^2 - 2(1-\alpha-u+\beta)]}\Bigg(C_i(k, j) \\ &- \frac{P_A(1-\alpha-u+\beta)}{2(\alpha-u)}\Bigg), \end{aligned} \tag{17.12}$$

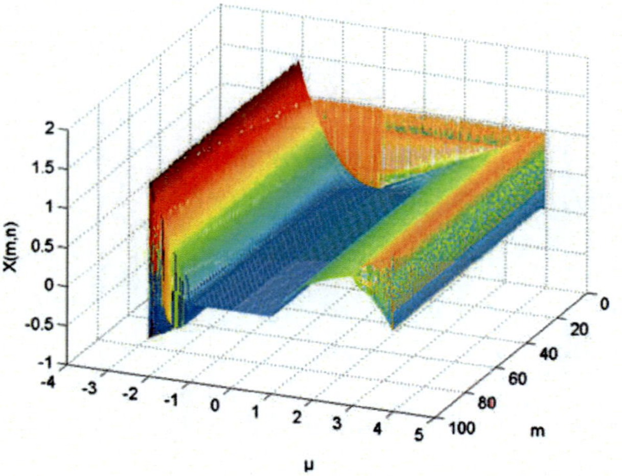

Fig. 17.1 The surface chaotic behavior of $D_i(k)$ as $\alpha + u = 0.7e^{-t} + 2.5, \beta = 0.4e^{-t}, t \in (0, 1)$, $N = 151$

system (17.6) becomes a special example for the following system as $i = 1$:

$$\begin{aligned} D_i(k, j) &= 1 - \frac{4(\alpha+u)^2}{P_A^2[(1-\alpha-u+\beta)^2 - 2(1-\alpha-u+\beta)]} \qquad (17.13)\\ &\cdot \frac{P_A^2[(1-\alpha-u+\beta)^2 - 2(1-\alpha-u+\beta)]^2}{(4(\alpha+u)^2} D_{i-1}(k, j)^2 \\ &= 1 - \frac{(\alpha+u-\beta)^2 - 1}{4} D_{i-1}(k, j)^2 \\ &= 1 - \mu D_{i-1}(k, j)^2. \end{aligned}$$

On the basis of surface chaos, the system (17.13) can display complex nonlinear behavior such as surface chaotic behavior shown in Fig. 17.1.

Moreover, the increasing effect of the periodic term on system (17.6) is obtained by the following two theorems.

Theorem 17.2 *The cell density P of the causative species in system (17.6) increases as time k increases when*

$$\begin{cases} \frac{\partial D_0(k,j)}{\partial k} > 0; \\ \frac{\partial(\alpha(k)+u(k))}{\partial k} < 0; \\ \alpha(k) + u(k) > \beta(k); \\ D_0(k, j)(\alpha + u(k) - \beta(k)) > 1; \\ (\alpha(k) + u(k))\frac{\partial \beta(k)}{\partial k} > \frac{\partial(\alpha(k)+u(k))}{\partial k}(\beta(k) + 1). \end{cases}$$

Proof According to the conditions (17.11) and (17.12), we have

$$
\begin{aligned}
P_{k,j} &= -D_0(k, j) \\
&\cdot \frac{P_A[(1-\alpha(k)-u(k)+\beta(k))^2 - 2(1-\alpha(k)-u(k)+\beta(k))]}{4(\alpha(k)+u(k))} \\
&- \frac{P_A(1-\alpha(k)-u(k)+\beta(k))}{2(\alpha(k)+u(k))}.
\end{aligned}
\tag{17.14}
$$

So we get

$$
\frac{\partial P(k, j)}{\partial k} = -\frac{\partial D_0(k, j)}{\partial k}\delta_1 - D_0(k, j)\frac{\partial \delta_1}{\partial k} - \frac{\partial \delta_2}{\partial k} \tag{17.15}
$$

owning to (17.14), where

$$
\delta_1 = \frac{P_A(\alpha(k)+u(k))}{4}\delta_0^2 - \frac{P_A}{2}\delta_0,
$$

$$
\delta_2 = \frac{P_a}{2}\frac{1-\alpha(k)-u(k)+\beta(k)}{\alpha(k)+u(k)},
$$

$$
\delta_0 = \frac{1-\alpha(k)-u(k)+\beta(k)}{\alpha(k)+u(k)},
$$

$$
\frac{\partial \delta_1}{\partial k} = \frac{P_A}{4}\frac{\partial(\alpha(k)+u(k))}{\partial k}\delta_0^2 + \frac{2P_A}{4}(\alpha(k)+u(k))\delta_0\frac{\partial \delta_0}{\partial k} - \frac{P_A}{2}\frac{\partial \delta_0}{\partial k},
$$

$$
\begin{aligned}
\frac{\partial \delta_2}{\partial k} &= \frac{P_A}{2}\frac{\delta_0}{\partial k}, \frac{\partial \delta_0}{\partial k} \\
&= \frac{(-\frac{\partial(\alpha(k)+u(k))}{\partial k} + \frac{\partial \beta(k)}{\partial k})(\alpha(k)+u(k)) - (1-\alpha(k)-u(k)+\beta(k))\frac{\partial(\alpha(k)+u(k))}{\partial k}}{(\alpha(k)+u(k))^2}.
\end{aligned}
$$

The relationship (17.15) means that

$$
\begin{aligned}
\frac{\partial P(k,j)}{\partial k} &= -\frac{P_A}{4}(\alpha(k)+u(k))\delta_0^2\frac{\partial D_0(k,j)}{\partial k} + \frac{P_A}{2}\delta_0\frac{\partial D_0(k,j)}{\partial k} \\
&- \frac{D_0(k,j)P_A}{4}\delta_0^2\frac{\partial(\alpha(k)+u(k))}{\partial k} \\
&+ \frac{P_A D_0(k,j)}{2(\alpha(k)+u(k))}(\alpha(k)+u(k))\delta_0\frac{\partial(\alpha(k)+u(k))}{\partial k} - \frac{P_A D_0(k,j)}{2(\alpha(k)+u(k))}(\alpha(k)+u(k))\delta_0\frac{\partial \beta}{\partial k} \\
&+ \frac{P_A D_0(k,j)}{2(\alpha(k)+u(k))^2}(\alpha(k)+u(k))(1-\alpha(k)-u(k)+\beta(k))\delta_0\frac{\partial(\alpha(k)+u(k))}{\partial k} \\
&- \frac{P_A D_0(k,j)}{2(\alpha(k)+u(k))}\frac{\partial(\alpha(k)+u(k))}{\partial k} + \frac{P_A D_0(k,j)}{2(\alpha(k)+u(k))}\frac{\partial \beta(k)}{\partial k} \\
&- \frac{P_A D_0(k,j)}{2(\alpha(k)+u(k))^2}(1-\alpha(k)-u(k)+\beta(k))\frac{\partial(\alpha(k)+u(k))}{\partial k} \\
&+ \frac{P_A}{2(\alpha(k)+u(k))}\frac{\partial(\alpha(k)+u(k))}{\partial k} \\
&- \frac{P_A}{2(\alpha(k)+u(k))}\frac{\partial \beta(k)}{\partial k} + \frac{P_A)}{2(\alpha(k)+u(k))^2}(1-\alpha(k)-u(k)+\beta(k))\frac{\partial(\alpha(k)+u(k))}{\partial k},
\end{aligned}
$$

and then

$$
\begin{aligned}
\frac{\partial P(k,j)}{\partial k} &= -\frac{\partial D_0(k,j)}{\partial k}\left(\frac{P_A(\alpha(k)+u(k))}{4}\delta_0^2 - \frac{P_A}{2}\delta_0\right) \\
&- D_0(k,j)\left[\frac{P_A}{4}\frac{\partial(\alpha(k)+u(k))}{\partial k}\delta_0^2 + \frac{2P_A}{4}(\alpha(k)+u(k))\delta_0\frac{\partial \delta_0}{\partial k} - \frac{P_A}{2}\frac{\partial \delta_0}{\partial k}\right] - \frac{P_A}{2}\frac{\partial \delta_0}{\partial k} \\
&= -\frac{P_A(\alpha(k)+u(k))}{4}\left[\frac{1-\alpha(k)-u(k)+\beta(k)}{(\alpha(k)+u(k))}\right]^2\frac{\partial D_0(k,j)}{\partial k} \\
&+ \frac{P_A}{2}\frac{1-\alpha(k)-u(k)+\beta(k)}{(\alpha(k)+u(k))}\frac{\partial D_0(k,j)}{\partial k} \\
&+ D_0(k,j)\frac{P_A}{4}\left[\frac{1-\alpha(k)-u(k)+\beta(k)}{(\alpha(k)+u(k))}\right]^2\frac{\partial(\alpha(k)+u(k))}{\partial k} - \frac{2P_A D_0(k,j)}{4}(\alpha(k) \\
&+ u(k))\frac{1-\alpha(k)-u(k)+\beta(k)}{\alpha(k)+u(k)} \\
&\cdot \frac{(-\frac{\partial(\alpha(k)+u(k))}{\partial k} + \frac{\partial \beta(k)}{\partial k})(\alpha(k)+u(k)) - (1-\alpha(k)-u(k)+\beta(k))\frac{\partial(\alpha(k)+u(k))}{\partial k}}{(\alpha(k)+u(k))^2} \\
&+ \frac{P_A D_0(k,j)}{2}\frac{(-\frac{\partial(\alpha(k)+u(k))}{\partial k} + \frac{\partial \beta(k)}{\partial k})(\alpha(k)+u(k)) - (1-\alpha(k)-u(k)+\beta(k))\frac{\partial(\alpha(k)+u(k))}{\partial k}}{(\alpha(k)+u(k))^2}\Bigg] \\
&- \frac{P_A}{2}\left[\frac{(-\frac{\partial(\alpha(k)+u(k))}{\partial k} + \frac{\partial \beta(k)}{\partial k})(\alpha(k)+u(k)) - (1-\alpha(k)-u(k)+\beta(k))\frac{\partial(\alpha(k)+u(k))}{\partial k}}{(\alpha(k)+u(k))^2}\right].
\end{aligned}
$$

At last, we obtain the following conditions for $\frac{\partial P(k,j)}{\partial k} > 0$:

$$\begin{cases} \frac{\partial D_0(k,j)}{\partial k} > 0; \\ \frac{\partial(\alpha(k)+u(k))}{\partial k} < 0; \\ (\alpha(k) + u(k)) > \beta(k); \\ D_0(k, j)(\alpha(k) + u(k) - \beta(k)) > 1; \\ (\alpha(k) + u(k))\frac{\partial \beta(k)}{\partial k} > \frac{\partial(\alpha(k)+u(k))}{\partial k}(\beta(k) + 1), \end{cases}$$

which can represent the increasing effect of the environmental factors on the cell density.

The growth rate defined as follows in Ref. [436]:

$$R_g = \frac{lnP_{t_1} - lnP_{t_0}}{t_1 - t_0},$$

is generally assumed to be constant. So we consider the more reasonable condition that the growth rate varies with time and then assume the following generalized growth rate

$$R_g(t, z) = \frac{lnP(t_1, z) - lnP(t_0, z)}{t_1 - t_0}.$$

Next theorem is about the control effect of the periodic factor on the growth rate in the discretized procee as $R_g(k, j) = lnP(k, j) - lnP(k - 1, j)$.

Theorem 17.3 *The growth rate $R_g(k, j)$ increases when the following conditions are satisfied:*

$$\begin{cases} \frac{\partial^2 D(k,j)}{\partial k^2} > 0; \\ \frac{\partial \beta(k)}{\partial k} > -\frac{1}{2}; \\ \frac{\partial^2(\alpha(k)+u(k))}{\partial k^2} < -\frac{[(\alpha(k)+u(k))\frac{\partial \beta(k)}{\partial k} - \frac{\partial(\alpha(k)+u(k))}{\partial k} - \beta(k)\frac{\partial(\alpha(k)+u(k))}{\partial k}]^2}{[(\alpha(k)+u(k))(1-\alpha(k)-u(k)+\beta(k))]^2}; \\ 1 + D(k, j)(\beta - \alpha(k) - u(k))) > 0; \\ (\alpha(k) + u(k))\frac{\partial \beta}{\partial k} > (\beta + 1)\frac{\partial(\alpha(k)+u(k))}{\partial k}. \end{cases}$$

Proof Due to the conditions (17.11) and (17.12), we have

$$\begin{aligned} R_g(k, j) &= lnP(k, j) - lnP(k - 1, j) \\ &= ln\frac{P(k,j)}{P(k-1,j)} \\ &= ln\frac{P(k,j)-P(k-1,j)+P(k-1,j)}{P(k-1,j)} \\ &\approx ln(\frac{\frac{\partial P(k,j)}{\partial k}}{P(k-1,j)} + 1). \end{aligned} \tag{17.16}$$

Then according to (17.16), it is clear that

$$\frac{\partial R_g(k,i)}{\partial k} \approx \frac{P(k,j)}{P(k,j)+\frac{\partial P(k,j)}{\partial k}} \frac{\frac{\partial^2 P(k,j)}{\partial k^2}P(k,j)-(\frac{\partial P(k,j)}{\partial k})^2}{P(k,j)^2}. \tag{17.17}$$

Because

$$\frac{\partial^2 P(k,j)}{\partial k^2} = -\frac{\partial^2 D(k,j)}{\partial k^2}\delta_1 - 2\frac{\partial D(k,j)}{\partial k}\frac{\partial \delta_1}{\partial k} - D(k,j)\frac{\partial^2 \delta_1}{\partial^2 k} - \frac{\partial^2 \delta_2}{\partial^2 k}$$
$$= -\frac{\partial^2 D(k,j)}{\partial k^2}\left[\frac{P_A(\alpha(k)+u(k))}{4}\delta_0^2 - \frac{P_A}{2}\delta_0\right]$$
$$-2\frac{\partial D(k,j)}{\partial k}\left[\frac{P_A\delta_0^2}{4}\frac{\partial(\alpha(k)+u(k))}{\partial k} + \frac{P_A(\alpha(k)+u(k))}{2}\delta_0\frac{\partial\delta_0}{\partial k} - \frac{P_A}{2}\frac{\partial\delta_0}{\partial k}\right]$$
$$-D(k,j)\left[P_A\delta_0\frac{\partial\delta_0}{\partial k}\frac{\partial(\alpha(k)+u(k))}{\partial k} + \frac{P_A}{4}\delta_0^2\frac{\partial^2(\alpha(k)+u(k))}{\partial k^2}\right.$$
$$+\frac{P_A}{2}(\alpha(k)+u(k))\left(\frac{\partial\delta_0}{\partial k}\right)^2$$
$$\left.+\frac{P_A}{2}(\alpha(k)+u(k))\delta_0\frac{\partial\delta_0^2}{\partial k^2} - \frac{P_A}{2}\frac{\partial\delta_0^2}{\partial k^2}\right] - \frac{P_A}{2}\frac{\partial\delta_0^2}{\partial k^2}$$

and after substituting it into (17.17), we get that

$$\frac{\partial R_g(k,j)}{\partial k} > 0,$$

when the following conditions are satisfied:

$$\begin{cases} \frac{\partial^2 D(k,j)}{\partial k^2} > 0; \\ \frac{\partial \beta}{\partial k} > -\frac{1}{2}; \\ \frac{\partial^2 \delta_0}{\partial k^2} < 0; \\ 1 + D(k,j)(\beta - \alpha(k) - u(k)) > 0; \\ 2(\alpha(k)+u(k))(\frac{\partial\delta_0}{\partial k})^2 + \delta^2\frac{\partial^2(\alpha(k)+u(k))}{\partial k^2} < 0, \end{cases}$$

and

$$\frac{\partial^2 P(k,j)}{\partial k^2}P(k,j) > \left(\frac{\partial P(k,j)}{\partial k}\right)^2, \tag{17.18}$$

which means that

$$\begin{cases} \frac{\partial^2 D(k,j)}{\partial k^2} > 0; \\ \frac{\partial \beta}{\partial k} > -\frac{1}{2}; \\ \frac{\partial^2(\alpha(k)+u(k))}{\partial k^2} < -\frac{[(\alpha(k)+u(k))\frac{\partial \beta}{\partial k} - \frac{\partial(\alpha(k)+u(k))}{\partial k} - \beta\frac{\partial(\alpha(k)+u(k))}{\partial k}]^2}{[(\alpha(k)+u(k))(1-\alpha(k)-u(k)+\beta)]^2}; \\ 1 + D(k, j)(\beta - \alpha(k) - u(k)) > 0; \\ (\alpha(k) + u(k))\frac{\partial \beta}{\partial k} > (\beta + 1)\frac{\partial(\alpha(k)+u(k))}{\partial k}. \end{cases}$$

Remark 17.3 Theorems 17.2 and 17.3 show the conclusion that the cell density and the growth rate of the causative species can be affected by the functional form of the birth rate, the controller and the death rate, meaning that if the conditions of Theorems 17.2 and 17.3 have been satisfied, the cell density and the growth rate will monotone increase. Furthermore, depending on the fact that the real growth process shows different characteristics for different algaes under different environments, we can conclude that our results play vital role in further understanding the complex characteristics of the growth process of the causative species.

17.4 Simulation Examples

Exponent growth is the famous form in the causative species growth phenomena so that we take the following form for the birth rate and the death rate as examples:

$$\alpha + u = A_1 e^{-\frac{t}{A_2}} + u = A_1 e^{-\frac{t}{A_2}} + A_3 sin\omega t,$$

$$\beta = A_4 e^{-\frac{t}{A_5}},$$

where $A_1 > 0$, $A_2 > 0$, $A_4 > 0$, $A_5 > 0$, A_3 and ω are constant.

In the light of Theorem 17.1, it is clear that the functional forms of α, u and β play important role in the variation of $||P||_2$, where

$$\delta = \frac{n(4 - \alpha - u + \beta)}{\alpha + u},$$

and the variation of δ is shown in Fig. 17.2. Besides this, the behavior of system (17.13) shows complex behavior as periodic factor changes illustrated in Fig. 17.3a, c and e and the chaotic characteristic which is calculated by Lyapunov exponent shown in Fig. 17.3b, d and f.

In addition, the cell density P and growth rate R_g can be controlled as increasing monotone by the variation of the periodic factor because of the relationships in Theorems 17.2 and 17.3. It is easy to obtain in terms of Theorem 17.2 that when we choose the time $t \in (0.1770, \pi/10)$, the increase of P can be controlled by the following condition:

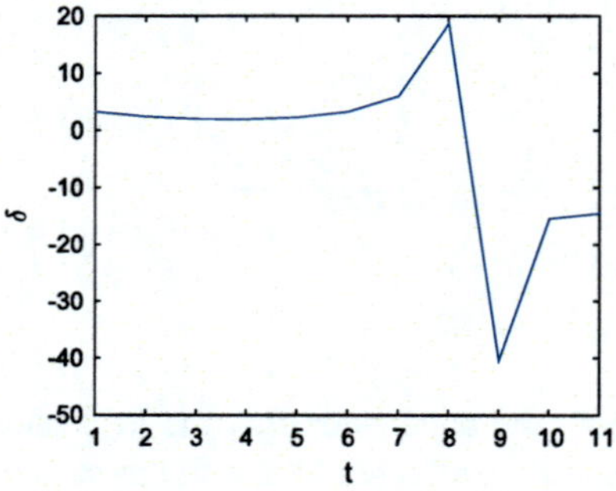

(a) $\alpha + u = 0.8e^{-t} + 0.5sin(5t)$, $\beta = 0.6e^{-t}$

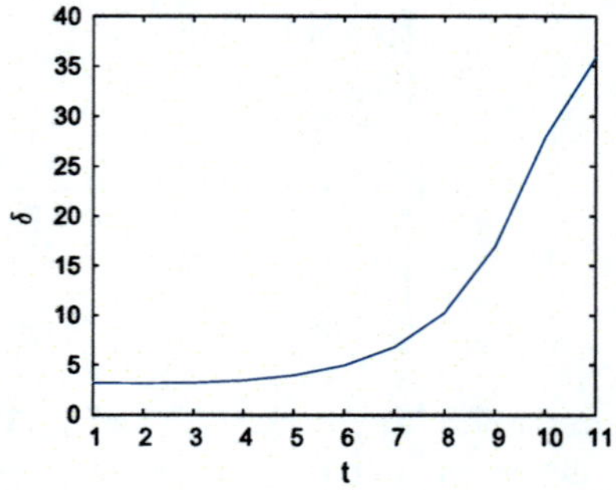

(b) $\alpha + u = 0.8e^{-t} + 0.2sin(5t)$, $\beta = 0.6e^{-t}$, $t \in (0, 1)$

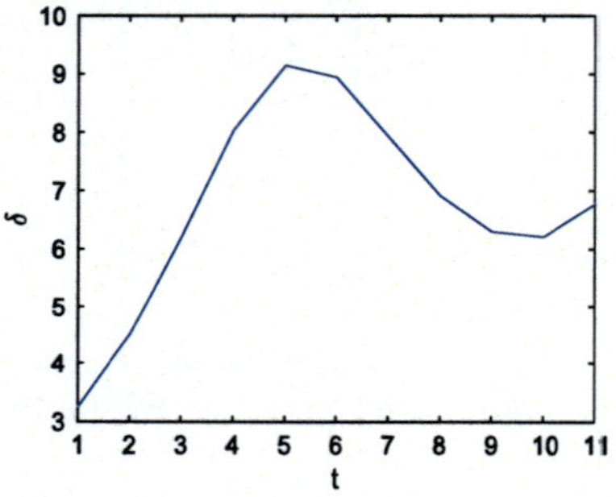

(c) $\alpha + u = 0.8e^{-t} - 0.2sin(5t)$, $\beta = 0.6e^{-t}$, $t \in (0, 1)$

Fig. 17.2 The variation of δ as the periodic factor changes

$$\frac{\partial(\alpha + u)}{\partial t} = -0.7e^{-t} + cos5t < 0,$$

where $cos5t > 0$, $sin5t > 0$, $\alpha + u = 0.7e^{-t} - 0.2sin5t, \beta = 0.4sin5t$,

$$\begin{aligned}&(\alpha + u)\tfrac{\partial \beta}{\partial t} - \tfrac{\partial(\alpha+u)}{\partial t}(\beta + 1)\\&= (0.7e^{-t} - 0.2sin5t)(-0.4e^{-t}) - (-0.7e^{-t} + cos5t)(0.4e^{-t} + 1)\\&= 0.08e^{-t}sin5t + 0.7e^{-t} + 0.4e^{-t}cos5t + cos5t > 0.\end{aligned}$$

For example, as the periodic factor increases, $D(k, j)$ decreases shown in Figs. 17.3a, b and c and 17.4. Meanwhile, the chaotic characteristic calculated by Lyapunov exponent of $D(k, j)$ increases illustrated by Fig. 17.3b, d and f.

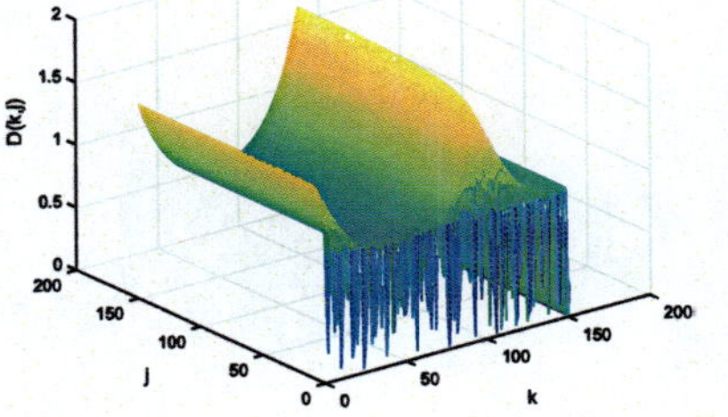

(a) $\alpha+u=0.7e^{-t}+1.1sin(5t)$, $\beta=0.4e^{-t}$, $t\in(0,1)$, $n=61$.

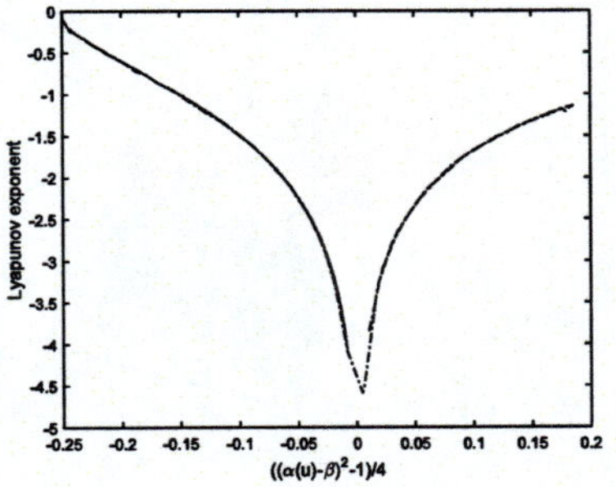

(b) $\alpha+u=0.7e^{-t}+1.1sin(5t)$, $\beta=0.4e^{-t}$, $t\in(0,1)$, $n=61$.

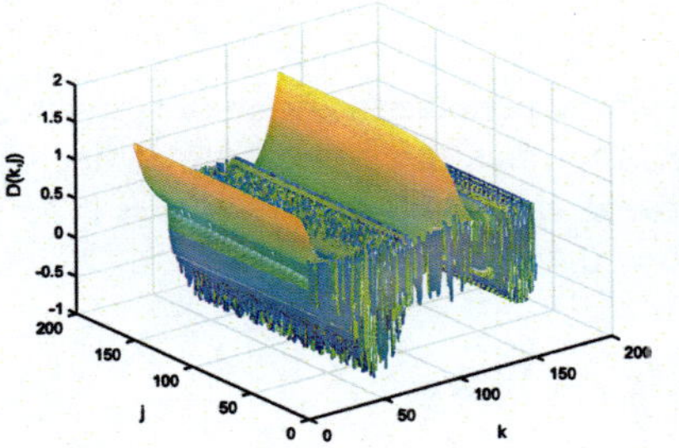

(c) $\alpha+u=0.7e^{-t}+2.6sin(5t)$, $\beta=0.4e^{-t}$, $t\in(0,1)$, $n=61$.

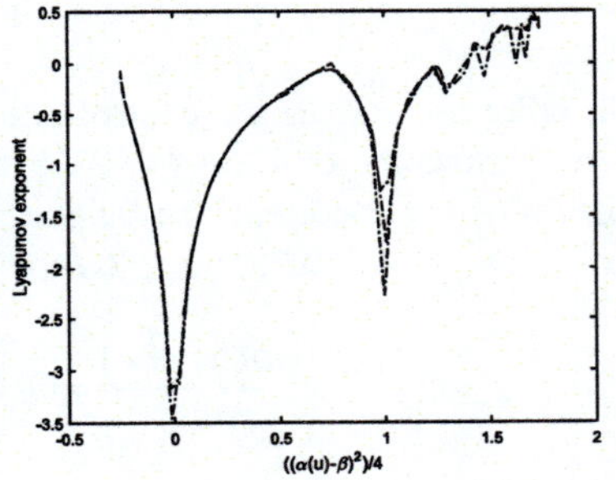

(d) $\alpha+u=0.7e^{-t}+2.6sin(5t)$, $\beta=0.4e^{-t}$, $t\in(0,1)$, $n=61$.

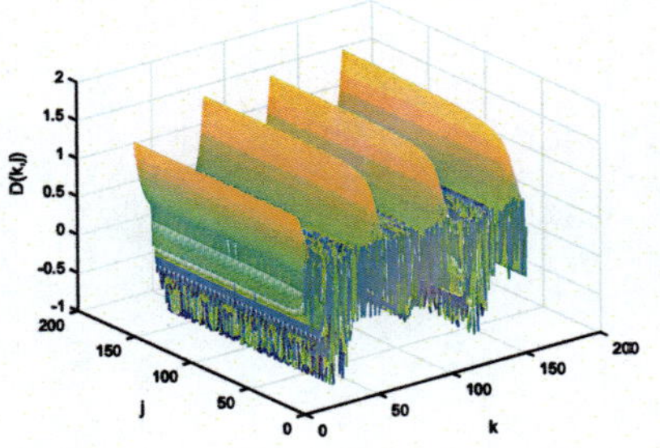

(e) $\alpha+u=0.7e^{-t}+2.6sin(10t)$, $\beta=0.4e^{-t}$, $t\in(0,1)$, $n=61$.

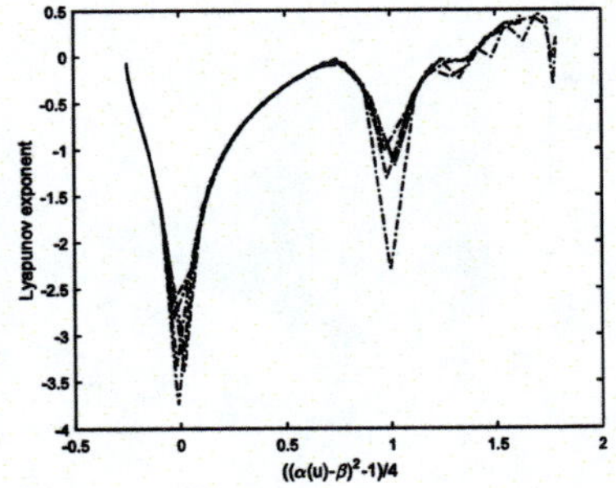

(f) $\alpha+u=0.7e^{-t}+2.6sin(10t)$, $\beta=0.4e^{-t}$, $t\in(0,1)$, $n=61$.

Fig. 17.3 The variation of $D(k,j)$ as the periodic factor changes

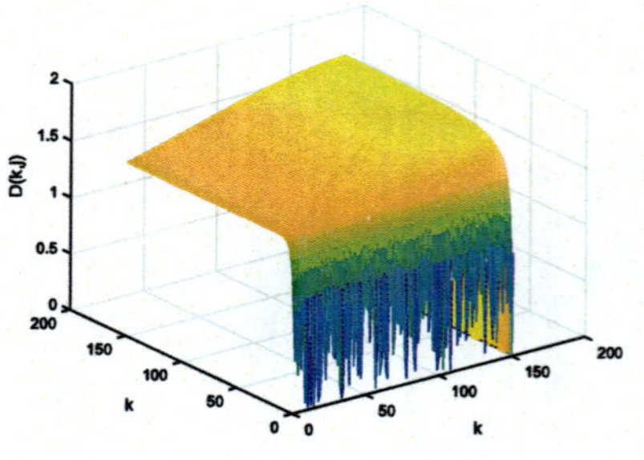

(a) $\alpha + u = 0.7e^{-t} - 0.2sin(5t)$, $\beta = 0.4e^{-t}$, $t \in (0.2, \pi/10)$

(b) $\alpha + u = 0.9e^{-t} - 0.2sin(5t)$, $\beta = 0.4e^{-t}$, $t \in (0.2, \pi/10)$

Fig. 17.4 The variation of $D(k, j)$ and its Lyapunov exponent as the periodic factor changes

In order to simulate the control effect of the periodic factor on the growth rate R_g, we choose the time $t \in (\pi/10, \pi/5)$. After that, the growth rate R_g increases as the following conditions are satisfied: $cos5t < 0$, $sin5t > 0$, $\alpha_1 + u = 0.7e^{-t} + 2.1sin5t$, $\alpha_2 + u = 0.7e^{-t} + 2.3sin5t$, $\beta = 0.4sin5t$,

$$\frac{\partial(\alpha_1 + u)}{\partial t} = -0.7e^{-t} + 10.5cos5t,$$

$$\frac{\partial^2(\alpha_1 + u)}{\partial t^2} = 0.7e^{-t} - 52.5sin5t < 0,$$

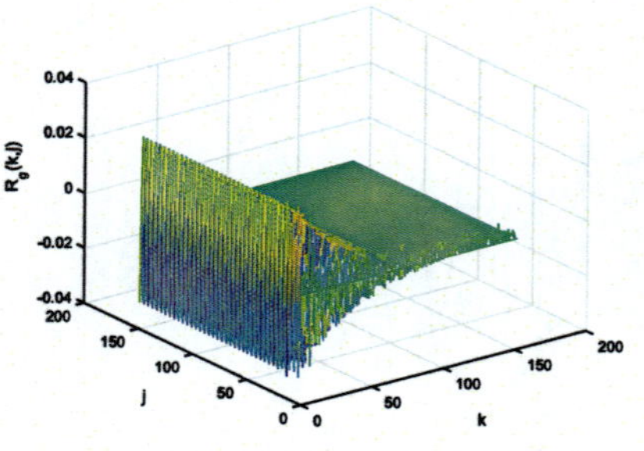

(a) $\alpha + u = 0.7e^{-t} + 2.3sin(5t)$, $\beta = 0.4e^{-t}$, $t \in (\pi/10, \pi/5)$

(b) $\alpha + u = 0.7e^{-t} + 2.1sin(5t)$, $\beta = 0.4e^{-t}$, $t \in (\pi/10, \pi/5)$

Fig. 17.5 The monotone increasing effect of the periodic factor u on $D(k, j)$

$$(\alpha_1 + u)\frac{\partial \beta}{\partial t} - \frac{\partial(\alpha_1 + u)}{\partial t}(\beta + 1)$$
$$= (0.7e^{-t} + 2.1sin5t)(-0.4e^{-t}) - (-0.7e^{-t} + 10.5cos5t)(0.4e^{-t} + 1)$$
$$= -0.84e^{-t}sin5t + 0.7e^{-t} - 4.2e^{-t}cos5t - 10.5cos5t > 0,$$

$$(\alpha_2 + u)\frac{\partial \beta}{\partial t} - \frac{\partial(\alpha_2 + u)}{\partial t}(\beta + 1)$$
$$= (0.7e^{-t} + 2.3sin5t)(-0.4e^{-t}) - (-0.7e^{-t} + 11.5cos5t)(0.4e^{-t} + 1)$$
$$= -0.92e^{-t}sin5t + 0.7e^{-t} - 4.6e^{-t}cos5t - 11.5cos5t > 0.$$

Figures 17.5 and 17.6 illustrate the monotone effect of the periodic factor on the growth rate.

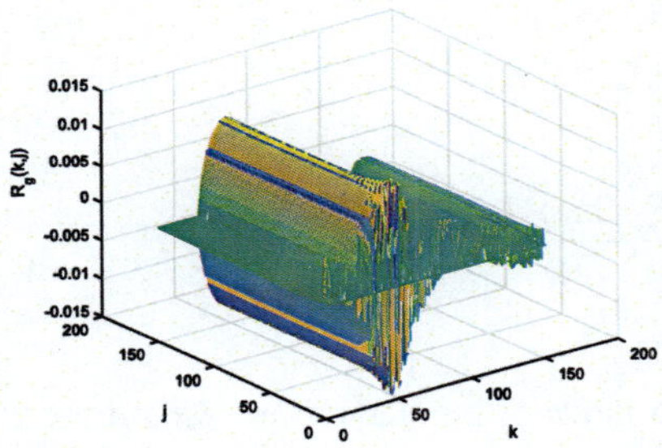

(a) $\alpha + u = 0.7e^{-t} + 2.2sin5t$, $\beta = 0.4e^{-t}$, $t \in (0, 1)$, $n = 61$.

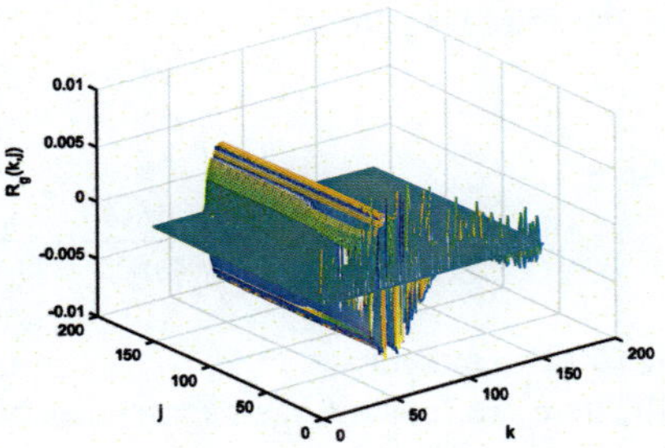

(b) $\alpha + u = 0.7e^{-t} + 1.9sin(5t)$, $\beta = 0.4e^{-t}$, $t \in (0, 1)$, $n = 61$.

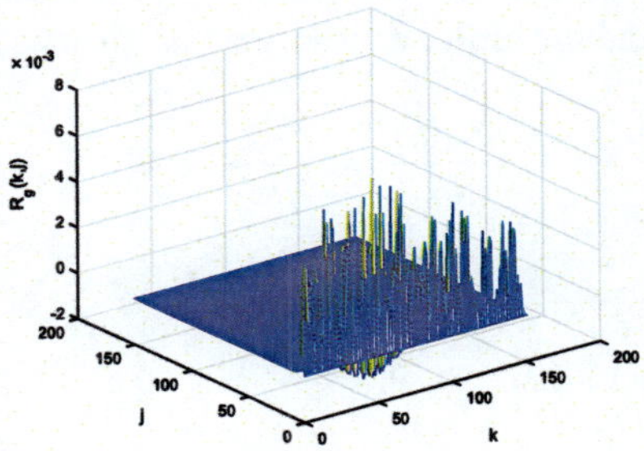

(c) $\alpha + u = 0.7e^{-t} + 1.1sin5t$, $\beta = 0.4e^{-t}$, $t \in (0, 1)$, $n = 61$.

Fig. 17.6 The monotone effect of the periodic factor u on the growth rate $\delta_g(k, j)$, where the line type named Control means the cell density without controller u and the other four line types stand for the cell density with different nitrogen source

17.5 Experimental Example

Besides, for the sake of showing the control effectiveness of our method, we use the real experimental data in Ref. [437] to test our model.

In the experiment, the conditions satisfying our model are as follows:
(1) the temperature, the illumination intensity, the ratio of light from the darkness, the salinity and the concentration of the different nitrogen sources are fixed as (24 ± 1)°C, 120 μmol · m^{-2} · s^{-1}, $12L : 12D$, 30.5 and 100 μmol · L^{-1}, respectively. Then the controller of the experiment is the different nitrogen source. The one-time deployment of nitrogen means the density of the different nitrogen sources decreases as time increases shown as $\frac{\partial u}{\partial t} < 0$.
(2) Because we consider the periodic control effect on the algal growth, we can assume that the monotone decreasing and positive function $u(t)$ in the experiment has the following form:

$$u(t) = Asin(\omega t + \theta),$$

where

$$t \in \left[\frac{2n + 0.5}{\omega}\pi - \frac{\theta}{\omega}, \frac{2n + 1}{\omega}\pi - \frac{\theta}{\omega}\right],$$

$n \in \mathbf{Z}$ as an integer set.
(3) There are four nitrogen sources as two kinds of inorganic nitrogen: nitrate and ammonium and two organic nitrogen: urea and mixed amino acids which have twenty basic amino acids and each is 5 μmol · L^{-1}. Moreover, the diffusion of the controller as the different nitrogen sources can be also regarded as the nitrogen uptake of the algae.
(4) Also, the process of the experiment holds on as the growth is the exponential growth so that we can assume that the variation of the birth rate and the death rate are positive constant and the birth rate is larger than the death rate shown as

$$\frac{\partial \alpha(k)}{\partial k} = \frac{\partial \beta(k)}{\partial k} = 0,$$

$$\alpha(k) > \beta(k) > 0.$$

Consequently we have that

$$\frac{\partial(\alpha(k) + u(k))}{\partial k} = \frac{\partial u(k)}{\partial k} < 0,$$

$$\alpha(k) + u(k) > \beta(k),$$

$$(\alpha(k) + u(k))\frac{\partial \beta(k)}{\partial k} = 0 > \frac{\partial u(k)}{\partial k}(\beta(k) + 1) = \frac{\partial(\alpha(k) + u(k))}{\partial k}(\beta(k) + 1)$$

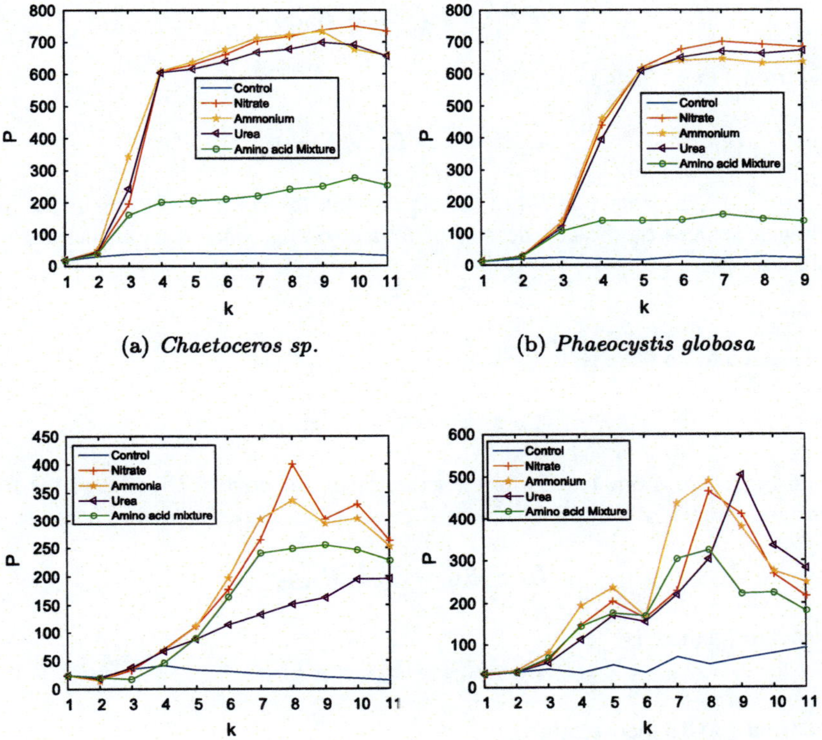

(a) *Chaetoceros sp.* (b) *Phaeocystis globosa*

(c) *Karenia sp.* (d) *Heterosigma akashiwo*

Fig. 17.7 The monotone effect of the periodic factor u on the cell density P which is simulated by the experimental data in Ref. [437], where the line type named as Control means the cell density without controller u and the other four line types stand for the cell density with different nitrogen sources

which satisfies the conditions of the Theorem 17.2.

Furthermore, the monotone increasing effect of the cell density P is shown clearly in Fig. 17.7a for 9 controlled days, in Fig. 17.7b for 7 controlled days, in Fig. 17.7c for 8 controlled days and in Fig. 17.7d for 5 controlled days. In Fig. 17.7, the density of the four algae with controllers as four different nitrogen sources is larger than that without these controllers. Moreover, the velocity of the increasing density of the four different algae and the controlled time are different for four nitrogen sources, respectively. And the growth mean velocity quantity V of the four algae in the controlled time is listed as follows:

$$V_{Chaetocerossp} > V_{Phaeocystisglobosa} > V_{Kareniasp} > V_{Heterosigmaakashiwo}.$$

In Fig. 17.7a, from 1 to 4 d in the experiment, the control effect E of the four nitrogen sources on the cell density of *Chaetoceros sp.* can be compared as follows:

$$E_A > E_U > E_N > E_{ACM}$$

and from 4 to 9 d there is

$$E_A > E_N > E_U > E_{ACM}.$$

In Fig. 17.7b, from 1 to 5 d in the experiment, the control effect E of the four nitrogen sources on the cell density of *Phaeocystis globosa* can be compared as follows:

$$E_A > E_N > E_U > E_{ACM}$$

and from 5 to 7 d there is

$$E_N > E_U > E_A > E_{ACM}.$$

In Fig. 17.7c, from 1 to 5 d in the experiment, the control effect E of the four nitrogen sources on the cell density of *Karenia sp.* can be compared as follows:

$$E_A > E_N > E_U > E_{ACM},$$

from 5 to 7 d there is

$$E_A > E_N > E_{ACM} > E_U$$

and from 7 to 8 d there exists

$$E_N > E_A > E_{ACM} > E_U.$$

In Fig. 17.7d, from 1 to 4 d in the experiment, the control effect E of the four nitrogen sources on the cell density of *Heterosigma akashiwo* can be compared as follows:

$$E_A > E_{ACM} > E_N > E_U$$

and from 4 to 5 d there is

$$E_A > E_N > E_{ACM} > E_U,$$

respectively.

Besides this, we consider the control effect of the controller u on the growth rate $R_g(k)$. Also because that the process of the experiment holds on as the growth is the exponential growth, we can assume that the variation of the birth rate and the death rate are positive constant and the birth rate is larger than the death rate. Then we have the following inequality:

$$\frac{\partial \beta(k)}{\partial k} = 0 > -\frac{1}{2},$$

$$\frac{\partial^2(\alpha(k)+u(k))}{\partial k^2}=\frac{\partial^2 u(k)}{\partial k^2}$$
$$<-\frac{[(\alpha(k)+u(k))\frac{\partial\beta(k)}{\partial k}-\frac{\partial(\alpha(k)+u(k))}{\partial k}-\beta(k)\frac{\partial(\alpha(k)+u(k))}{\partial k}]^2}{[(\alpha(k)+u(k))(1-\alpha(k)-u(k)+\beta(k))]^2},$$

$$(\alpha(k)+u(k))\frac{\partial\beta}{\partial k}=0>(\beta+1)\frac{\partial u(k)}{\partial k}=(\beta+1)\frac{\partial(\alpha(k)+u(k))}{\partial k},$$

which satisfies the conditions of the Theorem 17.3. Next, the monotone increasing effect of the growth rate $R_g(k)$ is shown clearly in Fig. 17.8a for 9 controlled days and in Fig. 17.8b for 7 controlled days. As for Fig. 17.8c and d, the monotone increasing effect of the controller on the growth rate is not obvious. But the control effect of the Ammonium, Amino acid Mixture and Urea on the growth rate of the *Karenia sp.* can be shown distinct from 3 to 6 d in Fig. 17.8c. And the control effect of the Ammonium and Amino acid Mixture on the growth rate of the *Heterosigma akashiwo*

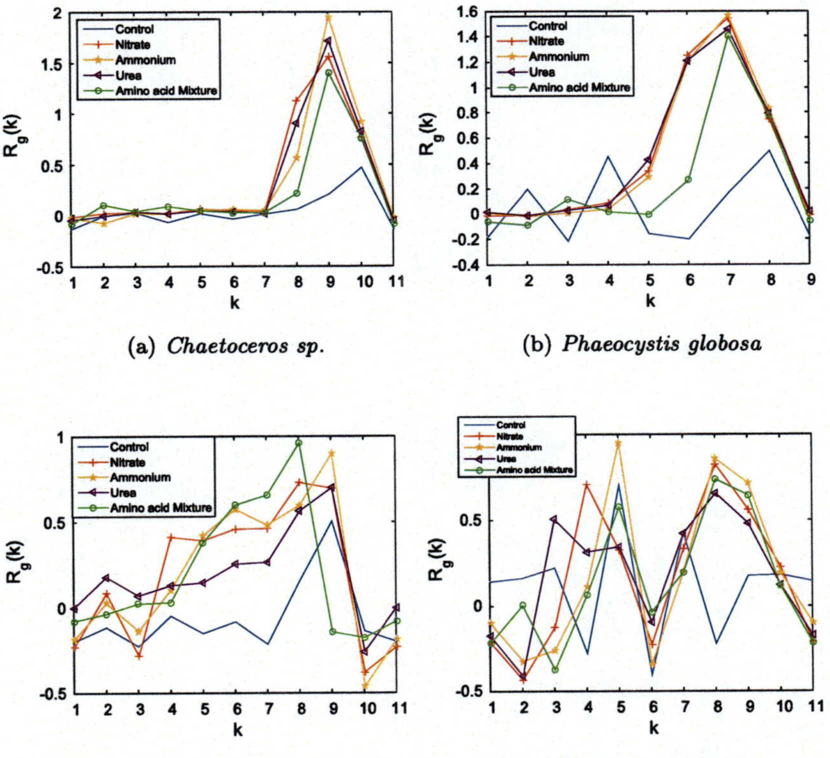

(a) *Chaetoceros sp.* (b) *Phaeocystis globosa*

(c) *Karenia sp.* (d) *Heterosigma akashiwo*

Fig. 17.8 The monotone effect of the periodic factor u on the growth rate $R_g(k)$

can be shown illustrated from 3 to 5 d in Fig. 17.8d. Moreover, the variation of the growth rate of *Heterosigma akashiwo* according to the four different nitrogen sources likes a shape of 'M'.

Chapter 18
Surface Chaos-Based Image Encryption Design

During the last decades, the use of chaos in cryptography is of great interest in many areas, including a database, Internet transaction banking, multimedia systems, medical imaging, because chaotic systems have several significant advantages favorable to secure communications [439, 440], such as aperiodicity (useful for one time pad cipher), sensitive dependence on initial conditions and system parameters(useful for confusion and diffusion processes), ergodicity and random-like behaviors(useful for producing output with satisfactory statistics). Chaotic communication schemes are based either on discrete or continuous systems. Many cryptosystems based on continuous systems utilize the idea of synchronization of chaos [442–444]. However, recent studies show that the performance of these communication schemes is very poor and most models of chaos communications are insecure [445]. Recently, much attention has been given to chaotic communication schemes based on discrete systems, many encryption schemes have been proposed based on different principles [446–465]. The basic ideas can be classified into the following major types: value transformation, position permutation, and their combining form. Moreover, multiple chaotic systems, high-dimensional chaotic systems, multiple iterations of chaotic systems and other techniques have been proposed. But some of them have been known to be insecure, various cryptanalysis have exposed some inherent drawbacks of chaotic cryptosystems. To improve the degree of the privacy, spatiotemporal chaotic systems are widely used in chaotic cryptography for the excellent chaotic dynamical properties. A spatiotemporal chaotic system is a spatially extended system that exhibits spatiotemporal chaos, i.e., nonlinear dynamics in both space and time. Coupled map lattices (CMLs) are often adopted as the basic model of a spatiotemporal chaos systems. Wang et al. have shown that the communication with the CMLs is more secure than the communication with a single map.

In this chapter, a new design of a class of chaotic cryptosystems is suggested to overcome the aforementioned drawbacks by using high dimensional chaotic map and

S. T. Liu and L. Zhang, *Surface Chaos and Its Applications*,
https://doi.org/10.1007/978-981-16-8229-2_18

some conventional cryptography techniques for obtaining high level security. Here we use high-dimensional chaos as the basic structure of the cryptography, which leads to the following significant advantages: due to the high-dimensionality and chaoticity, the output ciphertext has high complexity, long periodicity of computer realization of chaos, and effective byte confusion and diffusion in many directions in the variable space. All these properties are favorable to achieve high practical security.

The spatial generalized 2-D system difference form as follows:

$$x_{m+1,n} + \omega x_{m,n+1} = f(\mu, (1+\omega)x_{m,n}) \tag{18.1}$$

where $f(\mu, (1+\omega)x_{m,n})$ is nonlinear function, m, n, x_{mn} are three dimensional space geometric coordinates, μ is a real parameters, ω is a constant. And in fact, system (18.1) can be regarded as a discrete analog of the following functional partial differential system:

$$\frac{\partial v}{\partial x} + \omega \frac{\partial v}{\partial y} = f(\mu, (1+\omega)v). \tag{18.2}$$

System (18.2) is a *convection equation* with a forced term, quite classical in physics. Therefore, qualitative properties of system (18.1) may lead to some useful information for analyzing this companion partial differential system.

Particulary, when $f(\mu, (1+\omega)x_{m,n}) = 1 - (\mu(1+\omega)x_{m,n})^2$, the 2-D discrete dynamic system is described as:

$$x_{m+1,n} + \omega x_{m,n+1} = 1 - (\mu(1+\omega)x_{m,n})^2. \tag{18.3}$$

Research shows that when $2 > \mu \geq 1.55$, $\omega \in (-1, 1)$, the system is in chaotic state. Since there are two iterations variables, it can be used the surface generalized 2-D standard Logistic system.

The surface chaotic system is a generalization of one-dimension chaotic map. Since there are m and n two variables in spatial chaotic map, the encrypting sequences produced by this system is more complex and random-like than the one-dimension chaotic systems (only one variable m). It is more difficult to forecast such chaotic sequences. Therefore, The surface chaotic system is more advantage than the low-dimension chaotic system.

Compared with one-dimensional Logistic chaotic system, not only there are two control parameters μ, ω in urface chaotic system, but also simulation shows that the generalized spatial dynamic system orbit is extremely sensitive to the parameters.

The above surface chaotic system generates chaotic behaviors for a wide parameter range, so the parameters μ, ω and initial values can all be used as secret keys. Specially when the parameters $\mu = 1.75$, $\omega = -0.05$, the chaotic and bifurcation behavior of system (18.3) are shown in Figs. 18.1, 18.2 and 18.3.

In the proposed method, three different parameter values spatial chaotic maps are employed to achieve the goal of image encryption. One is when $\mu_1 = 1.6$, $\omega =$

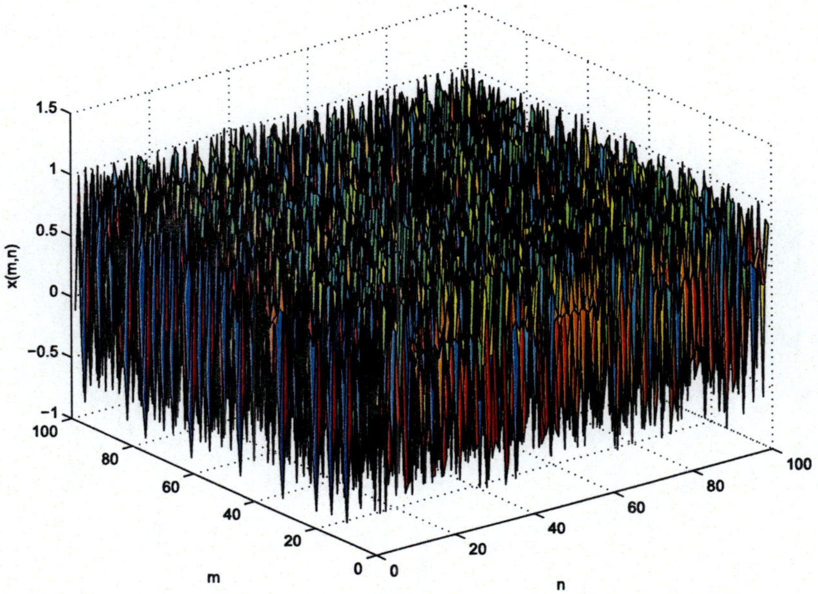

Fig. 18.1 Chaotic behavior of spatial chaotic system when $\mu = 1.75, \omega = -0.05$

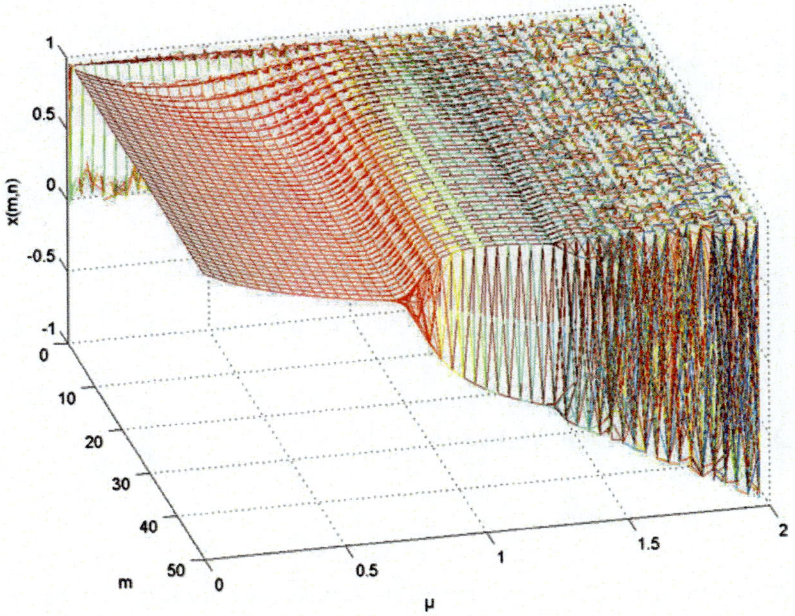

Fig. 18.2 Bifurcation behavior of system (18.3) in space

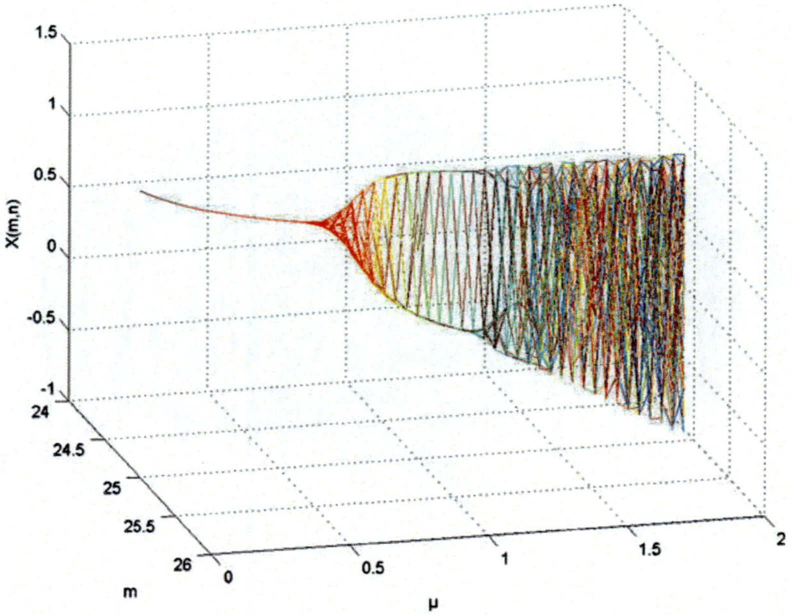

Fig. 18.3 A section of spatial bifurcation phenomena of system (18.3)

-0.05, the other is when $\mu_2 = 1.75$, $\omega = -0.05$, the third one is when $\mu_3 = 1.78$, $\omega = -0.05$. The corresponding equations are as follow:

$$x_{m+1,n} - 0.05x_{m,n+1} = 1 - 1.6[(1 - 0.05)x_{m,n}]^2, \tag{18.4}$$

$$y_{m+1,n} - 0.05y_{m,n+1} = 1 - 1.75[(1 - 0.05)y_{m,n}]^2, \tag{18.5}$$

$$z_{m+1,n} - 0.05z_{m,n+1} = 1 - 1.78[(1 - 0.05)z_{m,n}]^2. \tag{18.6}$$

18.1 The Proposed Method

18.1.1 Permutation

According to Golomb's three postulates for pseudo-random sequence, idea chaotic sequences should have such statistical characteristics as following: Its average value is zero, the autocorrelation is delta function, and the mutual correlation is zero. A demonstration of autocorrelation and cross-correlation properties of the achieved sequence $\{x_{i,j}\}$ by Eq. (18.4) is given in Fig. 18.4. Figure 18.4a shows the autocor-

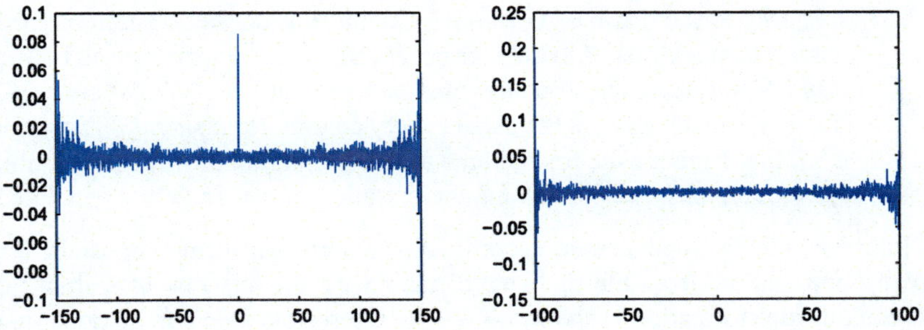

(a) (b)

Fig. 18.4 Autocorrelation characteristics and Cross-correlation characteristics

relation characteristics, Fig. 18.4b shows the cross-correlation characteristics of two sequences only with a low change($< 10^{-10}$) of initial value, which indicates $\{x_{i,j}\}$ is sensitive to the secret key. One can know that $\{y_{i,j}\}$ and $\{z_{i,j}\}$ are also sensitive to the secret key in the same way.

Image data have strong correlations among adjacent pixels, in order to disturb the high correlation among pixels, a higher-dimensional spatial chaotic map is used to shuffle the position of the plain-image. Without loss of generality, we assume that the dimension of the plain-image $M \times N$, the position matrix of pixels is $Q_{i,j}$, $i = 0, 1, \ldots, M-1$; $j = 0, 1, \ldots, N-1$ the procedure for shuffling the position of pixels is described as follows

Step 1: For system (18.4) and a given $x_{0,0}$, after do some iterations, $x_{m,n}$ is derived, as a new $x_{0,0}$, then let

$$h = mod(x_{0,0} \times 10^{14}, M). \tag{18.7}$$

Obviously, $h \in [0, M-1]$. Continue to do the iteration of spatial chaotic map and do (18.7) until one gets M different data which are all between 0 and $M-1$, these data can be recorder in the form of $h_i, i = 1, 2, \ldots, M$, where $h_i \neq h_j$ if $i \neq j$. Then rearrange the row of matrix $Q_{i,j}$ according to $h_i, i = 1, 2, \ldots, M$, that is, move the h_1 row to the first row, h_2 row to the second row, thus a new position matrix $Q^h_{i,j}$ is generated based on the transformation.

For the new matrix $Q^h_{i,j}$, we will produce column shuffling matrix column by column. The process is presented next.

Step 2: Using the present $x_{0,0}$ to do the iteration of (18.4), then let

$$c = mod(x_{0,0} \times 10^{14}, N). \tag{18.8}$$

It is easy to see that $c \in [0, N-1]$. Continue to do the iteration of spatial chaotic map and do (18.8) until we get N different data which are all between 0 and $N-1$, these data can be expressed $c_i, i = 1, 2, \ldots, N$, where $c_i \neq c_j$ if $i \neq j$. Then rearrange the data of every column for matrix $Q^h_{i,j}$ according to c_i, that is, move the first column to the c_1 column, the second column to c_2 column, thus a new column transformation of matrix $Q^h_{i,j}$ is generated.

In order to further enhance the security, the column transformation can be done line by line, that is, from the first row of matrix till the last row of matrix $Q^h_{i,j}$, the column transformation in the same way as the second step can be done under different $x_{0,0}$.

18.1.2 Substitution

The encryption process consist of three steps of operation.

Step 1: The each element of the pixels of the shuffled image is the decimal grey value, convert decimal pixel values to binary numbers and get a new $M \times N$ matrix $B(i, j)$.

Step 2: Choose two arbitrary initial conditions $y'_{0,0}, z'_{0,0}$, and quantize $y'_{0,0}, z'_{0,0}$ to binary values $y_{0,0}, z_{0,0}$. The two binary chaotic sequences $\{y_{i,j}\}, \{z_{i,j}\}$ are generated corresponding by the Eqs. (18.5) and (18.6). Respectively intercept finite sequences from $\{y_{i,j}\}$ and $\{z_{i,j}\}$ to construct two $M \times N$ matrix Y and Z.

Step 3: The substitution of $C(i, j)$ is defined by:

$$C(i, j) = (((B(i, j) + Y(i, j))mod2) \otimes Z(i, j)$$

where $0 \leq i \leq M-1, 0 \leq j \leq N-1$.

The operation is implemented until all the elements in the image are encrypted. The decryption process is evident. Conversely execute the encryption process and the decrypted image will be obtained.

18.2 Experimental Results and Security Analysis

18.2.1 Key Sensitivity Test

Experimental analysis of the proposed encryption scheme has been done with several images. Figure 18.5 is the 256 grey-scale plain-image and encrypted image of size 256×256 with the key $K = (x_{0,0} = y_{0,0} = z_{0,0} = 0.98; \omega = -0.05; \mu_1 =$

Fig. 18.5 Plain image and encrypted experimental result: **a** plain image, **b** encrypted image

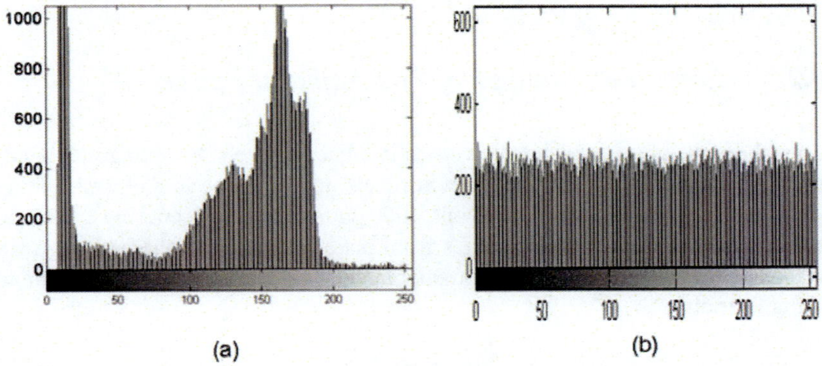

Fig. 18.6 Histograms of the plain-image and encrypted image: **a** histograms of plain-image, **b** histogram of encrypted image

1.6, $\mu_2 = 1.75$, $\mu_3 = 1.78$). The histograms of the plain-image and the corresponding encrypted image are shown in Fig. 18.6. As can be seen, the encrypted histogram is fairly uniform. Figure 18.7a is the decrypted image using the right keys. It can been seen that the decrypted image is clear and correct without any distortion. But if we use the wrong keys $K = (x_{0,0} = y_{0,0} = z_{0,0} = 0.98; \omega = -0.05; \mu_1 = 1.6, \mu_2 = 1.75, \mu_3 = 1.78000000000001)$ or $K = (x_{0,0} = y_{0,0} = z_{0,0} = 0.98000000000001; \omega = -0.05; \mu_1 = 1.6, \mu_2 = 1.75, \mu_3 = 1.78)$, two unexpected images will be got, Fig. 18.7b shows the decrypted image using the incorrectly key. So it can be concluded that the spatial chaotic encryption algorithm is sensitive to the key, a small change of the key will generate a completely different encryption result and can not get the correct plain image. As a result, differential attack would become very inefficient and practically useless.

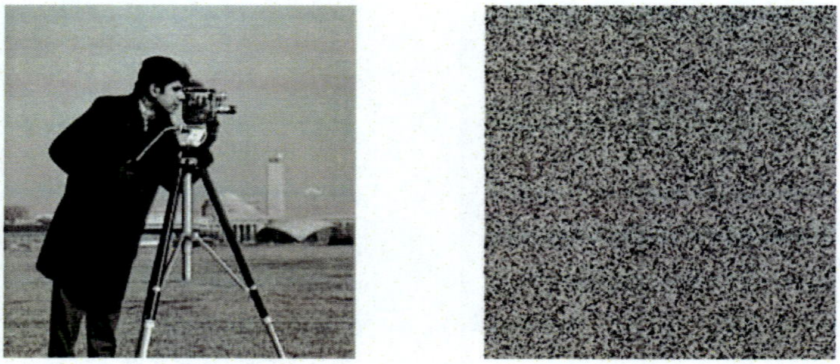

(a) (b)

Fig. 18.7 Decrypted image: **a** decrypted image by correct key, **b** decrypted image by wrong key $K = (x_{0,0} = y_{0,0} = z_{0,0} = 0.98; \omega = -0.05; \mu_1 = 1.6, \mu_2 = 1.75, \mu_3 = 1.78000000000001)$

18.2.2 Correlation Analysis of Two Adjacent Pixels

To test the correlation between two vertically adjacent pixels, two horizontally adjacent pixels, and two diagonally adjacent pixels, respectively, in a ciphered image, the following procedure was carried out. 1000 pairs of two adjacent (in vertical, horizontal, and diagonal direction) pixels from plain-image and ciphered image were randomly selected and the correlation coefficients were calculated by using the following formulas:

$$E(x) = \frac{1}{N}\sum_{i=1}^{N} x_i, \tag{18.9}$$

$$D(x) = \frac{1}{N}\sum_{i=1}^{N}(x_i - E(x))^2, \tag{18.10}$$

$$cov(x, y) = \frac{1}{N}\sum_{i=1}^{N}(x_i - E(x))(y_i - E(y)), \tag{18.11}$$

$$r_{xy} = \frac{cov(x, y)}{\sqrt{D(x)D(y)}}, \tag{18.12}$$

where x and y are grey-scale values of two adjacent pixel in the image. Figure 18.8a and b shows the correlation distribution of two horizontally adjacent pixels in the plain-image and that in the cipher-image seperately. The correlation coefficients are shown in Table 18.1. These correlation analysis prove that the chaotic encryption algorithm satisfy zero co-correlation.

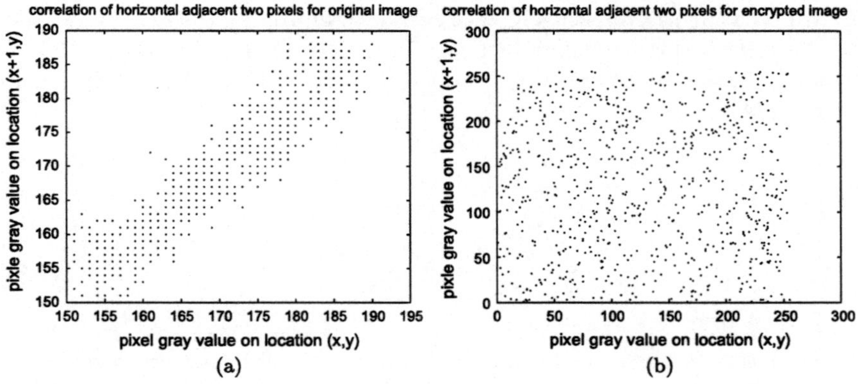

Fig. 18.8 Correlations of two horizontally adjacent pixels in the plain and cipher image

Table 18.1 Correlation coefficient of two adjacent pixels in plain-image and ciphered image

	Plain-image	Cipher image
Horizontal	0.994948	2.44573E-07
Vertical	0.961046	-9.1226E-08
Diagonal	0.956190	9.2465E-08

18.2.3 Key Space Analysis

A good encryption scheme should be sensitive to the secret keys, and the key space should be large enough to make brute-force attacks infeasible. In this algorithm, the initial conditions and parameters of three equations can be used as keys. If the precision is 10^{-14}, the key space can reach to 10^{126}. One can see the key space is large enough to resist the attacks.

18.2.4 Differential Attack

As a general requirement for all the image encryption schemes, the encrypted image should be greatly different from its original form. To satisfy this requirement, two common measures, including the number of pixel change rate (NPCR) and unified average changing intensity (UACI), can be adopted. NPCR stands for the number of pixels change rate while one-pixel of plain image is changed. UACI measures the average intensity of differences between the plain image and ciphered image. For calculation of NPCR and UACI, let us assume two ciphered images C_1 and C_2 whose corresponding plain images have only one-pixel difference. The grey-scale values of the pixels of the ciphered image C_1 and C_2 at grid (i, j) are labeled as

$C_1(i, j)$ and $C_2(i, j)$, respectively. Take a array D, with the same size as image C_1 or C_2. Then, $D(i, j)$ is determined by $C_1(i, j)$ and $C_2(i, j)$. So, if $C_1(i, j) = C_2(i, j)$ then $D(i, j) = 1$; otherwise, $D(i, j) = 0$. NPCR and UACI are defined through the following formulas:

$$NPCR = \frac{\sum_{ij} D(i, j)}{W \times H} \times 100, \tag{18.13}$$

$$UACI = \frac{1}{W \times H}\left[\sum_{ij} \frac{|C_1(i, j) - C_2(i, j)|}{255}\right] \times 100, \tag{18.14}$$

where W and H are the width and height of C_1 or C_2. Tests have been performed on the proposed scheme, about the one-pixel change influence on a 256 grey-scale image of size 256×256. We obtained NPCR = 0.3870% and UCAI = 0.3042%. The results show that a swiftly change in the original image will result in a significant change in the ciphered image, so the algorithm proposed has a good ability to anti differential attack.

18.2.5 Other Attacks

As for known-plaintext and chosen-plaintext attacks. In these two attacks, the illegal users are assumed to have obtained several plainimage and cipherimage pairs, and all of these pairs share a common key K. In these cases, the outlaws can analyze these pairs to obtain the common key K, and correctly decrypt the next cipherimage if the sender still encrypts his next original image by K. To prevent these attacks, we define that our private key is disposable, i.e., it is requested to change the secret key frequently between each session of communication. Since no common key exists in our cryptosystem, no one can break our cryptosystem by the known-plaintext or the chosen-text attack.

18.3 A Novel Image Encryption Scheme Based on Surface Chaos Map

High sensitivity of chaotic systems to initial conditions and parameters implies strong cryptographic properties of chaotic cryptosystems that makes them robust against any statistical attacks. Random-like behavior and unstable periodic orbits with long periods, which are quite advantageous to ciphers.

So far, for realizing secure image transmission over the Internet and through wireless networks, a large variety of chaos-based cipher have been proposed, but many of them are rather disappointing. There are some following reasons. Firstly, low security since there exists dynamical degradation of chaotic systems in their realization with digital computers. Secondly, useful information can be extracted from the chaotic orbits if chaotic systems with simple constructions are directly used to encrypt image. In addition, some chaos-based ciphers have slow performance speeds due to analytical floating-point computations, which makes the ciphers infeasible in practice. To overcome these drawbacks, multiple chaotic systems, multiple iterations of chaotic systems and other methods have been proposed to improve these situations. Especially, using high dimensional chaotic systems is a significant achievement in this aspect.

For

$$x_{m+1,n} = (1-\varepsilon) f(x_{mn}) + \frac{\varepsilon}{2}[f(x_{m,n-1}) + f(x_{m,n+1})], \tag{18.15}$$

and

$$x_{m+1} = f(\mu, x_m), \tag{18.16}$$

assume that x_{mn} is a state at time m. If there exists a disturbance $\{dx_0(n)\}$, then a new state $\overline{x}_{mn}$ is obtained. Define that

$$\overline{x}_{mn} = x_{mn} + dx_0(n), \quad n = 1, 2, \ldots, L$$

and iterate the 2-D coupled map lattice model (18.1). Then one has $\{x_{m+k,n}\}$, $\{\overline{x}_{m+k,n}\}$, and their difference is $\{dx_k(n)\}$. The largest Lyapunov exponent for the 2-D discrete system (18.15) is computed

$$\lambda = \lim_{N\to\infty} \frac{1}{N} \sum_{j=1}^{N} \log \frac{\|dx_j\|}{\|dx_{j-1}\|}, \tag{18.17}$$

where $\|dx_j\| = \sqrt{\sum_{i=1}^{N} [dx_k(i)]^2}$, in which $dx_k = [dx_k(1), dx_k(2), \ldots, dx_k(N)]^T$. On the other hand, the Lyapunov exponent of the 1-D discrete system (18.16) is given as the following:

$$\lambda = \lim_{k\to\infty} \frac{1}{k} \sum_{j=1}^{k-1} \ln |f'(x_j)|. \tag{18.18}$$

Notice, however, that (18.17) and (18.18) have no close relationship. Therefore, it is important and of significance to relax the above restrictions and to unify the 1-D and 2-D cases:

1. System (18.15) can be more complex, with a general nonlinear function $f(x)$.
2. The lattice site index n can have either a finite or an infinite boundary.
3. The periodic boundary condition can be either on n, $x_{m,n+L} = x_{mn}$; or on m, $x_{m+L,n} = x_{mn}$; or on (m, n); $x_{m+L,n+L} = x_{mn}$; and it can also be a non-periodic boundary.

4. Dynamical characteristics and behaviors of a 2-D system should be a natural generalization of that for a 1-D system.

This section deals with the following 2-D system:

$$x_{m+1,n} + \omega x_{m,n+1} = f(\mu, (1+\omega)x_{mn}), \tag{18.19}$$

where $f(\mu, (1+\omega)x_{mn})$ is nonlinear function, ω is a constant, μ is a real parameters. In fact, system (18.19) can also be regarded as a discrete analog of the following functional partial differential system:

$$\frac{\partial v}{\partial x} + \omega\frac{\partial v}{\partial y} + \omega v = f(\mu, (1+\omega)v). \tag{18.20}$$

System (18.20) is a convection equation with a forced term, quite classical in physics. Therefore, qualitative properties of system (18.19) may lead to some useful information for analyzing this companion partial differential system.

On the other hand, when $n = n_0, \omega = 0$, the system can be reduced to one-dimensional model

$$x_{m+1,n_0} = f(\mu, x_{m,n_0}), \tag{18.21}$$

because n_0 is a constant, the above equation can rewrite as:

$$x_{m+1} = f(\mu, x_m). \tag{18.22}$$

This is a familiar one-dimensional model. Specially, when $f(\mu, x_m) = \mu x_m(1 - x_m)$, the equation is described as:

$$x_{m+1} = \mu x_m(1 - x_m), x_m \in (0, 1), \tag{18.23}$$

which is the well-known classical one-dimensional Logistic system. Research shows that the system is in chaotic state under the condition $3.7 < \mu \leq 4$.

When $f(\mu, (1+\omega)x_{m,n}) = 1 - (\mu(1+\omega)x_{m,n})^2$, the 2-D discrete dynamic system is described as:

$$x_{m+1,n} + \omega x_{m,n+1} = 1 - (\mu(1+\omega)x_{m,n})^2. \tag{18.24}$$

Research shows that when $2 > \mu \geq 1.55, \omega \in (-1, 1)$, the system is in chaotic state. It can be called the spatial generalized 2D standard Logistic system. Compared with one-dimensional Logistic chaotic system, not only there are two control parameters and wide parameters range in spatial chaotic system, but also simulation shows that the generalized spatial dynamic system orbit is extremely sensitive to the parameters and initial conditions. The chaotic behavior of system (18.24) is demonstrated in Figs. 18.9 and 18.10. The fundamental theories and a lot of research results of spatial chaos system (18.24).

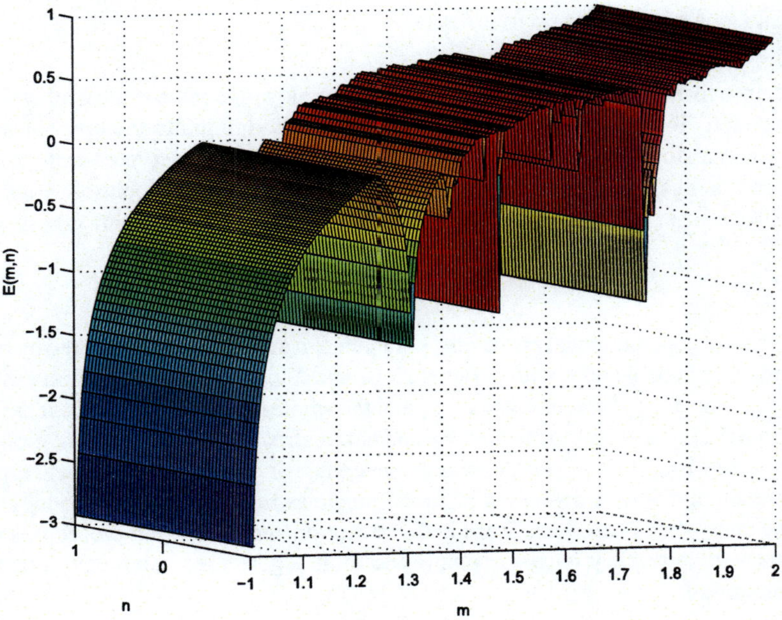

Fig. 18.9 The spatial Lyapunov exponent of system (18.24)

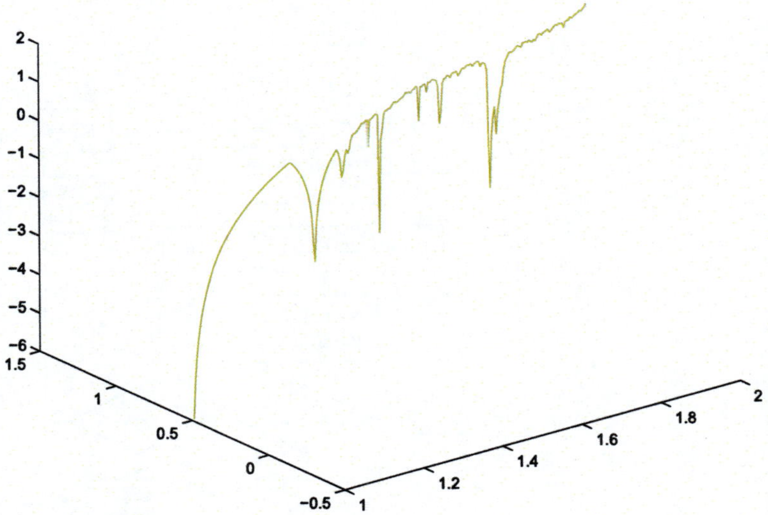

Fig. 18.10 The spatial Lyapunov exponent section of system (18.24)

18.3.1 The Proposed Method

In the proposed method, spatial chaotic map is employed to achieve the goal of image encryption. To apply our proposed encryption method, initially the original image $I_{m\times n}$ is transformed into $M_{(m\times n)\times 1}$, and then this matrix is encrypted using results of iteration of spatial chaos map. Using the initial conditions and control parameters of chaotic map, the spatial chaos map is iterated once and C_i is generated using

$$C_i = M_i\ XOR\ (X_{mn} \times 10^{14} mod\ 256).$$

At the next step, the spatial chaos map parameters μ and ω are both modified by using simple functions g and h with parameters C_i and X_{mn}. This process continues up to $M_{m\times n}$. Then $M_{m\times n}$ is set equal to $C_{m\times n}$ and the whole process is repeated for the new M from the last element to the first one and new matrix C is the output as the ciphered image (see Fig. 18.11). The decryption procedure is similar to that of encryption illustrated above with reverse of ciphered image as input instead of original image in the encryption procedure. Since both decryption and encryption procedures have similar structure, they have essentially the same algorithmic complexity and time consumption.

Fig. 18.11 Block diagram

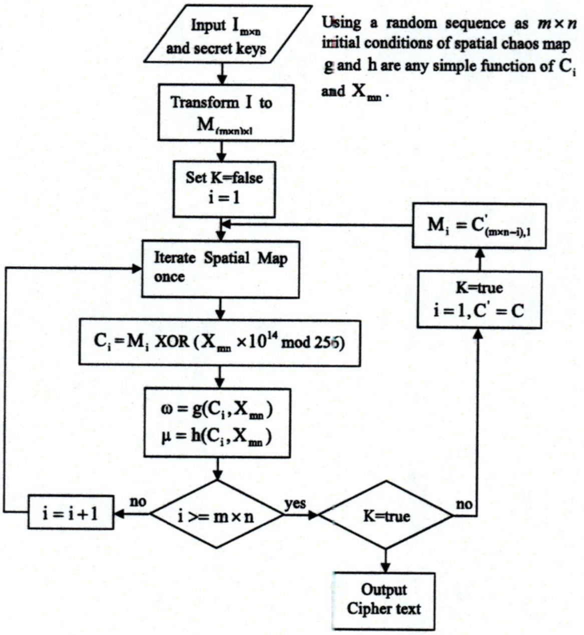

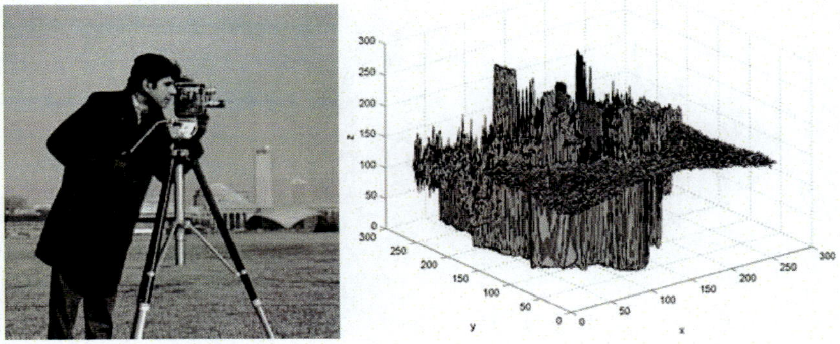

(a) (b)

Fig. 18.12 **a** Plain image. **b** Corresponding spatial image

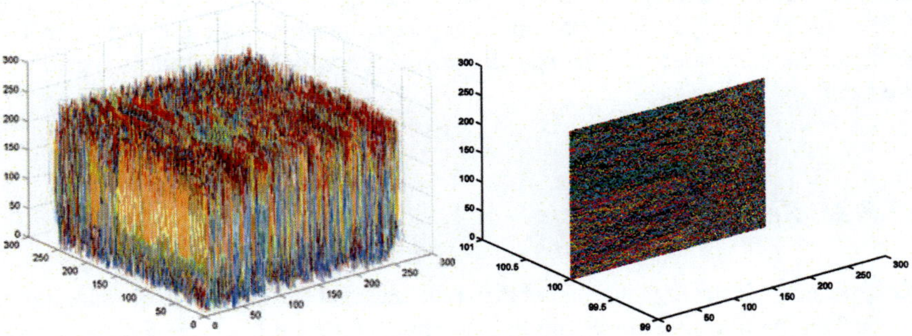

(a) (b)

Fig. 18.13 **a** Ciphered image. **b** A section of the ciphered image when $m = 100$

18.3.2 Experimental Results

Experimental analysis of the proposed encryption scheme has been done with several images. Figure 18.12a is the 256 × 256 grey-scale plain-image, its corresponding spatial image is displayed in Fig. 18.12b. Figure 18.13a is its encrypted image with the encryption key $K(\mu = 1.61234567891234, \omega = -0.05)$. Figure 18.13b is one section of the encrypted image. As one can see from Fig. 18.13, the ciphered image is as rough-and-tumble and unknowable as the others based on one dimensional chaos encryption algorithms, but the difference is that by using spatial chaos map to realize image encryption, the diffusion and confusion are conducted in multiple directions of high dimensional variable space, that is to say, the image pixels are distributed uniformly in multiple directions of space, which makes an unauthorized participant-an opponent-to more difficultly to gain the information. Figure 18.14a

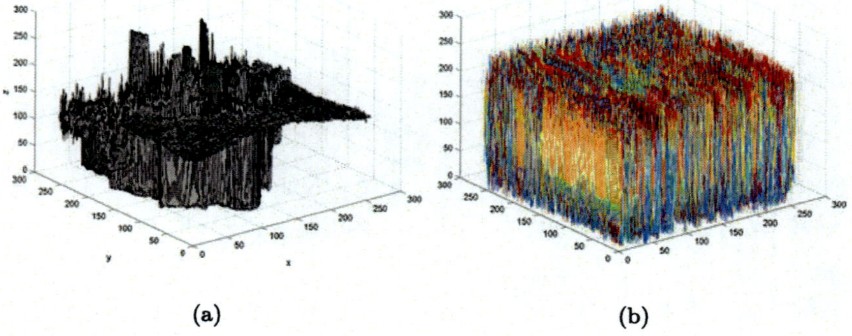

Fig. 18.14 a Decrypted image by correct key. b Decrypted image by wrong key

show the decrypted image using the right key. But if we use the wrong keys $K(\mu = 1.61234567891235, \omega = -0.05)$, an unexpected image will be got in Fig. 18.14b. So it can be concluded that the spatial chaotic encryption algorithm is sensitive to the keys.

18.3.3 Key Space Analysis

Key space size is the total number of different keys that can be used in the encryption. A good encryption scheme should be sensitive to the secret keys, and the key space should be large enough to make brute-force attacks infeasible. In this algorithm, the initial conditions and parameters can be used as keys. If the precision is 10^{-14}, the key space size can reach to $10^{28\times m\times n}$, the control parameters μ and ω modified in each iterate are not included. One can see the key space is incomparably large to resist the attacks.

18.3.4 Correlation Analysis of Two Adjacent Pixels

To test the correction between two adjacent pixels in plain-image and ciphered image, the following procedure was carried out. First, randomly select 1000 pairs of two adjacent (in horizontally, vertical, and diagonal direction)pixels from an image. Then, calculate the correction coefficient of each pair by using the following formulas:

$$E(x) = \frac{1}{N}\sum_{i=1}^{N} x_i, \tag{18.25}$$

$$D(x) = \frac{1}{N}\sum_{i=1}^{N}(x_i - E(x))^2, \tag{18.26}$$

$$cov(x, y) = \frac{1}{N}\sum_{i=1}^{N}(x_i - E(x))(y_i - E(y)), \tag{18.27}$$

$$r_{xy} = \frac{cov(x, y)}{\sqrt{D(x)D(y)}}, \tag{18.28}$$

where x and y are grey-scale values of two adjacent pixel in the image. Figure 18.15 shows the correlation distribution of two horizontally adjacent pixels in the plain-

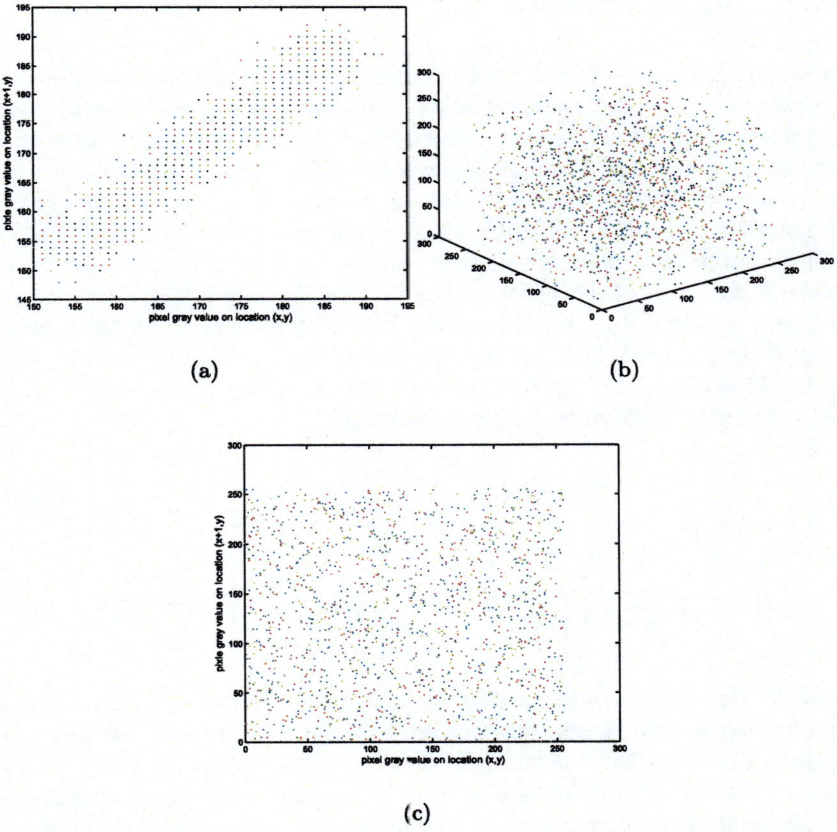

Fig. 18.15 **a** Correlations of two horizontally adjacent pixels in the plain-image. **b** Correlations of two horizontally adjacent pixels in the ciphered-image. **c** Section of two horizontally adjacent pixels correlations in the ciphered-image

Table 18.2 Correlation coefficient of two adjacent pixels in plain-image and ciphered image

	Plain-image	Cipher image
Horizontal	0.994948	0.00128
Vertical	0.961046	–0.00261
Diagonal	0.956190	0.00014

image and that in the cipher-image. The correlation coefficients are shown in Table 18.2. These correlation analysis prove that the chaotic encryption algorithm satisfy zero co-correlation.

18.3.5 Differential Attack

Two common measures, NPCR and UACI, are used to test the influence of changing a single pixel in the original image on the whole image encrypted by the proposed algorithm. NPCR stands for the number of pixels change rate while one-pixel of plain image is changed. The unified average changing intensity (UACI) measures the average intensity of differences between the plain image and ciphered image. For calculation of NPCR and UACI, let us assume two ciphered images C_1 and C_2 whose corresponding plain images have only one-pixel difference. The grey-scale values of the pixels of the ciphered image C_1 and C_2 at grid (i, j, k) are labeled as $C_1(i, j, k)$ and $C_2(i, j, k)$, respectively. Take a array D, with the same size as image C_1 or C_2 . Then, $D(i, j, k)$ is determined by $C_1(i, j, k)$ and $C_2(i, j, k)$. So, if $C_1(i, j, k) = C_2(i, j, k)$ then $D(i, j, k) = 1$; otherwise, $D(i, j, k) = 0$. NPCR and UACl are defined through the following formulas:

$$NPCR = \frac{\sum_{ijk} D(i, j, k)}{W \times H} \times 100, \tag{18.29}$$

$$UACI = \frac{1}{W \times H} \left[\sum_{ijk} \frac{|C_1(i, j, k) - C_2(i, j, k)|}{255} \right] \times 100, \tag{18.30}$$

where W and H are the width and height of C_1 or C_2. Tests have been performed on the proposed scheme, about the one-pixel change influence on a 256 grey-scale image of size 256 × 256. We obtained NPCR = 0.40% and UCAI = 0.3192%. The results show that a swiftly change in the original image will result in a significant change in the ciphered image, so the algorithm proposed has a good ability to anti differential attack.

Remark 18.1 This part presents a new encryption algorithm, it is to use spatial chaotic map to finish the encryption process. The proposed algorithm overcome the drawbacks of small key space and weak security in the widely used one-dimensional chaotic system. It is well known that the larger the cycles are, the higher is the security, yet, it is at the expense of processing speed. In this paper, the image becomes indistinguishable only with one cycle, and show the perfect encryption performance by using the proposed encryption method, which is advantageous to achieve faster implementation speed. It can be seen from the results that the proposed system offers a higher complexity. The high complexity of such chaotic dynamics system indicates that they could be advantageously used in chaotic cryptographic techniques with enhanced security. The experimental results and secure analysis demonstrate that the encryption algorithm is effective and highly secure.

Although the spatial chaotic system presented in this paper aims at image encryption, it is not just limited to this area and can be directly applied in encryption 3D images, such as holograms, which we believe will be used in communications in the future.

18.4 2-D Arnold Cat Graph

The classical Arnold cat graph can be described as the following

$$\begin{pmatrix} x_{n+1} \\ y_{n+1} \end{pmatrix} = \begin{pmatrix} 1 & 1 \\ 1 & 1 \end{pmatrix} \begin{pmatrix} x_n \\ y_n \end{pmatrix} mod\ 1, \tag{18.31}$$

which can be the special case of the system:

$$\begin{pmatrix} x_{n+1} \\ y_{n+1} \end{pmatrix} = \begin{pmatrix} 1 & a \\ b & ab+1 \end{pmatrix} \begin{pmatrix} x_n \\ y_n \end{pmatrix} mod\ 1, \tag{18.32}$$

where a and b are control parameters. Naturally, the generalized 2-D Cat graph is

$$\begin{pmatrix} m+1 \\ n+1 \end{pmatrix} = \begin{pmatrix} 1 & a \\ b & ab+1 \end{pmatrix} \begin{pmatrix} m \\ n \end{pmatrix} mod\ 1, \tag{18.33}$$

where $x_{m,n}$ is the pixel value at point (m, n), $m, n = 1, 2, \ldots, N$, N is the pixel level of 2-D image.

After discretizing the system (18.33), we obtain:

$$\begin{pmatrix} m+1 \\ n+1 \end{pmatrix} = \begin{pmatrix} 1 & a \\ b & ab+1 \end{pmatrix} \begin{pmatrix} m \\ n \end{pmatrix} mod\ N, \tag{18.34}$$

Example 18.1 Let take the Fig. 18.16 as an example. When $P(m, n)$ is the two dimensional coordinates at pixel point P and $x_{m,n}$ is its pixel value, the spatial plot of the pixel value for (18.34) is shown clearly in Fig. 18.17.

Fig. 18.16 Standard graph

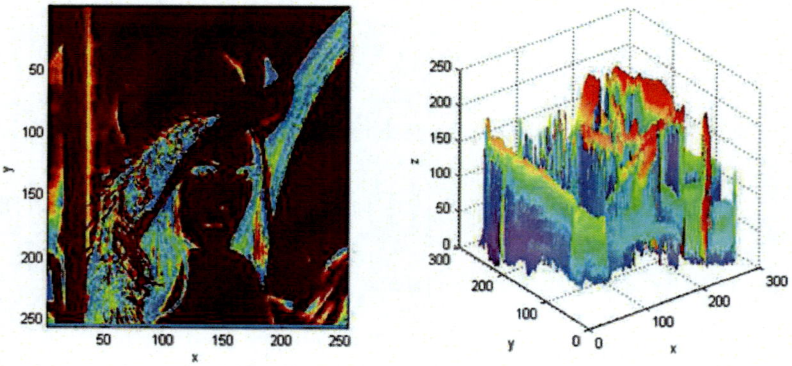

Fig. 18.17 The pixel value of the standard graph

Remark 18.2 Especially, the systems (18.32) and (18.33) are essential differences. For (18.32), (x_n, y_n) is the point in a plane graph, where n takes different values corresponding to different pixels on the image, traversing the whole image. In addition, $x_{m,n}$ is the pixel value at point (m, n), where m, n traverse the whole image.

18.4.1 2-D Cat Graph Image Encryption Based on Surface Chaos

Because of the encryption of image signal, it is actually the transformation of the midpoint of the set S of image pixels

$$S_{xy} = \{C(x, y)|(x, y)\ is\ the\ coordinate\ of\ the\ pixel\ point\ P\}.$$

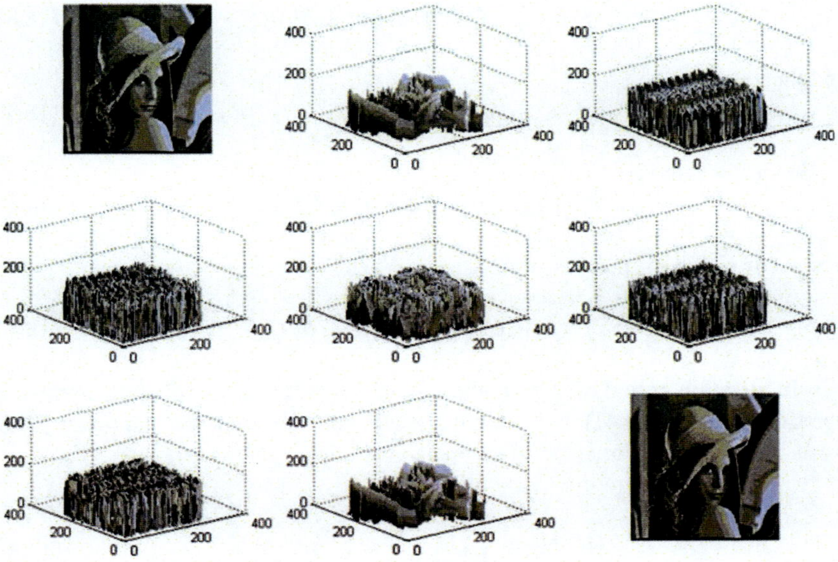

Fig. 18.18 Periodicity of standard graph encryption process

For a two-dimensional planar graph, each pixel $C(x, y)$ of Fig. 18.16 can be regarded as a continuous graph, while for a discrete graph $C_d(x, y)$ generated by (18.34), it is easy to get that:

$$\lim_{N\to\infty} \max_{0\leq i,j\leq N} |C(\frac{i}{N}, \frac{j}{N}) - C_d(i, j)| = 0. \tag{18.35}$$

However, similar to the two-dimensional chaotic graph, not all the mixed features can be preserved after discretization. For example, after discrete transformation, the non periodicity of chaotic graph will become periodicity, which will reduce the security of chaotic encryption. If the continuous graph (18.33) is described as (18.34) the 256 × 256 graph iterating 6 times would return to the original one illustrated in Fig. 18.18, where

$$x_{m+1,n} + \omega x_{m,n+1} \in \{0, 1, \ldots, N-1\},$$

and $a = 25, b = 5$.

Furthermore, if

$$C_{s+1,t+1} = x_{s+1,n} + \omega x_{m,t+1}, \tag{18.36}$$

we have

$$C_{s,t} = x_{s,n} + \omega x_{m,t}. \tag{18.37}$$

Let

$$f(\mu, a, b, C_{st}) = ai\mu(1 + ab)C_{st}^2, \tag{18.38}$$

and then we get

$$C_{s+1,t+1} = f(\mu, a, b, C_{st}). \tag{18.39}$$

Generally, there is

$$C_{s+r,t+r} = f^r(\mu, a, b, C_{st}), \tag{18.40}$$

where r is called the number of iterations. When

$$f^r(\mu, a, b, C_{st}) = f(\mu, a, b, C_{st}), \tag{18.41}$$

r is the iteration period of f and when r is the smallest, it is called the number of iteration orbits of (18.41). Note that the relationship between the maximum period r of a discrete three-dimensional graph and the integer n is as follows:

$$P(N, N) = \begin{cases} 3N : \ N = 2 \times 5^k, \\ 2N : \ N = 5^k \ or \ 6 \times 5^k, \\ \frac{12N}{7} : other \end{cases} \tag{18.42}$$

where $k = 1, 2, \ldots,$ and N is the width or height of the graph.

18.4.2 *Diffusion Process*

In order to solve the periodic problem of spatial image after discretization, a diffusion process is added as compensation to make the encryption system irreversible.

There are two reasons to introduce the diffusion process into the encryption algorithm: first, the diffusion process can make the chaotic system irreversible; second, the diffusion process can greatly change the histogram and statistical characteristics of the original image by changing the pixel value of each point in the original image. Therefore, for a secure encryption system, the diffusion process is necessary. Otherwise, the attacker can compare each pair of original text and encrypted text through several transformations to find some useful information and decode the image. For the purpose of diffusion, "XOR plus mod" operation is applied to the encryption system to operate every two adjacent pixels in the spatial image. First, take two numbers, one is a floating-point number on, as the initial value of the chaotic system. The other, denoted as, is an integer used as the initial value of the cryptosystem. Then, the nonlinear surface chaotic system is used:

$$x_{m+1,n} + \omega x_{m,n+1} = 1 - \mu[(1 + \omega)x_{m,m}]^2 \tag{18.43}$$

where $\mu = 1.65, \omega = -0.05, x_{0,0} = L(0, 0)$, and $(1 + \omega)x_{0,0} < 1.15$. It is noted that when the above conditions are satisfied, (18.43) is in the state of surface chaos. So

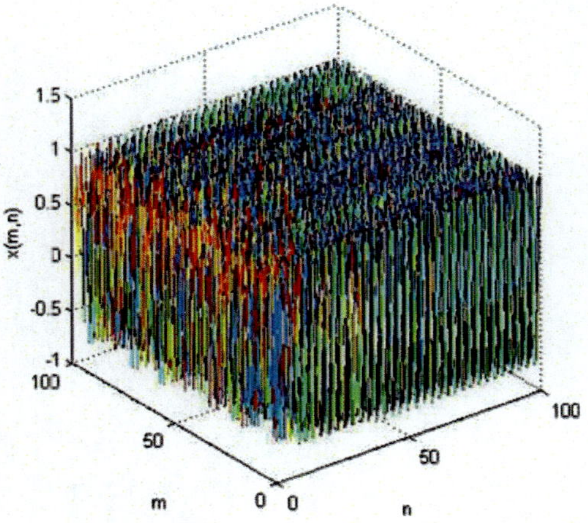

Fig. 18.19 Surface chaos as $\mu = 1.65, \omega = -0.05$

the system has a strong sensitivity dependence on the initial value, and the surface chaos diagram represented by it is shown in Fig. 18.19.

After calculating formula (18.43), If the next value is in the interval, then proceed to the next step. Once the appropriate value is obtained from formula (18.43), it can be digitized by enlarging the appropriate scale and sampling, where the digitized values are recorded as

$$x_{m+1,n} + \omega x_{m,n+1} = \phi(m+1, n+1).$$

The pixel value obtained after digitization is XOR calculated according to (18.44), which is expressed as follows:

$$C(m,n) = \phi(m,n) \oplus \{[I(m,n) + \phi(m,n)]\, mod\, N\} \oplus C(m-1, n-1), \quad (18.44)$$

where $C(m,n)$ is the pixel value obtained now, $C(m-1, n-1)$ is the output encrypted pixel of the previous step, the 256×256 vectors are connected into a sequence by this formula. $I(m,n)$ is the pixel value of the original 3-D image at the point (m,n), and N is the pixel level. We can set the initial value and finally get the encryption result. The decryption process is its inverse operation, and the decryption process is expressed as

$$I(m,n) = \{\phi(m,n) \oplus C(m,n) \oplus C(m-1, n-1) + N - \phi(m,n)\}mod\, N, \quad (18.45)$$

where the original value of step is known, the value of can be restored by Eq. (18.45).

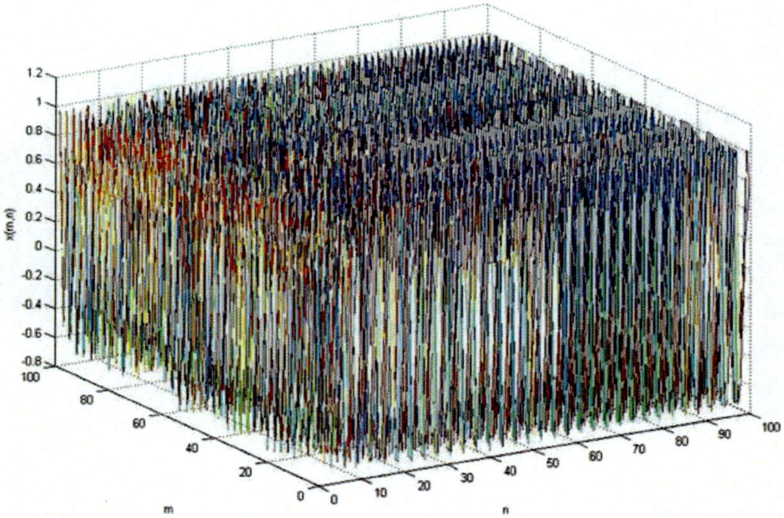

Fig. 18.20 Surface chaos as $\mu = 1.9, \omega = -0.72$

18.4.3 Two Step Generating Method of Cryptosystem

The basic principle of cryptosystem is that cryptosystem is sensitive to cryptosystem, and the sensitive dependence of chaos on initial value can meet this requirement. High security encryption system requires that the encrypted text and password are interrelated, but they cannot be decrypted by attackers.

In this part, the method has two steps: the first step is to thoroughly integrate the password into the original text through the encryption process; the second step is to have an ideal password generation mechanism.

Password recording directly applied to encryption system is regarded as k_m which is a parameter vector of chaotic graph, as floating point or integer. And the user input password k_u is a string, which is a column of decimal numbers. In this way, we need to design a password generation system to complete the transformation from k_u to k_m, which can protect the key from being stolen by password attackers.

We apply the following two surface chaotic systems:

$$\begin{cases} x_{m+1,n} + \omega_1 x_{m,n+1} = 1 - \mu_1[(1+\omega_1)x_{m,n}]^2, \\ y_{m+1,n} + \omega_2 y_{m,n+1} = 1 - \mu_2[(1+\omega_2)y_{m,n}]^2, \end{cases} \tag{18.46}$$

where μ_i, ω_i are real parameters, $i = 1, 2$. We introduce the system (18.46) into the cryptosystem, where the surface chaos of system (18.46) is illustrated clearly in Fig. 18.20.

If the initial value $x_{0,0}, y_{0,0} < 1.25$ is taken, the simulation shows that the orbit of the system is extremely sensitive to the parameters μ_i. So the generation of the key can be controlled. It is worth noting that the parameters in (18.46) are different from those in (18.43), so their chaotic characteristics are also different. The key used in the design of encryption system is a 96 bit binary sequence, which is divided into six parts as $k_a, k_b, k_x, k_y, k_l, k_s$, each part is 16 bits long, where the parameters k_a, k_b are the control parameters of two three-dimensional chaotic graphs, k_l, k_s are used to generate the initial value $L(0, 0)$ of chaotic graph and the initial value s of the encryption process, and k_x, k_y are two initial values of cryptosystem. In order to produce a and b, let

$$K_a = \sum_{i=0}^{15} k_a(i) \times 2^i, \tag{18.47}$$

where $k_a(i)$ is the ith in the sequence k_a. The two initial values of spatial Chaotic Cryptosystem (18.46) can be obtained by the following formula:

$$\begin{cases} K_x = \sum_{i=0}^{15} k_x(i) \times 2^i, \\ K_y = \sum_{i=0}^{15} k_y(i) \times 2^i. \end{cases} \tag{18.48}$$

The other parameters are obtained by 100 and 200 iterations of the spatial chaotic system (18.46), and two values $x_{100,100}$, $y_{200,200}$ are obtained. Then, the two control parameters a and b of the three-dimensional chaotic map are obtained by Eq. (18.49):

$$\begin{cases} a = round(x_{100,100}/60 \times N, \\ b = round(y_{200,200}/60 \times N, \end{cases} \tag{18.49}$$

where N is the pixel level of the original 2-D graph. The method of generating the initial value $L(0, 0)$ of chaotic graph is similar to that of encryption operation. The following two formulas are used to generate $L(0, 0)$ and s:

$$L(0, 0) = x_{100,100}/40, \tag{18.50}$$

$$s = round(y_{200,200}/60 \times 255. \tag{18.51}$$

In this way, the user password of chaotic encryption system is obtained.

Example 18.2 When the original 3-D image and password is 123456789012, the encryption result is shown in Fig. 18.21.

Remark 18.3 In this part, we apply spatial chaos to two-dimensional image encryption, which changes the traditional method. Its biggest advantage is that the carrier passed by is not the image, but the heap data in the space, which fully increases the attacker's cracking difficulty. In addition, because of its higher degree of chaos than one-dimensional, it can achieve better image encryption. In fact, the simulation results show that the new encryption method has high security and encryption speed.

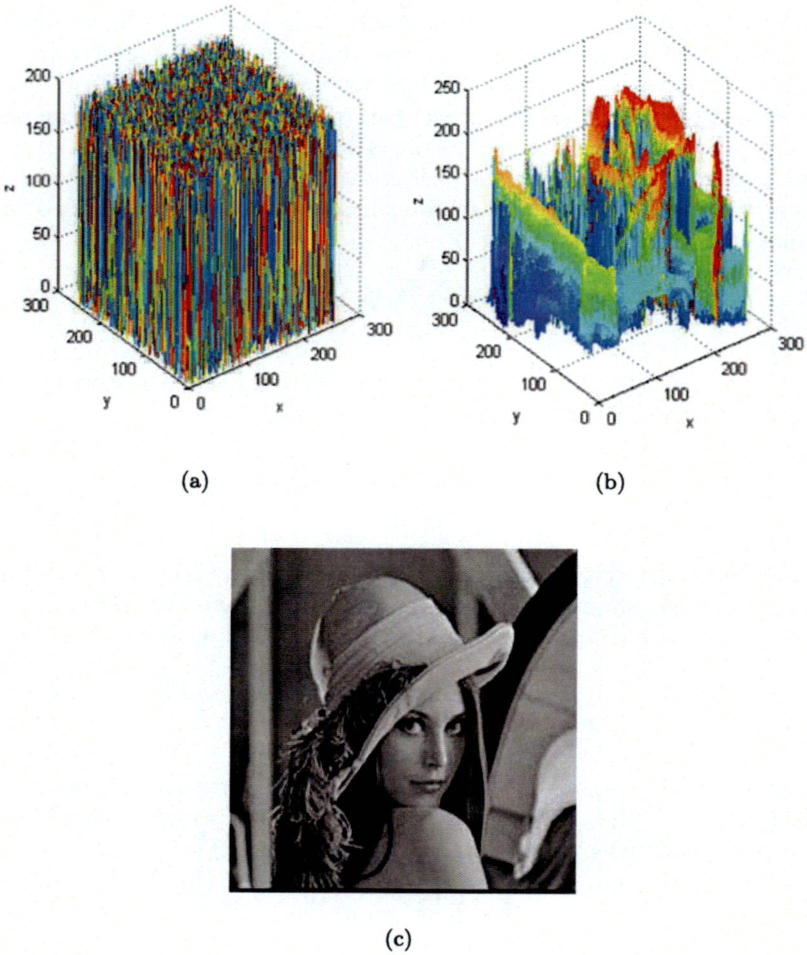

Fig. 18.21 Encryption and decryption results of surface chaos

Part V
Mathematical Mechanism and Fractal Analysis of Galaxy and Black Hole in the Universe

Chapter 19
The Relationships Between the Nonlinear Behavior of Star Motion in the Universe and the Black Hole of the Milky Way Galaxy

19.1 Nonlinear Motion of Stars in the Universe

Throughout the above contents, we can see that, for both the surface chaos and bifurcation, many interesting and important results have been developed. The topics involving theoretical and application in this field are still hot and crucial. For example, the nonlinear systems involved in a space body are as follows and have complicated chaos and bifurcation behavior:

$$F(u_x^{(n_1)}, u_y^{(n_2)}, u_z^{(n_3)}, \ldots, u_x^{(1)}, u_y^{(1)}, u_z^{(1)}, u(x, y, x, t), c) = 0, \tag{19.1}$$

where (n_i) is the order of partial derivatives, c is the constant parameter, $i = 1, 2, 3$, $u = u(x, y, z, t)$, and $F(\cdot)$ is a nonlinear function. In addition, we can get the generalization of system (19.1):

$$G(v_{x_1}^{(s_1)}, v_{x_2}^{(s_2)}, \ldots, v_{x_1}^{(s_1-1)}, v_{x_2}^{(s_2-1)}, \ldots, v_{x_1}^{(1)}, v_{x_2}^{(1)}, \ldots, v_{x_1}, v_{x_2}, \ldots, c) = 0, \tag{19.2}$$

where $s_i$is the order of partial derivatives, $i = 1, 2, \ldots, n$, $v = v(x_1, x_2, \cdots, x_n, t)$.

Specially, the nonlinear chaos and bifurcation behavior of the space body (19.2) are closely related to many characteristics in the universe [466–480]. Because in the vast space of the universe, the motion of the fragments and the planet are nonlinear, but the local definition domain is periodic, attractive and semi attractive [481–491]. These ones here can be usually related with the shape of the system (19.1). Naturally, it involves chaos and bifurcation, chaotic attractors, strange attractors of spatial bodies and semi strange attractors. For example, in the universe, there are countless galaxies, hyperbodies, and so on, among which there are nonlinear motion behaviors, while in the local area, it is a certain galaxy although it is still a nonlinear behavior. But the trend is bounded and attractive. These are orderly actions in disorderly movements. We can naturally use something similar to study the questions related by system

S. T. Liu and L. Zhang, *Surface Chaos and Its Applications*,
https://doi.org/10.1007/978-981-16-8229-2_19

Fig. 19.1 Galaxy picture not indicated source is downloaded from non copyright web pages on the Internet

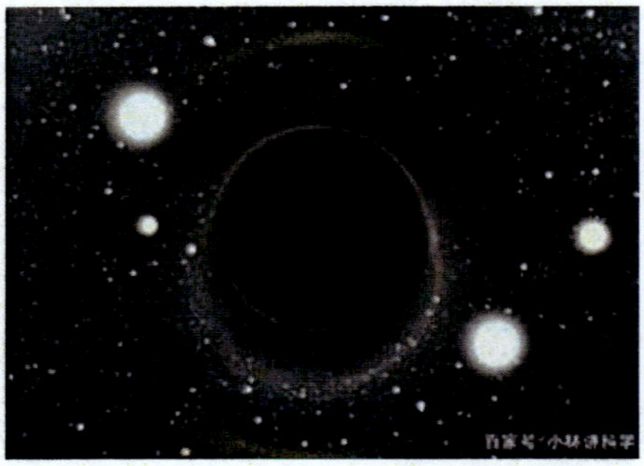

Fig. 19.2 Black hole picture not indicated source is downloaded from non copyright web pages on the Internet

(19.1). For instance, a galaxy in space shows its boundedness, attractiveness and the existence of semi strange attractors which relates closely with system (19.1). Similarly, for black holes in the universe, it is a typical characteristic of nonlinear, bounded and attractive [492–494]. In particular, it attracts any matter in the universe, including light, which is the similar behavior of the semi strange attractor in the nonlinear system (19.1) (Figs. 19.1, 19.2).

It can be seen that the nonlinear chaos and bifurcation behavior of the space body motion, the curve chaos and the curved surface chaos show more responsible and diverse behaviors. So the space body chaos and other nonlinear properties are also the content of rank exploration and research. This is a fertile soil, which contains rich things, hoping to arouse the vast number of science and technology high attention of workers (Fig. 19.3).

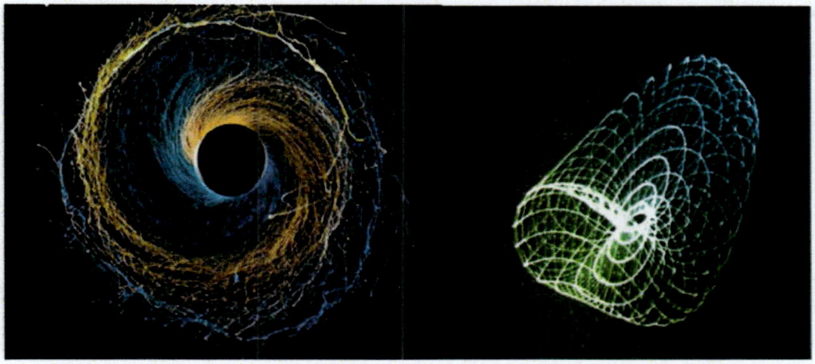

(a) Nonlinear motion behavior of space interstellar body

(b) Nonlinear motion of space body

(c) Chaotic motion of space body

(d) Galaxies caused by chaotic motion of space objects

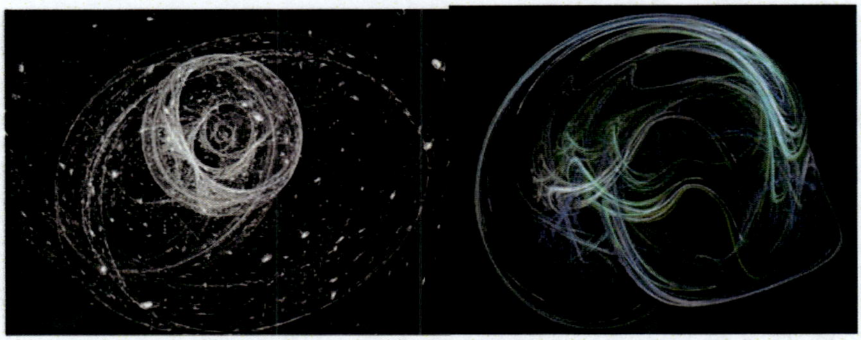

(e) Chaotic motion behavior of space body

(f) Chaotic motion behavior of space body

Fig. 19.3 Cosmic black hole and galaxy pictures not indicated source are downloaded from non copyright web pages on the Internet

19.2 Semi-Stable Limit Cycles and Semi-Strange Attractors

As the limit of human physics and the most mysterious celestial body in the universe [495–498], black holes have always been the focus of major science awards. “Understanding black holes may be necessary to understand the truth of the universe”, according to the chairman of the committee for the 2020 Nobel Prize in Physics. In the future, scientists who thoroughly understand the theory of the formation of supermassive black holes may also win the Nobel Prize in Physics [499], but the mathematical principles of the formation of black holes, galaxies and black hole singularities are still unresolved hot research issues in the world.

The Milky Way is one of an infinite number of galaxies in the universe, and black holes with different properties also form in galaxies. The core of this paper is to utilize the semi-stable limit cycle theory of nonlinear dynamical system to explain the mathematical principles of the formation of galaxies and black holes from the perspective of theoretical physics, and completely solve the problems that have been vexing people, including “black hole, void hole, dark energy, the existence of positive and anti matter universe”, as well as “physical black hole, real hole, radio black hole, white hole called” Sagittarius A “, fast radio burst, gamma ray burst”. Why is the center of the Milky Way a silent black hole? Why is the center of $M87$ an irascible black hole? And so on. For these unscientific explanations and guesses, we will give a thorough proof of mathematical principles and obtain the relationship between them.

Suppose $P(x, y)$, $Q(x, y)$ are nonlinear functions, then the nonlinear dynamic system is

$$\begin{cases} \frac{dx}{dt} = P(x, y), \\ \frac{dy}{dt} = Q(x, y) \end{cases},$$

set a closed track as C, in the sufficiently small neighborhood of C but except for C, the tracks are not closed, and when $t \to \infty$, these non-closed tracks all tend to C, then C is called isolated, besides, the closed track at this time is said to be a limit cycle of system, C is also called as a stable limit cycle. It can be seen that C divides the phase plane (phase space) into two regions, inner region and outer region. In particular, when $t \to +\infty(-\infty)$, the internal and external stability of C is opposite, then C is semi-stable limit cycle, refer to Fig. 19.4, [500–502]. Thus, there are two kinds of semi-stable limit cycles: externally stable but internally unstable, externally unstable but internally stable.

As for the limit cycle C, both inner and outer tracks of C tend to C, therefore C is attractive to other tracks, then the stable limit cycle is called the limit cycle attractor. As for a semi-stable limit cycle, the closed orbit divides the region into inner and outer parts, the attractor requires that the phase “volume” near a point decreases with time, while on both sides of the closed orbit, one side is stable and attractive, and the “volume” decreases with time. However, on the other side, it is unstable, unattractive and divergent, which means attraction and unattraction are interwoven at the inner and outer edges of C, and such attractive feature is called semi-strange attractor [503–509].

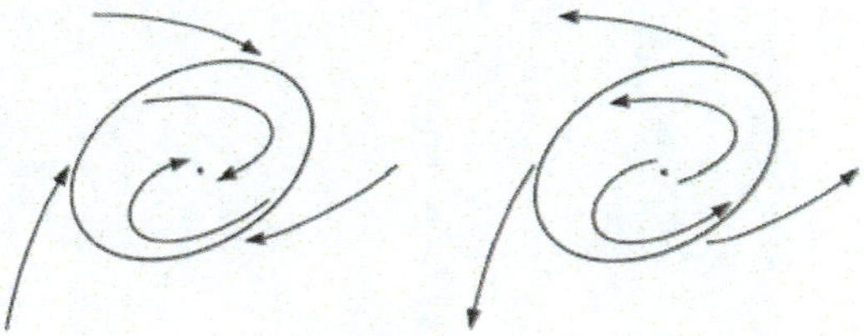

Fig. 19.4 Semi-stable limit cycle

(a) External stability and internal instability

(b) External stability and internal instability

Boundedness and Physical Properties of the Milky Way (Galaxies)

The Milky Way is one of the infinite galaxies in the universe, so its space-time structure is similar to that of other galaxies. Here we take this as an example to explain the mathematical mechanism of its formation, which accordingly explain the mathematical principle of the formation of other galaxies and black holes.

As is known to all, the Milky Way, it is an ellipse, rod-shaped, spiral galaxy with a disk-like spiral structure, its diameter up to a staggering 120,000 light years away. Looking down at the Milky Way, it is like a huge vortex, which is composed of four spiral arms, the arms are spaced 4500 light-years away, and the sun is on the Orion arm of one of the arms of Milky Way. From inside to outside, the Milky Way consists of a silver center, a silver core, a silver disk, a silver halo and a silver corona, its central diameter is about 20,000 light-years, and its thickness is about 10,000 light-years. Thus, it is a disk-like geometry with a bright and bulging center that constantly rotates in the cosmic space [499]. Suppose the geometry as E, its prominent physical feature is that it has the bounded behavior region. Let C be the largest edge line of the galactic random ellipse E on the plane, then C is a disk-like ellipse with random changes, and suppose $U(C)$ be the inner region of the galaxy surrounded by C, then

$$|U(C) \cup C| \leq M, \tag{19.3}$$

M is a very large number of light years (Fig. 19.5).

Denote any star in the Milky Way as p, and regard it as a particle in the Milky Way. Establishing a rectangular coordinate system $O - XYZ$ of the universe space, take the coordinates of the p as $p(x, y, z) \in E$. According to the physical form that the Milky Way is strongly attracted by the vortex of the galactic center, the motion behavior of particle p is asymptotic and uniformly tends to the rotational behavior. Furthermore, point p is any particle in the vortex of the Milky Way E, and its motion tends to be uniform, which determines that all stellar motion in the galaxy is strongly attracted by the galactic center.

The above two conditions are sufficient to determine the existence of the universal mathematical model, and the Milky Way is a large range of bounded motion which is dominated by infinite stars. Therefore, it is of profound historical and subversive

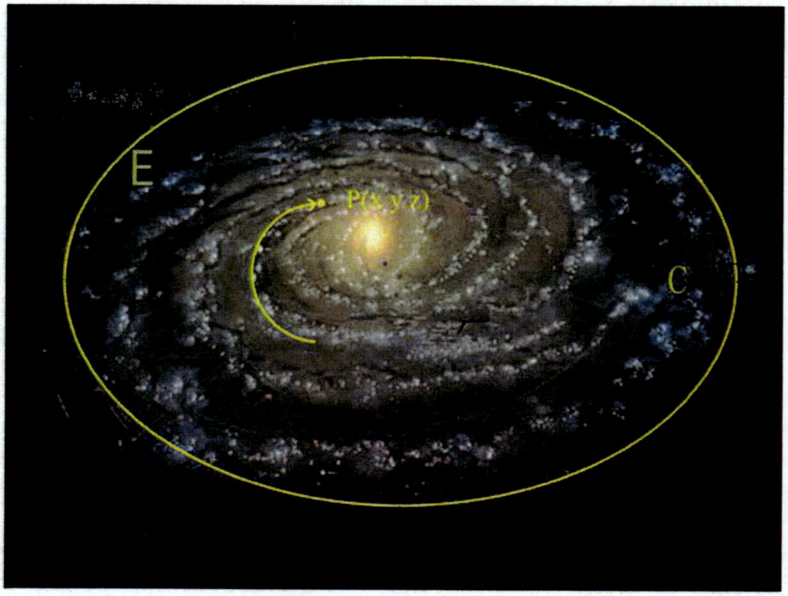

Fig. 19.5 Boundedness of the Milky Way and the asymptotic uniform tendency of rotational motion for any star as a particle

theoretical significance in the field of astrophysics to strengthen the study of its behavior mechanism and scientifically control the law of motion.

19.2.1 Mathematical Principles of the Formation of Black Holes and Galaxies

By the Sect. 19.2, in the universe, from the XY plane of the coordinate $O - XYZ$, the Milky Way disk geometry $E \in R^3$ is a random ellipse, whose long axis and short axis are constantly changing, and the maximum approximate ellipse circumference C and random elliptical radius R are both functions of time t, denote as $C\,(t)$, $R\,(t)$. Therefore, for the change of t at any time, the random ellipse can be expressed as

$$a(t)x^2 + b(t)y^2 = R(t), \tag{19.4}$$

thus the approximate random ellipse (19.4) of the the Milky Way on the OXY plane is the basic geometric element that forms the Milky Way. The following consideration is given to the motion momentum problem of a random ellipse (19.4) moving randomly in the universe. The uniform smooth ellipse $C(t)$, which is randomly enclosed by the unit material curve, is taken as a whole.

On the cosmic coordinate plane OXY, the equilibrium relationship between the circumference $C(t)$ of the random ellipse and the coordinates (x, y) on the coordinate axes OX and OY is established: $(x, y) \leftrightarrow C(x(t), y(t))$. Due to the constantly changing of $C(t)$ over time t, thus the momentum of unit quality random ellipse circumference $C(t)$ is $a(t)x^2 + b(t)y^2 - R^2(t)$. $C(t)$ is a random ellipse, then $(a(t)x^2 + b(t)y^2 - R(t)^2)^m$ is generalized random ellipse momentum. Similarly, in the space OX and OY, the momentum of the overall change motion of the unit mass material lines ox and oy formed by the coordinates x and y is respectively x', y', also, $\sqrt{x'}$, $\sqrt{y'}$. The coordinate x, y on the axis is the self-varying momentum of the numerical length stretching, under the regulation of momentum conservation factors $alpha_1(t)$, $\beta_1(t)$, then

$$\begin{cases} \sqrt{\frac{dx}{dt}} = \alpha_1(t)x, \\ \sqrt{\frac{dy}{dt}} = \beta_1(t)y \end{cases}.$$

Besides, unit mass random ellipse circumference $c(t)$, the motion momentum of unit material lines ox and oy, are both in the same coordinate space OXY, they are in the state of overall relative balance, thus under the regulation of momentum conservation factors $\alpha_2(t)$, $\beta_2(t)$, then

$$\begin{cases} \sqrt{\frac{dx}{dt}} = \alpha_2(t)(a(t)x^2 + b(t)y^2 - R(t)^2)^m, \\ \sqrt{\frac{dy}{dt}} = \beta_2(t)(a(t)x^2 + b(t)y^2 - R(t)^2)^m. \end{cases}$$

Due to the Milky Way is in a state of relative balance in the space OXY, take $\alpha(t) = \alpha_1(t)\alpha_2(t)$, $\beta(t) = \beta_1(t)\beta_2(t)$, comprehensively,

$$\begin{cases} \frac{dx}{dt} = \alpha(t)x(a(t)x^2 + b(t)y^2 - R(t)^2)^m, \\ \frac{dy}{dt} = \beta(t)y(a(t)x^2 + b(t)y^2 - R(t)^2)^m, \end{cases} \tag{19.5}$$

then (19.5) is the most basic mathematical principle of the Lagrange particle nonlinear dynamical system of the Milky Way.

As for (19.5), the interferences such as competition among gravitational force, gas pressure, magnetic field force and moment of inertia of interstellar motion in the universe are taken into account, let the noise disturbance be $p(x, y)$ and $q(x, y)$ respectively, by the Weierstrass polynomial approximation theorem, $p(x, y)$ and $q(x, y)$ can be approximated by polynomial function, and (19.5) can be reduced to

$$\begin{aligned} \frac{dx}{dt} &= \alpha(t)x(a(t)x^2 + b(t)y^2 - R(t)^2)^m + \omega\left(\sum_{i=0}^{r} a_i(t)x^i + \sum_{j=0}^{s} b_j(t)y^j + \sum_{k=0}^{u}\sum_{l=0}^{v} c_{ij}(t)x^k y^l\right), \\ \frac{dy}{dt} &= \beta(t)y(a(t)x^2 + b(t)y^2 - R(t)^2)^m + \omega\left(\sum_{i=0}^{\bar{r}} \bar{a}_0(t)x^i + \sum_{j=0}^{\bar{s}} \bar{b}_j(t)y^j + \sum_{\bar{k}=0}^{\bar{u}}\sum_{\bar{l}=0}^{\bar{v}} \bar{c}_{ij}(t)x^{\bar{k}} y\bar{l}\right), \end{aligned} \tag{19.6}$$

where ω is the gravitational interference coefficient between the motion of stars. This is the universal nonlinear dynamical system of galactic formatioon in the universe. Since (equblack4) is a 2-dimensional spatial model and the Milky Way is a random elliptical rodlike space-time body, and it's about 10,000 light years thick at its center, however, it's very short compared to the diameter of the plane. If considering its vertical structure $z = \varepsilon t, \varepsilon = \rho(f)$, f is the gravity coefficient, because of the immensity of the universe, for the convenience of dealing with the problem without loss of generality, the practical application form of (equblack4) is taken as

$$\begin{cases} \frac{dx}{dt} = \alpha x(ax^2 + by^2 - R^2)^m + \omega\lambda y, \\ \frac{dy}{dt} = \beta y(ax^2 + by^2 - R^2)^m - \omega\mu x, \\ \frac{dz}{dt} = \varepsilon(f), \end{cases} \tag{19.7}$$

because (19.6) is a general theoretical result, the nonlinear dynamical system (19.7) is the practical mathematical principle of the formation of the Milky Way (galaxy).

The Mathematical Classification of Milky Way (Galaxy)

In (19.7), α and β have a great influence on the formation of the semi-stable limit cycles of the galactic mathematical system [500–502], from the shooting method and the screening method [510, 511], we can be obtain, Mathematical Type of Class I Milky Way (Galaxy): Semi-stable limit cycle of the Milky Way (Galaxy) with internally stable and externally unstable must be generated when both α and β are positive. Mathematical Type of Class II Milky Way (Galaxy): Semi-stable limit cycle of the Milky Way (Galaxy) with internally unstable and externally stable must be generated when both α and β are negative. If α and β have different signs, it needs to be discussed and determined separately. a and b affect the horizontal and vertical sizes of the limit cycles formed by the Milky Way, while $\lambda\omega = \omega\sqrt{\frac{b}{a}}$, $\mu\omega = \omega\sqrt{\frac{a}{b}}$ affect the density of stellar orbital spacing in the Milky Way. In addition, according to [510, 511], the parameter distribution of (19.7) for the existence of semi-stable limit cycles is as follows (Table 19.2).

Refer to (19.7), take parameters as $(a, b, R, m, \alpha, \beta, \omega, \lambda, \mu) = (2, 1, 1, 2, 2, 1, 8, \frac{\sqrt{2}}{2}, \sqrt{2})$ to satisfy Table 19.1a, then the system turns into (19.8), which is a semi-stable limit cycle of class I galaxy that are internally stable but externally unstable. Then, take parameters as $(-1, 1, 1, 2, -1, -1, 8, 1, 1)$ to satisfy Table 19.1b, then the system converts into (19.9), which is a semi-stable limit cycle of class II galaxy that are internally unstable but externally stable.

$$\begin{cases} \frac{dx}{dt} = x(2x^2 + y^2 - 1)^2 + \frac{8\sqrt{2}}{2}y, \\ \frac{dy}{dt} = y(2x^2 + y^2 - 1)^2 - 8\sqrt{2}x, \\ \frac{dz}{dt} = \omega_1 f_1, \end{cases} \tag{19.8}$$

$$\begin{cases} \frac{dx}{dt} = -x(x^2 + y^2 - 1)^2 + 8y, \\ \frac{dy}{dt} = -y(x^2 + y^2 - 1)^2 - 8x, \\ \frac{dy}{dt} = \omega_2 f_2, \end{cases} \tag{19.9}$$

Table 19.1 Parameter range of internal stable and external unstable limit cycle for class I Milky Way (Galaxy) dynamical system

α	β	a	b	m	R	ω	$\lambda\omega$	$\mu\omega$
$[0, 3]$	$[-3, 3]$ $\alpha+\beta \neq 0$	$(0, 400]$	$(0, 400]$	$[1.5, 2.11]$ $[3.5, 4.11]$ $[5.5, 6.13]$	$[0.8, 1.1]$	$\omega \geq 0.8$	$\lambda\omega \geq 0.8\sqrt{\frac{b}{a}}$	$\mu\omega \geq 0.8\sqrt{\frac{a}{b}}$
$[-3, 0]$	$[0, 3]$ $\beta \neq \lvert\alpha\rvert$ $\lvert\alpha\rvert < 1.1\beta$	$(0, 400]$	$(0, 400]$	$[1.5, 2.11]$ $[3.5, 4.11]$ $[5.5, 6.13]$	$[0.8, 1.1]$	$\omega \geq 0.8$	$\lambda\omega \geq 0.8\sqrt{\frac{b}{a}}$	$\mu\omega \geq 0.8\sqrt{\frac{a}{b}}$

Table 19.2 Parameter range of internal unstable and external stable limit cycle for class II Milky Way (Galaxy) dynamical system

α	β	a	b	m	R	ω	$\lambda\omega$	$\mu\omega$
$[-3, 0]$	$[0, 3]$ $\lvert\alpha\rvert \geq 1.1\beta$	$(0, 400]$	$(0, 400]$	$[1.5, 2.11]$ $[3.5, 4.11]$ $[5.5, 6.13]$	$[0.8, 1.1]$	$\omega \geq 0.8$	$\lambda\omega \geq 0.8\sqrt{\frac{b}{a}}$	$\mu\omega \geq 0.8\sqrt{\frac{a}{b}}$
$[-3, 0]$	$[-3, 0]$	$(0, 400]$	$(0, 400]$	$[1.5, 2.11]$ $[3.5, 4.11]$ $[5.5, 6.13]$	$[0.8, 1.1]$	$\omega \geq 0.8$	$\lambda\omega \geq 0.8\sqrt{\frac{b}{a}}$	$\mu\omega \geq 0.8\sqrt{\frac{a}{b}}$

In the cosmic space-time (19.8)–(19.9), on the OXY plane of the two-dimensional flow field, for (19.8)–(19.9), according to the limit cycle theory [500–502], the motion behavior of stars as particles in the phase plane is considered. The planetary streamlines inside the semi-stable limit cycle tend to limit cycle C, while the streamlines outside are away from C, as shown in Fig. 19.6a. Similarly, for (19.9), in the two-dimensional flow field of cosmic space-time stellar motion, the stellar motion streamlines inside limit cycle C are all far away from C, and the external streamlines tend to close to the center of limit cycle C, as exhibited in Fig. 19.6c.

In universe, the inner stars of the semi-stable limit cycle C tend to limit cycle C and increase with the time t, the behavior of the semi-stable limit cycle is shown in Fig. 19.6b. Similarly, the stars motion streamlines inside the limit cycle C are all far away from C, while the stars outside C tend to converge on the central axis. With the increase of time t, the vortices rise and present a "vortex" shape, as shown in Fig. 19.6d.

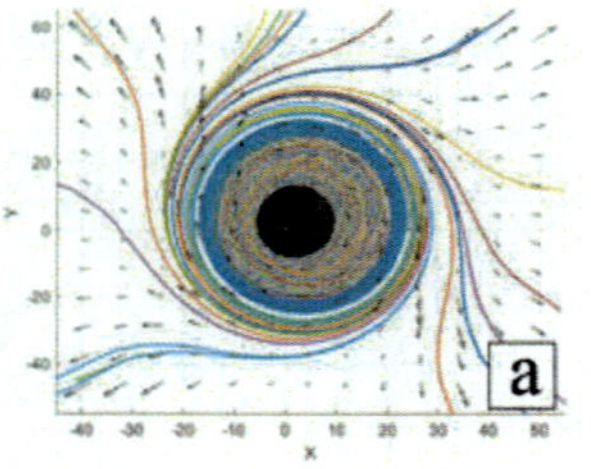

(a) Planar phase diagram of stellar motion semi-stable limit cycle with externally unstable and internally stable which attracts streamlines in the universe

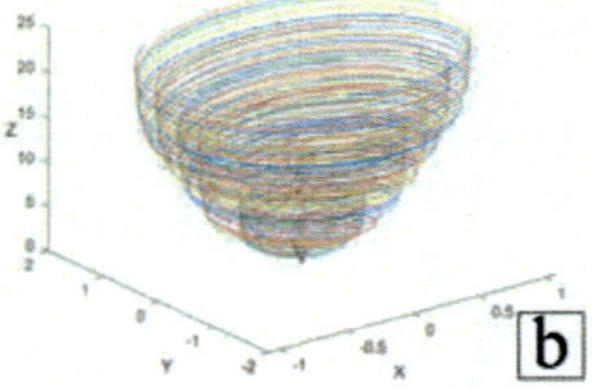

(b) Planar phase diagram of stellar motion semi-stable limit cycle with externally stable and internally unstable which attracts streamlines

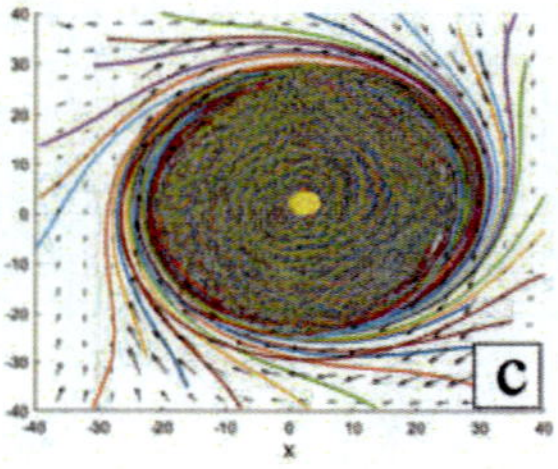

(c) Space-time behavior of stellar motion semi-stable limit cycle with externally unstable and internally stable

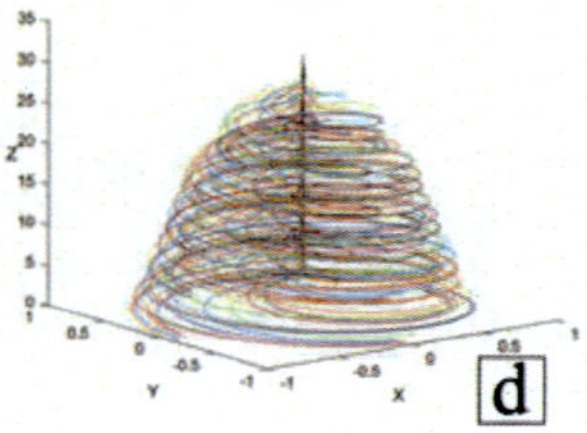

(d) Space-time behavior of stellar motion semi-stable limit cycle with externally stable and internally unstable

Fig. 19.6 The behavior of the stellar motion semi-stable limit cycle

19.2.2 The Space-Time Process of the Milky Way from Theoretical Principle to Cosmic Reality

The nonlinear dynamical system (19.7) of the Milky Way is generated by the attraction of the vortex motion of the interstellar stars in the universe. In view of the vortex attraction property, take any star within the stellar vortex flow field in the Milky Way as a particle $p(x, y, z)$, it is disturbed by interstellar multi-level gravity, gas pressure, magnetic force, and competition of rotational inertia, then it moves to $\bar{p}(\bar{x}, \bar{y}, \bar{z})$, which is

$$p(x, y, z) \xrightarrow{transform\ into} \bar{p}(\bar{x}, \bar{y}, \bar{z}).$$

Table 19.3 The transformation matrix

Translation	Rotation	Scale	Shear
$T = \begin{bmatrix} 1 & 0 & 0 \\ 0 & 1 & 0 \\ d & n & 1 \end{bmatrix}$	$R = \begin{bmatrix} cos\theta & sin\theta & 0 \\ -sin\theta & cos\theta & 0 \\ 0 & 0 & 1 \end{bmatrix}$	$S = \begin{bmatrix} k & 0 & 0 \\ 0 & k & 0 \\ 0 & 0 & k \end{bmatrix}$	$Shear = \begin{bmatrix} 1 & tan\phi & 0 \\ tan\phi & 1 & 0 \\ 0 & 0 & 1 \end{bmatrix}$

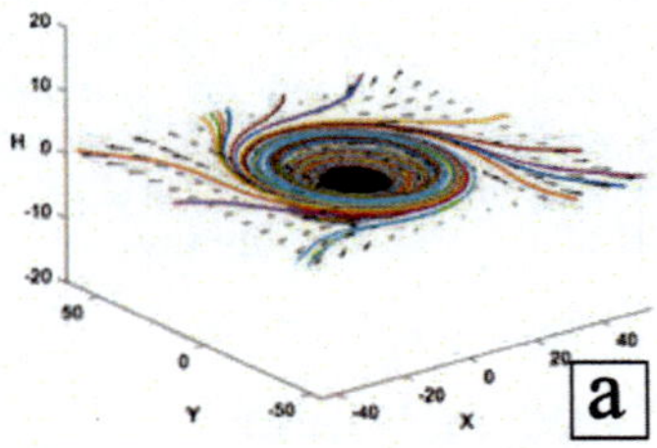

(a) The space-time behavior of class I galaxy that is externally unstable but internally stable

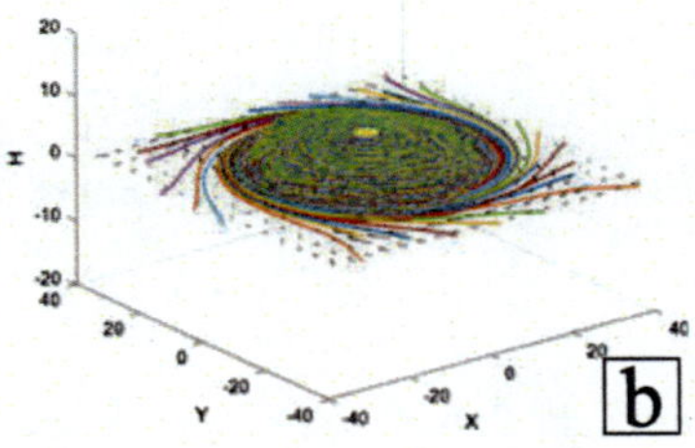

(b) The space-time behavior of class II galaxy that is externally stable but internally unstable

Fig. 19.7 The space-time behaviors of the two class galaxies

Usually after a finite or an infinite number of translation, rotation, scale, shear, flipping and other basic transformations, and their transformation matrix is shown in the Table 19.3.

The transformation from the original point $p(x, y, z)$ to the target point $\bar{p}(\bar{x}, \bar{y}, \bar{z})$ is obtained through the following transformation

$$(\bar{x}, \bar{y}, \bar{z})^{Tran} = \sigma_n \sigma_{n-1} \cdots \sigma_2 \sigma_1 (x, y, z)^{Tran}, \tag{19.10}$$

where the transformation $\sigma_n \sigma_{n-1} \cdots \sigma_2 \sigma_1$ represents the continuous transformation of translation, rotation, scale and shear. Note that transformation processes such as distortion, tilt, distortion nesting of different light years in the universe are sometimes used to form two types of the Milky Way. For example, the mathematical systems (19.8) and (19.9) of class I galaxy undergo a series of transformations of (19.10) and appropriate distortions, tilts, analogs, nesting and so on, then the space-time behaviors of the two classes of galaxies is shown in Fig. 19.7.

19.2.3 Production Principles, Classification, Singularities of Black Hole, Black Hole "M87" and "Sagittarius A"

According to the mathematical principle (19.7) of the formation of the Milky Way, the semi-stable limit cycles are divided into two classes, also, galaxies are divided into two classes. The following will confirm that black holes are contained in galaxies, therefore two classes of black holes are generated.

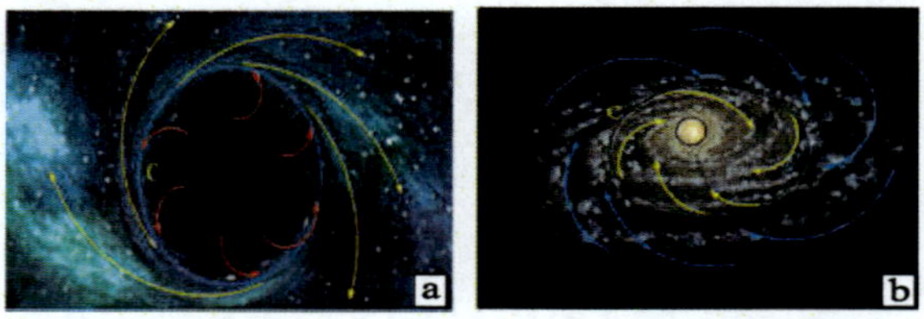

Fig. 19.8 The formation principle of the two class galaxies

(a) The formation principle of class I galaxy black hole formed by the semi-stable limit cycle with externally unstable but internally stable

(b) The formation principle of class II galaxy black hole formed by the semi-stable limit cycle with externally stable but internally unstable

Production Principle of Dark Matter Black Hole (Empty Hole)
Note that (19.7) can be transformed into the following system by polar coordinate transformation,

$$\begin{cases} r' = r(ar^2\cos^2\theta + br^2\sin^2\theta - R^2)^2(a\cos^2\theta + b\sin^2\theta) + r\sin\theta\cos\theta(\omega\lambda - \omega\mu), \\ \theta' = (\beta - \alpha)(ar^2\cos^2\theta + br^2\sin^2\theta - R^2)^2\sin\theta\cos\theta - (\omega\lambda\sin^2\theta + \omega\mu\cos^2\theta), \end{cases} \tag{19.11}$$

because of the vastness of stellar motion in the universe, take $a > b > 0$, $\omega\lambda \overset{\Delta}{=} \omega\mu$, from (19.11), it is semi-stable limit cycle, thus the Milky Way is a whirlpool of attractive behavior under the competition of gravity, gas pressure, magnetic forces, and moment of inertia. If C is externally unstable but internally stable, which belongs to the class I galaxy, thus $\alpha > 0$, $\beta > 0$, then $r^2a > R^2$, and $r' > 0$, which indicates that inside the domain C in the endless abyss, stars as particles, the radius of the whirlpool attraction movement becomes larger and larger, and finally, the whirlpool attracts and diverges approach the inner edge of C, and then lead to the vast void of infinite abyss inside C, this is the generation of class I black hole, as shown in Fig. 19.8a. By (19.7), in Fig. 19.8b, take

$$\begin{aligned} \omega\lambda p(x, y) &= y - 0.1(\tfrac{1}{3}x^3 - x) - \alpha(ax^2 + by^2 - R)^m, \\ \omega\mu q(x, y) &= -x - \beta(ax^2 + by^2 - R)^m, \end{aligned}$$

then the detailed process of black hole generated by mathematical system is shown in Fig. 19.8b.

Production Principle of Physical Black Hole (Real Hole)
From polar coordinates transformation (19.11), $a > b > 0$ and $\omega\lambda = \omega\mu$, and due to attractive behavior of vortex motion in semi-stable limit cycle, C is externally stable but inwardly unstable, which belongs to class II galaxy, thus $\alpha < 0$, $\beta < 0$, then $r^2a < R^2$, and $r' < 0$. From the viewpoint of the external of C, under the competition of gravity, gas pressure, magnetic force and moment of inertia, the stars are attracted in a whirlpool, which enable the motion radius smaller and smaller,

and they converge and condense on the central axis of the upper part of ring C. The convergence of stars leads to the thermonuclear reaction under violent collision, thus the formed space-time center above C is a boiling real hole, which is also a physical black hole. As shown in Fig. 19.9a, its process is the continuous state of instantaneous convergence and instantaneous combustion, which is the source of what is commonly called a massive quality black hole, while "massive quality" is just a continuous instantaneous process of convergence and combustion. Similarly, in the system of Fig. 19.9b, actually in the system (19.7), the vertical direction of the interference function is taken as

$$\begin{cases} \dot{z} = \varepsilon(f)t, \\ \omega\lambda p(x, y) = 2y - \alpha(ax^2 + by^2 - R)^m, \\ \omega\mu q(x, y) = 0.8y(x^3 - x^2 - 1) - 2x - 3x^2 - \beta(ax^2 + by^2 - R)^m, \end{cases}$$

the systematic process of generation of black hole is shown in Fig. 19.9b.

Perfect Unity of Black Holes and Mathematical Principles
To sum up, there are two forms of semi-stability for nonlinear system (19.7), black holes in the universe also have two forms, physical black holes and dark energy black holes, so they are exactly corresponding to each other. It can be seen from the above, (19.3) the semi-stable limit cycle of nonlinear dynamic system reveals the formation mechanism of black holes, (19.4) the random elliptical circle C of semi-stable limit cycle divides a layer of the universe into inner and outer regions, thus two forms are emerged, internally stable but externally unstable, internally unstable but externally stable. Two kinds of semi-stable limit cycles correspond to real holes and empty holes in the universe, which are physical black holes and dark energy black holes. This shows that mathematical theory and objective reality achieve perfect unity and symmetry, and it explains the classical application of semi-stable limit cycle theory in the formation and classification of galaxies and black hole formation in the universe.

19.2.3.1 The Singularity of Black Hole

In cosmic space $OXYZ$, both physical black holes (real holes) and dark matter black holes (empty holes) arise from the semi-stable limit cycles of nonlinear systems (19.7), and system (19.7) has equilibrium points, namely, the singularity problem. It has an ordinary fixed point (0, 0), and for the space of x and y, under the approximate equivalence of correlation coefficients, take $\xi = \frac{y}{x}$ in (19.7), $\mathbb{M}$ ={stars, star clusters, nebulae, interstellar gas, interstellar dust}, ξ is the equilibrium point at which any element of $\mathbb{M}$ moves and merges in the universe, which is the singularity of (19.7). By (19.7), the formula for nontrivial singularity is $\xi = \pm i$ (i is the imaginary unit).

For the singularity of class I black hole $\xi_I = \pm i$, astral motion is semi-stable limit cycle with externally unstable and internally stable, as for the motion of any star inside the semi-stable limit cycle, focus on (19.11), $r' > 0$, astral motion radius gets bigger and bigger, then tends to the inner edge of C, thermonuclear reaction

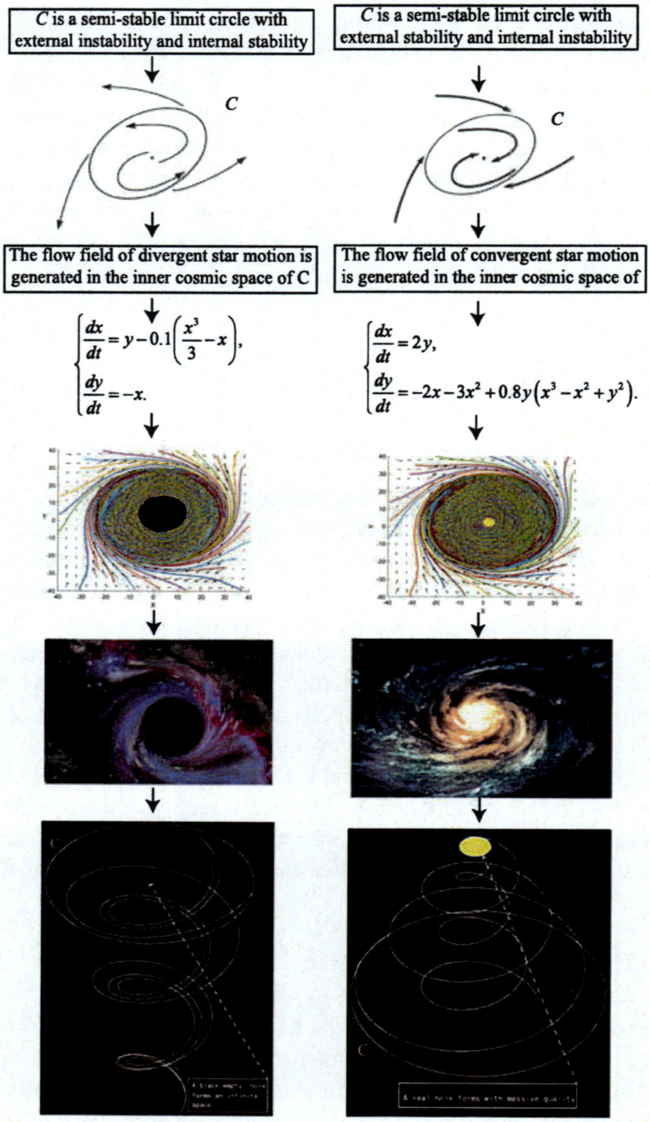

(a) The formation process of dark energy black hole formed by the semi-stable limit cycle with internally stable but externally unstable

(b) Physical black holes are formed by semi-stable limit cycle with internally unstable but externally stable

Fig. 19.9 The formation process of massive quality black hole and the semi-stable limit cycle

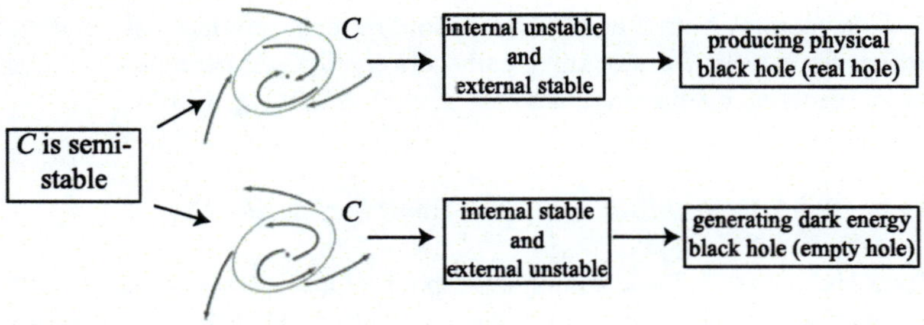

Fig. 19.10 Two black holes and semi-stable limit cycles

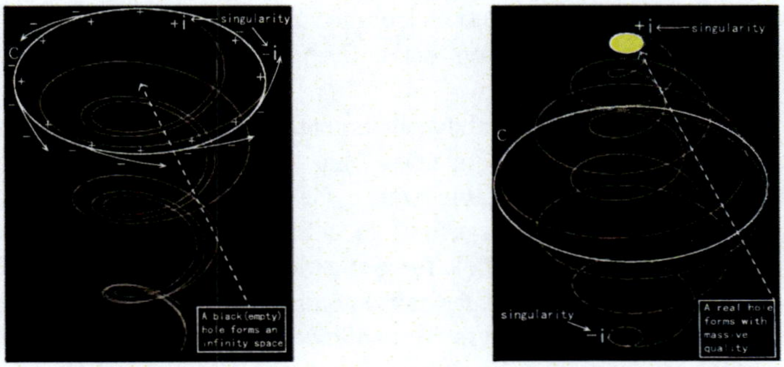

(a) Singularity of class I black hole $\xi_I = \pm i$

(b) Singularity of class II black hole $\xi_{II} = \pm i$

Fig. 19.11 Singularities of two class black holes

occurs, producing intense light beam, thus the closed inner edge of C is formed, and it keeps in balance with the outer stellar motion of C! The closed inner edge is the singularity of the system, denote it as $+i$. In addition, in the interior of C, the motion of the stars tends to the inner edge of C, and continue to swallow all the stars and light, thus forming empty holes, namely dark energy black holes. Outside of C, the stellar motion diverges along the outer edge of C. By physical and mathematical symmetry, the singularity of the outer edge of C is $-i$, as shown in Fig. 19.10a (Fig. 19.11).

For the singularity of class II black hole $\xi_{II} = \pm i$, astral motion is semi-stable limit cycle with externally stable and internally unstable, as for the motion of any star outside the semi-stable limit cycle, focus on (19.11), $r' < 0$, the radius of star attraction gets smaller and smaller, then converges towards the inner center of C, in form, it looks like a real star with super large mass and intense thermonuclear reaction, and produces a strong light beam. In essence, it is a process of continuous disappearance of fusion stars and continuous convergence under the action of gravity, thus forming a real hole, namely, a physical black hole! Its singularity is $+i$. In addition, the interior and exterior of C are in a stable state of equilibrium. As the stars inside C are far away from C and diverge, according to the physical and mathematical symmetry, the stars in C are symmetrically converged on the central axis of C in the opposite direction, and the singularity is $-i$, as shown in Fig. 19.10b.

19.2.3.2 Black Hole "M87" and Black Hole "Sagittarius A"

(I) Black Hole "M87": The common description of black holes is that (19.3) black holes are arisen from dying super massive stars, (19.4) when a supergiant star with more than 100 times the mass of the sun dies, the fuel in its core has run out and it cannot keep burning. The core collapses under the gravitational squeeze and forms a black hole, (19.5) It swallows up stars from the inside out, (19.6) As the black hole swallows objects, a bright accretion disk is formed around it, (19.7) The center of the $M87$ is an irascible black hole.

It is important to note that the above statement is just a guess, it is not scientific. From Sect. 19.2.1, the singularity of class I black hole is $\xi_I = \pm i$, it comes from the vortex attraction of a semi-stable limit cycle which is externally unstable but internally stable. From the significance of $\pm i$ in Sect. 19.2.1, under the competition of gravity, gas pressure, magnetic field force and moment of inertia, the stellar motion inside C tends to $+i$, while the stellar motion outside of C tends to $-i$. Thus within the ring C, all stars in endless abyss are swallowed up, light also can't escape, and the radius of the star's rotational motion becomes larger and larger, it diverges and collides towards the inner edge of C acutely and rotationally, which is the singularity $+i$, then thermonuclear reactions occur, producing the intense beams and spectacular jets that people have observed. This is the origin of gamma ray bursts or fast radio bursts. Correspondingly, in C, the black hole of the infinite abyss forms, that is, the empty hole! This is the origin of $M87$!

(II) Refer to the black hole "Sagittarius A" in the Milky Way: in the past, astrophysicists have always thought that, why the infinite stars are concentrated in the central region of the Milky way, so that the brightness of the central region is extremely amazing, then what is in the center of the Milky way that drives the infinite stars in the whole galaxy? In addition, a very strong radio source has been found in the center of the Galaxy! It's called "Sagittarius A".

These stars are found to come from a celestial body in Sagittarius A and are firmly "held" by strong gravity. To produce such a strong gravity, the mass of about 4 million suns is needed to concentrate, and the huge gravitational mass of the galactic center is concentrated in a very small range. How to have such a huge mass in such a small range? Astrophysical scholars believe that the center of the galaxy is a massive star cluster, but its influence is very limited. It can only produce enough gravitational effect on other nearby stars, and it can't drag the stars of the whole galaxy around it, so it is considered that there are mysterious forces in the universe that are not recognized by us, and scientists attribute it to dark matter [512–515].

All the above questions are also unscientific. According to the above section, the singularity of class II black hole is $\xi_{II} = \pm i$. Under the competition of gravity, gas pressure, magnetic force and moment of inertia, the key stars produce the vortex attraction of semi-stable limit cycle, which is externally stable and internally unstable. From the meaning of singularity $\pm i$, the stars motion outside of C tends to $+i$, and it converges on the central axis of the galaxy, it looks like a "huge mass" physical entity in form, but it is actually a quiet cluster. And it does not devour other stars, but is the result of convergence and condensation, so it does not form a spectacular jet like the $M87$! According to the discussion in Sect. 19.2.1, the motion of the outer stars of C tend to $-i$, which is the origin of black hole "Sagittarius A"!

Black Hole and the Existence of Positive and Anti Matter Universes
From Sect. 19.2.3, black holes are divided into two categories, physical black holes and dark matter black holes. And physical black hole = quiet black hole = real hole, dark energy black hole = irascible black hole = empty hole, thus, black hole = physical black hole + dark energy black hole = quiet black hole + irascible black hole (black hole highlighted by dark matter) = real hole + empty hole.

Therefore, there are two types of galaxies in the universe, and the physical black hole corresponds to the positive material galaxy, also known as the positive material universe, and the dark energy black hole corresponds to the anti matter galaxy, also known as the anti matter universe. Then, we have, universe = galaxies with semi-stable limit cycle (the Milky Way is one of them) + $\mathbb{M}$ = galaxies of the limit cycle with externally unstable but internally stable + galaxies of the limit cycle with externally stable but internally unstable + $\mathbb{M}$ =infinite physical black hole galaxies + infinite dark energy black hole galaxies + $\mathbb{M}$ =positive material universe + anti matter universe + $\mathbb{M}$ =infinite positive material galaxies + infinite anti matter galaxies + $\mathbb{M}$ =infinite positive material galaxies + infinite dark matter galaxies + $\mathbb{M}$.

Black Hole and Galaxy Under the Noise Interference
Stars, star clusters, nebulae, interstellar gas and dust in the universe contract and flatten under the competition of gravity, gas pressure, magnetic field force and moment of inertia, forming an astronomical unit. According to the mathematical principle of black hole and galaxy formation, this process tends to the semi-stable limit cycle, they interfere with each other and the trajectory may not be smooth. According to the nonlinear system (19.8) in Sect. 19.2.1, in interstellar motion, different stars produce different levels of tilt and twist values such as rotation angle, tangent angle, expansion coefficient and twist degree under the interference of different gravitation, which is similar to the behavior under the interference of transformation (19.10), as shown in Fig. 19.12a–c. It is also shown in Fig. 19.12d–f from nonlinear system (19.9) and under interference similar to transform (19.10).

19.2.4 Spatiotemporal Structure (Self-Similar Fractal Behavior) of Black Holes

(I) Spatiotemporal structure of physical black holes: in the nonlinear system (19.7), as for the motion of n stars, move from the initial time $(x_{i0}, y_{i0}), i = 1, 2, \ldots, n$, for each (x_{i0}, y_{i0}), n vortex trajectories are generated from the nonlinear system (19.7), and each vortex trajectory is set as $C(x_{i0}, y_{i0}), i = 1, 2, \ldots, n$, then the trajectories of n stars are recorded as E_n, thus

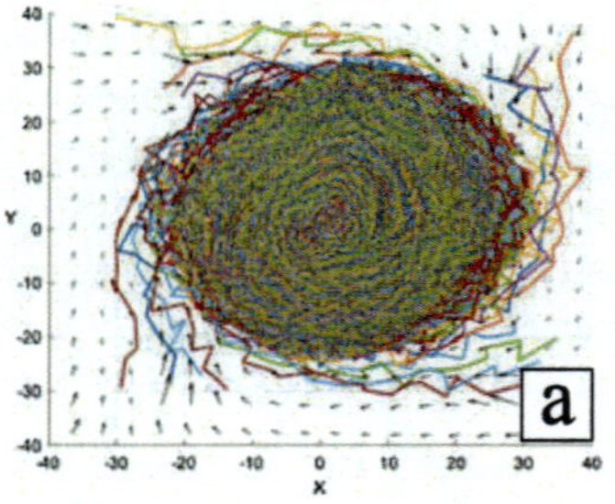

(a) plane phase diagram under interference of system (19.10)

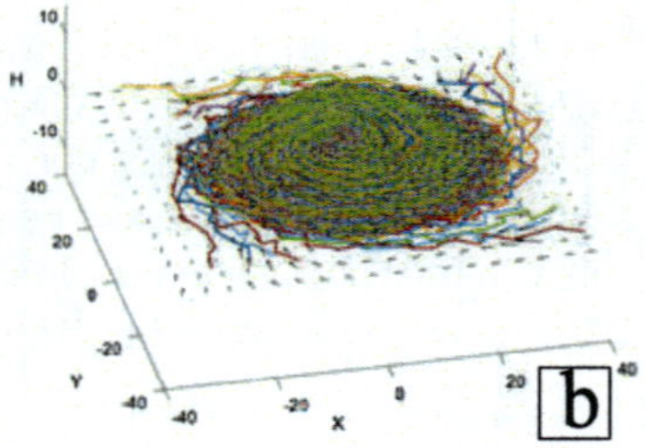

(b) space-time behavior under interference of system (19.10)

(c) physical black hole under interstellar gravitational interference

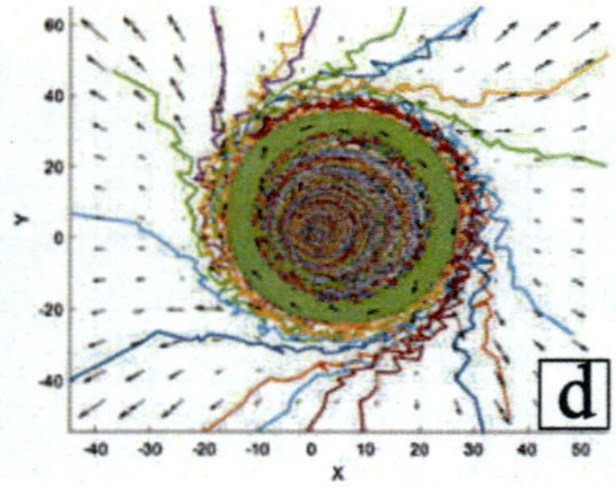

(d) plane phase diagram of system (19.9)

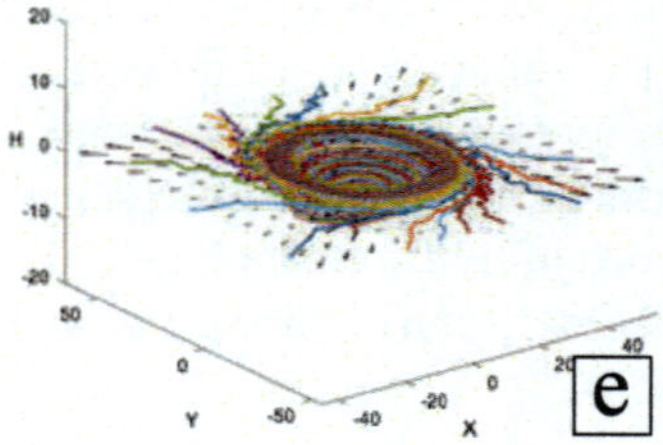

(e) space-time behavior under interference of system (19.9)

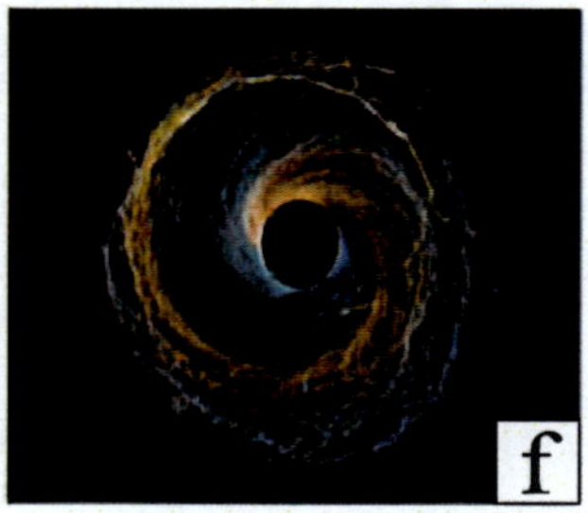

(f) dark energy black hole under interstellar gravitational interference

Fig. 19.12 Spatiotemporal structure of system (19.9) and the black holes

$$E_n = \bigcup_{i=1}^{n} C(x_{i0}, y_{i0}), \tag{19.12}$$

then (19.12) is the case of the spatial vortex attraction of the motion of *n* stars formed by system (19.7). Practically, the physical black hole formed by the space vortex is composed by the uniform gravitational rules that each star moves from east to west as a particle, and the gravity is also consistent with itself. Therefore, the physical black hole in the universe is

$$E_\infty = \bigcup_{i=1}^{\infty} C(x_{i0}, y_{i0}), \tag{19.13}$$

from the vorticity of black holes, (19.13) and [516], the space-time structure of the universe is infinitely nested self-similar fractal behavior, as shown in Fig. 19.12a, b. Besides, from the limit cycle theory [500], (19.7) is externally stable and internally unstable, which means (19.7) can form a black hole.

(II) The space-time structure of dark energy black hole: refer to the system (19.7), the motion of *n* stars at the initial time $(\bar{x}_{i0},\ \bar{y}_{i0}),\ i = 1, 2,\ \ldots, n$ in the universe is the spatial vortex

$$\bar{E}_n = \bigcup_{i=1}^{n} \bar{C}(\bar{x}_{i0},\ \bar{y}_{i0})$$

formed by *n* particles, where $\bar{C}(\bar{x}_{i0},\ \bar{y}_{i0})$ is the i-th vortex trajectory generated by $(\bar{x}_{i0},\ \bar{y}_{i0})$. Similarly, the dark energy black hole in the universe is

$$\bar{E}_\infty = \bigcup_{i=1}^{\infty} \bar{C}(\bar{x}_{i0},\ \bar{y}_{i0}),$$

focusing on the attractiveness of black hole flow field, from $\bar{E}_\infty$ and [516], the space-time structure of dark energy black hole is infinitely nested self-similar fractal behavior, as shown in Fig. 19.13c, d.

The Existence of the Black Hole Formulated by Symmetry

It can be seen from Sect. 19.2.1 that both physical black (real) holes and empty holes are the equilibrium states maintained by the motion of stars in the universe. In these two different equilibrium states, it is apparent that, firstly, the formation of the fixed circle *C* of black (empty) holes is due to the fact that in the nonlinear system (19.11) represented by polar coordinates, when $r' > 0$, the motion of the internal stars of *C* tends to the singularity $+i$, while the motion of the external stars diverges and tends to the singularity $-i$, resulting in such a continuously changing physical scene of closed circle *C* exists and does not exist in form. Moreover, the equilibrium physical state with such characteristics is formulated by the symmetry of internal singularities $+i$ and external singularities $-i$. Secondly, from the polar coordinate system (19.11), $r' < 0$, due to the different direction of attraction or tendency between the external and internal stars of the closed circle C, the formation of physical black (real) hole is just appears at the singularity $+i$ on the central axis of C. And the existence of the equilibrium state of the physical real hole is formulated by the symmetry of these two singularities.

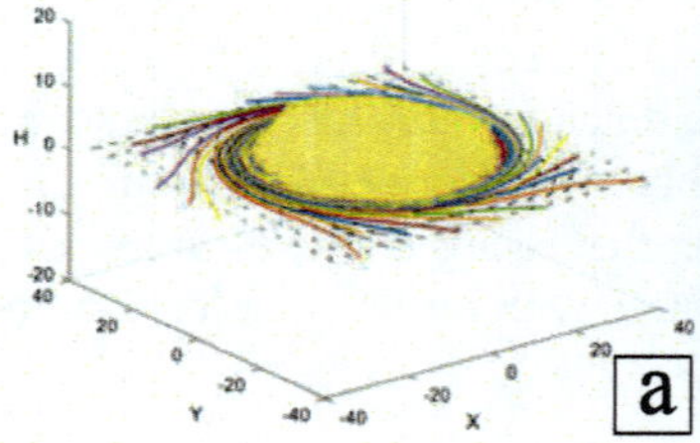

(a) The space-time structure of the physical black hole generated by the limit cycle with internally unstable and externally stable

(b) The space-time structure of the galactic physical black hole in the universe

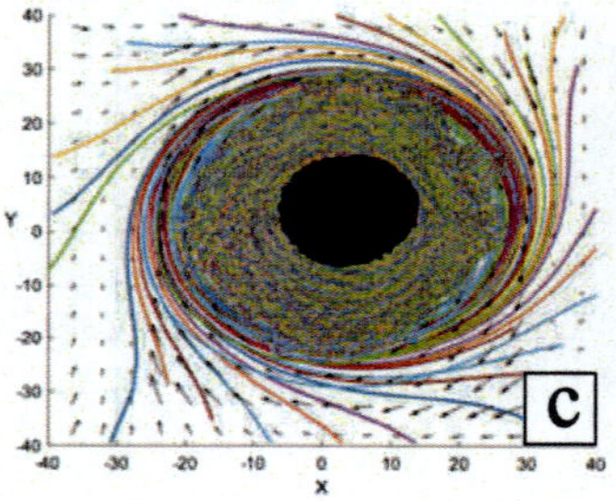

(c) The space-time structure of the dark energy black hole generated by the limit cycle with internally stable and externally unstable

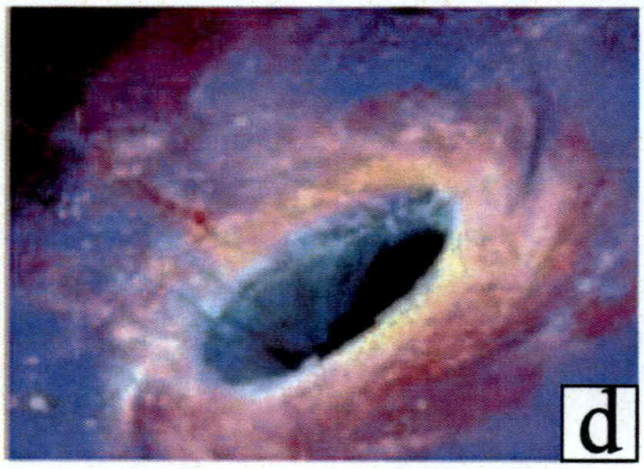

(d) The space-time structure of the galactic dark energy black hole in the universe

Fig. 19.13 The space-time structure of the physical black hole and structure of the galactic dark energy black hole

Fractal Prediction of the Complexity of the Milky Way

From [496, 497, 508–513], it can be proven that the space-time structure of the Milky Way has Cantor self-similar fractal structure behavior. If the cross section of an attractor is a set of Cantor type, it is called a strange attractor, and the dynamic system is called fractal.

(I) Fractal prediction of the complexity of class I Milky Way with dark energy black hole.

For the nonlinear system (19.7), two types of galaxies with physical black hole and dark energy black hole in the universe are obtained by finite affine transformation, as shown in Fig. 19.9. Figure 19.9a is taken as an example to study the complexity

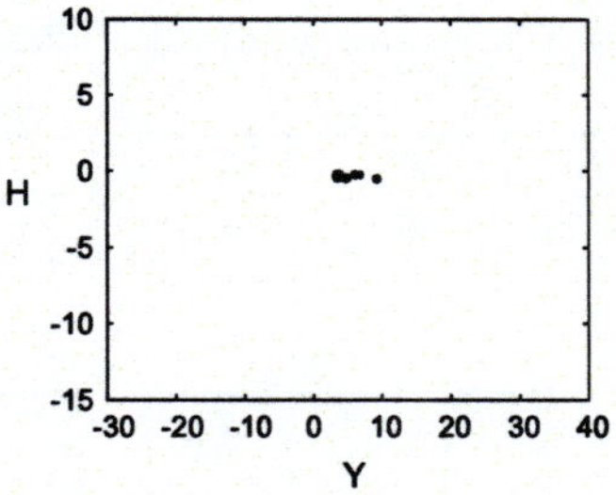

(a) The cross section of $x = 24.5$ light year of dark energy black hole

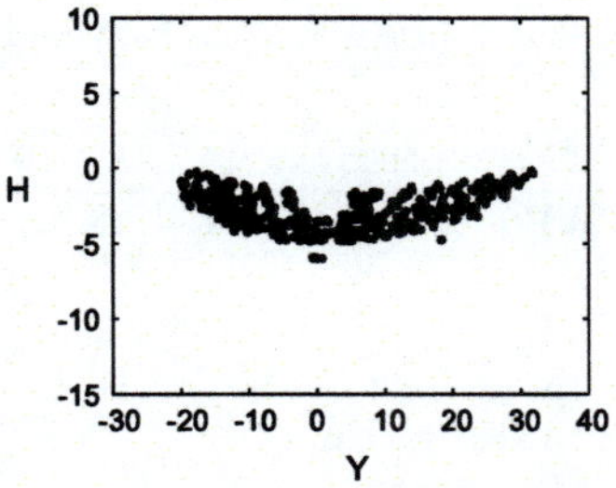

(b) The cross section of $x = 15.6$ light year of dark energy black hole

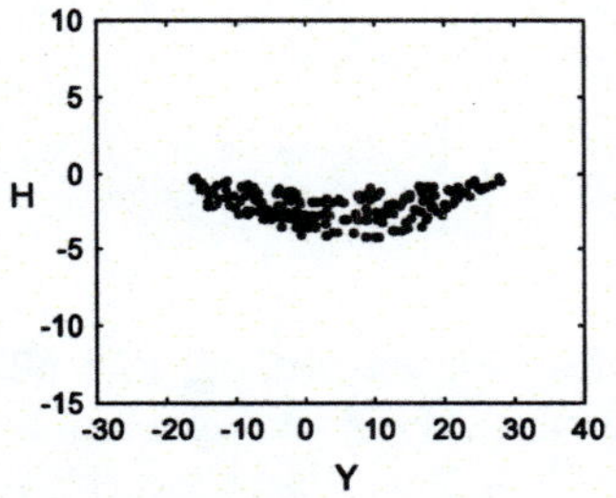

(c) The cross section of $x = 7.8$ light year of dark energy black hole

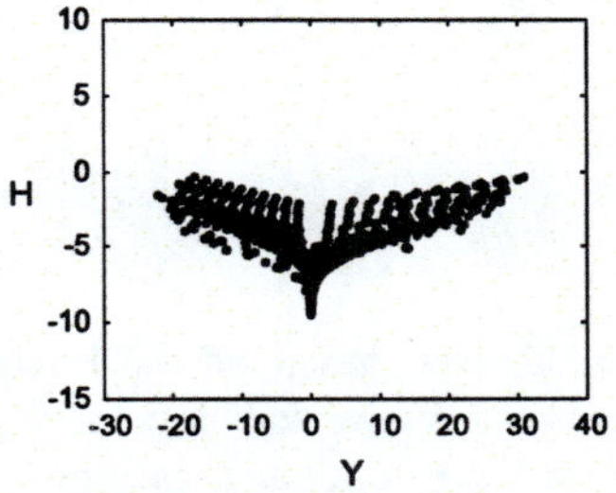

(d) The cross section of $x = 0$ light year of dark energy black hole

Fig. 19.14 Moving particle diagram of stars in vertical section of dark energy black hole

of the space-time structure of galaxies with dark energy black holes. It is known from Sect. 19.2.1, as for class I galaxy, the Milky way with such black holes is a self-similar fractal structure [496, 497, 508–513], so the fractal dimension can be utilized to describe its complexity. Firstly, take the vertical cross section of the Milky way, and a series of cross sections are obtained by finite vertical interception. When the interval between them is infinitely small, the Milky way is composed of many sections with infinitesimal intervals, so the complexity of the space-time structure of the Milky way can be revealed according to the fractal dimensions of the sections.

Focus on the axisymmetry of the Milky Way, consider the situation in $|x| \leq 25$ light year region, it is necessary to intercept the Milky Way along the x axis because the geometric structures of $x < 0$ is same as $x > 0$. Without losing generality, we only show the $(0 \leq x < 25)$ section. Take the vertical section and we obtain the Fig. 19.14.

From the perspective of fractal, fractal dimension can describe the variation of the section, the change of cross section graphics can cause the change of the fractal dimension, fractal dimension corresponds to cross section graph, through the adjustment of the parameters, the fractal diagram of different galactic sections can

Table 19.4 Fractal dimension D and box number $\log_2(N)$ of dark energy black hole in the universe

	1	2	3	4	5	6	7	...	61
D	1.5121	1.5095	1.5056	1.5129	1.4847	1.4909	1.5105	...	1.0909
$\log_2 N$	18.319	18.378	18.383	18.505	18.178	18.187	18.368	...	14.370

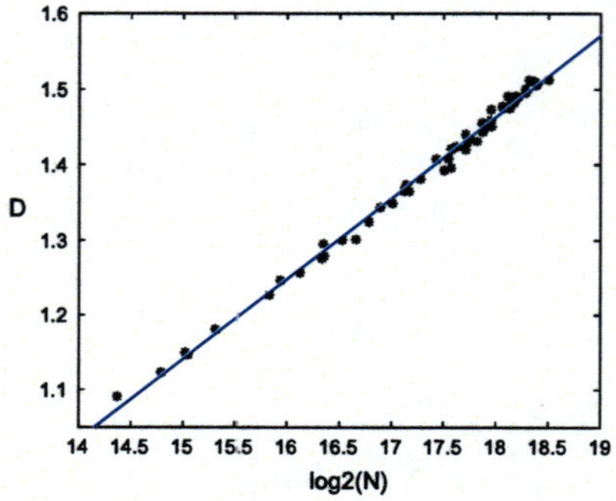

Fig. 19.15 The curve of fractal dimension D and box numbers $\log_2(N)$ of class I Milky Way ("*": data points; "–": Fitting curve)

be obtained. And based on the fractal box dimension method, take 60 sections (Fig. 19.14a–d shows four legends in 60 groups), the fractal dimensions and the box numbers are computed, as shown in the Table 19.4. Where $\log_2(N)$ represents the numbers of small squares with a side length of 0.001 covering the cross-section edge of the Milky way, besides, the complexity of class I Milky way is depicted by the fractal dimension D.

Figure 19.14a–d correspond to the data in group 1, group 35, group 47 and group 60 in the table respectively. These four graphics describe that the fractal dimension and the number of boxes increase one by one, and also reflect that the fractal dimension and the number of boxes of the 60 groups of the cross-section of the Milky way from the edge to the center gradually increase, and the complexity of the cross-section augments from weak to strong.

The data in Table 19.4 were fitted to obtain the numerical curve as shown in Fig. 19.15. In Fig. 19.15 we can see that the 60 cross section data points in Table 19.4 from the "negative" near-vortex edge to the vortex center and then to the "positive" near-vortex edge ($-16 \le x \le 25$) are uniformly distributed on the fitting curve. The fractal dimension D of the Milky way is positively correlated with the number of boxes $\log_2(N)$. The fractal dimension D increases with the number of boxes $\log_2(N)$ increasing, and the complexity of the space-time structure of the Milky Way also increases gradually. The fractal dimension of the vortex cross-section is mostly in the range (1.1, 1.5) and the complexity of the Milky Way is also in this range. The relationship between the fractal dimension D of the Milky Way and the number of boxes $\log_2(N)$ is $D = 0.1074 \log_2 N - 0.4687$, it gives a prediction of the complexity of class I Milky Way.

(II) Fractal prediction of the complexity of class II Milky Way with physical black hole.

Table 19.5 Fractal dimension D and box numbers $\log_2(N)$ of physical black hole in the universe

	1	2	3	4	5	6	7	…	61
D	1.5072	1.4953	1.4912	1.4932	1.5001	1.4180	1.4265	…	1.4239
$\log_2 N$	18.465	18.427	18.363	18.363	18.442	17.728	17.796	…	17.830

Similarly, after affine transformation, (19.7) is still in the region of $-20 \leq x \leq 25$ light year. Moreover, 60 sections are taken as 60 groups to calculate their fractal dimension and the number of boxes, as shown in the Table 19.5. Where $\log_2(N)$ represents the numbers of small squares with a side length of 0.001 covering the cross-section edge of the Milky way, besides, the complexity of the attraction of space-time gravitational vortices is depicted by the fractal dimension D. These 60 groups also reflect the trend that the fractal dimension D and box numbers $\log_2(N)$ of the cross-section from the edge to the center of the vortex increase gradually, what's more, the complexity of the cross-section of the Milky way with physical black hole increases from weak to strong.

Fitting the data in Table 19.5 [500, 501], it can be seen that the 60 cross section data points in Table 19.5 from the "negative" near-vortex edge to the vortex center and then to the "positive" near-vortex edge ($-20 \leq x \leq 25$ light year) are uniformly distributed on the fitting curve. The fractal dimension D of the Milky way is positively correlated with the number of boxes $\log_2(N)$. The fractal dimension D increases with the number of boxes $\log_2(N)$ increasing, and the complexity of structure of the Milky way also increases gradually. The fractal dimension of the vortex cross-section is mostly in the range (1.41 light year, 1.52 light year) and the complexity of the Milky way is also in this range. The relationship between the fractal dimension D of class II Milky way and the number of boxes $\log_2(N)$ is $D = 0.1097\log_2 N - 0.526$. It describes the fractal prediction of the complexity of the galaxy with physical black hole in the circle.

References

1. C Tunnard and B Pushkarev. 1963. Man-made america: Chaos or control? *Geographical Journal*, 136(2).
2. R M May. 1975. Biological populations obeying difference equations: stable points, stable cycles, and chaos. *Journal of Theoretical Biology*, 51(2):511–524.
3. M Mackeyl and L Glass. 1977. Oscillation and chaos in physiological control systems. *Science*, 197(4300):287–289.
4. J P Crutchfield and B A Huberman. 1980. Fluctuations and the onset of chaos. *Physics Letters A*, 77(6):407–410.
5. J L Hudson. 1981. Chaos in the belousov-zhabotinskii reaction. *Journal of Chemical Physics*, 74(11):6171–6177.
6. S Fishman, D R Grempel and R E Prange. 1982. Chaos, quantum recurrences, and anderson localization. *Physical Review Letters*, 49(8):509–512.
7. M R Guevara and L Glass. 1982. Phase locking, period doubling bifurcations and chaos in a mathematical model of a periodically driven oscillator: A theory for the entrainment of biological oscillators and the generation of cardiac dysrhythmias. *Journal of Mathematical Biology*, 14(1):1–23.
8. C Grebogi and E Ott. 1983. Crisis, sudden changes in chaotic attractors and transient chaos. *Physica D Nonlinear Phenomena*, 7(1):181–200.
9. A Brandstater, J Swift, H L Swinney, et al. 1983. Low-dimensional chaos in a hydrodynamic system. *Physical Review Letters*, 51(16):1442–1445.
10. H L Swinney. 1983. Observations of order and chaos in nonlinear systems. *Physica D-nonlinear Phenomena*, 7(1–3):3–15.
11. M Jensen, P Bak and T Bohr. 1984. Transition to chaos by interaction of resonances in dissipative systems. i. circle maps. *Physical Review A*, 30(4):1970–1981.
12. M D Feit and Jr J A Fleck. 1984. Wave packet dynamics and chaos in the henon-heiles system. *Journal of Chemical Physics*, 80(6).
13. J P Eckmann and D Ruelle. 1985. Ergodic theory of chaos and strange attractors. *Reviews of Modern Physics*, 273–312.
14. T R Chay and J Rinzel. 1985. Bursting, beating, and chaos in an excitable membrane model. *Biophysical Journal*, 47(3):357–366.
15. H G Schuster, J Buchner, J Kurths, et al. 1986. Book-Review - Deterministic Chaos - an Introduction. Astronomische Nachrichten, 307:259.
16. J Büchner and L M Zeleny. 1986. Deterministic chaos in the dynamics of charged particles near a magnetic field reversal. *Physics Letters A*, 118(8):395–399.

S. T. Liu and L. Zhang, *Surface Chaos and Its Applications*,
https://doi.org/10.1007/978-981-16-8229-2

17. J Gleick. 1987. Chaos: Making a new science. *Physics Today*, 41(11):79–79.
18. C Grebogi, E Ott and J A Yorke. 1987. Chaos, strange attractors, and fractal basin boundaries in nonlinear dynamics. *Science*, 238(4827):632–638.
19. P Coullet, C Elphick and D Repaux. 1987. Nature of spatial chaos. *Physical Review Letters*, 58(5):431.
20. S W Mcdonald and A N Kaufman. 1987. Wave chaos in the stadium: Statistical properties of short-wave solutions of the helmholtz equation. *Physical Review A*, 37(8):3067.
21. J Gleick and R C Hilborn. 1988. Chaos: Making a new science. *Physics Today*, 41(11):79–79.
22. S Wiggins. 1988. Global bifurcations and chaos: Analytical methods. Springer, London.
23. H Sompolinsky, A Crisanti and H J Sommers. 1988. Chaos in random neural networks. *Physical Review Letters*, 61(3):259–262.
24. L F Olsen, G L Truty and W M Schaffer. 1988. Oscillations and chaos in epidemics: a nonlinear dynamic study of six childhood diseases in copenhagen, denmark. *Theoretical Population Biology*, 33(3):344–370.
25. L A Bunimovich and Y G Sinai. 1988. Spacetime chaos in coupled map lattices. *Nonlinearity*, 1(4):491.
26. G Leon, M C Mackey and P F Zweifel. 1989. From clocks to chaos: The rhythms of life. *Physics Today*, 42(7):72–72.
27. R Z Sagdeev, D A Usikov and G M Zaslavsky. 1989. Nonlinear physics : from the pendulum to turbulence and chaos. *Physics Today*, 42(4):61–62.
28. K Kaneko. 1989. Spatiotemporal chaos in one- and two-dimensional coupled map lattices. *Physica D-nonlinear Phenomena*, 37(1–3):60–82.
29. W Stephen and D S. Mazel. 1990. Introduction to applied nonlinear dynamical systems and chaos. *Computers in Physics*, 4(5):843.
30. E Ott, C Grebogi and J A Yorke. 1990. Controlling chaos. *Physical Review Letters*, 64(11):1196.
31. G Sughihara and R M May. 1990. Nonlinear forecasting as a way of distinguishing chaos from measurement error in time series. *Nature*, 344(6268):734–41.
32. R Lima and M Pettini. 1990. Suppression of chaos by resonant parametric perturbations. *Physical Review A*, 41(2):726.
33. B J West and P F Zweifel. 1990. Fractal physiology and chaos in medicine. *Physics Today*, 45(3).
34. A L Goldberger, D R Rigney and B J West. 1990. Chaos and fractals in human physiology. *Scientific American*, 262(2):42.
35. T A Denton, G A Diamond, R H Helfant, et al. 1990. Fascinating rhythm: A primer on chaos theory and its application to cardiology. *American Heart Journal*, 120(6 Pt 1):1419–1440.
36. S Wiggins and J P Mucket. 1991. Book-review - global bifurcations and chaos - analytical methods. *Astronomische Nachrichten*, 312(5):340.
37. M P Hassell, H N Comins and R M Mayt. 1991. Spatial structure and chaos in insect population dynamics. *Nature*, 353(6341):255–258.
38. M Schroeder. 1991. Fractals, chaos, power laws: Minutes from an infinite paradise. *Physics Today*, 44(11):91–91.
39. M C Gutzwiller and J V José. 1991. Chaos in classical and quantum mechanics. *Physics Today*, 44(11):94–96.
40. A Hastings and T Powell. 1991. Chaos in a three-species food chain. *Ecology*, 72(3).
41. T Puu. 1991. Chaos in duopoly pricing. *Chaos Solitons & Fractals*, 1(6):573–581.
42. F T Arecchi, G Giacomelli, P L Ramazza, et al. 1991. Vortices and defect statistics in two-dimensional optical chaos. *Physical Review Letters*, 67(27):3749.
43. D Tilman and D Wedin. 1991. Oscillations and chaos in the dynamics of a perennial grass. *Nature*, 353(6345):653–655.
44. M M Waldrop. 1992. Complexity: The emerging science at the edge of order and chaos. *Physics Today*, 45(12):83–83.
45. T M Addiscott. 1998. Modelling concepts and their relation to the scale of the problem. Nutrient Cycling in Agroecosystems, 50(1–3):239–245.

46. K Pyragas. 1992. Continuous control of chaos by self-controlling feedback. *Physics Letters A*, 170(6):421–428.
47. B Hoop, M D Burton, H Kazemi, et al. 1996. Fractal Noise in Breathing. Bioengineering Approaches to Pulmonary Physiology and Medicine. Springer US, 1996.
48. B I Shraiman, A Pumir, W V Saarloos, et al. 1992. Spatiotemporal chaos in the one-dimensional complex ginzburg-landau equation. *Physica D-nonlinear Phenomena*, 57(3–4):241–248.
49. A Garfinkel, M L Spano, W L Ditto, et al. 1992. Controlling cardiac chaos. *Science*, 257(5074):1230–1235.
50. W A Brock, D A Hsieh and B Lebaron. 1992. Nonlinear dynamics, chaos, and instability. *Journal of the Operational Research Society*, 48(1):202–203.
51. H Nozawa. 1992. A neural network model as a globally coupled map and applications based on chaos. *Chaos*, 2(3):377–386.
52. A A Tsonis and J B Elsner. 1992. Nonlinear prediction as a way of distinguishing chaos from random fractal sequences. *Nature*, 358(6383):217–220.
53. D Auerbach, C Grebogi and E Ott, et al. 1992. Controlling chaos in high dimensional systems. *Physical Review Letters*, 69(24):3479–3482.
54. K Pyragas and A Tamaevicius. 1993. Experimental control of chaos by delayed self-controlling feedback. *Physics Letters A*, 180(1–2):99–102.
55. G Chen and X Dong. 1993. Control of chaos-a survey. In *IEEE Conference on Decision & Control.*
56. D L Turcotte. 1993. Fractals and chaos in geology and geophysics. *Physics Today*, 46(5):68–68.
57. I Hanski, P Turchin and E Korpimaki, et al. 1993. Population oscillations of boreal rodents: regulation by mustelid predators leads to chaos. *Nature*, 364:232–235.
58. J C Allen, W M Schaffer and D Rosko. 1993. Chaos reduces species extinction by amplifying local population noise. *Nature*, 364(6434):229–32.
59. H Gang and H Kaifen. 1993. Controlling chaos in systems described by partial differential equations. *Physical Review Letters*, 71(23):3794–3797.
60. V Petrov, G Vilmos and J Masere, et al. 1993. Controlling chaos in the belousov-zhabotinsky reaction. *Nature*, 361:240–243.
61. S Rinaldi and M Y Kuznetsov. 1993. Multiple attractors, catastrophes and chaos in seasonally perturbed predator-prey communities. *Bulletin of Mathematical Biology*, 55(1):15–35.
62. A M Lyapunov and M I Rabinovich. 1993. Chaos theory. (book reviews: The general problem of the stability of motion.; nonlinearities in action. oscillations, chaos, order, fractals.). *Science*, 260.
63. S Hayes, C Grebogi and E Ott. 1993. Communicating with chaos. *Physical Review Letters*, 70(20):3031.
64. E N Lorenz. 1993. The essence of chaos. *Physics Today*, 63(2).
65. X Zeng, R A Pielke and R Eykholt. 1993. Chaos theory and its applications to the atmosphere. *Bulletin of the American Meteorological Society*, 74(4):631–644.
66. T Shinbrot, C Grebogi and J A Yorke, et al. Using small perturbations to control chaos. *Nature*, 363(6428):411–417.
67. J B Elsner and A A Tsonis. 1993. Corrigendum to "nonlinear prediction, chaos, and noise". *Bulletin of the American Meteorological Society*, 74(2):243–354.
68. R Lewin. 1993. Complexity: Life at the edge of chaos. *American Journal of Physics*, 61(8):L627–L633(7).
69. H D Chiang and C W Liu. 1993. Chaos in a simple power system. *IEEE Transactions on Power Systems*, 8(4):1407–1417.
70. E Ott and K Wiesenfeld. 1994. Chaos in dynamical systems. *Bulletin of Mathematical Biology*, 47(1):45–45.
71. S J Schiff, K Jerger, D H Duong, et al. 1994. Controlling chaos in the brain. *Nature*, 370(6491):615–620.

72. R C Hilborn. 1994. Chaos and nonlinear dynamics: An introduction for scientists and engineers. *American Journal of Physics*, 62(9):654.
73. G Hu and Z Qu. 1994. Controlling spatiotemporal chaos in coupled map lattice systems. *Physical Review Letters*, 72(1):68–71.
74. H Waelbroeck. 1994. Deterministic chaos in tropical atmospheric dynamics. *Journal of the Atmospheric Sciences*, 52(6):320–324.
75. T Elbert, W J Ray, Z J Kowalik, et al. 1994. Chaos and physiology: deterministic chaos in excitable cell assemblies. *Physiological Reviews*, 74(1):1.
76. G M Zaslavsky. 1994. Fractional kinetic equation for hamiltonian chaos. *Physica D-nonlinear Phenomena*, 76(1-3):110–122.
77. A W Jayawardena and F Lai. 1994. Analysis and prediction of chaos in rainfall and stream flow time series. *Journal of Hydrology*, 153(1-4):23–52.
78. A A Tsonis. 1994. Chaos: From theory to applications. *American Journal of Physics*, 62(2):562–570.
79. S Hayes, C Grebogi, E Ott, et al. 1994. Experimental control of chaos for communication. *Physical Review Letters*, 73(13):1781.
80. T Sugawara, M Tachikawa, T Tsukamoto, et al. 1994. Observation of synchronization in laser chaos. *Physical Review Letters*, 72(22):3502.
81. A Maritan and J R Banavar. 1994. Chaos, noise, and synchronization. *Physical Review Letters*, 72(10):1451.
82. F L Moore, J C Robinson, C Bharucha, et al. 1994. Observation of dynamical localization in atomic momentum transfer: A new testing ground for quantum chaos. *Physical Review Letters*, 73(22):2974.
83. S H Strogatz. 1995. Nonlinear dynamics and chaos: With applications to physics, biology, chemistry and engineering. *Physics Today*.
84. C Fuh and P Tung. 1995. Controlling chaos using differential geometric method. *Physical Review Letters*, 75(16):2952–2955.
85. P Glendinning and E Ott. 1995. Book-review - stability instability and chaos - an introduction to the theory of nonlinear differential equations. *Nature*, 374(2):402–406.
86. C K Aihara. 1995. Chaotic simulated annealing by a neural network model with transient chaos. *Neural Networks*, 8(6):915–930.
87. L Chen and K Aihara. 1995. Chaotic simulated annealing by a neural network model with transient chaos. *Neural Networks*, 8(6):915–930.
88. A Hastings. 1995. Chaos in dynamical systems. *Bulletin of Mathematical Biology*, 57(6):943–944.
89. K Pyragas. 1995. Control of chaos via extended delay feedback. *Physics Letters A*, 206(5–6):323–330.
90. S Ellner, P Turchin. 1995. Chaos in a noisy world: New methods and evidence from time-series analysis. *American Naturalist*. 145(3):343–375.
91. G Perez and H A Cerdeira. 1995. Extracting messages masked by chaos. *Physical Review Letters*, 74(11):1970–1973.
92. E Ott and M Spano. 1995. Controlling chaos. *Physics Today*, 48(5):34–40.
93. H G Schuster. Deterministic chaos: An introduction. *Quarterly Review of Biology*, 1995.
94. S H Strogatz. 1994. Stability, instability and chaos: An introduction to the theory of nonlinear differential equations.*Glendinning, Cambridge University Press*, Cambridge, UK.
95. V Vreeswijk and C Sompolinsky. 1996. Chaos in neuronal networks with balanced excitatory and inhibitory activity. *Science*, 274(5293).
96. L A Goldberger. 1996. Non-linear dynamics for clinicians: chaos theory, fractals, and complexity at the bedside. *Lancet*, 347(9011):1312–1314.
97. J Schopman and M G Mann. 1996. The Quark and the Jaguar; Adventures in the Simple and the Complex. *Annales De Linstitut Pasteur*, 102(1):153–173.
98. J H Xiao, G Hu and Z Qu. 1996. Synchronization of spatiotemporal chaos and its application to multichannel spread-spectrum communication. *Physical Review Letters*, 77(20):4162.

99. A Kudrolli and J P Gollub. 1996. Patterns and spatiotemporal chaos in parametrically forced surface waves: a systematic survey at large aspect ratio. *Physica D-nonlinear Phenomena*, 97(1–3):133–154.
100. N F Rulkov. 1996. Images of synchronized chaos: Experiments with circuits. *Chaos*, 6(3):262–279.
101. M Ding, E J Ding, W L Ditto, et al. 1997. Control and synchronization of chaos in high dimensional systems: Review of some recent results. *Chaos*, 7(4):644–652.
102. R O Grigoriev, M C Cross and H G Schuster. 1997. Pinning control of spatiotemporal chaos. *Physical Review Letters*, 79(15):2795–2798.
103. K T Alligood, T D Sauer and J A Yorke. 1997. Chaos : an introduction to dynamical systems. *Physics Today*, 50(11):67–68.
104. K Efetov and B Simons. 1997. Supersymmetry in disorder and chaos. *Physics Today*, 51(3):90–94.
105. A Garfinkel, P S Chen, D O Walter, et al. 1997. Quasiperiodicity and chaos in cardiac fibrillation. *Journal of Clinical Investigation*, 99(2):305–314.
106. D S Coffey. 1998. Self-organization, complexity and chaos: The new biology for medicine. *Nature Medicine*, 4(8):882–885.
107. K Mischaikow and M Mrozek. 1998. Chaos in the lorenz equations: A computer assisted proof. partii: Details. *Mathematics of Computation*.
108. M I Rabinovich and H D I Abarbanel. 1998. The role of chaos in neural systems. *Neuroscience*, 87(1):5–14.
109. S Abbott. 1999. Fractals and chaos: An illustrated course. *Mathematical Gazette*, 83(496):256–186.
110. J Huisman and F Weissing. 1999. Biodiversity of plankton by species oscillations and chaos. *Nature*, 402:407–410.
111. I R Epstein, J A Pojman and G Nicolis. 1999. An introduction to nonlinear chemical dynamics: Oscillations, waves, patterns, and chaos. *Physical Today*.
112. B H K Lee, S J Price and Y S Wong. 1999. Nonlinear aeroelastic analysis of airfoils: bifurcation and chaos. *Progress in Aerospace Sciences*, 35(3):205–334.
113. J N Weiss, A Garfinkel, H S Karagueuzian, et al. 1999. Chaos and the transition to ventricular fibrillation: a new approach to antiarrhythmic drug evaluation. *Circulation*, 99(21):2819.
114. P Gaspard and J Bricmont. 1999. Chaos, scattering and statistical mechanics. *Physics Today*, 52(8):62–64.
115. G J Bazzoli, S M Shortell, N L Dubbs, et al. 1999. A taxonomy of health networks and systems: bringing order out of chaos. *Health Services Research*, 33(6):1683–717.
116. L J Dubé and P Després. 2000. The control of dynamical systems - recovering order from chaos -. *AIP Conference Proceedings*, 500:551–570.
117. T Izuyama, D J Kim and R Kubo. 2000. Pollen coupling of forest trees: Forming synchronized and periodic reproduction out of chaos. *Journal of Theoretical Biology*, 203(2):63–84.
118. J M T Thompson and H B Stewart. 2000. Nonlinear dynamics and chaos. *Journal of Statistical Physics*, 78(5-6):1635–1636.
119. I Fischer, Y Liu and P Davis. 2000. Synchronization of chaotic semiconductor laser dynamics on subnanosecond time scales and its potential for chaos communication. *Physical Review A*, 62(1):11801.
120. T Damour and M Henneaux. 2000. Chaos in superstring cosmology. *Physical Review Letters*, 85(5):920–923.
121. T R Kim, Y H Joo, J B Park, et al. 2000 Anticontrol of chaos in continuous-time systems via time-delay feedback. *Chaos*, 2000.
122. P W O Hoskin. 2000. Patterns of chaos: fractal statistics and the oscillatory chemistry of zircon. *Geochimica Et Cosmochimica Acta*, 64(11):1905–1923.
123. A Y Loskutov. 2001. Chaos and control in dynamical systems. *Computational Mathematics & Modeling*, 12(4):314–352.
124. E J Heller. 2001. Quantum chaos (book review). *Physics Today*.
125. F Mathias. 2001. Chaos and time-reversed acoustics. *Physica Scripta*, 2001(1):268.

126. K Pyragas. 2001. Control of chaos via an unstable delayed feedback controller. *Physical Review Letters*, 86(11):2265–8.
127. B Dennis, R A Desharnais, J M Cushing, et al. 2001. Estimating chaos and complex dynamics in an insect population. *Ecological Monographs*, 71(2):277–303.
128. K Mischaikow, M Mrozek and A Szymczak. 2001. Chaos in the lorenz equations: a computer-assisted proof. *Journal of Differential Equations*, 169(1):17–56.
129. D A Steck, W H Oskay and M G Raizen. 2001. Observation of chaos-assisted tunneling between islands of stability. *Science*, 293(5528):274–278.
130. S Tang and J M Liu. 2001. Chaotic pulsing and quasi-periodic route to chaos in a semiconductor laser with delayed opto-electronic feedback. *IEEE Journal of Quantum Electronics*, 37(3):329–336.
131. H Stckmann and M Gutzwiller. 2001. Quantum chaos: An introduction. *American Journal of Physics*, 54(8):49–50.
132. G Jakimoski and L Kocarev. 2001. Analysis of some recently proposed chaos-based encryption algorithms. *Physics Letters A*, 291(6):381–384.
133. S V Petrovskii and H Malchow. 2001. Wave of chaos: new mechanism of pattern formation in spatio-temporal population dynamics. *Theoretical Population Biology*, 59(2):157–174.
134. D Xiu and G Karniadakis. 2002. The wiener–askey polynomial chaos for stochastic differential equations. *SIAM Journal on Scientific Computing*.
135. A L Fradkov and R J Evans. 2002. Control of chaos: Survey 1997–2000. *IFAC Proceedings Volumes*, 35(1):131–142.
136. E S Selleck. 2002. Order out of chaos: Assembly of ligand binding sites in heparan sulfate1. *Annual Review of Biochemistry*.
137. J Lü, T Zhou and S Zhang. 2002. Chaos synchronization between linearly coupled chaotic systems. *Chaos Solitons & Fractals*, 14(4):529–541.
138. G Chen, X Yu. 2003. Chaos control theory and applications *Germany, Springer*.
139. J D Scargle. 2003. Chaos and time-series analysis. *Technometrics*, 47(3):373–373.
140. L J Pike. 2003. Lipid rafts bringing order to chaos. *Journal of Lipid Research*, 44(4):655–67.
141. G Contopoulos and P Saha. 2003. Order and chaos in dynamical astronomy. *Physics Today*, 56(11):68–69.
142. W M Ahmad, J C Sprott. 2003. Chaos in fractional-order autonomous nonlinear systems. *Chaos Solitons & Fractals*, 16(2):339–351.
143. D Xiu, G E Karniadakis. 2003. Modeling uncertainty in flow simulations via generalized polynomial chaos. *Journal of Computational Physics*, 187(1):137–167.
144. M Feki. 2003. An adaptive chaos synchronization scheme applied to secure communication. *Chaos Solitons & Fractals*, 18(1):141–148.
145. G S Duane and J J Tribbia. 2004. Weak atlantic pacific teleconnections as synchronized chaos. *Journal of the Atmospheric Sciences*, 61(17):2149–2168.
146. F Yizhak. 2004. Nonlinear dynamics and chaos in the sea and land breeze. *Journal of the Atmospheric Sciences*, 61(17):2169–2187.
147. C Li and G Peng. 2004. Chaos in chen's system with a fractional order. *Chaos Solitons & Fractals*, 22(2):443–450.
148. N Bertschinger and T Natschlager. 2004. Real-time computation at the edge of chaos in recurrent neural networks. *Neural Computation*, 16(7):1413–1436.
149. H T Yau. 2004. Design of adaptive sliding mode controller for chaos synchronization with uncertainties. *Chaos Solitons & Fractals*, 22(2):341–347.
150. C Li and G Chen. 2004. Chaos in the fractional order chen system and its control. *Chaos Solitons & Fractals*.
151. M Ji and H Tang. 2004. Application of chaos in simulated annealing. *Chaos Solitons & Fractals*, 21(4):933–941.
152. B J Debusschere, H N Najm, P P Pébay, et al. 2004. Numerical challenges in the use of polynomial chaos representations for stochastic processes. *Siam Journal on Scientific Computing*, 26(2):698–719.

153. B Sivakumar. 2004. Chaos theory in geophysics: past, present and future. *Chaos Solitons & Fractals*, 19(2):441–462.
154. C Soize and R Ghanem. 2005. Physical systems with random uncertainties: Chaos representations with arbitrary probability measure. *Siam Journal on Scientific Computing*, 26(2):395–410.
155. A L Fradkov and R J Evans. 2005. Control of chaos: Methods and applications in engineering. *Annual Reviews in Control*, 29(1):33–56.
156. A Argyris, D Syvridis, L Larger, et al. 2005. Chaos-based communications at high bit rates using commercial fibre-optic links. *Nature*, 438(7066):343–346.
157. B Liu, L Wang, Y H Jin, et al. 2005. Improved particle swarm optimization combined with chaos. *Chaos Solitons & Fractals*, 25(5):1261–1271.
158. H K Chen. 2005. Global chaos synchronization of new chaotic systems via nonlinear control. *Chaos Solitons & Fractals*, 23(4):1245–1251.
159. Z H Guan, F Huang and W Guan. 2005. Chaos-based image encryption algorithm. *Physics Letters A*, 346(1-3):153–157.
160. X Wan and G E Karniadakis. 2005. An adaptive multi-element generalized polynomial chaos method for stochastic differential equations. *Journal of Computational Physics*, 209(2):617–642.
161. M T Yassen. 2005. Chaos synchronization between two different chaotic systems using active control. *Chaos Solitons & Fractals*, 23(1):131–140.
162. H P Ju. 2005. Chaos synchronization of a chaotic system via nonlinear control. *Chaos Solitons & Fractals*, 25(3):579–584.
163. X Gao and J Yu. 2005. Chaos in the fractional order periodically forced complex duffing's oscillators. *Chaos Solitons & Fractals*, 26(4):1125–1133.
164. L Becks, F M Hilker, H Malchow, et al. 2005. Experimental demonstration of chaos in a microbial food web. *Nature*, 435(7046):1226–9.
165. J D Phillips. 2006. Deterministic chaos and historical geomorphology: A review and look forward. *Geomorphology*, 76(1–2):109–121.
166. A L Fradkov, R J Evans and B R Andrievsky. 2006. Control of chaos: methods and applications in mechanics. *Annual Reviews in Control*, 364(1846):2279–2307.
167. B R Hunt, E J Kostelich and I Szunyogh. 2006. Efficient data assimilation for spatiotemporal chaos: A local ensemble transform kalman filter. *Physica D-nonlinear Phenomena*, 230(1–2):112–126.
168. J Huisman, N N P Thi and D M Karl, et al. 2006. Reduced mixing generates oscillations and chaos in the oceanic deep chlorophyll maximum. *Nature*, 439(7074):322–325.
169. S Gnutzmann and U Smilansky. 2006. Quantum graphs: Applications to quantum chaos and universal spectral statistics. *Advances in Physics*, 55(5-6):527–625.
170. H P Ju. 2006. Chaos synchronization between two different chaotic dynamical systems. *Chaos Solitons & Fractals*, 27(2):549–554.
171. M T Yassen. 2006. Chaos control of chaotic dynamical systems using backstepping design. *Chaos Solitons & Fractals*, 27(2):537–548.
172. Q Y Wang, S L Qi, R C Guan, et al. 2006. Chaos synchronization of coupled neurons with gap junctions. *Physics Letters A*, 356(1):17–25.
173. P L Knight. 2007. Deterministic chaos: An introduction. *Optica Acta International Journal of Optics*, 35(7):1114–1114.
174. A L Goldberger, D R Rigney and B J West. 2007. Science in pictures: Chaos and fractals in human physiology. *Scientific American*, 262(262):42–49.
175. H Adeli, S Ghosh-Dastidar, N Dadmehr. 2007. A wavelet-chaos methodology for analysis of eegs and eeg subbands to detect seizure and epilepsy. *IEEE Transactions on Biomedical Engineering*, 54(2):205–211.
176. L W Maass. 2007. Edge of chaos and prediction of computational performance for neural circuit models. *Neural Networks*, 20(3):323–334.
177. R A Todor and C Schwab. 2007. Convergence rates for sparse chaos approximations of elliptic problems with stochastic coefficients. *Ima Journal of Numerical Analysis*, 27(2):232–261.

178. R Femat and G SolíS-Perales. 2007. On the chaos synchronization phenomena. *Physics Letters A*, 262(1):50–60.
179. D Yang, G Li and G Cheng. 2007. On the efficiency of chaos optimization algorithms for global optimization. *Chaos Solitons & Fractals*, 34(4):1366–1375.
180. S S E H Elnashaie and J R Grace. 2007. Complexity, bifurcation and chaos in natural and man-made lumped and distributed systems. *Chemical Engineering Science*, 62(13):3295–3325.
181. O A Rosso, H A Larrondo, M T Martin, et al. 2007. Distinguishing noise from chaos. *Physical Review Letters*, 99(15):154102.
182. C Jarvis. 2007. Order out of chaos: Linnaean plant names and their types. *Nature*.
183. R A Todor and C Schwab. 2007. Convergence rates for sparse chaos approximations of elliptic problems with stochastic coefficients. *Ima Journal of Numerical Analysis*, 27(2):232–261.
184. B Sudret. 2008. Global sensitivity analysis using polynomial chaos expansions. *Reliability Engineering & System Safety*, 93(7):964–979.
185. T Gao and Z Chen. 2008. A new image encryption algorithm based on hyper-chaos. *Physics Letters A*, 372(4):394–400.
186. W C Chen. 2008. Nonlinear dynamics and chaos in a fractional-order financial system. *Chaos Solitons & Fractals*, 36(5):1305–1314.
187. Q Zhang and J Lu. 2008. Chaos synchronization of a new chaotic system via nonlinear control. *Chaos Solitons & Fractals*, 37(1):175–179.
188. H Adeli, S Ghosh-Dastidar and N Dadmehr. 2008. A spatio-temporal wavelet-chaos methodology for eeg-based diagnosis of alzheimer's disease. *Neuroscience Letters*, 444(2):190–194.
189. S Zhou, H Li and Z Zhu. 2008. Chaos control and synchronization in a fractional neuron network system. *Chaos Solitons & Fractals*, 36(4):973–984.
190. J Rinne and H Jarvinen. 2009. Numerical approximations can create chaos-related errors in blocking forecasts. *Monthly Weather Review*, 120(5):884.
191. B Alatas, E Akin and A B Ozer. 2009. Chaos embedded particle swarm optimization algorithms. *Chaos Solitons & Fractals*, 40(4):1715–1734.
192. Y M Marzouk and H N Najm. 2009. Dimensionality reduction and polynomial chaos acceleration of bayesian inference in inverse problems. *Journal of Computational Physics*, 228(6):1862–1902.
193. H N Najm. 2009. Uncertainty quantification and polynomial chaos techniques in computational fluid dynamics. *Annual Review of Fluid Mechanics*, 41(1):35–52.
194. T Crestaux, O L Mattre and J M Martinez. 2009. Polynomial chaos expansion for sensitivity analysis. *Reliability Engineering & System Safety*, 94(7):1161–1172.
195. Y Wang, K W Wong, X Liao, et al. 2009. A chaos-based image encryption algorithm with variable control parameters. *Chaos Solitons & Fractals*, 41(4):1773–1783.
196. D Xiao, X Liao and P Wei. 2009. Analysis and improvement of a chaos-based image encryption algorithm. *Chaos Solitons & Fractals*, 40(5):2191–2199.
197. L Jia and D Xiu. 2009. A generalized polynomial chaos based ensemble kalman filter with high accuracy. *Journal of Computational Physics*, 228(15):5454–5469.
198. Z Duan, J Wang, Ying Yang, et al. 2009. Frequency-domain and time-domain methods for feedback nonlinear systems and applications to chaos control. *Chaos Solitons & Fractals*, 40(2):848–861.
199. H Zhu, S Zhou and Z He. 2009. Chaos synchronization of the fractional-order chen's system. *Chaos Solitons & Fractals*, 41(5):2733–2740.
200. W S Pritchard, D W Duke and K K Krieble. 2010. Dimensional analysis of resting human eeg. ii: Surrogate-data testing indicates nonlinearity but not low-dimensional chaos. *Psychophysiology*, 32(5):486–491.
201. G Blatman and B Sudret. 2010. An adaptive algorithm to build up sparse polynomial chaos expansions for stochastic finite element analysis. *Probabilistic Engineering Mechanics*, 25(2):183–197.
202. L W Powell. 2010. The nature of cell death in piecemeal necrosis: Is order emerging from chaos? *Hepatology*, 7(4).

203. G Blatman and B Sudret. 2010. Efficient computation of global sensitivity indices using sparse polynomial chaos expansions. *Reliability Engineering & System Safety*, 95(11):1216–1229.
204. O E Rossler. 2010. Continuous chaos–four prototype equations. *Annals of the New York Academy of Sciences*, 316(1):376–392.
205. M B Sharifi, K P Georgakakos and I Rodrigueziturbe. 2010. Evidence of deterministic chaos in the pulse of storm rainfall. *Journal of the Atmospheric Sciences*, 47(7):888–893.
206. S Cang, G Qi and Z Chen. 2010. A four-wing hyper-chaotic attractor and transient chaos generated from a new 4-d quadratic autonomous system. *Nonlinear Dynamics*, 59(3):515–527.
207. J Bricmont. 2010. Science of chaos or chaos in science? *Annals of the New York Academy of Sciences*, 775(1):131–175.
208. M Arnst, R Ghanem and C Soize. 2010. Identification of bayesian posteriors for coefficients of chaos expansions. *Journal of Computational Physics*, 229(9):3134–3154.
209. V Daftardar-Gejji and S Bhalekar. 2010. Chaos in fractional ordered liu system. *Computers & Mathematics with Applications*, 59(3):1117–1127.
210. H Degn, L F Olsen and J W Perram. 2010. Bistability, oscillation and chaos in an enzyme reaction. *Annals of the New York Academy of Sciences*, 316(1):623–637.
211. C Desceliers, R Ghanem and C Soize. 2010. Maximum likelihood estimation of stochastic chaos representations from experimental data. *International Journal for Numerical Methods in Engineering*, 66(6):978–1001.
212. D Davendra, I Zelinka and R Senkerik. 2010. Chaos driven evolutionary algorithms for the task of pid control. *Computers & Mathematics with Applications*, 60(4):1088–1104.
213. G Blatman and B Sudret. 2011. Adaptive sparse polynomial chaos expansion based on least angle regression. *Journal of Computational Physics*, 230(6):2345–2367.
214. Z Qu. 2011. Chaos in the genesis and maintenance of cardiac arrhythmias. *Progress in Biophysics & Molecular Biology*, 105(3):247–257.
215. Z Wang, X Huang and G Shi. 2011. Analysis of nonlinear dynamics and chaos in a fractional order financial system with time delay. *Computers & Mathematics with Applications*, 62(3):1531–1539.
216. R Ye. 2011. A novel chaos-based image encryption scheme with an efficient permutation-diffusion mechanism. *Optics Communications*, 284(22):5290–5298.
217. I Omelchenko, Y Maistrenko, P Hovel, et al. 2011. Loss of coherence in dynamical networks: spatial chaos and chimera states. *Physical Review Letters*, 106(23):234102–234102.
218. S Li, Y Li, B Liu, et al. 2012. Model-free control of lorenz chaos using an approximate optimal control strategy. *Communications in Nonlinear Science & Numerical Simulation*, 17(12):4891–4900.
219. M Banerjee and S Banerjee. 2012. Turing instabilities and spatio-temporal chaos in ratio-dependent holling-tanner model. *Mathematical Biosciences*, 236(1):64–76.
220. F Christiansen, P Cvitanovic and V Putkaradze. 2012. Hopf's last hope: spatiotemporal chaos in terms of unstable recurrent patterns. *Nonlinearity*, volume 10(1):55–70(16).
221. R Senkerik. 2013. On the evolutionary optimization of chaos control - a brief survey. *Advances in Intelligent Systems & Computing*, 192:35–48.
222. R Rhodes and V Vargas. 2013. Gaussian multiplicative chaos and applications: A review. *Probability Surveys*, 11(2014).
223. Van Straaten, C W Elisabeth and C J Stam. 2013. Structure out of chaos: Functional brain network analysis with eeg, meg, and functional mri. *European Neuropsychopharmacology*, 23(1):7–18.
224. R Laje and D V Buonomano. 2013. Robust timing and motor patterns by taming chaos in recurrent neural networks. *Nature Neuroscience*, 16(7):925–33.
225. J Banks, T T D Nguyen, P Oprocha, et al. 2013. Dynamics of spacing shifts. *Discrete & Continuous Dynamical Systems*, 33(9):4207–4232.
226. L A Aguirre and L Christophe. 2014. Modeling nonlinear dynamics and chaos: A review. *Mathematical Problems in Engineering*, 2009(1024–123X).

227. Y Q Zhang and X Y Wang. 2014. Analysis and improvement of a chaos-based symmetric image encryption scheme using a bit-level permutation. *Nonlinear Dynamics*, 77(3):687–698.
228. F Iztok, P Matja, K Salahuddin, et al. 2015. A review of chaos-based firefly algorithms: Perspectives and research challenges. *Applied Mathematics & Computation*, 252:155–165.
229. J Maldacena, S H Shenker and D Stanford. 2016. A bound on chaos. *Journal of High Energy Physics*, 2016(8):1–17.
230. M Hanada, H Shimada and M Tezuka. 2018. Universality in chaos: Lyapunov spectrum and random matrix theory. *Physical Review E*, 97(2):022224.
231. I Garcia-Mata, M Saraceno, R A Jalabert, et al. 2018. Chaos signatures in the short and long time behavior of the out-of-time ordered correlator. *Physical Review Letters*, 121(21):210601.1–210601.5.
232. C Hainaut, P Fang, A Raneon, et al. 2018. Experimental observation of a time-driven phase transition in quantum chaos. *Physical Review Letters*, 2018, 121(13):13401.1-134101.6.
233. B Senyange, B M Manda and C Skokos. 2018. Characteristics of chaos evolution in one-dimensional disordered nonlinear lattices. *Physical Review E*, 98(5).
234. R H Du, S X Qu and Y C Lai. 2018. Transition to high-dimensional chaos in nonsmooth dynamical systems. *Physical Review E*, 98(5).
235. P O Vandervoort. 2018. On order and chaos in the mergers of galaxies. *Monthly Notices of the Royal Astronomical Society*, (3):4186–4210.
236. Z Hua, B Zhou and Y Zhou. 2018. Sine chaotification model for enhancing chaos and its hardware implementation. *IEEE Transactions on Industrial Electronics*, 99:1-1.
237. A Gusso, W G Dantas and S Ujevic. 2019. Prediction of robust chaos in micro and nanoresonators under two-frequency excitation. *Chaos*, 29(3):033112.
238. X Wang, Y Wang, S Wang, et al. 2019. A novel pseudo-random coupled lp spatiotemporal chaos and its application in image encryption. *Chinese Physics B*, 27(11):423–433.
239. N Stankevich, A Kuznetsov, E Popova, et al. 2019. Chaos and hyperchaos via secondary neimark-sacker bifurcation in a model of radiophysical generator. *Nonlinear Dynamics*, 97(4).
240. T Xu, T Scaffidi and X Cao. 2020. Does scrambling equal chaos? *Physical Review Letters*, 124(14).
241. M Kumar, A Kundu, M Kulkarni, et al. 2020. Transport, correlations, and chaos in a classical disordered anharmonic chain. *Physical Review E*, 102(2).
242. D Mondal, S Sinha and S Sinha. 2020. Chaos and quantum scars in coupled top model. *Physical Review E*, 102(2).
243. G Chen and X Dong. 1998. *From Chaos to Order:Methodologies, Perspectives and Applications*. World Scientifc.
244. G Chen, J Lu, B Nicholas, et al. 1999. Bifurcation dynamics in discrete-time delayed-feedback control systems. *International Journal of Bifurcation and Chaos*, 9(1).
245. S T Liu and G Chen. 2003. On spatial lyapunov exponents and spatial chaos. *International Journal of Bifurcation & Chaos*, 13(05):1163–1181.
246. S T Liu, Y Q Liu and F Q Deng. 2002. Oscillation for nonlinear delay partial difference equations with positive and negative coefficient. *Computers & Mathematics with Applications*, 43(10–11):1219–1230.
247. S T Liu and G Chen. Oscillations of nonlinear partial difference systems. 2003. *Journal of Mathematical Analysis and Applications*, 277.
248. S T Liu and P Jin. 2003. Oscillatory behavior of delay partial difference equations. *Periodica Mathematica Hungarica*, 47(1–2):151–167.
249. G Chen, C Tian and Y Shi. 2005. Stability and chaos in 2-d discrete systems. *Chaos Solitons & Fractals*, 25(3):637–647.
250. F Sun, S Liu, Z Li, et al. 2008. A novel image encryption scheme based on spatial chaos map. *Chaos Solitons & Fractals*, 38(3):631–640.
251. Y Zhang, W Sun and S Liu. 2009. Control of generalized julia sets. *Chaos Solitons & Fractals*, 42(3):1738–1744.
252. F Sun, Z Lü and S Liu. 2010. A new cryptosystem based on spatial chaotic system. *Optics Communications*, 283(10):2066–2073.

253. S Liu, F Sun and J Sun. 2010. Uniformity and spatial chaos of spatial physics kinematic system. *International Journal of Modern Physics B*, 24(28):5495–5503.
254. S Liu and L Ping. 2011. Adaptive anti-synchronization of chaotic complex nonlinear systems with unknown parameters. *Nonlinear Analysis Real World Applications*, 12(6):3046–3055.
255. J Sun, S T Liu and W Qiao. 2011. Parameter identification of generalized julia sets. *Acta Physica Sinica*, 60(7):70510–070510.
256. L Ping and S Liu. 2012. Robust adaptive full state hybrid synchronization of chaotic complex systems with unknown parameters and external disturbances. *Nonlinear Dynamics*, 70(1):585–599.
257. Y Zhang, S Liu and Z Jia. 2012. Localization using joint distance and angle information for 3d wireless sensor networks. *IEEE Communications Letters*, 16(6):809–811.
258. F Zhang and S Liu. 2013. Full state hybrid projective synchronization and parameters identification for uncertain chaotic (hyperchaotic) complex systems. *Journal of Computational & Nonlinear Dynamics*.
259. F Zhang and S Liu. 2014. Self-time-delay synchronization of time-delay coupled complex chaotic system and its applications to communication. *International Journal of Modern Physics C*, 25(03):1350102.
260. D Wang and S T Liu. 2015. Synchronization between the spatial julia sets of complex lorenz system and complex henon map. *Nonlinear Dynamics*, 81(3):1197–1205.
261. C Jiang, S Liu and D Wang. 2015. Generalized combination complex synchronization for fractional-order chaotic complex systems. *Entropy*, 17(8):5199–5217.
262. J Liu, S Liu and C Yuan. 2015. Adaptive complex modified projective synchronization of complex chaotic (hyperchaotic) systems with uncertain complex parameters. *Nonlinear Dynamics*, 79(2):1035–1047.
263. P Wang and S Liu. 2015. The gradient control of spatial-alternated julia sets. *Nonlinear Dynamics*, 80(3):1291–1302.
264. J Liu, S Liu and J C Sprott. 2016. Adaptive complex modified hybrid function projective synchronization of different dimensional complex chaos with uncertain complex parameters. *Nonlinear Dynamics*, 83:1109–1121.
265. D Wang, S T Liu, Y Zhao, et al. Control of the spatial mandelbrot set generated in coupled map lattice. *Nonlinear Dynamics*, 84(3):1795–1803.
266. W Wang, S Liu, D Tian, et al. 2017. Permanence and extinction of a nonautonomous impulsive plankton model with help. *Mathematical Methods in the Applied Sciences*.
267. C Jiang and S Liu. 2017. Synchronization and antisynchronization of n -coupled complex permanent magnet synchronous motor systems with ring connection. *Complexity*, 2017(4):1–15.
268. J Liu, K Liu, S Liu, et al. 2017. Adaptive control for a class of nonlinear complex dynamical systems with uncertain complex parameters and perturbations. *Plos One*, 12(5):e0175730.
269. C Jiang and S Liu. 2017. Synchronization and antisynchronization of n -coupled complex permanent magnet synchronous motor systems with ring connection. *Complexity*, 2017(4):1–15.
270. Y Zhang, C Liu, Shutang Liu, et al. 2017. Fractal dimension and synchronization of the controlled julia sets of a reaction-diffusion system. *Asian Journal of Control*, 19(4).
271. D Tian and S Liu. 2017. Exponential stability of switched positive homogeneous systems. *Complexity*, 2017:1–8.
272. Y Wang and S Liu. 2018. Fractal analysis and control of the fractional lotka-volterra model. *Nonlinear Dynamics*, 95:1457–1470.
273. Q Y Zhao, S T Liu and D Tian. 2018. Dynamic behavior analysis of phytoplankton-zooplankton system with cell size and time delay. *Chaos, Solitons & Fractals*, 113:160–168.
274. W Wang, S T Liu, D Tian, et al. 2018. Pattern dynamics in a toxin-producing phytoplankton-zooplankton model with additional food. *Nonlinear Dynamics*, 94:1–18.
275. D Tian and S Liu. 2018. Stability analysis for a class of switched positive nonlinear systems under dwell-time constraint. *Advances in Difference Equations*, 2018(1):95.

276. Q Zhao, S Liu and X Niu. 2019. Dynamic behavior analysis of a diffusive plankton model with defensive and offensive effects. *Chaos Solitons & Fractals*, 129.
277. Q Hai, S Liu and C Hu. 2019. Linear generalized synchronization of spatial chaotic systems. *Asian Journal of Control*, 21(2):649–659.
278. Y Wang and S Liu. 2019. Fractal analysis and control of the fractional lotka-volterra model. *Nonlinear Dynamics*, 95:1457–1470.
279. W Wang, S Liu and Z Liu. 2019. Spatiotemporal dynamics near the turing-hopf bifurcation in a toxic-phytoplankton-zooplankton model with cross-diffusion. *Nonlinear Dynamics*, 98(1):27–37.
280. Q Zhao, S Liu and X Niu. 2020. Turing instability and hopf bifurcation for a diffusion-plankton system with cell size. *International Journal of Computer Mathematics*, (3):1–23.
281. F Sun, S Liu, Z Li, et al. 2008. A novel image encryption scheme based on spatial chaos map. *Chaos Solitons & Fractals*, 38(3):631–640.
282. F Zhang and S Liu. 2014. Complex function projective synchronization of complex chaotic system and its applications in secure communication. *Nonlinear Dynamics*, 76(2):1087–1097.
283. Jian Liu and S Liu. 2017. Complex modified function projective synchronization of complex chaotic systems with known and unknown complex parameters. *Applied Mathematical Modelling*, 48(aug.):440–450.
284. Z Liu, W Wang and S Liu. 2019. Spatiotemporal dynamics near the turing-hopf bifurcation in a toxic-phytoplankton-zooplankton model with cross-diffusion. *Nonlinear Dynamics*, 98(1):27–37.
285. M Yoshimoto, S Kurosawa and H Nagashima. 1999. Effect of noise on chaos in a one-dimensional map. *Journal of the Physical Society of Japan*, 67(6):1924–1929.
286. M Yoshimoto, T Yamaguchi and H Nagashima. 1996. New aspect of noise-induced order in the one-dimensional chaos of the belousov-zhabotinsky reaction. *Chemical Physics Letters*, 257(3–4):397–400.
287. S Codreanu and L Matyas. 1999. An analytical study of bifurcations generated by some iterated function systems. *Chaos Solitons & Fractals*, 10(8):1343–1348.
288. Y Minoru, K Shigeru, N Hiroyuki. 1998. Effect of noise on chaos in a one-dimensional map. *Journal of the Physical Society of Japan*, 67(6):1924–1929.
289. T Shinbrot, E Ott, C Grebogi, et al. 1992. Using chaos to direct orbits to targets in systems describable by a one-dimensional map. *Physical Review A*, 45(6):4165–4168.
290. K A Jukka. 1992. Universal criterion for the breakup of invariant tori in dissipative systems. *Physical Review Letters*, 69(15):2180–2183.
291. J Güémez and M A Matías. 1993. Control of chaos in unidimensional maps. *Physics Letters A*, 181(1):29–32.
292. A Sen and D Mukherjee. 2009. Chaos in the delay logistic equation with discontinuous delays. *Chaos Solitons & Fractals*, 40(5):2126–2132.
293. V Gintautas, G Foster and A W Hübler. 2008. Resonant forcing of chaotic dynamics. *Journal of Statistical Physics*, 130(3):617–629.
294. H O Peitgen, H Jürgens and D Saupe. 2004. Chaos and fractals : new frontiers of science. Springer-Verlag.
295. F Hofbauer and G Keller. 1990. Quadratic maps without asymptotic measure. *Communications in Mathematical Physics*, 127(2):319–337.
296. R Coutinho and B Fernandez. 2004. Spatially extended circle maps: Monotone periodic dynamics of functions with linear growth. *Journal of Nonlinear Science*, 14(1):93–118.
297. H Rauhut and C Schwab. 2016. Compressive sensing petrov-galerkin approximation of high-dimensional parametric operator equations. *Mathematics of Computation*, 86(304):661–700.
298. E A. Pekoz, A Rollin and N Ross. 2013. Generalized gamma approximation with rates for urns, walks and trees. *Annals of Probability*, 44(3): 1776–1816.
299. W G Mitchener. 2017. A stochastic model of language change through social structure and prediction-driven instability. *Siam Journal on Applied Mathematics*, 77(6).
300. T Prosen. 1999. Map from one-dimensional quantum field theory to quantum chaos on a two-dimensional torus. *Physical Review E*, 60(2):1658.

301. M Yoshimoto. 2003. Noise effects on the one-dimensional return map of the high flow rate chaos of the belousov-zhabotinsky reaction. *Physics Letters A*, 312(1–2):59–64.
302. W H Steeb and F Solms. 1996. Complexity, chaos and one-dimensional maps. South African Journal of Science, 161–178.
303. J M. Aguirregabiria. 2009. Robust chaos with variable lyapunov exponent in smooth one-dimensional maps. *Chaos Solitons & Fractals*, 42(4):2531–2539.
304. P Favati, G Lotti and L Margara. 1997. Additive one-dimensional cellular automata are chaotic according to devaney's definition of chaos. *Theoretical Computer Science*, 174(1–2):157–170.
305. D A Smirnov, V S Vlaskin and V I Ponomarenko. 2005. Estimation of the parameters of one-dimensional maps from chaotic time series. *Technical Physics Letters*, 31(2):97–100.
306. H Kawai and S H H Tye. 1984. Approach to chaos: Universal quantitative properties of one-dimensional maps. *Physical Review A*, 30(4):2005–2023.
307. A Csordás, G Gyorgyi, P Szépfalusy, et al. 1993. Statistical properties of chaos demonstrated in a class of one-dimensional maps. *Chaos*, 3(1):31–49.
308. N Zandi-Mehran, S Panahi, Z Hosseini, et al. 2020. One dimensional map-based neuron model: A phase space interpretation. *Chaos Solitons & Fractals*, 132:109558.
309. G Yuan and J A Yorke. 2000. Collapsing of chaos in one dimensional maps. *Physica D-nonlinear Phenomena*, 136(1–2):18–30.
310. D Jorge, R Carla, J Cristina, et al. 2015. How complex, probable, and predictable is genetically driven red queen chaos? *Acta Biotheoretica*, 63(4):341–361.
311. D L Hitzl and F Zele. 1985. A three-dimensional dissipative map with three routes to chaos. *Physics Letters A*, 110(4):181–187.
312. L Gardini, R Lupini and M G Messia. 1987. Bifurcations and transitions to chaos in the three-dimensional lotka-volterra map. *Siam Journal on Applied Mathematics*, 47(3):455–482.
313. G Chen, Y Chen and X Liao. 2007. An extended method for obtaining s-boxes based on three-dimensional chaotic baker maps. *Chaos Solitons & Fractals*, 31(3):571–579.
314. R L Marichal, E J González and G N Marichal. 2012. Hopf bifurcation stability in hopfield neural networks. *Neural Networks*, 36:51–58.
315. R E Gillilan and G S Ezra. 1991. Transport and turnstiles in multidimensional hamiltonian mappings for unimolecular fragmentation: Application to van der waals predissociation. *Journal of Chemical Physics*, 94(4):2648–2668.
316. A S Gonchenko and S V Gonchenko. 2014. Lorenz-like attractors in nonholonomic models of celtic stone. *Nonlinearity*, 28(9).
317. T Shnerb and K G Kay. 2005. Relationship between singularities in classical mechanics for complex initial conditions and for complex time. *Nonlinear Dynamics*, 42(2):165–173.
318. X Zhuo, M Bi, Z Hu, et al. 2020. Secure scheme for ofdm-pon system using tr based on modified henon chaos. *Optics Communications*, 462:125304.
319. N Parekh, S Parthasarathy and S Sinha. 1998. Global and local control of spatiotemporal chaos in coupled map lattices. *Physical Review Letters*, 81(7):1401–1404.
320. C Primo, I G Szendro, M A Rodríguez, et al. 2007. Error growth patterns in systems with spatial chaos: From coupled map lattices to global weather models. *Physical Review Letters*, 98(10):108501.
321. D York and A Harald. 2005. Characterizing spontaneous irregular behavior in coupled map lattices. *Chaos Solitons & Fractals*, 24(1):313–327.
322. H Ma, K E Zhu and T L Chen. 2006. A cryptographic scheme based on spatiotemporal chaos of coupled map lattices. *Communications in Theoretical Physics*, 45(3):477.
323. L M Pecora and T L Carroll. 1990. Synchronization in chaotic systems. *Physical Review Letters*, 64(8):821–824.
324. G Hu and J Xiao. 1997. Synchronization of spatiotemporal chaos its applications. *Phys Rev E*, 56(3):2738–2746.
325. Z Qu and H Gang. 1994. Spatiotemporally periodic states, periodic windows, and intermittency in coupled-map lattices. *Physical Review E Statal Physics Plasmas Fluids & Related Interdiplinary Topics*, 49(2):1099.

326. D H Zanette and A S Mikhailov. 1997. Condensation in globally coupled populations of chaotic dynamical systems. *Physical Interdiplinary Topics*, 57:276–281.
327. P Muruganandam, G Francisco, M Menezes, et al. 2009. Low dimensional behavior in three-dimensional coupled map lattices. *Chaos Solitons & Fractals*, 41(2):997–1004.
328. M Wang, X Wang, C Wang, et al. 2020. Spatiotemporal chaos in cross coupled map lattice with dynamic coupling coefficient and its application in bit-level color image encryption. *Chaos Solitons & Fractals*, 139:110028.
329. J Gravner and K Johnson. 2018. Coupled map lattices as musical instruments. *Computer Music Journal*, 42(2):22–34.
330. D Faranda, H Ghoudi, P Guiraud, et al. 2018. Extreme value theory for synchronization of coupled map lattices. *Nonlinearity*, 31(7):3326–3358.
331. M A Khan, D Maity and S D Jabeen. 2019. Spatiotemporal synchronization in asymmetrically bidirectionally coupled neurons over a complex network. *International Journal of Modern Physics B*, 33(19):1950213.
332. S T Liu, Y Liu, X P Guan, et al. 2000. Existence of monotone positive solution of neutral partial difference equation. *Journal of Mathematical Analysis and Applications*, 247(2):384–396.
333. X P Guan, J Yang and S T Liu. 2000. Nonexistence of positive solution of a class of nonlinear delay partial difference equation. *Acta Mathematica Sinica*, 43:233–238.
334. B G Zhang and S T Liu. 1997. On the oscillation of two partial difference equations. *Journal of Mathematical Analysis & Applications*, 206(2):480–492.
335. P J Y Wong and R P Agarwal. 1998. On the oscillation of partial difference equations generated by deviating arguments. *Acta Mathematica Hungarica*, 79(1–2):1–29.
336. J A Yorke, T Y Li. Period three implies chaos. *The American Mathematical Monthly*, 82.
337. R E Amritkar and N Gupte. 1993. Synchronization of chaotic orbits: The effect of a finite time step. *Physical review. E, Statal physics, plasmas, fluids, and related interdiplinary topics*, 47(6):3889–3895.
338. G Chen and X Dong. 1999. *From Chaos to Order:Methodologies, Perspectives and Applications*. World Scientific.
339. F R Marotto. 1978. Snap-back repellers imply chaos in rn. *Journal of Mathematical Analysis & Applications*, 63(1):199–223.
340. G Chen, S B Hsu and J Zhou. 1998. Snapback repellers as a cause of chaotic vibration of the wave equation with a van der pol boundary condition and energy injection at the middle of the span. *Journal of Mathematical Physics*, 39(12):6459–6489.
341. X F Wang and G Chen. 2000. Letter to the editor; chaotifying a stable map via smooth small-amplitude high-frequency feedback control. *International journal of circuit theory and applications*, 28(3):305–312.
342. R F Marotto. 2005. On redefining a snap-back repeller. *Chaos Solitons & Fractals*, 25(1):25–28.
343. P Glendinning and H W Chi. 2009. Border collision bifurcations, snap-back repellers, and chaos. *Phys Rev E Stat Nonlin Soft Matter Phys*, 79:025202.
344. W Lin, J Ruan and W Zhao. 2002. On the mathematical clarification of the snap-back-repeller in high-dimensional systems and chaos in a discrete neural network model. *International Journal of Bifurcation & Chaos in Applied Sciences & Engineering*, 12(05):1129-1139.
345. L Gardini, I Sushko, V Avrutin, M Schanz. 2011. Critical homoclinic orbits lead to snap-back repellers. *Chaos Solitons & Fractals*, 44(6):433–449.
346. M C Li, M J Lyu and P Zgliczyński. 2008. Topological entropy for multidimensional perturbations of snap-back repellers and one-dimensional maps. *Nonlinearity*, 21(11):2555.
347. S M Chang and H H Chen. 2011. Applying snapback repellers in resource budget models. *Chaos*, 21(4):199.
348. W M Yang. 1994. *Spatiotemporal Chaos and Coupled Map Lattices*. (Shanghai Scientific and Technological Education, Publishing House, Shanghai) (in Chinese).
349. H Bailin. 1989. *Elementary symbolic dynamics and chaos in dissipative systems*. World Scientific, 1989.
350. R L Devaney. *An Introduction to Chaoric Dynamical Systems*. Addision-Wesley, 1989.

351. Y K Li. 1975. *Period three implies chaos*. American Journal of Mathematics, 82: 481–485
352. J Frøyland. 1992. *Introduction to Chaos and Coherence*. Bristol: Institute of Physics Publishing.
353. F R Marotto. 1978. Snap-back repellers implies chaos in r. *Journal of Mathematical Analysis*, 63.
354. F R Marotto. 1979. Chaotic behavior in the hénon mapping. *Communications in Mathematical Physics*, 68(2):187–194.
355. X F Wang and G Chen. 2000. Chaotification via arbitrarily small feedback controls: Theory, method, and applications. *International Journal of Bifurcation & Chaos*, 10(03):0000037.
356. G Chen and S T Liu. 2003. On spatial periodic orbits and spatial chaos. *International Journal of Bifurcation & Chaos*, 13(04):935–941.
357. S T Liu and G Chen. 2002. Asymptotic behavior of delay 2-d discrete logistic systems. *IEEE Transactions on Circuits and Systems I: Fundamental Theory and Applications*, 49(11):1677–1682.
358. M Hénon. 1976. A two-dimensional mapping with a strange attractor. *Communications in Mathematical Physics*, 50(1):69–77.
359. F Sun and Z Lv. 2016. Stability and spatial chaos in 2d he non system. *Applied Mathematics & Information Sciences*, 10(2):739–746.
360. L Kocarev and U Parlitz. 1996. Generalized synchronization, predictability, and equivalence of unidirectionally coupled dynamical systems. *Physical Review Letters*, 76(11):1816–1819.
361. S S Cheng. Partial difference equations. 2003. *Advances in Discrete Mathematics and Applications*, 3, Taylor & Francis, London.
362. B Zhang and S Cheng. 1994. Qualitative theory of partial difference equations. i: Oscillation of nonlinear partial difference equations. *Tamkang Journal of Mathematics*, 25(3).
363. B G Zhang and Y Zhou. *Qualitative Analysis of Delay Partial Difference Equations*, Hindawi Publishing Corporation.
364. B G Zhang and R P Agarwal. 2003. The oscillation and stability of delay partial difference equations. *Computers & Mathematics with Applications*, 45(6–9):1253–1295.
365. B G Zhang and B M Liu. 1999. Necessary and sufficient conditions for oscillations of partial difference equations with continuous variables. *Computers & Mathematics with Applications*, 38(5–6):163–167.
366. B G Zhang and S T Liu. 1995. Necessary and sufficient conditions for oscillations of delay partial difference equations. *Discrete Dynamics in Nature & Society*, 1(4):265–268.
367. S T Liu, B G Zhang and G Chen. 2003. Asymptotic behavior and oscillation of delay partial difference equations with positive and negative coefficients. *Rocky Mountain Journal of Mathematics*, 43(8–9): 953–970.
368. C Yuan and S Liu. 2014. An envelope surface method for determining oscillation of a delay 2-d discrete convection system. *Journal of Dynamics & Differential Equations*, 29(1):1–16.
369. C Yuan and S Liu. 2014. Exact regions of oscillation for a mixed 2d discrete convection system. *Journal of Difference Equations and Applications*, 21(1):37–52.
370. S Cheng and Y Z Lin. 2009. *Dual Sets of Envelopes and Characteristic Regions of Quasi-Polynomials*. World Scientific, Singapore.
371. R P Agarwal and Z Yong. 2000. Oscillation of partial difference equations with continuous variables. *Mathematical & Computer Modelling*, 31(2-3):17–29.
372. G B Zhang, S J Yu. 1998. Linearized oscillation theorems for certain nonlinear delay partial difference equations. *Computers & Mathematics with Applications*, 35(4):111–116.
373. R P Agarwal and P J Y Wong. 1996. Oscillation criteria for nonlinear partial difference equations. *Computers & Mathematics with Applications*, 32(none):57–86.
374. B G Zhang and S T Liu. 1995. Necessary and sufficient conditions for oscillations of delay partial difference equations. *Discrete Dynamics in Nature & Society*, 1(4):265–268.
375. A F Henry and E Melkonian. 1977. Nuclear reactor analysis. *Nuclear Science IEEE Transactions on*, 30(6):53–55.
376. M S E Naschie. 1999. Radioactive decay and the structure of E (∞) quantum spacetime. *Chaos Solitons & Fractals*, 10(1):17–23.

377. A G Palmer, C D Kroenke and J P Loria. 2001. Nuclear magnetic resonance methods for quantifying microsecond-to-millisecond motions in biological macromolecules. *Methods in Enzymology*, 339:204–238.
378. J R Lamarsh and G I Bell. 1967. Introduction to nuclear reactor theory. *Physics Today*, 20(3):110–111.
379. M S E Naschie. 2004. A review of e infinity theory and the mass spectrum of high energy particle physics. *Chaos Solitons & Fractals*, 19(1):209–236.
380. H V Dam. 1992. Physics of nuclear reactor safety. *Reports on Progress in Physics*, 55(11):2025.
381. J B Dragt, J W M Dekker, H Gruppelaar, et al. 1977. Methods of adjustment and error evaluation of neutron capture cross sections; application to fission product nuclides. *Nuclear Science & Engineering the Journal of the American Nuclear Society*, 62:1(1):117–129.
382. W G Kelley and A C Peterson. 1991. Difference Equations, *Academic Press, New York.*
383. I Gyori and G Ladas. 1991. Oscillation Theorem of Delay Differential Equations with Applications.*Clearndon Press, Oxford.*
384. J C Strikwerda.1989. Finite Difference Schemes and Partial Differential Equations. *Wadsworth and Brooks, California.*
385. B G Zhang, S T Liu and S S Cheng. 1995. Oscillation of a class of delay partial difference equation.*Journal of Difference Equations and Its Applications*. 1: 215–226.
386. B G Zhang and S T Liu. 1997. On the oscillation of two partial difference equation.*Journal of Mathematical Analysis and Applications*. 206: 480–492.
387. B G Zhang and S T Liu 1995. Necessary and sufficient conditions for oscillations of delay partial difference equation, Discussiones Mathematicae.*Differential Inclusions*. 15: 213–219.
388. S T Liu and B G Zhang. 1998. Oscillation of a class of partial difference equation. *Pan American Mathematical Journal* 8: 93–100.
389. S T Liu, X P Guan and J Yang. 1999. Nonexistence of positive solution of a class of nonlinear delay partial difference equations. *Journal of Mathematical Analysis and Applications* 234: 361–371.
390. S T Liu and H Wang. 1998. Necessary and sufficient conditions for oscillations of a class of delay partial difference equations. *Dynamic Systems and Applications*. 7: 495–500.
391. B G Zhang and S T Liu. 1998. Oscillation of partial difference equation with variable coefficients. *Computers Math. Applic.* 36–38: 235–242.
392. B. G. Zhang and S. T. Liu, Oscillation of partial difference equations, Pan America Mathematical Journal, **5** (1995), 60–71.
393. W R Rudin. 1964. *Principles of mathematical analysis*. McGraw-Hill.
394. P Liu, S Liu, J Wang, et al. 2015. Spatial chaos on surface and its associated bifurcation and feigenbaum problem. *Nonlinear Dynamics*, 81:283–298.
395. J J Duderstadt. 1976. *Nuclear Reactor Analysis*. Wiley.
396. B G Zhang, S T Liu and S S Cheng. 1995. Oscillation of a class of delay partial difference equations. *Journal of Difference Equations & Applications*, 1(3):215–226.
397. J M Ottino, F J Muzzio, M Tjahjadi, et al. 1992. Chaos, symmetry, and self-similarity: exploiting order and disorder in mixing processes. *Science*, 257(5071):754–760.
398. S M Henson. 2001. Lattice effects observed in chaotic dynamics of experimental populations,. *Science*, 294(5542):602–605.
399. R L Devaney. 1987. Chaotic bursts in nonlinear dynamical systems. *Science*, 235(4786):342–345.
400. S J Schiff, K Jerger, D H Duong, et al. 1994. Controlling chaos in the brain. *Nature*, 370(6491):615–620.
401. G Klambauer. 1975. *Mathematical Analysis*. Mercel Dekker,Inc.York, 1975.
402. S Glasstone and G I Bell. 1970. *Nuclear Reactor Theory.*
403. J P Keener. 1980. Chaotic behavior in piecewise continuous difference equations. *Transactions of the American Mathematical Society*, 261:589–604.
404. Michael Robert Herman. *Mesure de Lebesgue et Nombre de Rotation*. Springer Berlin Heidelberg, 1977.

405. S Smale and R F Williams. 1976. The qualitative analysis of a difference equation of population growth. *Journal of Mathematical Biology*, 3(1):1–4.
406. A Denjoy. 1932. Sur les courbes definies par les equations differentiables a la surface du tore. *Journal de Mathematiques Pures et Appliquees*, 11.
407. J Pople and D Beveridge. 1970. *Approximate Molecular Orbital Theory*. Mcgraw-Hill, 1–6.
408. E Heilbronner. 1976. The HMO model and its application, *Journal of Molecular Structure*, 38:87136.
409. X Li. 1982. Partial difference equation used in the study of molecular orbits. *Acta Chimica Sinica*, 40:688–698.
410. B Kirkpatrick, L E Fleming, D Squicciarini, et al. 2004. Literature review of florida red tide: implications for human health effects. *Harmful Algae*, 3(2):99–115.
411. N H Ye, X W Zhang, Y Z Mao, et al. 2011. 'green tides' are overwhelming the coastline of our blue planet : taking the world's largest example. *Ecological Research*, 26(3):477–485.
412. G M Hallegraeff. 2010. Ocean climate change, phytoplankton community responses, and harmful algal blooms: A formidable predictive challenge. *Journal of Phycology*, 46(2):220–235.
413. E E Prepas and A M Trimbee. 1987. Evaluation of total phosphorus as a predictor of the relative biomass of blue-green algae with emphasis on alberta lakes. *Canadian Journal of Fisheries and Aquatic ences*, 44(7):1337–1342.
414. V H Smith. 1983. Low nitrogen to phosphorus ratios favor dominance by blue-green algae in lake phytoplankton. *ence*, 221(4611):669–671.
415. A Bennett. 1973. Complementary chromatic adaptation in a filamentous blue-green alga. *The Journal of cell biology*, 58:419–435.
416. A F Talebi, M Tohidfar, M Tabatabaei, et al. 2003. Genetic manipulation, a feasible tool to enhance unique characteristic of chlorella vulgaris as a feedstock for biodiesel production. *Molecular Biology Reports*, 40:4421–4428.
417. M A Reiter. 2001. Colonization time and substratum relief as factors in the uptake of algae from the water column. *Journal of Freshwater Ecology*, 16(1):57–65.
418. S Schilling and C Oesterhelt. 2007. Structurally reduced monosaccharide transporters in an evolutionarily conserved red alga. *Biochemical Journal*, 406(2):325.
419. T I Pisman and Y V Galayda. 2006. Experimental and mathematical model of the interactions in the mixed culture of links in the "producer-consumer" cycle. In *36th Cospar Scientific Assembly*, 16–23.
420. J Wang and W Jiang. 2014. Impulsive perturbations in a predator-prey model with dormancy of predators. *Applied Mathematical Modelling*, 38(9–10):2533–2542.
421. H Serizawa, T Amemiya, T Enomoto, et al. 2008. Mathematical modeling of colony formation in algal blooms: phenotypic plasticity in cyanobacteria. *Ecological Research*, 23(5):841–850.
422. B Mukhopadhyay and R Bhattacharyya. 2006. Modelling phytoplankton allelopathy in a nutrient-plankton model with spatial heterogeneity. *Ecological Modelling*, 198(1–2):163–173.
423. S Chakraborty, R Shovonlai and J Chattopadhyay. 2008. Nutrient-limited toxin production and the dynamics of two phytoplankton in culture media: A mathematical model. *Ecological Modelling*, 213(2):191–201.
424. J B Shukla, A K Misra and P Chandra. 2008. Mathematical modeling and analysis of the depletion of dissolved oxygen in eutrophied water bodies affected by organic pollutants. *Nonlinear Analysis Real World Applications*, 9(5):1851–1865.
425. B Beran and F Kargi. 2005. A dynamic mathematical model for wastewater stabilization ponds. *Ecological Modelling*, 181(1):39–57.
426. K Sugiura, Y Kawasaki, M Kinoshita, et al. 2003. A mathematical model for microcosms: formation of the colonies and coupled oscillation in population densities of bacteria. *Ecological Modelling*.
427. Y Yamasaki, Y Zou, J Go, et al. 2011. Cell contact-dependent lethal effect of the dinoflagellate heterocapsa circularisquama on phytoplankton-phytoplankton interactions. *Journal of Sea Research*, 65(1):76–83.

428. P Fergola, M Cerasuolo, A Pollio, et al. 2007. Allelopathy and competition between chlorella vulgaris and pseudokirchneriella subcapitata: Experiments and mathematical model. *Ecological Modelling*, 208(2-4):205–214.
429. J Solé, E G Ladona, P Ruardij, et al. 2005. Modelling allelopathy among marine algae. *Ecological Modelling*, 183(4):373–384.
430. A C Fassoni and M L Martins. 2014. Mathematical analysis of a model for plant invasion mediated by allelopathy. *Ecological Complexity*, 18:49–58.
431. L A Lewis and R M Mccourt. 2004. Green algae and the origin of land plants. *American Journal of Botany*, 91(10).
432. Q Yang, B F Pando, G Dong, et al. 2010. Circadian gating of the cell cycle revealed in single cyanobacterial cells. *Science*, 327(5972):1522–1526.
433. F B Wang. 2015. A pde system modeling the competition and inhibition of harmful algae with seasonal variations. *Nonlinear Analysis Real World Applications*, 25:258–275.
434. M E Baird and I M Suthers. 2007. A size-resolved pelagic ecosystem model. *Ecological Modelling*, 203(3-4):185–203.
435. L Zhang and S T Liu. 2010. Spatial dynamical behavior of narrow composite beams. *Applied Mathematics and Computation*, 216(12):3627–3633.
436. G W Fairchild. 2010. Birth and death rates of a littoral filter feeding microcrustacean, sida crystallina (cladocera), in cochran lake, michigan. *Internationale Revue der gesamten Hydrobiologie und Hydrographie*, 68(3):339–350.
437. H U Zhangxi and X U Ning. 2010. Effects of nitrogen sources on the growth of heterosigma akashiw,karenia sp.,phaeocystis globosa and chaetoceros sp. *Ecology and Environmental ences*, 19(10):2452–2457.
438. B L Hao. 1993. Starting with Parabolas: An Introduction to Chaotic Dynamics. *Shanghai Scientific and Technological Education Publishing House, Shanghai, China*.
439. R Brown and L O Chua. 1996. Clarifying chaos: examples and counterexamples. *International Journal of Bifurcation and Chaos*, 6(2):219–242.
440. A Lasota and M C Mackey. 1994. Chaos, Fractals, and Noise: Stochastic Aspects of Dynamics. *Springer, New York*.
441. B L Hao. 1989. Elementary Symbolic Dynamics and Chaos in Dissipative Systems. *Singapore: World Scientific*.
442. S L Yan, Y Y Chi and W J Chen. 2004. Chaotic laser synchronization and its application in optical fiber secure communication. *Science in China Series E-Technological Sciences*, 34.(4):467–480.
443. J H Dai. 1996. Chaotic application in information encryption. *Chinese Science Bulletin*, 41(5):402–405.
444. Y Zhang, J M Yu and G H Du. 1998. Continuous feedback chaotic synchronization and its application in secure communication. *Chinese Science Bulletin*, 43(17):1831–1835
445. T Yang, L B Yang, C M Yang. 1998. Cryptanalyzing chaotic secure communications using return maps. *Physics Letters A* 245:495–510.
446. H K C Chang and J L Liu. 1997. A linear quadtree compression scheme for image encryption. *Signal Process. Image Communication*, 10(4):279–290.
447. N K Pareek, V Patidar and K K Sud. 2005. Cryptography using multiple one-dimensional chaotic maps.*Communications in Nonlinear Science and Numberical Simulation*, 10(7):715–723.
448. H Zhou and X T Ling. 1997. Problems with the chaotic inverse system encryption approach. *IEEE Trans Circ System I:*, 44(3):268–271.
449. N K Pareek, V Patidar and K K Sud. 2006. Image encryption using chaotic logistic map. *Image and vision compution*, 24(9):926–934.
450. K Wang, W J Pei, L H Zou LH, et al. 2006. Security of public key encryption technique based on multiple chaotic systems.*Physics Letters A* 360:259–262.
451. K M Short. 1997. Signal extraction from chaotic communication. *International Journal of Bifurcation and Chaos* 7(7):1579–97.
452. M S Baptista. 1999. Cryptography with chaos. *Physics Letters A* 240:50–4.

453. N K Pareek, V Patidar, K K Sud. 2005. Cryptography using multiple one-dimensional chaotic maps. *Communications in Nonlinear Science and Numberical Simulation* 10(7):715–23.
454. T Xiang, X Liao, G Tang, et al. 2006. A novel block cryptosystem based on iterating a chaotic map. Physics Letters A 349:109–15.
455. P Garcia, A Parravano, M G Cosenza, et al. 2002. Coupled map networks as communication schemes. *Phys Review E* 65:045201.
456. H P Lu, S H Wang, X W Li, et al. 2004. A new spatiotemporally chaotic cryptosystem and its security and performance analyses. *Chaos* 14(3):617–29.
457. P Li, Z Li, W A Halang, et al. 2006. A multiple pseudorandom-bit generator based on a spatiotemporal chaotic map. *Physics Letters A* 349:467–73.
458. P Li, Z Li,W A Halang, et al. 2007. A stream cipher based on a spatiotemporal chaotic system. *Chaos, Solitons & Fractals* 32:1867–76.
459. C Grebogi,E Ott, and A Yorke. 1988. Roundoff-induced periodicity and the correlation dimension of chaotic attractors.*Physical Review A* 38:3688–92.
460. U Bastolla, G Parisi. 1997. Attraction basins in discretized maps. *Journal of Physics A-Mathematical and Theoretical* 30:3757–3769.
461. S H Wang, W R Liu, H P Lu, et al. 2004. Periodicity of chaotic trajectories of single and coupled maps in realizations of finite computer precisions. *International Journal of Modern Physics B* 18(17–19):2617–2622.
462. G Tang,S Wang, H Lu, et al. 2003. Chaos-based cryptography incorporated with S-box algebraic operation. *Physics Letters A* 318:388–398.
463. K Kaneko, I Tsuda. 2001. Complex Systems: Chaos and Beyond: A Constructive Approach with Applications in Life Sciences. *Berlin: Springer-Verlag*.
464. S H Wang, J Y Kuang, J Li, et al. 2002. Chaos-based secure communication in a large community. *Physical Revview E* 66:065202(R).
465. W M Yang. 1991. On the largest exponent for coupled surjective map lattice with weak diffusive coupling. *Chaos Solition & Fractal* 1:389–396.
466. T Furusawa. 1986. Quantum chaos of mixmaster universe. *Progress of Theoretical Physics*, 76(1):67–74.
467. T C BountisL and B Drossos. 1997. Evidence of a natural boundary and nonintegrability of the mixmaster universe model. *Journal of Nonlinear Science*, 7:45–55.
468. F M Lewis. 2002. Cardassian expansion: a model in which the universe is flat, matter dominated, and accelerating. *Physics Letters B*, 540:1-8.
469. M R Setare. 2006. Interacting holographic dark energy model in non-flat universe. *Physics Letters B*, 642(1–2):1–4.
470. H Alfven. 1986. Model of the plasma universe. *IEEE Transactions on Plasma Science*, 14(6):629–638.
471. V P Korobeinikov, P I Chushkin and L V Shurshalov. 1976. Mathematical model and computation of the tunguska meteorite explosion. *Acta Astronautica*, 3(7–8):615–622.
472. P M Bauleo and J R Martino. 2009. The dawn of the particle astronomy era in ultra-high-energy cosmic rays. *Nature*, 458(7240):847.
473. J Thilmany. 2010. Universe pictured. *Mechanical Engineering*, 132(3):18–18.
474. R A Burger and M S Potgieter. 1989. The calculation of neutral sheet drift in two-dimensional cosmic-ray modulation models. *Astrophysical Journal*, 339(1):501–511.
475. O A Bessonovv and I Polezhaev. 2001. Mathematical modeling of convection in the dacon sensor under conditions of real space flight. *Cosmic Research*, 39:170–178.
476. M M M Low, D V Buren, D O S Wood, et al. 1991. Bow shock models of ultracompact h ii regions. *Astrophysical Journal*, 369(2):395–409.
477. A G Sgro. 1975. The collision of a strong shock with a gas cloud - a model for cassiopeia a. *Astrophysical Journal*, 197(3):621–634.
478. Y G Markov, I N Sinitsyn and M L Kiselev. 2009. Effect of parametric and nonlinear perturbations on statistical dynamics of the earth's pole. *Cosmic Research*, 47(3):251–258.
479. V I Abrashkin, K E Voronov, I V Piyakov, et al. 2016. A simplified technique for determining the rotational motion of a satellite based on the onboard measurements of the angular velocity and magnetic field of the earth. *Cosmic Research*, 54(5):375–387.

480. M Y Belyaev, T V Matveeva, M I Monakhov, et al. 2018. Modes of uncontrolled rotational motion of the progress m-29m spacecraft. *Cosmic Research*, 56(1):54–67.
481. N Y Gnedin. 2009. Hidden chaos in cosmic order. *Nature*, 461:43–44.
482. T Kitamura, S Ohara, T Konishi, et al. 1997. Chaos in cosmic ray air showers. *Astroparticle Physics*, 8(3-4):279–291.
483. A Provenzale, E A Spiegel and R Thieberger. 1997. Cosmic lacunarity. *Chaos*, 7(1):82–88.
484. J A Lezniak and W R Webber. 1978. The charge composition and energy spectra of cosmic-ray nuclei from 3000 mev per nucleon to 50 gev per nucleon. *Chaos*, 14(4):1123–1129.
485. F S Labini and L Pietronero. 2006. Statistical physics for cosmic structures. *European Physical Journal B*, 59(12):60–61.
486. B G Sidharth. 2000. A brief note on cosmic background radiation. *Chaos Solitons & Fractals*, 11(9):1471–1472.
487. V V Uchaikin and R T Sibatov. 2017. Fractional derivatives on cosmic scales. *Chaos Solitons & Fractals*, 197–209.
488. B A Carreras, V E Lynch, I Dobson, et al. 2002. Critical points and transitions in an electric power transmission model for cascading failure blackouts. *Chaos*, 12(4):985–994.
489. S Behnia, A Akhshani, H Mahmodi, et al. 2008. A novel algorithm for image encryption based on mixture of chaotic maps. *Chaos Solitons & Fractals*, 35(2):408–419.
490. S F Dermott, S Jayaraman, Y L Xu, et al. 1994. A circumsolar ring of asteroidal dust in resonant lock with the earth. *Nature*, 369(6483):719–723.
491. H Murayama, H Suzuki, T Yanagida, et al. 1992. Chaotic inflation and baryogenesis by right-handed sneutrinos. *Physical Review Letters*, 70(13):1912–1915.
492. P Sukováand O Semerák. 2013. Free motion around black holes with discs or rings: between integrability and chaos-i. *Monthly Notices of the Royal Astronomical Society*, 545–574.
493. N Bobev, F F Gautason and K Parmentier. 2020. Holographic uniformization and black hole attractors. *Journal of High Energy Physics*, 2020(6).
494. T Andrade, R Emparan, A Jansen, et al. 2020. Entropy production and entropic attractors in black hole fusion and fission. *Journal of High Energy Physics*, 2020(8).
495. Y F Cai, C Chen, X Tong, et al. 2019. When primordial black holes from sound speed resonance meet a stochastic background of gravitational waves. *Physical Review D*, 100:043518.
496. B Carr, S M Kühnel F. 2016. Primordial black holes as dark matter. *Physical Review D*, 94:083504.
497. Y F Cai, X Tong, D G Wang, et al. 2018. Primordial black holes from sound speed resonance during inflation. *Physical Review Letters*, 121:081306.
498. P A R Ade, Z Ahmed, R W Aikin, et al. 2018. Constraints on primordial gravitational waves using planck, wmap, and new bicep2/keck observations through the 2015 season. *Physical Review Letters*, 121:221301.
499. The nobel prize in physics is awarded to physical celestial bodies again, how mysterious is black hole? *Chinese Academy of Sciences*, 2020.10.9.
500. Y Q Ye. 1965. *Theory of limit cycle. Shanghai Scientific & Technical Publishers.*
501. Z Zhang, H Wenzao and Z Dong. 1985. *Qualitative theory of differential equations*. Science Press.
502. Y X Qin. 1954. The monotonically approaching and semi-stable line of limit cycle. *Acta Mathematica Sinica*, 4:445–465.
503. Y Z Liu and L Q Chen. 2001. *Nonlinear vibration*. Shanghai Jiao Tong University Press.
504. B H Wang. 1992. *The introduction of chaotic dynamics*. Press of University of Science and Technology of China.
505. G R Wang. 1995. *Chaotic motion of non-linear system*. Press of Guangxi Science and Technology, 1995.
506. S Y Zhang and B L Hao. 1983. The research of frequency division sampling method about period doubling bifurcation and "chaos" phenomenon in forced nonlinear oscillator. *Acta Physica Sinica*, 32:198–208.
507. W Q Zeng and F Kenneth. 2007. *Fractal geometry: mathematical foundations and applications*. Posts & Telecom Press.

508. S S Liu and S D Liu. 2014. *Fractal in physics*. Peking University Press.
509. L B Ao L. 1996. *The introduction of fractal*. Inner Mongolia People's Publishing House.
510. L B Rall. 2006. Nonlinear two point boundary value problems (paul b. bailey, lawrence f. shampine and paul e. waltman). *Siam Review*, 12:160–161.
511. C B Pan and C D Pan. 1998. *The concise introduction about number theory*. Peking University Press.
512. R Penrose. 1965. Gravitational collapse and space-time singularities. *Physical Review Letters*, 14:57.
513. R Penrose. 1973. Naked singularities. *Annals of the New York Academy of Sciences*, 224(1):125–134.
514. R Penrose. 1994. *Shadows of the Mind*. Oxford University Press.
515. W Rindler and R Penrose. 1984. *Spinors and space-time: Volume 2, Spinor and twistor methods in space-time geometry*. Cambridge University Press.
516. G Z Chen and G R Wang. 1995. *Chaotic motion in nonlinear system*. Press of Guangxi Science and Technology.